Praise for *3D Data Science with Python*

Florent has a unique ability to break down complex real-world problems, allowing the reader to confidently understand and address the issue. This book provides a solid foundation for working solutions and real-world applications.

—Adam Tindall, consultant/problem solver

This book quite literally adds a whole new dimension to data science. Why settle for flat insights when you can think in 3D? A word from the Author reminds us that a fresh perspective can change everything. With its hands-on approach and real-world examples, this book doesn't just teach—you experience firsthand how complex problems are solved through the lens of 3D data science. From healthcare to autonomous vehicles, it unveils how the fusion of 3D technology and data science is shaping the future.

—Lipi Deepaakshi Patnaik, senior software developer, Zeta

The processing of a 3D point cloud is a complex digital encoding process that begins with an uncritical instrumental acquisition and must ultimately produce a representative product capable of preserving meanings, relationships, and properties within a newly defined digital ontology. This book serves as an excellent fundamental reference, providing a methodological breakdown for selecting and applying techniques and workflows designed to maintain the semantics of the recorded environment. Through clear and accessible language, Florent offers valuable support for potential research applications in both scientific and operational fields.

—Riccardo Tavolare, head of MAULab,
Polytechnic University of Bari, Italy

Dr. Florent Poux's work has been instrumental in my academic journey, particularly during the development of my final paper. His clear and engaging teaching style, combined with the depth of his content, not only makes complex topics accessible but also inspires new ideas and supports the development of practical applications.

—*Davi Ribeiro, AI researcher, Ford Motor Company*

3D Data Science with Python demystifies working with 3D data by blending clear theory, hands-on projects, and step-by-step guidance throughout. I've rarely seen a book make advanced 3D data workflows feel this approachable. It's a must-have for anyone serious about 3D.

—*Tarun Narayanan Venkatachalam, AI engineer, Unstructured*

3D Data Science with Python
*Building Accurate Digital Environments
with 3D Point Cloud Workflows*

Florent Poux
Foreword by Dr. Uwe Bacher

O'REILLY®

3D Data Science with Python

by Florent Poux

Published by O'Reilly Media, Inc., 1005 Gravenstein Highway North, Sebastopol, CA 95472.

O'Reilly books may be purchased for educational, business, or sales promotional use. Online editions are also available for most titles (*http://oreilly.com*). For more information, contact our corporate/institutional sales department: 800-998-9938 or *corporate@oreilly.com*.

Acquisitions Editor: Aaron Black
Development Editor: Jill Leonard
Production Editor: Elizabeth Faerm
Copyeditor: Charles Roumeliotis
Proofreader: Krsta Technology Solutions

Indexer: BIM Creatives, LLC
Interior Designer: David Futato
Cover Designer: José Marzan Jr.
Illustrator: Kate Dullea

April 2025:   First Edition

Revision History for the First Edition

2025-04-09:   First Release

See *http://oreilly.com/catalog/errata.csp?isbn=9781098161330* for release details.

978-1-098-16133-0

[LSI]

Table of Contents

Foreword

The world around us is incredibly diverse and, most importantly, three-dimensional. Everything has a position and a meaning. To better understand our environment and answer current questions, we need accurate representations of the real world to be represented in the virtual world. These representations are often referred to as *digital twins*. While digital twins usually encompass more than just three-dimensional images, they include as much additional data and information as possible. However, one thing is common in almost all digital twins: they require a foundational dataset that is three-dimensional, or often even four-dimensional, when we include the temporal component. Hence, 3D data is transforming our world in many areas, whether in geosciences, self-driving cars, or medicine.

In *3D Data Science with Python*, Florent demonstrates why this data is so valuable and how it can be effectively utilized and leveraged. There are numerous ways to generate 3D data, such as photogrammetric methods to generate point clouds from images, laser scanners that "scan" the environment, or MRI machines that create an image of the body, to name a few. What all these methods have in common is the fact that, although they generate highly accurate data, this data lacks semantic information. The real value is created through the processing and analysis of the data. To achieve value, nowadays one cannot avoid using AI methods and their implementation in Python.

I've worked on projects for capturing and analyzing high-precision spatial data from airborne systems like laser scanners and cameras. These data are often used as the basis for digital twins and initially provide valuable information about the spatial context. The real value, however, is created through the processing, analysis, and visualization of the data for the end user. For instance, AI-based methods can automatically infer land use, extract individual objects like buildings or trees, or create 3D models of the scene. In this book, and with practical applications in mind, Florent illustrates how data can be manipulated, analyzed, and visualized to achieve impressive results.

The significance of 3D data goes far beyond merely capturing spatial information. It enables deeper understanding and a more thorough analysis of the world. In healthcare, 3D data contributes to more precise diagnoses and personalized treatment plans. In manufacturing, complex products are designed and optimized, efficiency is increased, and production costs are reduced. In cultural heritage preservation, 3D data helps document and preserve historical sites and artifacts.

A key aspect of 3D data science is the ability to manipulate and visualize data. Python offers a variety of libraries and tools specifically developed for these tasks. From basic data manipulation with NumPy and pandas to advanced visualization with Open3D and PyVista, Florent demonstrates how it's done and provides understandable tutorials for an easy and successful entry into the world of 3D data analysis.

With over ten years of research experience, Florent has made a name for himself and shares his extensive knowledge in this book. He has focused on using modern methods and tools, such as Python as a programming language and AI for the analysis of 3D data. He introduces readers to various topics through easily understandable tutorials and provides simple, reproducible examples to follow the solution paths themselves. He regularly publishes on platforms like LinkedIn, the 3D Geodata Academy he founded, and Medium.com. These publications complement and deepen the topics covered in his book.

The field of AI is becoming increasingly important for the highly efficient analysis of 3D data. By applying techniques such as machine learning and deep neural networks, patterns in 3D data can be detected that are invisible to the human eye. This enables new insights to be gained and innovative solutions to be developed.

The book offers practical examples and exercises that help you understand and apply these techniques and more, including the core concepts of 3D data, load, manipulate, and visualize 3D data, perform advanced pattern recognition, generate 3D datasets, and create digital environments for spatial AI. Florent shares his personal journey from surveyor to expert in 3D data science, emphasizing the importance of curiosity and continuous learning in this fascinating field.

Florent and I share a passion for the analysis of 3D data and strive to share this enthusiasm with as many interested individuals as possible. I invite you to read this book and discover the world of 3D data science. Be inspired, and use the possibilities that 3D data offers to develop innovative solutions and shape the future.

— Dr. Uwe Bacher
Technical director, Hexagon AB
Munich, March 2025

Preface

The physical world we live in is grounded in three dimensions. With new technologies, we can digitize or emulate its geometry, which opens up marvelous spatial playgrounds. From medical scans used to create detailed visualizations of internal organs to self-driving cars, where 3D data is used to create detailed road network maps, 3D digital assets are becoming the backbone of critical decisions.

This book, *3D Data Science with Python,* stands out with its hands-on approach. It actively involves you by sharing the ideas, tools, and methods to unlock the power of spatial AI.

But why not exclusively depend on computer vision and computer graphics? Why are traditional non-3D data science methods inadequate for effectively handling 3D data?

These initial questions beautifully convey how we approach 3D data science in our modern world. Come with me as I take you on a journey that will change the way you think about and deal with complicated, three-dimensional data. Be wary, though, because it's difficult to stop analyzing everything that your senses (sight, sound, and touch) bring back to you after reading.

3D data science is the pinnacle of mathematical excellence, computational rigor, and 3D data expertise. Combined, these fields create a new branch of data science with wide-ranging applications integrated into the most advanced workflows, ensuring you stay relevant and connected to the forefront of modern technology.

Here are some key applications from six domains where 3D data science is making a significant impact:

Prosthetics (healthcare and medicine)
　　Designing custom prosthetics that fit seamlessly with the patient's anatomy

Product design (manufacturing and engineering)
　　Creating and testing product prototypes virtually before physical production

Heritage preservation (architecture and urban planning)
　　Digitizing historic sites to preserve their cultural heritage and facilitate restoration

Autonomous vehicles (robotics and automation)
　　Enabling cars to perceive, map, and navigate their surroundings

Immersive reality (entertainment and gaming)
　　Designing virtual worlds that users can interact with in real time (games, virtual reality, augmented reality, mixed reality)

Geospatial analysis (Earth science and environmental monitoring)
　　Studying the Earth's surface and natural resources using 3D data from satellites, drones, or laser scanners

Who Should Read This Book?

This book is a practical reference for data scientists, engineers, and anyone curious about working with 3D data. It assumes very little, and you'll find value even without any understanding of Python programming and little familiarity with fundamental data science concepts.

Moreover, no prior experience with 3D data processing is necessary. I will guide you through the essential *libraries* and *techniques* step by step, ensuring that you can apply the knowledge to real-world scenarios in a 0-to-1 fashion.

What Will You Learn?

In this book, we dive deep into 3D data science with Python, building 3D workflows and wielding Python to implement solutions. You will learn:

- The core concepts and representations of 3D data (meshes, point clouds, voxels, CAD)
- How to load, manipulate, and visualize 3D data using powerful Python libraries
- Advanced techniques for 3D pattern recognition (supervised and unsupervised)
- How to generate 3D datasets and leverage 3D automated modeling workflows

- Practical applications of 3D data science in areas like computer vision/graphics, geospatial intelligence, scientific computing, robotics, and autonomous driving
- How to build accurate digital environments that spatial AI solutions (perception and reasoning) can leverage through real-world projects

Why This Book?

"Why this book?" may have been the first question I asked myself. Why should I write a book on this topic?

Well, if you have your eyes on these lines, there is no doubt that you still value human knowledge input in the age of AI. And this is the core of my motivation. I want to ensure that we perfectly understand what we wield. Future 3D systems should build onto a sound codebase layer and be implemented with as little bias as possible. But most importantly, I want this book to trigger your creativity and ability to innovate in a space where opportunities open every week.

The challenge, of course, is that the field of 3D data science is rapidly evolving, with new tools and techniques constantly emerging. These build upon a fundamental backbone, which has a transformative power when it comes to designing new solutions. Where I can, I provide links and references within the book that will give the most current information.

My goal is to provide you with the most comprehensive and practical guide to navigating this exciting landscape. With its focus on Python, you'll be able to tackle a broad range of 3D data problems for many years to come.

A Word from the Author

Confession time: my journey into 3D data science began not with mind-blowing algorithms but with mud-caked boots. I was a land surveyor, meticulously translating the real world into digital plans, one point at a time. Every day felt like a rerun on TV—the thrill of discovery was long gone, replaced by a sense of repetitive routine. It was in this mundane setting that I had my first encounter with the potential of 3D data science.

I remember precisely one evening: I was alone in the office, staring at a *.csv* file of 2,587 lines, each holding records of a specific data point surveyed on site. I had to deliver the topographic map the next day. I felt this stressful rush of not knowing how I would be able to find the will and time to do it. Going through these monotonous data points, I had my spark moment: "Couldn't this be automated?" That question, fueled by a late-night cup of coffee, changed everything. It sent me down a rabbit hole of research, leading me to the world of 3D data science.

The learning curve was steep—steeper than the hills I climb every time I hop on my bike in my favorite locations in the Pyrenees. Textbooks became my bible, and online forums my lifeline. The transition from a blue-collar job to a PhD program was a wild ride but fueled by a newfound passion. After years of research, a postdoc, and even becoming a professor, the journey unfolded one exciting discovery after another.

But my passion wasn't confined to academia. I craved to see these tools put to real-world use, so I ventured into the industry, helping top companies leverage 3D data for innovation. Witnessing the impact—from optimizing factory layouts to designing life-saving medical implants—was truly inspiring.

Today, as the Director of an Academy, I dedicate my time to empowering professionals with the magic of 3D data science. And guess what? The most rewarding part isn't seeing them ace a project but rather witnessing that spark of 3D creativity ignite in their eyes— the same spark I felt when I first dared to ask, "Can this be automated?"

This book is for you, the curious mind who wants to break free from routine and unlock the potential of 3D data. Remember, groundbreaking discoveries often begin with a simple question. Are you ready to ask yours?

Navigating This Book

I created this book to help you acquire all the knowledge needed to become a 3D expert (or specialize in certain areas, depending on your goals). Throughout my experience in the 3D space, I have identified several profiles that work within different yet complementary boundaries.

The book is structured to cater to different learning styles, with theoretical explanations, practical examples, and hands-on exercises. It ensures that you can learn at your own pace and in the way that best suits you.

I recommend reading the book linearly, as the learning curve is aligned with the challenges addressed. Also, previous chapters are referenced as you progress through the book. Nevertheless, to ensure maximum flexibility, I have designed the book so that you can grasp each chapter (or set of chapters) independently based on your profile, as shown in Figure P-1.

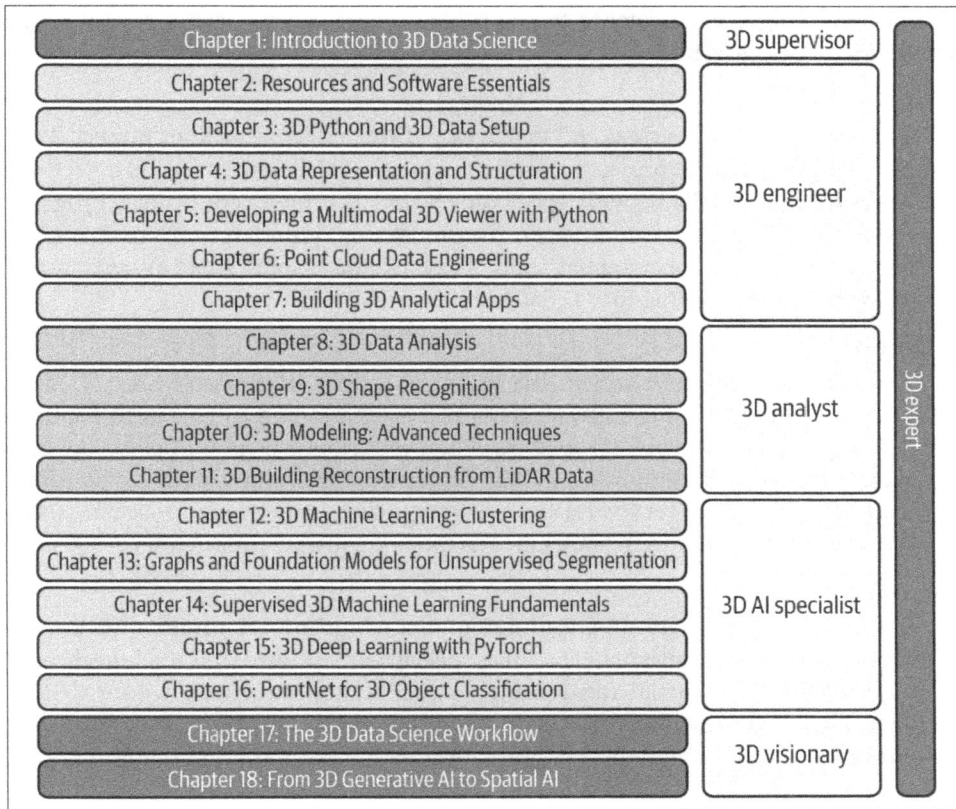

Figure P-1. Structure of the book and accompanying profiles

This book presents five tracks (3D supervisor, 3D engineer, 3D analyst, 3D AI specialist, 3D visionary) that can be followed independently, or combined to form the ultimate combination (3D expert), as shown in Figure P-2.

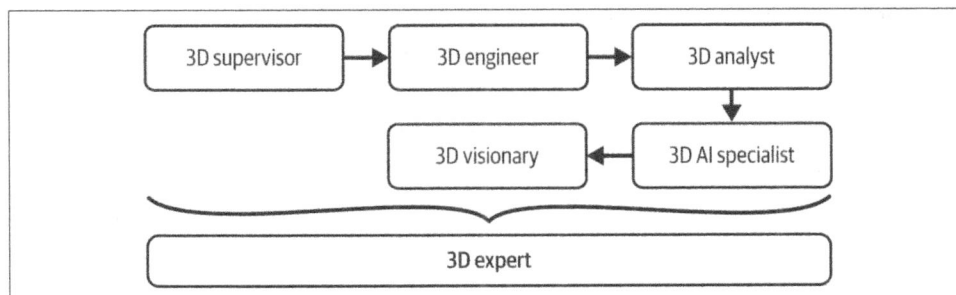

Figure P-2. The road to 3D expertise through various learning profiles

For the rest of this book, I refer to these profiles as personas you can unlock through practice! Let's define them:

3D supervisor

The 3D supervisor oversees and manages a 3D data science team or project. They provide guidance, set project priorities, and ensure the team meets its objectives. Additionally, they liaise with stakeholders and communicate project progress. Aside from being a great manager, they must clearly understand possible challenges and how to form a great team. They also have some basic understanding of Python programming for 3D data science.

3D engineer

The 3D engineer specializes in the technical aspects of working with 3D data. They are skilled in using software tools and programming languages to manipulate, process, and analyze 3D datasets. Their expertise is crucial in tasks like data acquisition, feature engineering, 3D reconstruction, and clearly establishing workflows. They are very versatile and can quickly adapt to many scenarios, but they do not specialize in 3D machine learning or software production.

3D analyst

The 3D analyst is focused on examining, modeling, and interpreting 3D data to extract meaningful insights. They use specialized software and analytical techniques to analyze spatial relationships, patterns, and structures within the 3D data. Their findings contribute to decision making and problem solving in various industries.

3D AI specialist

The 3D AI specialist is an expert in leveraging artificial intelligence techniques and algorithms tailored explicitly for 3D data applications. They profoundly understand AI and 3D data science, allowing them to develop and implement advanced AI models operating in three-dimensional space. Their expertise is crucial for 3D object recognition, scene understanding, and other complex cognitive tasks within a 3D context. They work on cutting-edge applications integrating AI capabilities with 3D data to achieve innovative solutions in various industries.

3D visionary

The 3D visionary is an individual with a deep understanding of the potential applications and innovations that can be achieved using 3D data science. They conceptualize and envision projects, often pushing the boundaries of what can be accomplished with three-dimensional information.

3D expert

The 3D expert is a highly skilled professional with comprehensive knowledge and experience in working with 3D data science. They are proficient in all the aspects mentioned previously, including data acquisition techniques, software tools, analytical methods, and industry-specific applications. 3D Experts often serve as the go-to resources for the most complex 3D projects.

Prerequisites

To best prepare to progress through this book, you need:

- A computer with Windows, macOS, or Linux
- 100 GB of free HDD (to download the data) and 16 GB of RAM (to be comfortable)
- A decent CPU (for performance) and an NVIDIA GPU (optional but recommended)
- To be ready for a transformative experience

Code, Data, and Resources

Working with 3D data means that we rely heavily on visualization modules. While we can work remotely with access to computing servers, it is usually much more efficient for 3D innovation to work and design first-hand solutions locally, leveraging your local machine capabilities.

This means that we are going to rely on a local Python setup accompanied by several well-known open source tools such as CloudCompare, MeshLab, Blender, and Meshroom. The ideal "starter pack" is laid out in Chapter 3, where I made sure to avoid any extra costs other than your computer, ideally equipped with a recent CUDA-enabled GPU (more on that later).

Regarding datasets, I am thrilled to provide everything necessary from my collection, which you will be free to use following the license guidelines (mostly open). Finally, all the code is also given to you, either in a notebook *(.ipynb)* file format or as a Python script *(.py)*. It is also part of the text to ensure maximum clarity, e.g.:

```
import numpy as np
x = np.array([1,2,3])
```

Figure P-3 presents a QR code that allows you to access the resources offered in this book, hosted thanks to the 3D Geodata Academy (*https://learngeodata.eu/3d-data-science-with-python*). You will find a complimentary course, code, datasets, and any relevant resources. This is your learning hub. All in all, this is a standalone experience with no extra dependencies other than the few Python libraries that we are going to use (less than 20), the open source software mentioned, and code with data provided in the book.

Figure P-3. QR code allowing access to the resources offered in this book

At the start of each chapter, all you have to do is download the chapter folder that contains everything necessary for hands-on learning with 3D Python. If you want to deepen your expertise, I also created several courses to complement the book at the 3D Geodata Academy (*https://learngeodata.eu*).

Conventions Used in This Book

The following typographical conventions are used in this book:

Italic
: Indicates new terms, URLs, email addresses, filenames, and file extensions.

`Constant width`
: Used for program listings, as well as within paragraphs to refer to program elements such as variable or function names, databases, data types, environment variables, statements, and keywords.

`Constant width bold`
: Shows commands or other text that should be typed literally by the user.

`Constant width italic`
: Shows text that should be replaced with user-supplied values or by values determined by context.

This element signifies a tip or suggestion.

This element signifies a general note.

This element indicates a warning or caution.

Using Code Examples

Supplemental material (code examples, exercises, etc.) is available for download at the 3D Geodata Academy (*https://learngeodata.eu/3d-data-science-with-python*).

If you have a technical question or a problem using the code examples, please send email to *support@oreilly.com*.

This book is here to help you get your job done. In general, if example code is offered with this book, you may use it in your programs and documentation. You do not need to contact us for permission unless you're reproducing a significant portion of the code. For example, writing a program that uses several chunks of code from this book does not require permission. Selling or distributing examples from O'Reilly books does require permission. Answering a question by citing this book and quoting example code does not require permission. Incorporating a significant amount of example code from this book into your product's documentation does require permission.

We appreciate, but generally do not require, attribution. An attribution usually includes the title, author, publisher, and ISBN. For example: "*3D Data Science with Python* by Florent Poux (O'Reilly). Copyright 2025 Florent Poux, 978-1-098-16133-0."

If you feel your use of code examples falls outside fair use or the permission given above, feel free to contact us at *permissions@oreilly.com*.

O'Reilly Online Learning

O'REILLY® For more than 40 years, *O'Reilly Media* has provided technology and business training, knowledge, and insight to help companies succeed.

Our unique network of experts and innovators share their knowledge and expertise through books, articles, and our online learning platform. O'Reilly's online learning platform gives you on-demand access to live training courses, in-depth learning paths, interactive coding environments, and a vast collection of text and video from O'Reilly and 200+ other publishers. For more information, visit *https://oreilly.com*.

How to Contact Us

Please address comments and questions concerning this book to the publisher:

O'Reilly Media, Inc.
1005 Gravenstein Highway North
Sebastopol, CA 95472
800-889-8969 (in the United States or Canada)
707-827-7019 (international or local)
707-829-0104 (fax)
support@oreilly.com
https://oreilly.com/about/contact.html

We have a web page for this book, where we list errata, examples, and any additional information. You can access this page at *https://oreil.ly/3d-data-science*.

For news and information about our books and courses, visit *https://oreilly.com*.

Find us on LinkedIn: *https://linkedin.com/company/oreilly-media*.

Watch us on YouTube: *https://youtube.com/oreillymedia*.

Acknowledgments

Embarking on this journey into 3D data science wouldn't have been possible without the support of remarkable people. First, my genuine gratitude goes to my past and present students. Their curiosity, unwavering determination, and "Aha!" moments fueled my passion for this field and made teaching an unending source of joy.

I want to give special thanks to my industry collaborators, who not only provided invaluable insights into real-world applications but also trusted me to be a partner in their innovative endeavors. Witnessing the impact of 3D data science in action was a truly humbling experience.

To my fellow researchers and professors, thank you for the stimulating discussions, shared knowledge, and unwavering support. Your expertise and camaraderie propelled me forward in my academic journey.

I also want to say how grateful I am to my family and friends for putting up with my late nights, enthusiastic rants about 3D point clouds, and unshakable faith in my ability to finish this project. Your love and encouragement were the wind beneath my wings.

I'd also be remiss not to express my sincere gratitude to the dedicated editors and reviewers who meticulously combed through this manuscript. Their insightful feedback, suggestions for improvement, and a keen eye for detail significantly enhanced the clarity and comprehensiveness of this book. Their expertise has undoubtedly made this a valuable resource for all aspiring 3D data scientists.

## CHAPTER 1

Introduction to 3D Data Science

Have you seen the movie *The Matrix*? Can you picture the thousands of moving green characters set against a black background? These represent "the world" for a computer program, with a set of rules and algorithms that determine how things work. Without stretching the fundamentals too much, our setting for 3D data science is not too far from this idea.

Imagine having the ability to create accurate digital settings of real-world phenomena. From an existing factory, you could study the impact of a new virtual assembly line with robots and conveyor belts. In healthcare, you could simulate the effect of different medical treatments on a patient virtual model. Based on a city model, you could predict the impact of natural disasters and propose the best emergency response scenario. We have so much to gain by digitizing our world. But where do we start?

In 3D data science, we generate and work with spreadsheets that describe environments such as your city. We then combine interdisciplinary techniques to extract valuable insights and design tools that can augment the three-dimensional groundwork. This revolutionizes our decision-making capabilities by working toward an accurate digital link with reality.

The best part? You do not need a PhD to get started. A basic curiosity and a sprinkle of programming skills (think very basic Python) are all you need to dive into this exciting field.

This chapter establishes the foundation for our exploration of 3D data science. I explain the importance of dimensions in science and how spatial AI uses geometry to mimic how our brains analyze 3D data. We examine the definition of 3D data science and why it is experiencing rapid growth. I then introduce the 3D data science modular workflow and discuss its associated challenges, including 3D data acquisition, preprocessing, augmentation, the scarcity of annotated datasets, computational demands, model building, explainability, performance, and stability. To ensure a strong understanding of these concepts, I've included a hands-on exercise where we explore and manipulate point cloud data from the OpenTopography repository.

3D Data Science in Brief

Science is the process of asking questions about the world around us, gathering evidence to answer those questions, and using that evidence to develop explanations that can be tested and refined. It is based on observation, measurement, experimentation, and logical reasoning. When it leverages 3D data, it forms an exploding field: 3D data science (Figure 1-1).

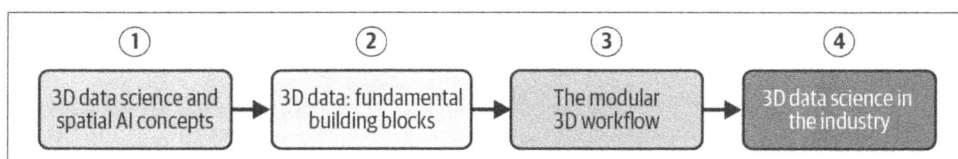

Figure 1-1. The four core components of 3D data science

This field bridges the gap between our three-dimensional world and the science of extracting knowledge and insights from data. It encompasses a range of techniques, from processing complex 3D models to crafting stunning visualizations that reveal hidden patterns.

To get started, we need to understand that three-dimensional world. The next section will break down what a dimension is and what makes it 3D.

Dimensions and 3D Data Science

What does the word "dimension" evoke in you? Is it something you tie to science fiction, like the Dark Dimension of some ethereal world? While not unrelated to fiction, in the context of science, specifically 3D data science, the dimensions we explore have a whole other meaning.

In essence, a dimension is a measurable aspect or characteristic of something. From a mathematical point of view, dimensions refer to the number of variables needed to describe a system or object. From a 3D data science perspective, dimensions often refer to spatial extent. Now, let me give you some examples.

A one-dimensional object can be described by a single variable, like the temperature of boiling water or the number of pages in this chapter. 2D data refers to information that is represented in two dimensions, typically with x and y coordinates. Using those x and y coordinates, it's possible to create digital images consisting of pixels that are analogous to geographical maps that depict places and features on a planar area utilizing latitude and longitude coordinates.

Extending this logic, we're able to comprehend that the real world we inhabit is considered three-dimensional, with points in space defined by x, y, and z coordinates. This is why when we relate dimensions to our visual perception system, we essentially refer to three-dimensional spaces that tie objects in contextual surroundings. For instance, a chair is a 3D object because it has length, width, and height, allowing us to distinguish it from a picture of a chair, which is a 2D representation.

> We can push the dimension concept further. An n-dimensional object or space can be described by n variables. While it's difficult to visualize, mathematical concepts like higher-dimensional spaces are often explored in fields like physics and geometry. For instance, string theory postulates that the universe might exist in 10 or 11 dimensions. But of course, as we increase the number of dimensions, we add more complexity and possibilities to describe objects and systems.

Dimensions become the base that permits us to anchor the reality we experience. With 3D data science, we have a newfound way to try and mimic the analytical parts of what our brain does with this "3D data." So, suppose we can somehow get a hold of a virtual replica of observed three-dimensional shapes that can serve as some data groundwork on which we can do science. Spatial AI opens the digital door to incredible brain-inspired possibilities.

Spatial AI: From Reality to Virtuality

Spatial AI is a powerful tool that allows machines to perceive, reason about, and interact with the world in a way that mimics human spatial intelligence. It is like giving computers a pair of eyes that can see beyond the flat screen, understanding the nuances of depth, distance, and orientation. It is comprised of foundational components including:

Perception
 How do machines "see" the world in 3D? Sensors like LiDAR, cameras, and depth sensors capture spatial information.

Reasoning
 Once machines have perceived the world, how do they make sense of it? Algorithms and techniques enable spatial reasoning and decision making.

Action
 How can machines translate their understanding of the 3D world into meaningful actions? Accurate digital environments form a robust base.

Let's break this down further. 3D shapes dominate our physical world. If someone asks you to describe the room you're in, you would likely describe your environment as an assembly of 3D shapes and geometries. As I write these lines, I identify my keyboard as the primary 3D subject and delineate each key, having a shape relating closely to deformed cubes. As I lift my head, my screen is standing still one meter in front of me in my current local spatial representation of my office.

When you break down this kind of positioning and observation behavior, it is amazing what our brain is capable of. A. M. Treisman explained in 1980 the complex mechanism of human visual perception in simple terms:

> When we open our eyes on a familiar scene, we form an immediate impression of recognizable objects, organized coherently in a spatial framework.[1]

For nonimpaired human beings, it is often the primary source of information that our cognitive decision system acts on. Our brain extends itself and adapts quickly to new surroundings, retaining the most crucial information from our vision. If we look deeper, the brain receives just three "images" every second. These are then sorted and combined with the available prior knowledge to create the reality we experience.[2]

This highly efficient mechanism allows us to brake at a red light or read and understand text. Our vision can also adapt to different attention modes, such as *orientation attention,* which conserves energy by not fully processing the surroundings, and *discover attention*, which functions in a "slow mode" while the brain collects data from our memory to gain a complete understanding of the scene.

With the help of advanced computational power and dematerialization, it is now possible to virtually replicate complex processes. Although replicating certain operations is challenging, studying how humans interact with the environment can help us better understand the limits and mechanisms that can be used effectively. Let me share an example.

Figure 1-2 depicts the relationship between the physical and digital worlds, specifically in the context of digital twins. The process begins in the real world, where sensors capture various physical objects or system aspects. These sensors can be anything from a camera capturing a building facade to a thermometer recording temperature. The captured data represents the as-is/as-built state of the physical world. This captured data is then processed and used to create a digital replica of the physical object or system.

---

1 A.M. Treisman and G. Gelade, "A Feature-Integration Theory of Attention" (*https://oreil.ly/UwV99*), *Cognitive Psychology*, vol. 136, pp. 97–136, 1980.

2 F. Poux, "The Smart Point Cloud: Structuring 3D Intelligent Point Data" (*https://oreil.ly/86Gii*), doctoral dissertation, Université de Liège, 2019.

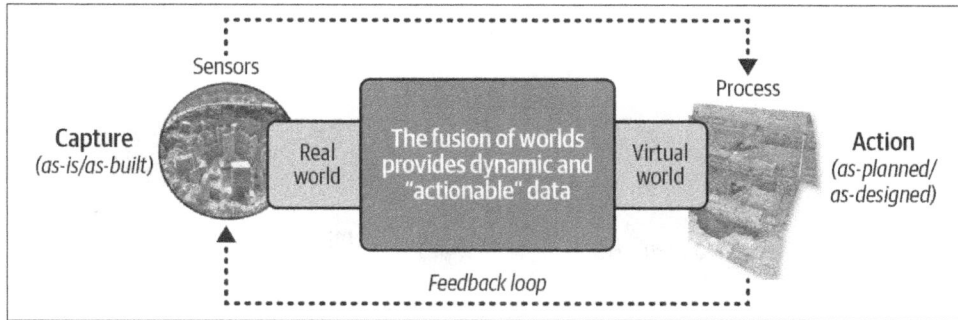

Figure 1-2. The process of going from reality to virtuality in the context of digital twins

This digital twin resides in the virtual world. The fusion process permits the linking of the data from the sensors in the real world, which can be fed back into the digital twin. The simulations or models can be updated to reflect the changing conditions in the real world. This creates a feedback loop between the two worlds.

Even at the molecular level, the constituting particles use specific geometries that give a functional meaning to what they describe. This fascinating trait can be primarily studied through 3D data science: the perfect combination of mathematics, computer science, and 3D data expertise (Figure 1-3). It capitalizes on this synergy and brings 3D machine learning, 3D data processing, and geometry and space analysis to new frontiers. Mixed, we obtain a mighty new branch that has widespread applications and is intertwined within the most cutting-edge workflows. It is a powerful area of study that is leading to spatial AI advances leveraging 3D machine learning, 3D data processing, geometry, and space.

3D data science provides a great chance to explore advanced technology and help solve real-world problems. There are a lot of 3D data applications, and the need for skilled professionals is growing, making this a secure career choice for the future. 3D data science drives innovation and creativity, helping create immersive virtual experiences, new product designs, and advanced AI systems. By mastering this field, you can help with important research, solve difficult problems, and influence the future of technology. Chapter 2 reviews the principal concepts of mathematics, computer science, and 3D data expertise to help break down the main constituents.

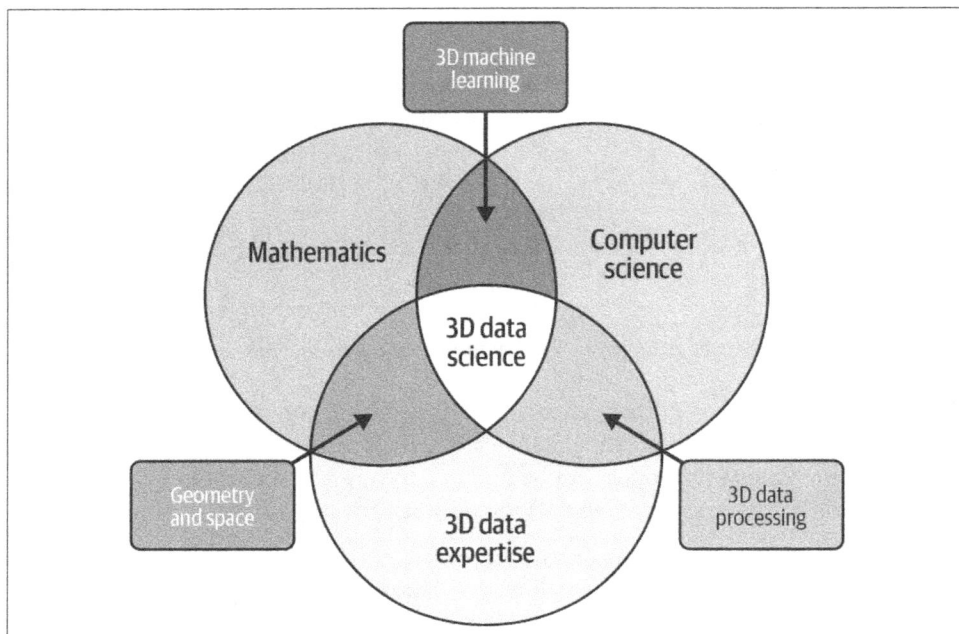

Figure 1-3. 3D data science is the intersection of mathematics, computer science, and 3D data expertise

Let us shift gears now and get into the specifics: establishing the workflows for 3D data science.

3D Data: Fundamental Building Blocks

The primary objective of 3D data science is to define the boundaries of a "digital environment." In essence, we want to create a replica of the world we inhabit, which will serve as a robust foundation for our digital applications. This precise digital representation of the real world will then enable intelligent agents to perform tasks in a manner that resembles human behavior. Once these foundations are established, we can connect to decision-making elements that can be triggered through voice commands, manipulations, and other means.

The first step is to understand that we have a base, which is the real world. Then, we take this base and create what we call the low-level digital model (Figure 1-4). It's essentially a geometric depiction of what we observe. Usually, most workflows stop here. But to get to these so-called "smart systems" or digital environments on which you can act and make decisions, you need to go an extra step and create a high-level digital model. This is a model that can be used for reasoning services.

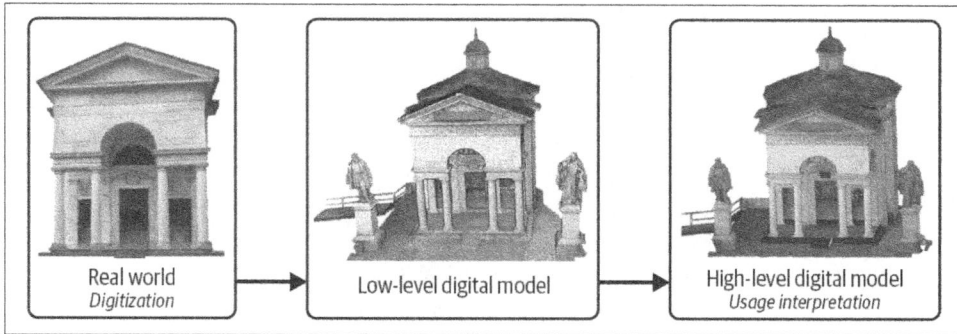

Figure 1-4. From the real world to a high-level digital model

What does this mean in practical terms? We need to inject additional information on top of geometry, precisely semantics and topology. Taking Figure 1-4 as an example, it could be knowing which portion of your model belongs to the pillars, which portion belongs to the roofs, what the statue is used for, how all these elements interact with each other, what their relationships are with one another, what would happen if this element had a fallback, or what physics are involved.

We aim to create a simple procedure that uses available data and information repositories to produce a semantic representation of a scene, which integrates concepts and their meanings. How do we get there? In this scenario, a spatial sensor acts as our eyes to capture a digital spatial asset that is refined into a semantic representation using available knowledge, either directly or indirectly. Availability is often the first complication we encounter when it comes to developing online cognitive perception. Our memory-based structure is optimized to access required information quickly. However, replicating this process using a computer is incredibly complex, and creating a general solution that works for all situations is a significant challenge.

The second bottleneck when virtualizing a cognitive decision system is creating a semantic representation. Gathering domain knowledge and attaching it to spatial data can lead to complex integration and mining challenges due to data types, sources, and representations.

So, basically, you want to model everything as it stands or extend from it to make some kind of simulation. But let's not build the system with the roof first. Let us lay the bricks, then move onto the mortar to hold them, and only after optimizing the tools to build faster will we achieve a more robust 3D data science workflow. To start small, at a conceptual level, we need to combine the three main elements that result in accurate semantic representations: geometry, semantics, and topology.

Geometry, Topology, and Semantics

Understanding the interplay between geometry, topology, and semantics is crucial for extracting meaningful information from three-dimensional data. Each of these concepts provides distinct insights into the structure and meaning of spatial information.

Geometry

Geometry encompasses the study of spatial properties, characterizing 3D shapes according to their size, position, orientation, and scale. To better illustrate this, let's consider a house (Figure 1-5). In this context, geometry defines the physical attributes of the house, including its dimensions, shape, and spatial arrangement.

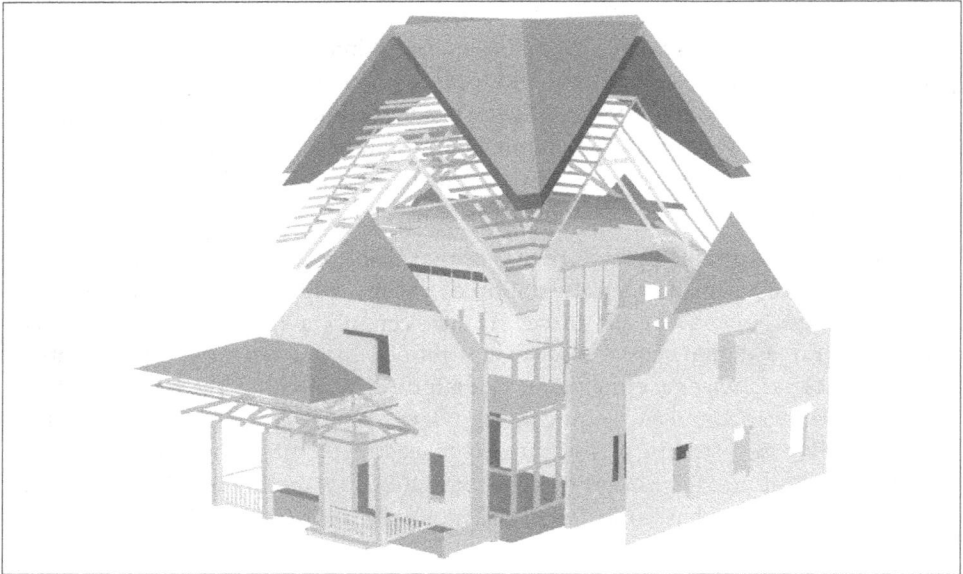

Figure 1-5. An example of a geometric house: note the many different planes and surfaces that need to be represented in a model

It permits extracting quantifiable aspects such as the length and width of the walls, the height of the roof, and the angles at which different surfaces intersect. The house's geometry can be described with various methods. For example, you can generate many discrete points sampled on the surface of the house's structural elements. Or you could see an assembly of polygons defined by points, lines, and surfaces that describe the physical footprint. You can also combine several basic 3D shapes (see Figure 1-6), then deform and combine them to obtain a final 3D geometry.

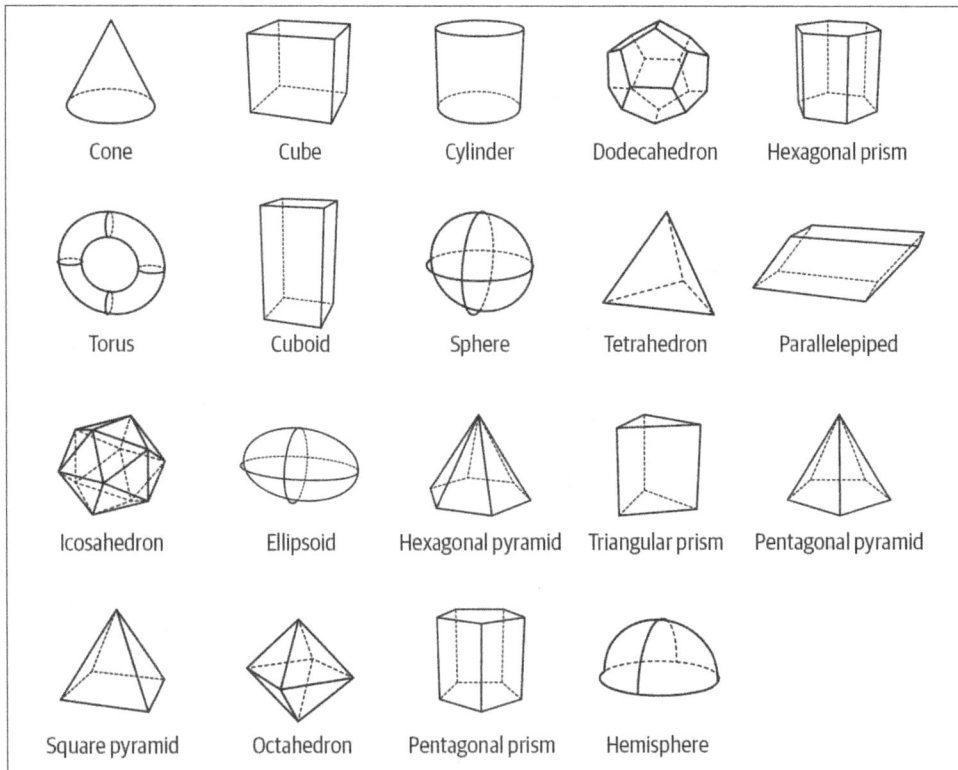

Figure 1-6. Basic 3D shapes that can be combined and deformed to create complex 3D geometries

Overall, obtaining the geometry is often the first step in 3D data science workflows. But if we stop at this stage, we fail to gather the profound insights needed for complex decision-making scenarios. This is why you need to define the topology once you acquire a sound geometry.

Topology

Topology focuses on the intrinsic properties of space that remain unchanged under continuous transformations, regardless of stretching, bending, or tearing. It deals with concepts like connectivity, adjacency, containment, and continuity.

It may be harder to grasp, so let me try to clarify with our house example (Figure 1-7). Topology helps us understand the relationships between its constituent parts. For instance, it identifies that the walls are connected to form rooms, and doorways connect rooms. Topology allows us to analyze how different elements of the house are spatially related, providing insights into the layout and connectivity of the structure.

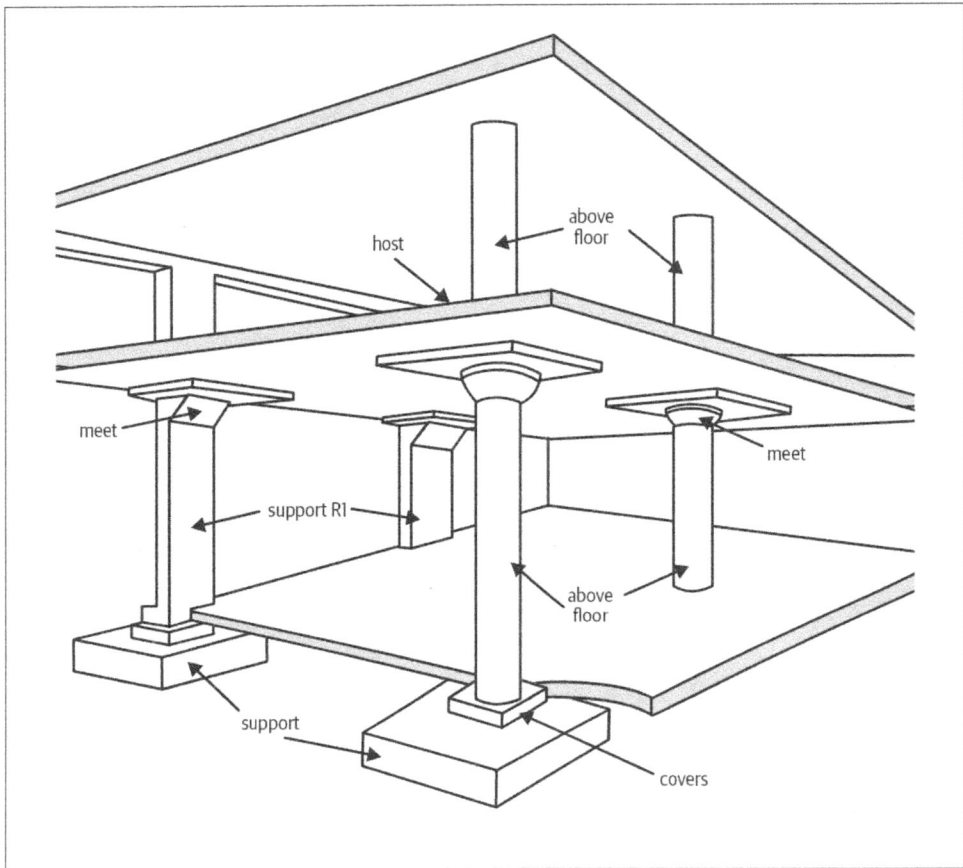

Figure 1-7. Inside a house, several elements have topological relationships: the column supports the floor, the floor stands on supports, and the columns are aligned

It is a way to contextualize objects of interest within a 3D geometrical scene. This all sounds great in practice, but how do we bring meaning to the entities we are manipulating and defining topologically? How do we know something is a floor, and something else a column? Well, with semantics.

Semantics

Semantics refers to the meaning, interpretation, or significance associated with elements or structures in a dataset. It involves assigning labels, attributes, or classifications that provide context and understanding to the data. Again, in the context of a house, semantics enable us to describe and categorize its components (Figure 1-8).

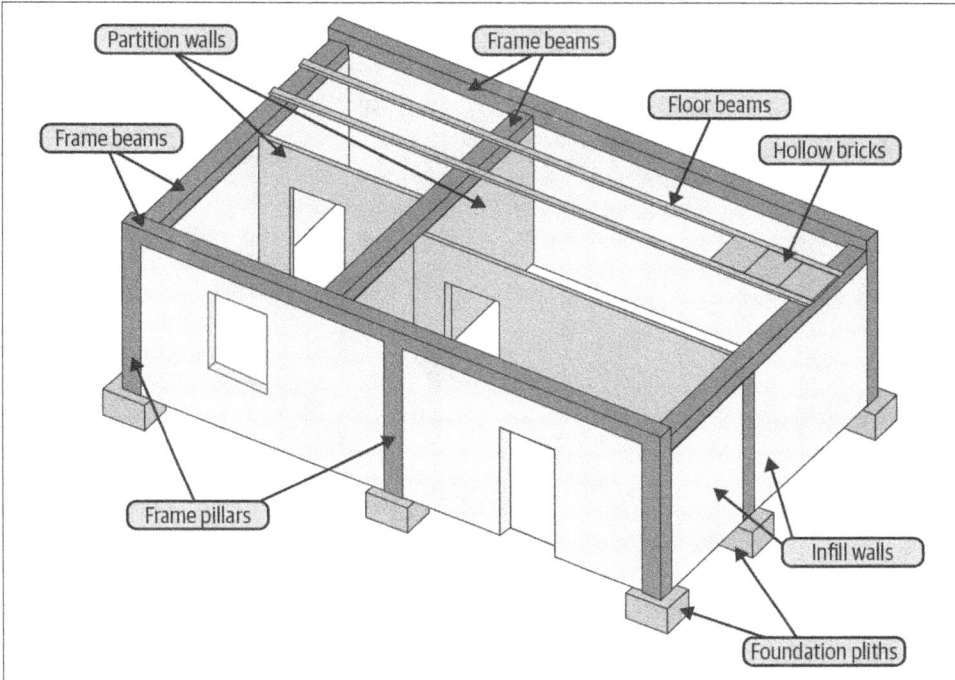

Figure 1-8. A conceptual label identifies a scene or object's main constituents which we can later tie to deeper knowledge sources

Semantics allows us to identify and label the kitchen, bedrooms, and living room within your house. This additional layer of information goes beyond the geometric representation and provides a higher-level understanding of each part of the house's purpose and function.

Naturally, each one of these pillars can be considered independently. However, to develop the most robust 3D data science solutions, integrating geometry, topology, and semantics is a necessary component.

Integrating Geometry, Topology, and Semantics

The synergy between geometry, topology, and semantics is crucial for comprehensive 3D data analysis. By combining these concepts, we gain a holistic understanding of spatial data. For example, in a 3D urban environment, geometry provides information about the physical structures, topology defines how these structures are connected, and semantics give meaning to various elements like buildings, roads, and parks.

But we can even describe what we consider a base object, such as a chair. We can refine this object into its constituent parts and define how each interacts with the others. In Figure 1-9 we have a backrest, a seat, and four legs (legs 1 to 4) that are located in a kitchen.

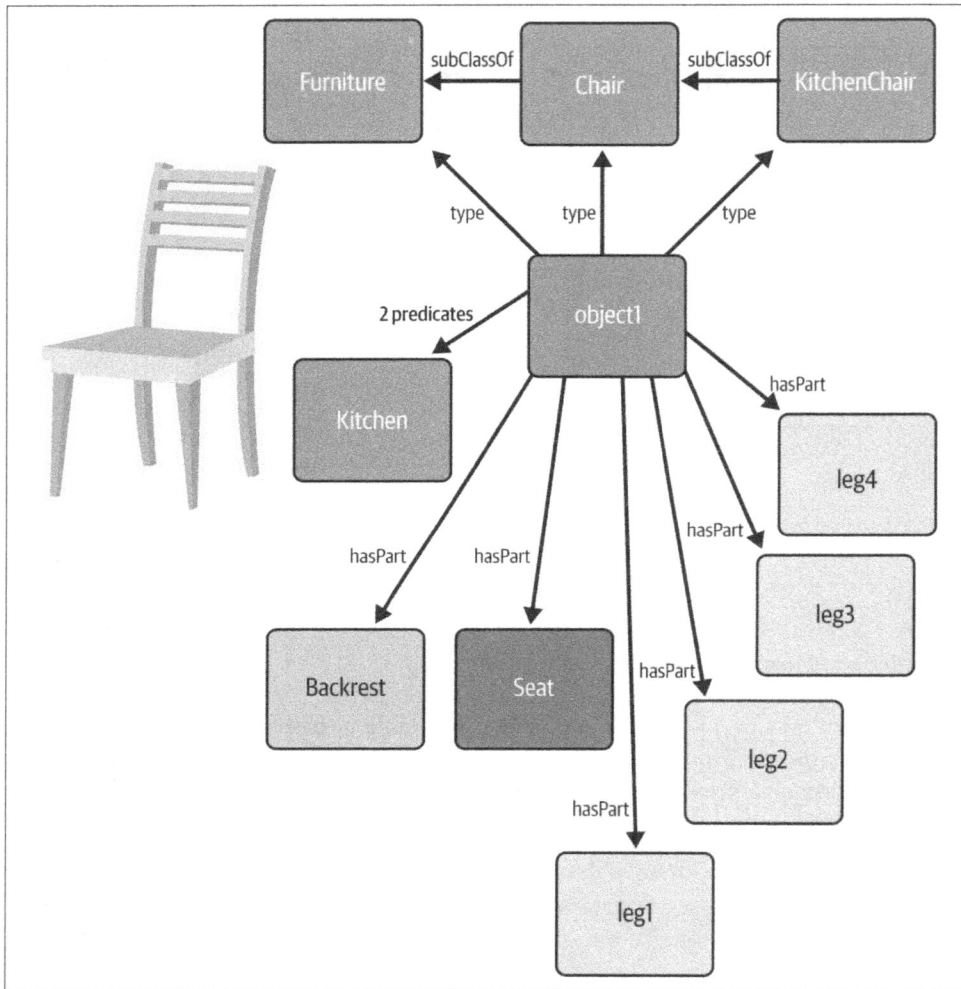

Figure 1-9. A chair, described with its geometry, topology, and semantics

Now that we understand what we are searching for and why, let me quickly illustrate a central data modality that often forms the base layer of R&D solutions: 3D point cloud data.

Introduction to 3D Point Clouds

Entities and objects can be described at varying levels of detail. For instance, a point can represent a city, a person, or a country. However, in the field of 3D data science, data points are typically collected using a set of sensors. While single raster images or video streams can be useful in certain cases, they may only provide the necessary depth of information to emulate our 3D visual perception accurately. Therefore, a more comprehensive and diverse data foundation is often required.

Reality capture devices allow for the capture of comprehensive 3D spatial information in the form of a point cloud, a spatial ensemble consisting of {X, Y, Z} coordinates, along with attributes that represent the recorded environment based on the strengths and limitations of the sensor.

The landscape of suitable hardware, software, and methodological choice has matured to the extent where it is possible to create digital replicas of the real world, ranging from micro+ level scale to country scale, as illustrated in Figure 1-10.

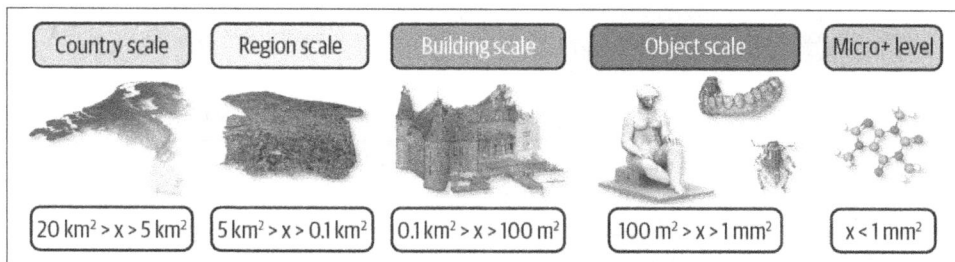

Figure 1-10. Multiscale point cloud of different datasets at various scales

Acquiring point clouds has become faster and more accessible with low-cost solutions. However, the software that supports these hardware advancements has not kept pace and is still facing challenges due to the five Vs of big data, namely volume, velocity, value, veracity, and variety, as illustrated in Figure 1-11.

To explain this concept in simple terms, when we use different sensors and methods to collect data, we end up with diverse point cloud datasets that are large in size. This creates challenges in terms of processing speed and accuracy, as we need to ensure that the data is reliable and can be translated into useful information. Therefore, it is crucial to verify the truthfulness of the data and extract actionable insights from it in a timely and efficient manner.

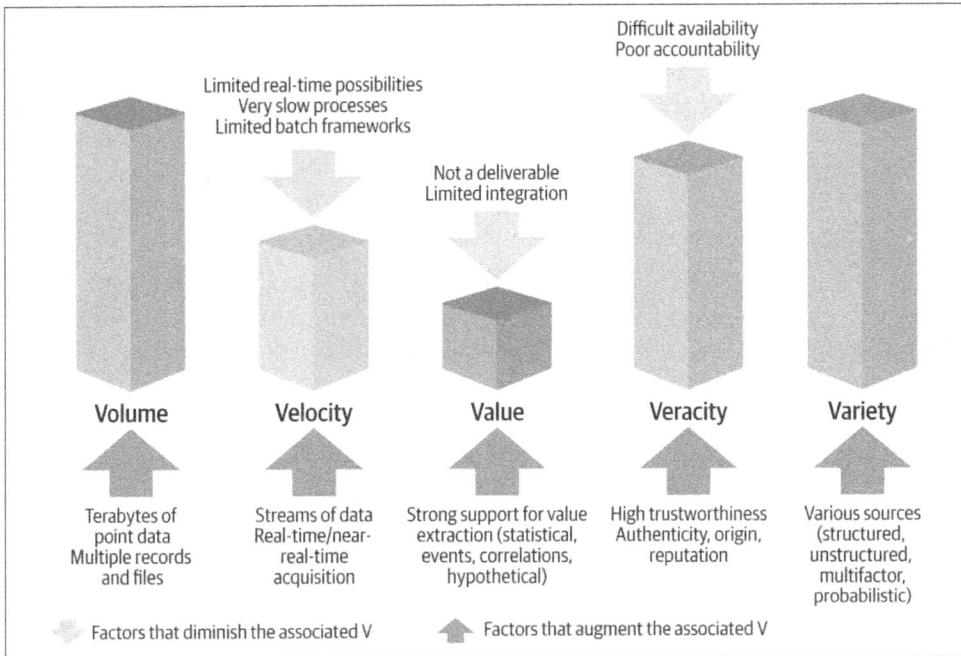

Figure 1-11. The five Vs of big data in the context of point clouds

It's worth noting that the 3D data science workflow, which relies on point cloud data-sets, is usually application specific. The workflow follows a classic progression from gathering relevant data to creating deliverables. While the data acquisition step may be specific to the sensors and platforms available, point-cloud-as-a-deliverable is becoming more popular and is now the preferred choice for many industries. In this task-oriented scenario, point clouds serve as a spatial reference, which experts use to create other deliverables. As such, it is the project's closest link to reality, providing accurate real-world information to facilitate decision making based on digital reality instead of outdated or interpreted information.

Experts manually process point cloud data in many production operations to extract useful information. This specific process is time-consuming and usually prone to human errors. One pillar idea behind 3D data science is to integrate as much expert knowledge into the data itself, giving spatial entities a semantic and topological meaning in an automated manner. This saves time, reduces errors, and improves the overall efficiency of such operations.

Integrating components to allow automated decision-making scenarios on a 3D point cloud model refers to the concept of smart point clouds, first introduced in 2016 by Poux et al (*https://oreil.ly/wl6zY*). If this concept triggers your curiosity, you can also refer to the "Smart Point Cloud" thesis (*https://oreil.ly/OL7Vw*), which is openly accessible, to structure 3D intelligent point data.

Furthermore, transforming point clouds into application-specific deliverables can be quite time-consuming and often requires significant manual intervention. Because the volume and complexity of the information we have to process continues to grow, it is becoming increasingly challenging for human experts to manage it effectively, especially when dealing with conflicting data dispersed among various project stakeholders and platforms.

As a result, translating big point cloud data into more efficient processes is key for a sustainable system. This opens up a new generation of products and services that help decision making and information extraction. In the next section, we'll delve into efficient workflows to avoid task-specific and nonsustainable systems.

Before we move on, there is one important note to add. I purposefully illustrated 3D challenges with 3D point clouds to provide a visual representation. But of course, as with any 3D data science workflow, viewing visuals independently from the 3D data modality doesn't show the complete picture. Throughout the rest of this chapter, I'll provide initial workflow definitions that will set up a stable base for various applications.

The 3D Data Science Modular Workflow

Have you ever wondered how research and development transitions into real-world applications? When building production-ready systems or tackling engineering challenges, understanding the significance of workflows in 3D data science is crucial. If you refer back to Figure 1-1, this is stage three out of four.

Indeed, having a system for managing repetitive processes and tasks in a particular order helps us understand the whole picture while clearly delineating each step. In 3D data science, workflows often share the same structure but provide different sequences or combinations based on the needs of an application. The word "application" is significant, as countless workflows can be used for data science applications. Depending on your specific goals, you may need to go in a particular direction.

However, I propose taking a higher-level approach, identifying families of workflows that can be used more broadly and holistically. This way, you can select any of these workflows and tailor them to your needs. There are eight families of defined workflows that fall into this category (Figure 1-12).

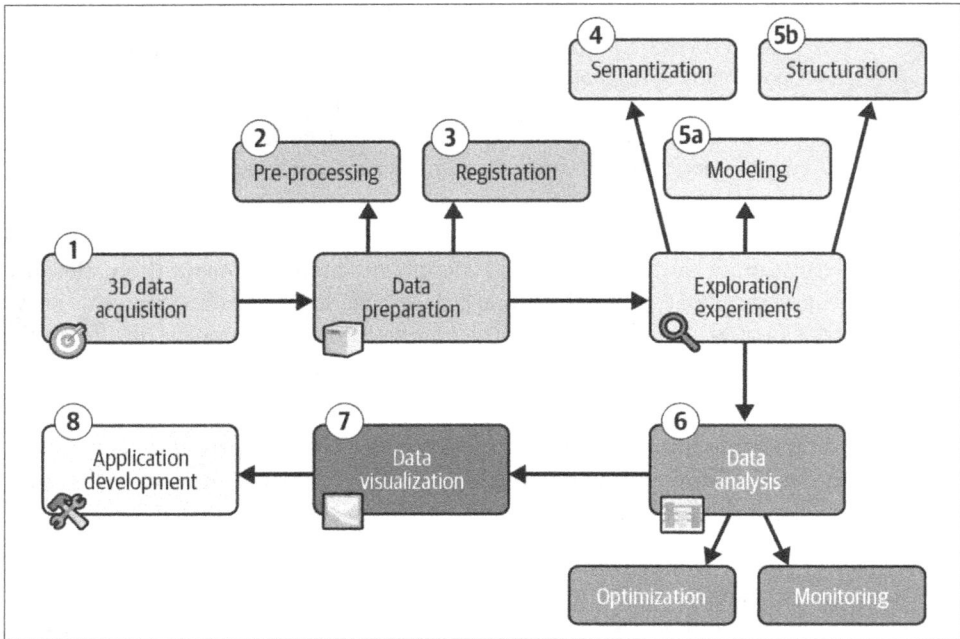

Figure 1-12. A high-level view of the 3D data science workflow

Let's explore each of these critical components of a 3D data science workflow and how they work together to create valuable solutions.

Data Acquisition

The first step in acquiring this extra digital representation is to have a way to capture it. This is where we rely essentially (but not only) on 3D sensors that allow us to capture the three-dimensional shape and sometimes even other properties of objects and scenes. Now, 3D sensors are not a monolithic entity. They come in various flavors, each with their strengths and weaknesses. If you want an extensive view of these systems, I wrote a complimentary guide that you can find on the 3D Data Science Resource Hub (*https://learngeodata.eu/3d-data-science-with-python*). It specifically looks at stereo vision, structured light, time of flight (ToF), and LiDAR sensors.

> We have other ways to capture data that do not stem from real-world scenarios but from pure creativity, such as simulations and video games. Chapters 2 and 3 include explicit ways to get these datasets or how to best source the information needed. Additionally, Chapter 18 will showcase how you can use generative AI tools for 3D dataset generation.

Essentially, this means that capturing data in three dimensions allows us to create precise representations of real-world objects and environments. We can create 3D data representations from real-world objects by using data capture vectors and methods to assess a given object, such as a city. Then, the data is processed and 3D representations can be created.

These methods are essential for creating accurate digital environments that provide a base for critical applications in urban planning, infrastructure management, and disaster response. Incorporating 3D data improves our capacity to depend on accurate and trustworthy data representations for decision making. Furthermore, integrating this data into digital environments can produce remarkably lifelike replicas of real-world assets and systems, thereby increasing the efficiency of simulation and analysis.

3D Digital Twin

There's another important concept to talk about with 3D data representations: 3D digital twins. This is a term that you may have heard or used, but it is important to understand what that encompasses. I prefer to avoid vague definitions, so instead of simply calling it a "virtual representation," let's break down what that truly entails. A 3D digital twin is a dynamic, data-driven replica of a physical object or system existing in the real world. This replica isn't just a static 3D model; it's constantly updated with real-time data, mirroring the state and behavior of its physical counterpart. Think of it as a living, breathing digital mirror.

This real-time connection is key. Imagine a wind turbine. A 3D digital twin of this turbine wouldn't just show its shape and location. Sensors on the real turbine would feed information about its rotational speed, temperature, vibration levels, and even the surrounding wind conditions to the digital twin. This allows us to monitor the turbine's performance remotely, predict potential failures, and optimize its operation—all within the digital realm.

The "3D" aspect is crucial as well. While digital twins can exist in 2D or even as purely numerical representations, the 3D aspect allows for spatial understanding and analysis. We can visualize the turbine's blades interacting with the wind, simulate stress points under different loads, or even plan maintenance operations in a virtual environment before applying them in the real world. This spatial context adds immense value, particularly in engineering, manufacturing, and urban planning.

Now, how do we actually build these digital twins? It's a multistage process that starts with a highly accurate 3D representation of the physical object. Next, we integrate this 3D model within our 3D data science pipeline. Finally, we need a platform to visualize and interact with the digital twin, allowing us to extract insights and make informed decisions.

Therefore, consider the digital twin as a dynamic, data-driven virtual representation of a specific physical entity or system, continuously synchronized with its real-world counterpart. A digital environment, while also a virtual space, represents a broader context or setting—like a factory floor, a city, or a natural landscape—and doesn't necessarily maintain a live, individualized connection with a physical counterpart.

This should help clarify the whole concept and explain why I specifically employ the term "digital environment" within this book, avoiding inaccurate hype taxonomies.

Now, we are ready to see what 3D datasets coming from 3D data acquisition techniques look like in practice. Let us go through a six-step process that leverages open data repositories as illustrated in Figure 1-13.

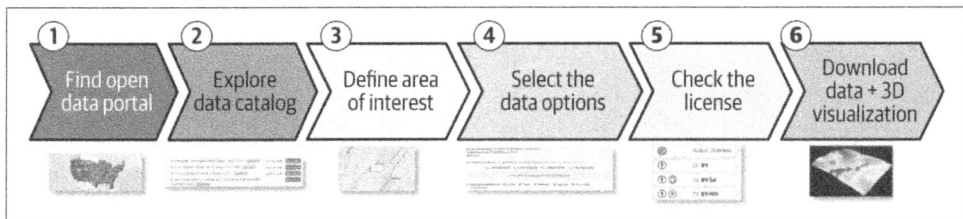

Figure 1-13. An example of a data curation workflow that leverages open data repositories rather than relying on on-site 3D data acquisition

The first step is to identify an open data portal. This example uses open data because, for these purposes, we do not want to go on-site to do a data acquisition, and this gives us a chance to assess what is openly available. Let's start with an excellent resource for getting our hands on 3D point cloud datasets: OpenTopography (*https:// opentopography.org*).

From this site, we can explore the data catalog (step 2) and choose from there what we would like to explore. For this example, let's choose Manawatu - Whanganui, New Zealand 2022-2023 (*https://oreil.ly/8MM6H*). In step 3, select a tiny region to access point cloud data (*https://oreil.ly/0FZuV*). Next, select the option to generate a 3D point cloud web browser and remove the generation of TIN. Finally, check the use license (CC BY 4.0) and the data citation[3] to submit our job and obtain our 3D results. If satisfied, we can download our dataset and the license for clarity.

Now that we have our 3D dataset, let's consider the next step in the overall pipeline.

---

3 Horizons Regional Council, Toitū Te Whenua Land Information New Zealand (LINZ) (2023). Manawatu - Whanganui, New Zealand 2022-2023 (*https://oreil.ly/s-yT3*). Collected by Landpro, distributed by OpenTopography and LINZ. Accessed: 2023-11-14.

This section was designed to ensure you can quickly access pragmatic matters. To avoid overflowing our cognitive capacities at this stage, we will move on to the next stages of the workflow. You'll find additional methodologies and tools for 3D data acquisition in Chapter 2 (software), Chapter 3 (methods and tools), and Chapter 18 (3D generative AI with spatial AI).

Preprocessing

In a 3D data science workflow, going through a data preprocessing step is essential. Clean, noise-free, and manageable datasets are needed to form the base for extracting significant features for further processing, such as registration, segmentation, and classification.

This process is divided into six main categories (Figure 1-14): data profiling, data cleaning, data reduction, data transformation, data enrichment, and data validation. Preprocessing pipelines almost systematically focus on addressing noise reduction and missing data issues.

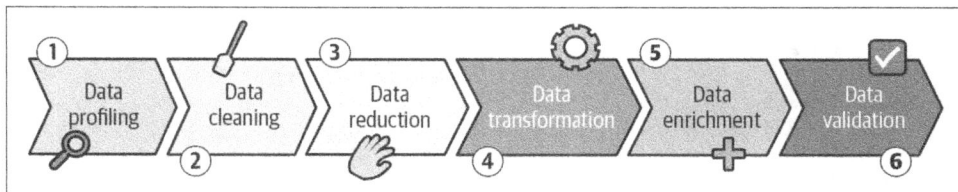

Figure 1-14. A high-level view of the six main components of data preprocessing workflows

We will explore each step in greater detail in Chapters 5 and 6, but first, let me synthesize each step very simply. Data profiling is straightforward: it involves gaining a good understanding of the data at hand. Data cleaning then focuses on dealing with noise and missing data, particularly in scenarios like collecting data in adverse weather conditions. Data reduction involves sampling strategies and dimensionality reduction techniques, while data transformation includes normalizing, scaling, and selecting features.

These steps can be implemented with various strategies, leveraging programming languages. As you know, we'll focus on Python in this book. Let us execute some code to demystify data preprocessing. To do that without setting up a local environment (we'll cover this in Chapter 3), we can use a cloud solution, such as Google Colab (*https://colab.research.google.com*).

Let me give some notes on Google Colab, drawing a parallel to something familiar—a fully configured workstation in the cloud. It is close to a computer accessible through your web browser, preinstalled with data science tools like Python libraries, a Jupyter Notebook (*https://jupyter.org*) interface, and access to processing unit (GPUs (*https://oreil.ly/IlMEN*), CPUs (*https://oreil.ly/HBIau*), and TPUs (*https://oreil.ly/RLNSQ*)). In essence, that is Google Colab. It lets you write and execute Python code in your browser by leveraging hardware acceleration without local installation or resource management. This is very handy to try out new ideas, share interactive analyses, or dive into resource-intensive tasks without substantial up-front investment. Think of it as your on-demand coding environment. I like to use it for quick demos that do not need heavy 3D visualization, which is the case in this section.

To follow along, you can open and execute the Python script hosted on a Colab I've created, First 3D Steps (*https://oreil.ly/kSQLP*), or use the following code if you are already well versed in Python:

```python
# Import the first library
import numpy as np

# Import the second library
import matplotlib.pyplot as plt

# Define the coordinates of the cube vertices
vertices = np.array([
  [-1, -1, -1], [1, -1, -1], [1, 1, -1], [-1, 1, -1], [-1, -1, 1], [1, -1, 1],
  [1, 1, 1], [-1, 1, 1]
])

# Define the edges of the cube
edges = np.array([
  [0, 1], [1, 2], [2, 3], [3, 0], [4, 5], [5, 6], [6, 7], [7, 4], [0, 4], [1, 5],
  [2, 6], [3, 7]
])

# Create a figure and axes object
fig = plt.figure()
ax = fig.add_subplot(111, projection='3d')

# Plot the cube vertices
ax.scatter(vertices[:, 0], vertices[:, 1], vertices[:, 2])

# Plot the cube edges
for edge in edges:
  ax.plot(vertices[edge, 0], vertices[edge, 1], vertices[edge, 2], color='black')

# Set the axis labels
ax.set_xlabel('X')
```

```
ax.set_ylabel('Y')
ax.set_zlabel('Z')

# Show the plot
plt.show()
```

Here, we simply create a 3D cube composed of vertices (points) and edges (lines between points) and then draw it in a Euclidean space. After this, we could transform the cube by translating and scaling it:

```
# Translate the cube by (1, 1, 1)
vertices_translated = vertices + np.array([1, 1, 1])

# Scale the cube by a factor of 2
vertices_scaled = vertices * 2

# Create a figure and axes object
fig = plt.figure()
ax = fig.add_subplot(111, projection='3d')

# Plot the original cube vertices
ax.scatter(vertices[:, 0], vertices[:, 1], vertices[:, 2], label='Original Cube')

# Plot the translated cube vertices
ax.scatter(vertices_translated[:, 0], vertices_translated[:, 1],
vertices_translated[:, 2], label='Translated Cube')

# Plot the scaled cube vertices
ax.scatter(vertices_scaled[:, 0], vertices_scaled[:, 1], vertices_scaled[:, 2],
label='Scaled Cube')

# Set the axis labels
ax.set_xlabel('X')
ax.set_ylabel('Y')
ax.set_zlabel('Z')

# Add a legend
ax.legend()

# Show the plot
plt.show()
```

In the preceding example, you should obtain the corners of the original cube in blue, the scaled cube in green, and the translated cube in orange. Transformations extend the case of translation and scaling, and we dive deep into this aspect throughout the rest of the book. Once this is well constrained, we usually move on to the data enrichment phase. It is linked to what is called feature extraction. We aim to derive values from the data that enrich overall understanding.

These values are *features*, and they need to address three main properties:

- They need to be informative. There is no use in deriving things that do not give us more information than we have.
- They need not be redundant. If we can, we must limit them to things that provide substantial information, not very limited information.
- Finally, they need to play the role of facilitator in downward processes.

Let me illustrate a classic 3D object classification task: you want to know if a 3D object is a chair or a table. You need to extract some information as features from the data so that it can be helpful for your model (e.g., size, point distribution along the z-axis, height). Various approaches exist, whether part of feature engineering phases (see Chapters 6 and 7) or included within major 3D machine learning architectures (covered in Chapters 12 to 16).

At this stage, we can already note the influence and importance of understanding the larger goal of the targeted application. Different applications, such as structural deviation analysis, semantic segmentation, classification, and 3D modeling, require varying preprocessing and feature computation levels. Finally, we go through the data validation phase to ensure we answer the specific criteria we established.

> Additional steps are often needed when processing data. These may include preparing a zone of interest that provides greater detail, creating acceleration structure data, computing prefeatures, and transforming the 3D dataset into a helpful format for our objective(s). Therefore, it is important to remember that the serialization of the workflow modules may vary in practice.

3D data preprocessing is extensively used in every chapter, with various depths depending on the end goal. Once we have achieved our goal of transforming the initial messy data into a refined and valuable dataset, we are ready for registration challenges between datasets.

Registration

When dealing with multiple datasets from various perspectives, aligning them into a unified frame of reference becomes crucial. Whether bringing together scans of the same object or photographs taken from different angles, registration workflows enable holistic analysis and understanding. Additionally, data registration deals with scenarios where you have incomplete data and need to be able to translate, rotate, and scale components that are inside.

A typical scenario is if you have two "scans" from two different perspectives and you want to bring them both into one perspective (e.g., Figure 1-15). In this case, we want to merge two scans from photogrammetry assets. We first extract some features of interest and then apply a registration technique that leverages these features to align the two datasets in one common frame of reference.

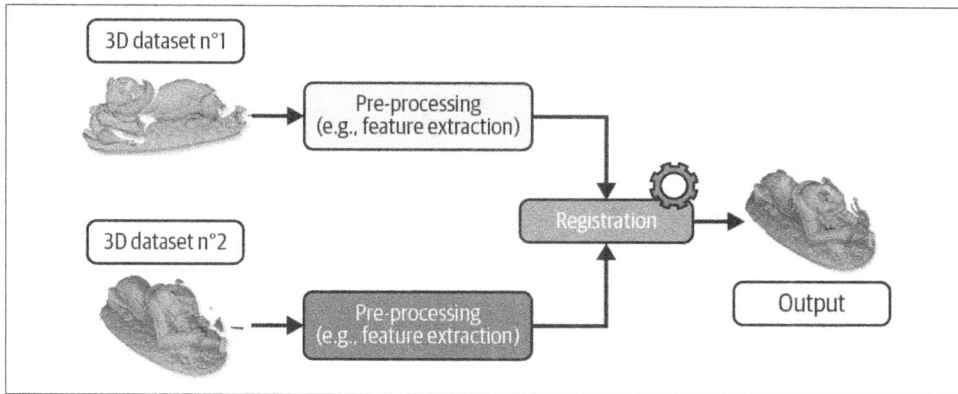

Figure 1-15. An example of the data registration process

We refer to this task as data registration, where the goal is to go from two local coordinate systems to one common frame of reference, where the datasets are coherently aligned. This also refers to interplatform registration, where you have different perspectives, different scales of acquisition, and different data typologies down the line.

However, the process is a bit more complex and can involve data calibration (*https://oreil.ly/Y8hQz*) or data fusion (*https://oreil.ly/NXw9R*) subtasks, which would extend the book's scope. These are covered in depth in the 3D Resource Hub (*https://learngeodata.eu/3d-data-science-with-python*) for anyone interested.

> Data registration is fundamental for creating accurate digital environments. A complete solution is provided in Chapter 6 with a global to local registration solution with Python that you can replicate and then extend to your own projects.

Now, let us move on to the most predominant module in 3D data science workflows: semantization.

3D Data Classification (Semantic Injection)

What we classically cover in the 3D scene understanding framework is the task of semantization, which we will refer to as classification. The main goal of this step is to understand the input data and to interpret the different parts of the sensor data. For example, we have a point cloud of an outdoor scene, such as a highway, collected by an autonomous robot or a car.

The goal of classification is to figure out the main parts of the scene. In other words, we need to know which parts of this point cloud are roads, which parts are buildings, or where the humans are. In this sense, it is an overall category that aims to extract specific semantic meaning from our sensor data. And from there, we want to add the semantics at different granularities. Let me describe each of these techniques (represented in Figure 1-16) with a bit more texture:

3D object detection

3D object detection techniques, or bounding-box detection techniques, enclose the object. They are a vital component for a lot of applications. Basically, they enable the system to capture objects' sizes, orientations, and positions in the world. As a result, we can use these 3D detections in real-world scenarios such as augmented reality applications, self-driving cars, or robots that perceive the world through limited spatial/visual cues. Nice 3D cubes that contain different objects. But what if we want to fine-tune the contours of objects?

3D semantic segmentation

With semantic segmentation techniques, we are able to attack one of the most challenging tasks: assigning semantic labels to every base unit (i.e., every point in a point cloud) that belongs to the objects of interest. Essentially, 3D semantic segmentation aims at better delineation of objects present in a scene. A 3D bounding-box detection on steroids, if you will. This method provides semantic information per point, allowing us to go deep in our analysis. However, a limitation still remains: we cannot directly handle different objects per category (class) we attack. Do we have techniques for this as well?

3D instance segmentation

Yes! And it is called 3D instance segmentation. It has even broader applications, from 3D perception in autonomous systems to 3D reconstruction in mapping and digital twinning. For example, we could imagine an inventory robot that identifies chairs, is able to count how many there are, and then moves them by grasping them by the fourth leg. Achieving this goal requires distinguishing different semantic labels and instances with the same semantic label. Think about instance segmentation as a semantic segmentation step on mega-steroids.

3D geometries and semantics are extracted and attached to the entities at this stage. Let's consider the structuration and modeling module next.

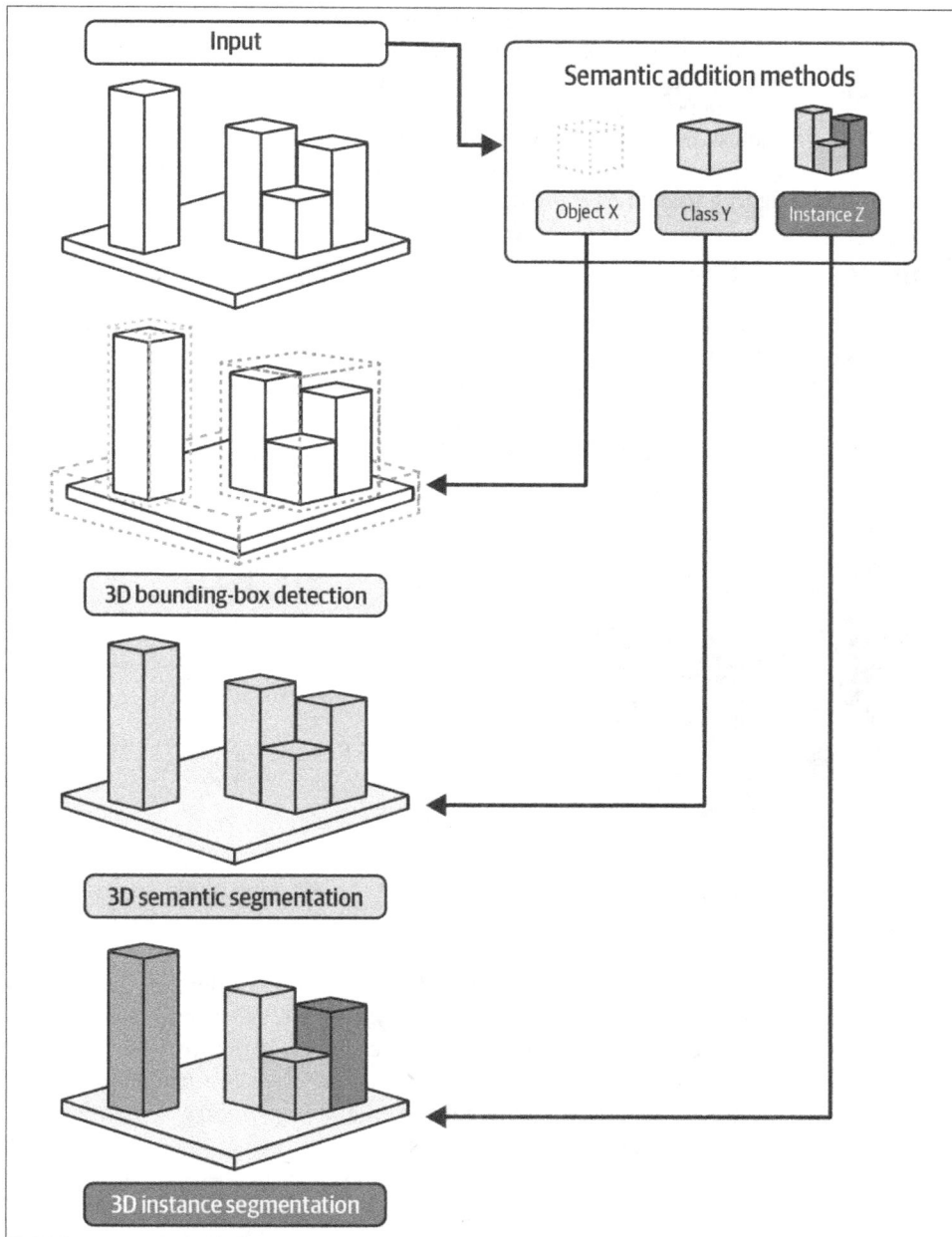

Figure 1-16. The three main semantization techniques: 3D object detection (bounding-box detection), 3D semantic segmentation, and 3D instance segmentation

The semantization module is heavily targeted by new AI approaches and sees the interplay between 3D machine learning and 3D deep learning methodologies. A large chunk of the book is dedicated to these topics to ensure you have the most cutting-edge understanding and methods to deal with these complex tasks. You can find methodologies and Python solutions leveraging principles, algorithms, and AI approaches in Chapters 9 and 12 through 18.

Structuration/Modeling

Structuring 3D datasets is often overlooked, but this is a crucial step. It ensures that the data is organized in a way that allows for easy manipulation and understanding. It also ensures that you can create coherent models that perfectly suit simulations or the underlying phenomenon that the data is trying to capture. We can leverage structuration techniques like constructing hierarchical trees, among others, to contribute to faster processing and analysis (Figure 1-17).

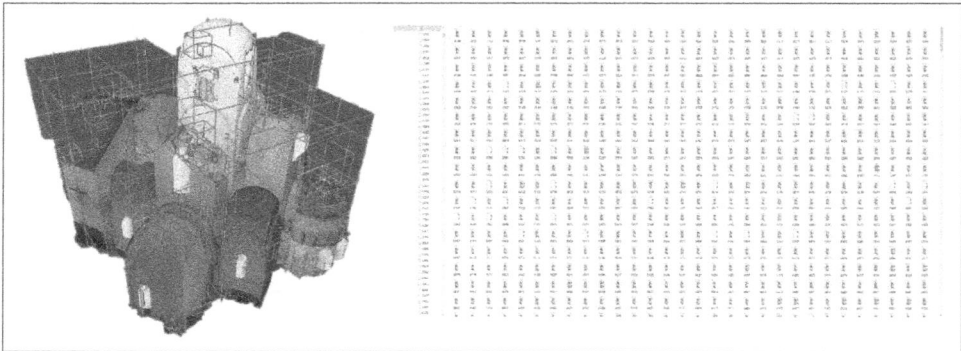

Figure 1-17. Example of an octree construction on top of a 3D point cloud and its associated out-of-core storage filesystem

Indeed, we often need to switch between the data structure we deliver and what is directly underlying it. When we deal with 3D data, we have an ensemble of geometries that we need to connect through a topological data structure with semantic pointers extracted. We need to overcome this obstacle, and we can usually achieve a data structure that coherently organizes our three pillars of geometry, semantics, and topology for the application or many applications.

At this stage, if we keep the linear progression of modules, we will have a dataset with a sound 3D geometry, semantics, and topology. As a result, we have all the elements needed to proceed to the analytical step.

> When dealing with 3D geometry, we often resort to acceleration structures, such as k-d trees, binary trees, and octrees, to speed up processes. Chapter 4 will introduce how to use Python and methodologies to best structure your 3D datasets. Then, Chapter 10 complements Chapter 4 by giving you advanced 3D modeling techniques.

3D Data Analysis

Once the structure is clear, analytical workflows come into play. Here, the goal is to describe the data accurately, understand trends, and make inferences. This phase forms a significant part of the overall workflow, as it uncovers insights hidden within vast amounts of data (Figure 1-18).

Figure 1-18. Example of a decomposition of a 3D point cloud dataset of a library into its main constituents

For example, we can extract valuable insights to gain a better understanding of the underlying 3D data by analyzing the spatial relationships between objects, their geometries, and their semantic meaning. These analytical tasks enable us to identify patterns, detect anomalies, and predict future trends, which can be incredibly useful across various fields.

Moreover, by leveraging 3D geometries and topology, we can perform complex analyses that would not be feasible using traditional non-3D data science methods. Ultimately, the ability to extract useful information from 3D data opens up new possibilities for innovation and problem-solving, making analytical tasks a critical component of the 3D data science workflow.

> Analytical tasks are intertwined with the application. However, we often rely on a standard subworkflow that leverages models to make inferences and predictions. As this is a large part of the challenges we encounter with 3D modeling, I cover this stage in depth in Chapter 8. Also, we extend the use of 3D data analysis through Chapter 9, which focuses on 3D shape recognition.

3D Data Visualization

Building visualizations and rendering pipelines adds more understanding to the data. Visualization workflows make it easier for people to grasp complex information by enhancing the capacity to present the data effectively. It's a vital component that ties together the other workflow modules. When you think about something, do you feel that you create some kind of visual representation? If we were to talk about rare species, like the narwhal (*Monodon monoceros*), if you already have some data or knowledge points on it, you would immediately picture these "unicorns of the sea," known for their long spiral tusks.[4]

This example shows how instinctive data visualization is for us. We have tremendous capabilities to make robust sense and synthesize or represent various topics with visual support. Figure 1-19 breaks down the analytics, discovery, presentation, perception, and interaction that all flow into the science behind data visualization.

---

4 I got a bit into narwhal recently. I have never seen any of them, but now, I know that I need to go to the Arctic waters near Canada and Greenland to get a chance to spot some of these beautiful creatures. I learned they are social creatures, often found in "pods" and with impressive diving abilities. I mean, I am happy if I can hold 1.5 minutes underwater; narwhals can dive to depths of up to 1,500 meters (4,921 feet) and remain submerged for extended periods, often over 25 minutes. What a marvelous power!

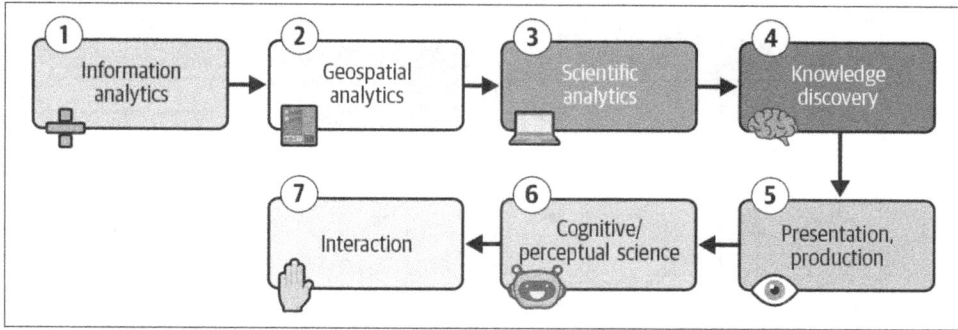

Figure 1-19. The main components of data visualization for analytics

Even more powerful than our instincts is the ability to handle 3D geometries and attributes that represent real objects with our modern computing systems. Indeed, this forms an essential tool for us to create better visuals that can serve any goal we set ourselves with more precision.

And speaking about pertinency, have you heard about Jacques? Not the Belgian singer Jacques Brel, not the French ex-president Jacques Chirac, not the Naval Officer Jacques Cousteau, but the French cartographer Jacques Bertin (1918–2010) (*https:// oreil.ly/w8ew8*)?

It is hard to speak about data visualization without mentioning Bertin's visual variables. They were introduced by Bertin and because they encode information visually, clearly, and effectively, they continue to be fundamental attributes to represent data graphically today. Figure 1-20 illustrates these variables, which include orientation, size, shape, color, texture, value, and position.

By strategically selecting and combining visual variables, we can create effective visualizations that simplify complex information for viewers to interpret (for cartography and 2D applications, these visualizations are most often found in the form of maps). And now, it is time to tie these modules to the application layer.

You will find some inspiration for your 3D visualization goals in Chapters 5 and 7. In Chapter 5, we will create a multimodal 3D viewer to handle various 3D modalities. In Chapter 7, we will develop some analytical tools like thresholding sliders to interact with our datasets in real time and bring an immersive component to the 3D experience.

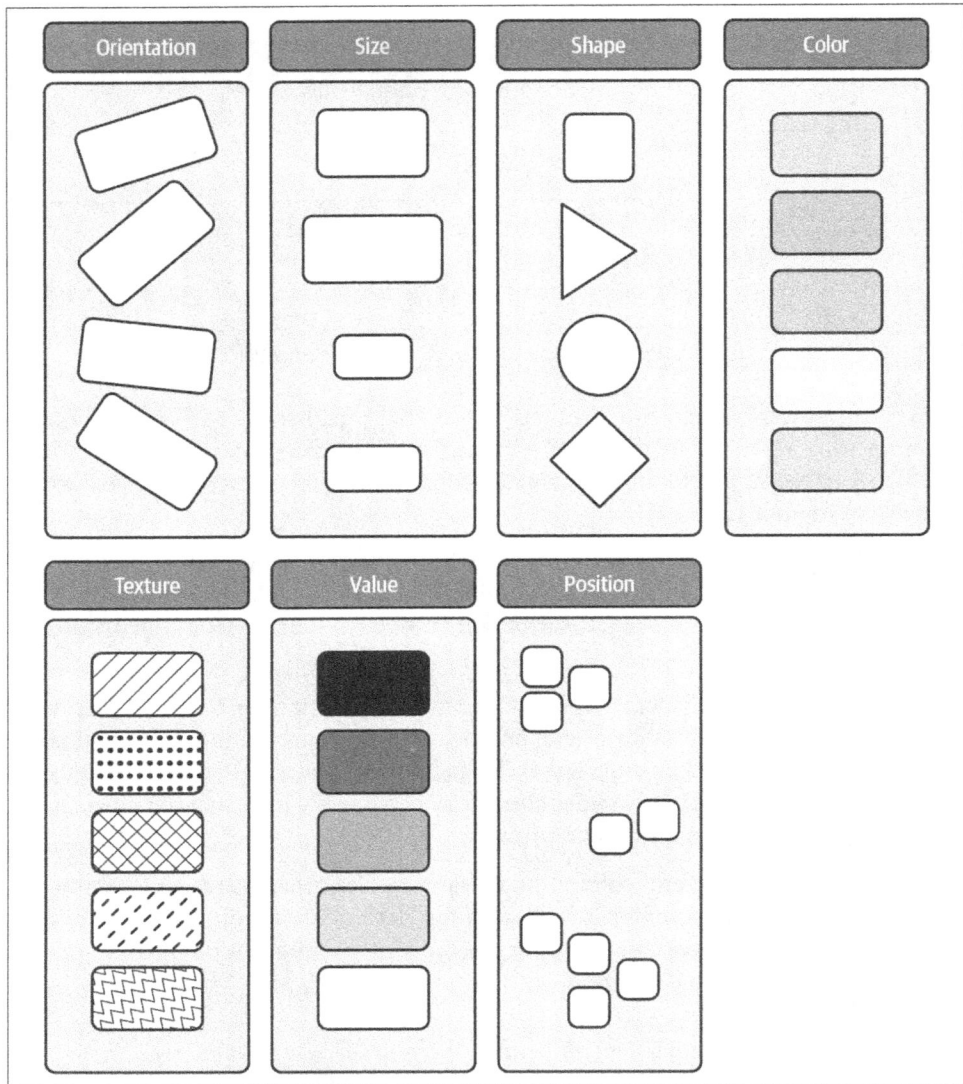

Figure 1-20. Bertin's visual variables

Application (Software) Development

This is an optional step that focuses on the meticulous development and deployment of our developed models and algorithms. While this book does not intend to give you a complete solution to application development, let me share a simple view of the critical phase as shown in Figure 1-21.

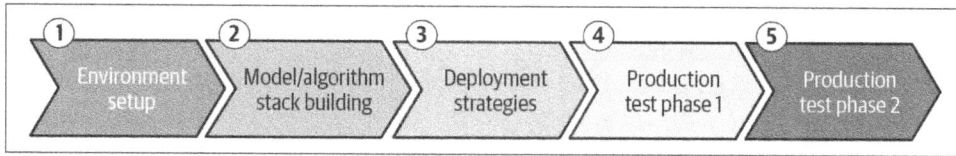

Figure 1-21. The main steps to create an application module that delivers production-ready solutions

As you may notice, by that point, you should already have stage 1, environment setup, and stage 2, model/algorithm stack, done. This will have been completed through the previous steps. However, during this production cycle, we usually refactor the code and make sure it is optimized to run as smoothly as possible. Also, compatibility checks follow, providing seamless integration with the target deployment environment, be it a cloud-based infrastructure, edge computing devices, or an existing system. This means that, especially for the models, we have to consider dependencies and craft a data processing pipeline that seamlessly processes 3D datasets. The focus here is on optimization—ensuring efficiency and processing capabilities.

> Imagine you are developing a 3D computer-aided design (CAD) application for architectural design. For this example, the environment setup involves installing and configuring software libraries such as OpenCASCADE (*https://www.opencascade.com*) for geometric modeling and PyTorch (*https://pytorch.org*) for deep learning integration and ensuring compatibility with the target operating systems used by architects and designers.

This is where we usually decide on a deployment strategy that aligns with real-world scenarios. The idea is to define the approach taken to deploy software changes from development environments to production environments. This is part of your software architecture definition. I encourage you to read *Fundamentals of Software Architecture* by Mark Richards and Neal Ford (O'Reilly) if you intend to make your 3D data science journey a real product.

> If you intend to transform your computer code solution into a tangible, marketable product, then you have the choice of several productization methods. You could create a standalone application, software as a service (SaaS), embedded software, a software component, or even offer a professional service. Creativity does not stop at the science behind the data only. However, the choice of offering naturally impacts the architecture of your software.

Nowadays, you can benefit from cloud deployment, which provides scalability and accessibility and may help you minimize downtime, reduce risks, and ensure successful and efficient implementations. Let's consider another example. In the domain of geospatial analysis, utilizing cloud services like Amazon Web Services (*https://oreil.ly/7eoYI*), Microsoft Azure (*https://oreil.ly/vlpkb*), or Google Cloud Platform (*https://oreil.ly/XpEJ9*) ensures the scalability and accessibility of users globally for analysis and visualization of large-scale 3D spatial datasets.

Once your solution is live, there are still some important stages: rigorous production testing is essential to validate the robustness and scalability of your deployed 3D data science application. This entails performance testing to evaluate response times and resource utilization, stress testing to assess peak load handling, and security testing to fortify against vulnerabilities. This is Phase 1, which mostly involves internal testing.

Imagine deploying a 3D medical imaging application for tumor detection. In the production testing phase, performance testing assesses the application's ability to process high-resolution 3D medical images efficiently. Stress testing evaluates how the system handles simultaneous requests from multiple medical professionals, while security testing ensures patient data confidentiality.

However, there is still a major component missing, which is taken care of by what I call Production Test Phase 2. It usually involves a pilot phase that initiates the rollout to a limited user base and collects initial feedback and insights. Staged deployment follows, with a gradual expansion to larger user groups.

This specific process is a "world of its own." It largely extends the scope of this book but is also super interesting when you are ready to turn your ideas into products. Therefore, I can recommend a series of seven books that will guide you step by step through the process of realizing a great software product:

- *The Pragmatic Programmer* by David Thomas and Andrew Hunt (Addison Wesley Longman)
- *Modern Software Engineering* by David Farley (Pearson Education)
- *Code Complete: A Practical Handbook of Software Construction* by Steve McConnell (Microsoft Press)
- *Software Engineering at Google* by Titus Winters, Tom Manshreck, and Hyrum Wright (O'Reilly)
- *Fundamentals Of Software Architecture* by Mark Richards and Neal Ford (O'Reilly)
- *Software Architecture: The Hard Parts* by Neal Ford, Mark Richards, Pramod Sadalage, and Zhamak Dehghani (O'Reilly)
- *Designing Machine Learning Systems* by Chip Huyen (O'Reilly)

At this stage, you should have a clearer idea of what lies behind the 3D data science workflow. Understanding these processes paves the way for groundbreaking advancements in data capture, preprocessing, registration, extraction of semantics, optimization of datasets, analysis, visualization, and tailoring to specific applications.

But understanding the basics doesn't end here. In the world of 3D data science, automation is critical. While automating these workflows is no easy feat, achieving a certain level of automation brings immense benefits. Let's explore this next.

The Case for Automation

Streamlining our team's tasks and workflows in production can greatly improve project execution. Indeed, effective workflow management allows us to build, automate, and manage workflows, gaining a clear understanding of the entire process. It helps eliminate confusion and chaos and simplifies processes. One aspect is especially central: automation. Automating tasks provides benefits, as illustrated in Figure 1-22.

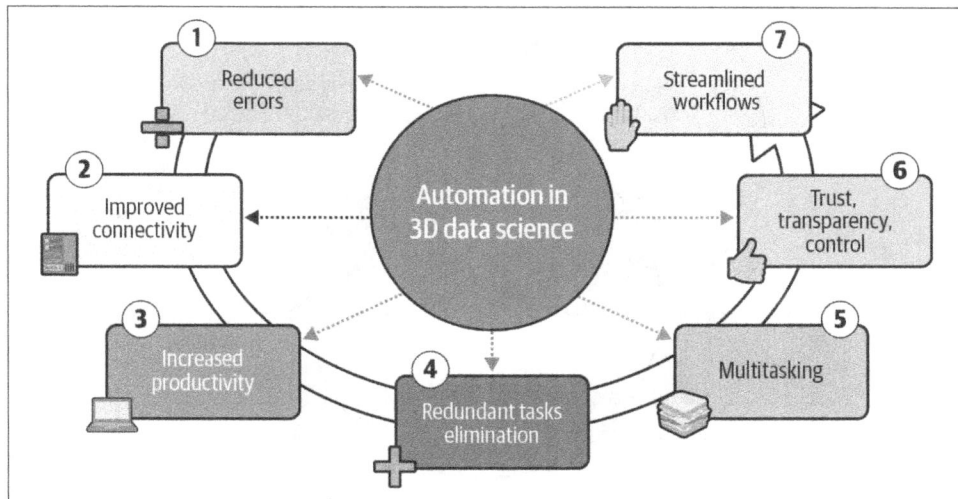

Figure 1-22. The benefits of automation in 3D data science

Related to 3D data science, we can extend our thoughts from a human-centered process to an autonomous workflow. This orients our goal to develop automation and spatial AI to speed up inference processes. It is crucial when developing workflows that encompass semantization tasks, such as where objects must be identified.

When we look at the industry's perspective, we see that robotics research has made significant progress in providing autonomous 3D recording systems. These systems can capture a 3D point cloud of environments without any human intervention. However, just collecting datasets without context is not enough to make a valid decision. We need the knowledge of experts to extract the necessary information and create viable data support for decision making.

Automating the process of fully autonomous cognitive decision systems is an attractive prospect. However, it poses significant challenges, primarily related to knowledge extraction, integration, and representation of 3D data. Therefore, the structuring of 3D datasets must be designed in a way that allows computers to use them as a basis for information extraction through reasoning and agent-based systems. This paves the way for understanding how 3D data science relies on 3D data expertise, which requires proficient knowledge management. Achieving this is the ultimate stage of automated workflows (Figure 1-23).

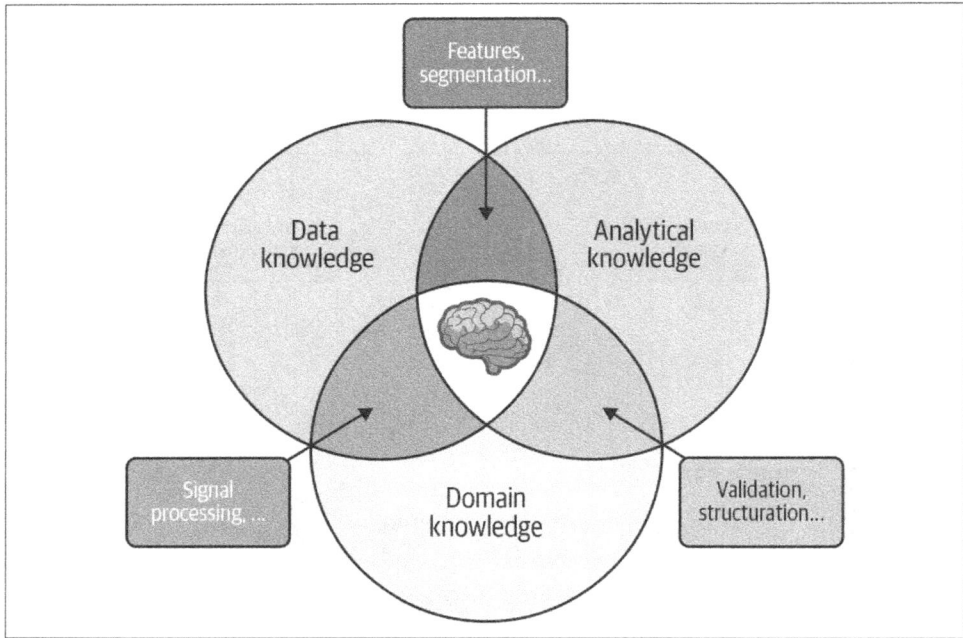

Figure 1-23. The organization of knowledge for reaching 100% automation and reasoning with 3D data science

Imagine if we attempt to understand the origin of data characteristics by understanding the influences of the characteristics of the acquisition process and the different factors influencing it. Such an understanding allows for the anticipation of data characteristics. This data-centric process requires explicit knowledge of the domains of data, scene, and data processing and the influence between these domains. Therefore, we identify three main areas:

Data knowledge

The data domain is composed of knowledge about the data's characteristics and knowledge about the acquisition process (e.g., acquisition methodology, acquisition technology, acquisition instrument).

Domain knowledge

The domain of the digitalized scene is composed of knowledge about objects, their distribution in the scene, the scene context, and external factors.

Analytical knowledge

Finally, we need knowledge of algorithms (e.g., conditions of use, purpose of the algorithms, configuration, preferred usage conditions) used in our workflows.

If we can design our systems with these in mind, we can be sure that moving to full automation will not lead to devilish retro-engineering of our modules.

Now that we have lined up the various workflow modules and the case for automation, are we on a straightforward path? Well, there are a fair number of challenges that you should be aware of to better grasp the path that lies ahead.

Workflow Challenges in 3D Data Science

We are interested in real-world applications, so it's very important to understand the challenges that we're going to face. What should we be careful about? What are the limitations of 3D data science? Let's explore in a synthetic way what I compiled for you here.

Basically, whenever we have a real-world data science workflow, we'll always have three main phases: data preparation, model(s) building to answer a specific task, and inference and analysis. With these steps in mind, I compiled eight main challenges that we are going to face (Figure 1-24).

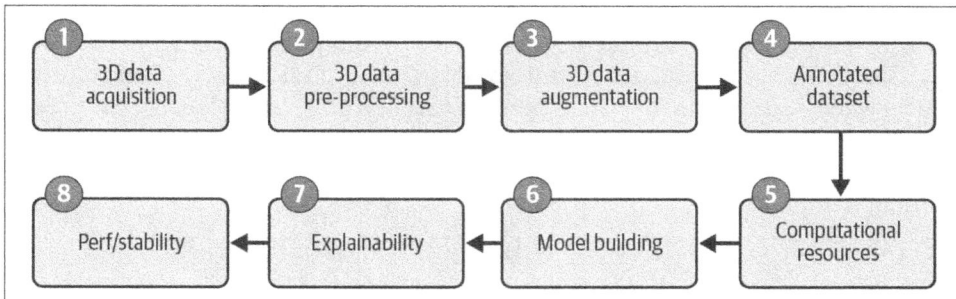

Figure 1-24. The main workflow challenges in 3D data science

These challenges do not have a specific order, but I linked them in a logical way inspired by the main workflow presented in Figure 1-12.

3D data acquisition challenges

It's very important to understand where the data comes from, especially when you try building the architecture, so that you are sure that it will be tuned to your application. Remember, this application of data science is very specialized and high-performing on specific real-world applications. In this case, you have different sensors, for example, LiDAR sensors, active sensors, and passive sensors. You have different platforms, which could be mobile or static. You also have other methods of acquisition, which makes the process of gathering data less straightforward, and it's good to frame how the data was collected before using it.

3D data preprocessing

Before pushing a dataset into an architecture, there is almost always a need to preprocess the data to make sure that it will work, as shown previously. It can be as simple as just translating, rotating the data, normalizing to a specific range, or tiling and sampling to be sure that you always have the same number of points that will pass through the network. Those are examples of preprocessing 3D data in general for 3D deep learning applications. This stage poses some challenges about which preprocessing we should do and how we should do it.

3D data augmentation

Usually, we will not have a sufficient number of annotated datasets, which means that to ensure that we encompass the full range of variation within an application, we need to do some augmentation. This could be noise injection, data synthetics, or other methods that will execute rotations, translations, or scaling on your data, all to extend the possible configuration of your datasets, for example.

Lack of annotated datasets

One of the biggest challenges in 3D data science is the need for annotated datasets. Indeed, we often rely on models that learn from annotated data. To train such a model in a supervised manner, you need data that is labeled. This specific point is critical when dealing with 3D data. There are very few annotated 3D scenes with both data and semantics (labels) available, and it is even harder to find them for specific applications. This makes it challenging to train models accurately using supervised learning techniques.

Let's consider an example. I used to work for cultural heritage projects. In this line of work, having a dataset that you gather on-site means that you will usually not find another specimen of this specific object. You need to provide another layer of abstraction about an encompassing class that you want to classify your data onto. All applications need to find enough objects per class to sufficiently annotate the segmentation workflow. It usually comes down to having a lot of data, which is not

---

really possible. That is a big problem. This complexity extends even more when trying to provide multimodal datasets that are nicely annotated. There is a famous saying: garbage in, garbage out. When you want to push something in a production-ready environment, you need to make sure what you train on is super top-notch label data. Usually, that means a lot of human supervision and a limited amount of unsupervised mechanisms to try and facilitate the labeling process because it can be quite time consuming. This is a big challenge.

Computational resources

This is a massive challenge as well. We are playing with huge datasets and that will trigger our need to go into deep learning rather than classical machine learning. However, if you have a very small dataset, machine learning usually performs if you are just getting the top result rather than the intersection of a union. This is the first challenge. Then, there is another one: the computational resources you have at your disposal.

We deal with extremely complex tasks. We need to do a lot of successive operations in the most efficient manner. Usually, training deep learning models on a central processing unit (CPU) is not recommended for production-ready applications (though inference on a CPU is possible). For training, we will need to be careful to make sure that our computation executes on a GPU (graphical processing unit), which is much more performance-oriented, or a TPU (tensor processing unit). This means that we often need to move to cloud computing or parallel computing. If you're using a GPU with CUDA, we will cover this a bit later. In this scenario you usually have all your parallel computing handled for you, making it much more efficient to use all the CUDA cores at your disposal.

Model building

Model building is a significant challenge because, first, we will need to define which paradigm we will choose and which mechanism to use. Then, we must select a specific task followed by choosing a specific method to achieve the functions that we want. Then, we need to make sure that we define the training, the inference, and how we generalize it with the right proportions for solid performance. That means also monitoring the performance and making sure that we get the highest performance possible. This is the biggest challenge.

Explainability

What do interpretability and explainability mean? There are several model layers, and the model will make a lot of complex computations to find the optimal parameters for performing a specific task, which means that it's usually pretty hard to understand exactly the rules or the logic behind how the models work. This is something that you will often discover whenever you deal with an artificial neural network. Interesting work is being done in deep learning, explainability, and explainability AI, which will be covered in Chapters 8 and 10.

Performance and stability

The final challenge is performance and stability. You want to have the best performance for your model on a specific task or multiple tasks, and you also want stability, which means that either during your training or inference, or both, you want your model to perform smartly and robustly. For training, this means that you wish to have a manageable number of variations per epoch. In the case of inference, what you aim at having is something that can generalize well. Make sure that you prepare your data in a way that really represents what you will want to have as a real-world application.

Remember, we are not looking at something from an academic perspective or research perspective, where it's fine to build a theory with no direct application. We want to solve real-world pragmatic problems. That's why we need to make sure that all of these challenges are addressed whenever we solve the issue at hand. It's very good to have these eight points in mind as you think about 3D data science. It will guide you through the full course of the book and beyond to ensure that whenever you build something, it rightly addresses all these challenges.

Table 1-1 summarizes workflow challenges in 3D data science and should help guide your efforts in future developments.

Table 1-1. Workflow challenges in 3D data science

Challenge category	Description	Mitigation strategy
3D data acquisition challenges	Understanding the origin and characteristics of 3D data is crucial, especially considering factors like sensor types (LiDAR, active, passive), platforms (mobile, static), and acquisition methodologies.	Carefully consider sensor characteristics, platform limitations, and acquisition methodologies during data collection. The choice of sensor, platform, and methodology should align with application requirements.
3D data preprocessing	This step ensures data quality and compatibility with subsequent processing steps. Key operations include translation, rotation, normalization, tiling, and sampling.	Apply appropriate preprocessing techniques based on data characteristics and the requirements of subsequent processing steps. Techniques like noise reduction, outlier removal, and data normalization enhance data quality and model performance.
3D data augmentation	This step helps to address the challenge of limited annotated datasets. It involves techniques like noise injection, synthetic data generation, rotation, translation, and scaling to increase data variability and improve model generalization.	Utilize various data augmentation techniques to expand the training dataset and improve model robustness. Augmentation techniques, such as rotation, scaling, and adding noise, create variations of existing data, reducing the risk of overfitting and improving model generalization.

Challenge category	Description	Mitigation strategy
Lack of annotated datasets	Obtaining sufficient labeled 3D data for supervised learning is a significant challenge. This is particularly true for specialized applications where annotated datasets are scarce.	Explore alternative approaches like semisupervised or unsupervised learning when labeled data is limited. Leverage transfer learning techniques, utilizing pretrained models on related datasets to bootstrap the training process. Invest in data annotation efforts to create high-quality labeled datasets for specific applications.
Computational resources	3D deep learning often requires substantial computational resources. Training large models on extensive datasets necessitates powerful hardware like GPUs or TPUs, potentially requiring cloud computing solutions for efficient processing.	Optimize code and algorithms for efficient computation, considering parallel processing and GPU utilization. Explore cloud computing platforms that offer scalable computational resources on demand, enabling the training and deployment of large 3D models.
Model building	Selecting an appropriate model architecture, training methodology, and performance evaluation metrics is critical. Balancing model complexity with generalization capability while ensuring robust training and inference procedures is a key challenge.	Carefully choose model architecture based on the specific 3D data analysis task and dataset characteristics. Experiment with different training strategies, hyperparameter tuning, and regularization techniques to achieve optimal model performance. Continuously monitor training progress and validate model performance on independent datasets to prevent overfitting and ensure generalization.
Explainability	Understanding the reasoning behind 3D deep learning model predictions can be challenging. Interpreting complex model decisions and providing clear explanations for predictions is essential for building trust and ensuring responsible AI development.	Investigate explainability techniques such as feature visualization or attention maps to provide insights into model decision making. Develop methods for generating human-interpretable explanations of model predictions, enhancing transparency and accountability.
Performance and stability	Achieving strong model performance while maintaining stability during training and inference is crucial. Managing training variability and ensuring consistent, reliable model predictions in real-world applications is a continuous challenge.	Implement robust training procedures with appropriate regularization techniques and learning rate schedules to mitigate training instability. Thoroughly evaluate model performance on diverse datasets to ensure stable and reliable predictions in real-world scenarios.

3D Data Science in the Industry

Before wrapping up this chapter, I would like to offer you an optional view of how 3D data science is used in practice (stage 4 in Figure 1-1). You will find an extended view on the 3D Data Science Resource Hub (*https://learngeodata.eu/3d-data-science-with-python*), which delineates the major fields that likely encompass most of the applications you will encounter when dealing with 3D data science.

There are many tasks in each field, supported by roles in 3D engineering, 3D analysis, 3D reconstruction, and 3D visualization. Then, for each of these tasks and roles, you can define a workflow around various knowledge pillars. Framed this way, a clear path emerges, enabling us to navigate how we go into these different categories (Figure 1-25).

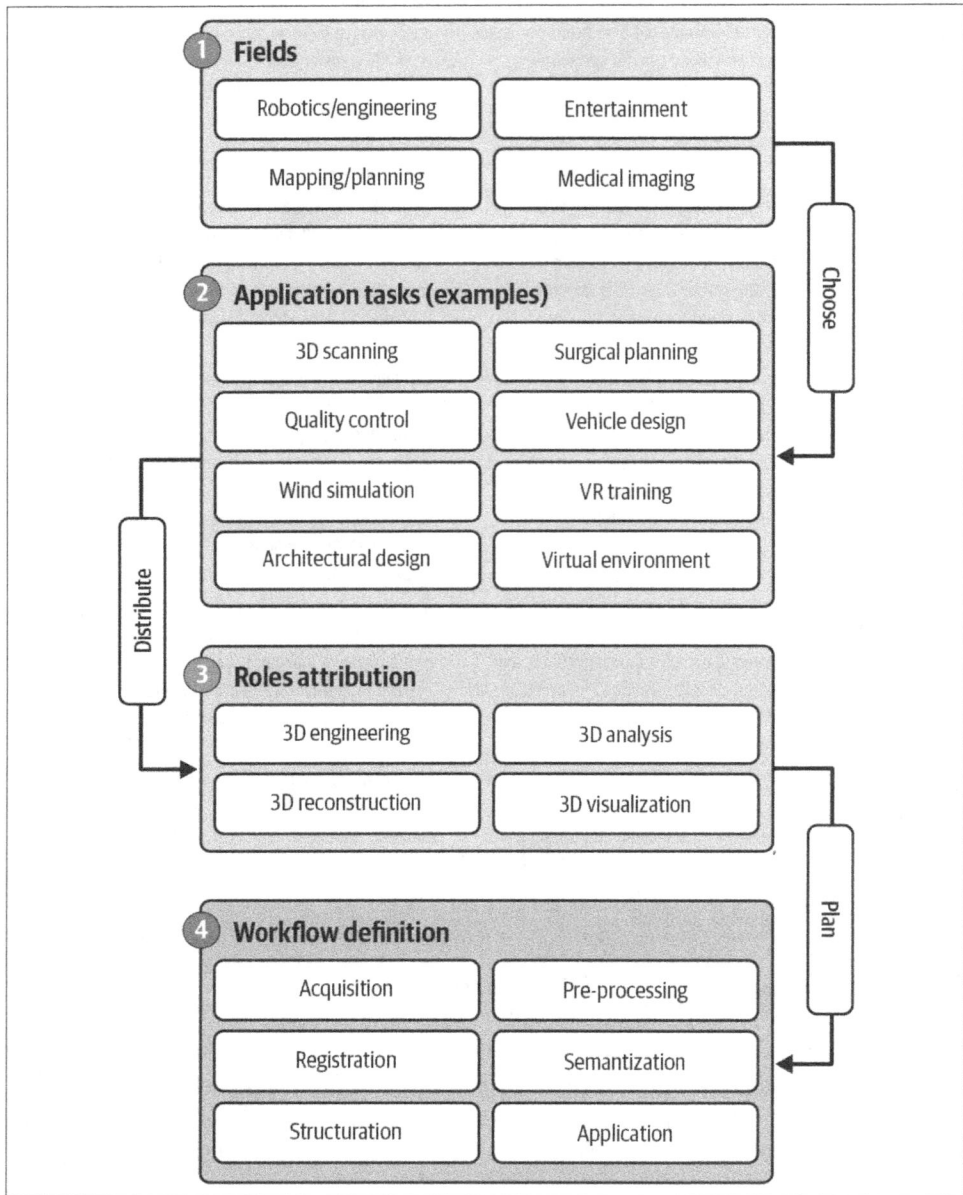

Figure 1-25. 3D data science in the industry

Identifying fields, applications, and roles can give you an idea about the process and targeting needed before moving on to the definition of a workflow. Such a workflow can be derived from careful observation and research curation of current industry players.

> Aside from academia, there is a dynamic landscape of 3D data science industry leaders such as NVIDIA, Microsoft, Google, Amazon, and Facebook. They have propelled innovation with sophisticated solutions. These companies, renowned for their technological prowess, contribute significantly to advancing 3D data science applications.

Summary

By this point, you have taken your first steps into 3D data science. We studied the role of dimensions in science and how spatial AI uses geometries as a base to mimic the analytical parts of what our brain does with 3D data. There are nine key learning points that are important to understand before continuing to the next chapter:

- 3D data science analyzes three-dimensional data to extract knowledge and insights. It bridges the gap between our three-dimensional world and the science of data.

- Spatial AI is a specialization of artificial intelligence that focuses on understanding and interacting with the world in 3D space. It is the outcome of the best 3D data science workflows.

- A dimension is a measurable aspect or characteristic of something. In 3D data science, dimensions often refer to spatial extent.

- 3D data science offers a lot of opportunities, such as low entry cost, high versatility, and a future-proofed career path.

- Geometry, topology, and semantics are crucial for extracting meaningful information from three-dimensional data. Geometry encompasses the study of spatial properties. Topology focuses on the intrinsic properties of space. Semantics refers to the meaning, interpretation, or significance associated with elements or structures.

- Point cloud data is a central data modality in 3D data science. Point clouds are discrete spatial ensembles that represent the recorded environment.

- The 3D data science modular workflow includes data acquisition, preprocessing, registration, 3D data classification, structuration, 3D modeling, 3D data analysis, 3D data visualization, and application development.

- Automating 3D data science workflows is crucial for efficiency and scalability.
- Workflow challenges in 3D data science include 3D data acquisition, 3D data preprocessing, 3D data augmentation, lack of annotated datasets, computational resources, model building, explainability, performance, and stability.

The workflow definition established in this chapter constitutes the basis of our endeavors in this book. Next, we'll explore the essential resources and tools to help you set up your 3D data science journey.

Hands-on Project

As you progress through this book, you will find that each chapter contains an exercise that you can complete to self-assess your ability to apply what you have learned through a small hands-on project. For this chapter, familiarize yourself with point cloud data, a core component of 3D data science. You can do this in six steps:

1. OpenTopography provides a variety of 3D point cloud datasets. Visit the website (*https://opentopography.org*) and select a dataset that interests you. Consider factors like location, size, and point density.

2. Download a small portion of the chosen dataset.

3. Visualize the point cloud: numerous tools are available for point cloud visualization, which we will cover in the next chapter. As a first layer, feel free to explore your point cloud with the generated viewer from OpenTopography to avoid installing software at this stage.

4. Observe the point cloud and try to identify distinct features or objects. What shapes or patterns can you discern? Can you make any inferences about the environment or objects captured in the point cloud?

5. While you're working with a small point cloud subset, reflect on how the five Vs (volume, velocity, value, veracity, and variety) might pose challenges when dealing with large-scale point cloud data.

6. Consider the preprocessing steps that might be necessary to clean and prepare this point cloud data for further analysis or model training.

   This exercise focuses on hands-on exploration of point cloud data using external tools and resources. It should take 30 to 45 minutes to complete.

## CHAPTER 2

Resources and Software Essentials

Starting from a blank canvas is both exciting and intimidating. It is exciting because you start fresh, and everything seems possible. I love taking on a project from scratch, where I can find solutions with a first-principle approach. But I also know that I can quickly fall into builder's block: staring at my screen or the white piece of paper until I feel lost.

Getting into a new discipline can be pretty intimidating because everything is new and you're building from scratch. It can be hard to know where to start, especially when we try to leverage the full load of information available on the internet as we get up to speed.

This chapter will help you avoid these common pitfalls and establish a healthy groundwork for building 3D products. The first strategic piece in a 3D data science project is ensuring proper access to the right resources and tools. This stage precedes setting up systems, processes, and methods that can scale and benefit from automation.

Figure 2-1 illustrates the fundamental knowledge resources, hardware recommendations for 3D development, and software solutions that provide the foundation for 3D data science. I call this framework the 3D Data Science Starter Pack.

"Fundamental Resources" on page 45 will investigate the resources available, including the essential mathematical, computer science, and specialized AI concepts needed for 3D data science. This extends to established disciplines, methods, workflows, and industries that heavily benefit from its application.

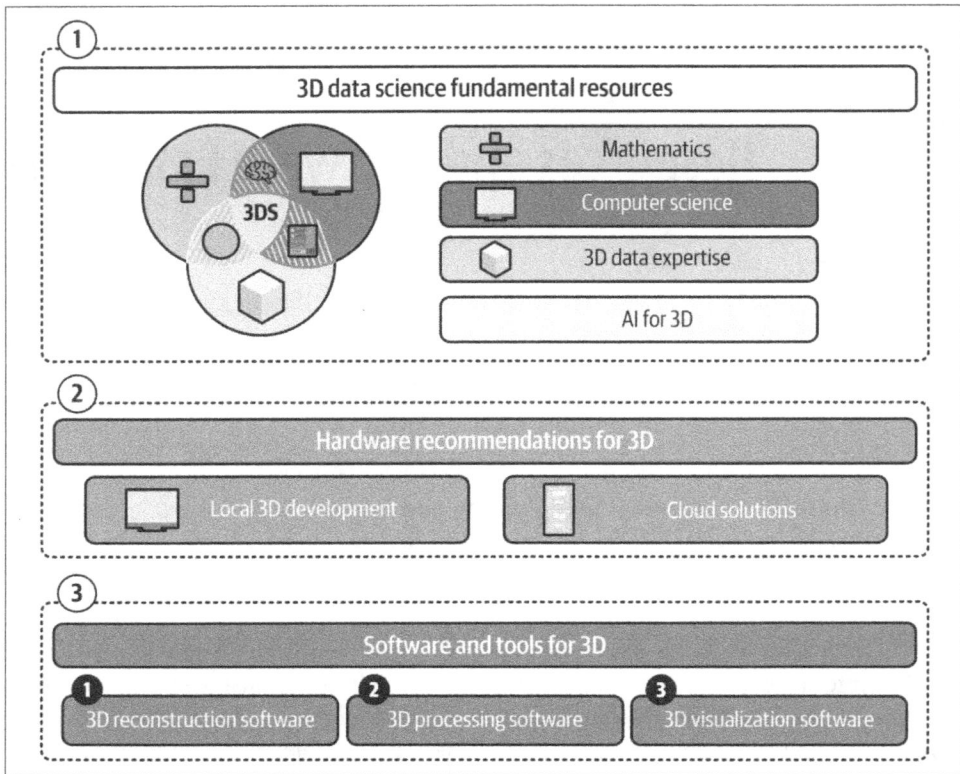

Figure 2-1. An overview of the 3D Data Science Starter Pack

The next stage is to examine your specific hardware setup. You will need a computer to create 3D data tools. But what is the best configuration for your case, and how can you ensure a lean setup for flexible 3D development? These are relevant questions, and I'll walk you through some recommendations. I've also included a list of my recommended software for 3D reconstruction, 3D processing, and 3D visualization tasks.

> This book allows you to create 3D data tools and solutions from scratch with Python. However, leveraging existing software and algorithms to quickly iterate and test your ideas before going into extensive 3D coding is a good idea. Knowing which software can help you cut iteration time is a great asset.

Fundamental Resources

There are many essential 3D data science concepts. Because this is foundational information that many readers may be familiar with, I've provided an overview here, and more complete coverage lives on the 3D repository hub (*https://learngeodata.eu/3d-data-science-with-python*). My goal here is to quickly delineate the concepts and taxonomy used in this book. I recommend bookmarking this section to refer to later if you want to strengthen your understanding of the building blocks for successful 3D data science projects.

Mathematics

"That which is learned" is how Greek etymology views mathematics. It is a beautiful way to discover the properties of abstract objects by using pure reasoning to prove them. Today, mathematics steers most of our scientific principles to the data branch. In this section, I delineate the branches of mathematics that are useful for creating powerful innovations using 3D data science.

When studied in the context of 3D data science, the four main mathematical concepts are linear algebra, calculus, statistics, and combinatorics. These concepts are used to represent 3D data, analyze it, and extract insights from it. While you do not need to hold a PhD in mathematics to follow along, it is recommended that you grasp some of the fundamentals. To ensure you receive adequate guidance, I have compiled a list of my preferred resources in Table 2-1.

It is tough to be exhaustive when it comes to curating resources. Therefore, I limited myself to providing you with free resources. I included three mediums to best suit your learning affinity: books, videos, and articles. Again, these are my recommendations to complement the knowledge of this book at no additional cost but know that there are lovely resources outside of these, for example, within the O'Reilly collection (*Essential Math for AI* and *Essential Math for Data Science*).

> It is perfectly OK to dive head-on into the material and teachings of this book, even if you are not a skilled mathematician. Just know that if you feel constrained by your understanding at any point, you can fall back on the available resources.

Table 2-1. Additional probability and statistics resources

Medium	Resource
Books	Statistics: *Think Stats: Probability and Statistics for Programmers* (*https://oreil.ly/9XnYN*), Allen B. Downey (Green Tea Press)
	Linear Algebra: *Practical Linear Algebra for Data Science: From Core Concepts to Applications Using Python*, Mike X. Cohen (O'Reilly)
	Combinatorics: *Introduction to Graph Theory* (*https://oreil.ly/k0OdL*), Richard J. Trudeau (Dover)
	Calculus: *Calculus* (*https://www.stewartcalculus.com*), James Stewart (Cengage Learning)
Videos/ courses	Statistics: "Statistics and Probability" (*https://oreil.ly/2pYmC*) (Khan Academy)
	Linear Algebra: "Linear Algebra" (*https://oreil.ly/arMwS*) (Khan Academy), "Essence of Linear Algebra" (*https://oreil.ly/5BGjW*)
	Combinatorics: "Graph Theory Explained" (*https://oreil.ly/VfavL*)
	Calculus: "Integral Calculus" (*https://oreil.ly/4_RNo*) (Khan Academy)
Articles	Statistics: "What Are Variables in Statistics?" (*https://oreil.ly/sTcwK*), " What is Hypothesis Testing?" (*https://oreil.ly/Q0EaX*) (Stat Trek)
	Linear Algebra: "Linear Algebra" (*https://oreil.ly/d0nQb*) (MIT OpenCourseWave)
	Combinatorics: Graph Theory Notes ("Introduction to Programming Contests") (*https://oreil.ly/aKBcg*) (Stanford University)
	Calculus: "Integration: Why Do We Need to Study Integration?" (*https://oreil.ly/0CHkP*) (Interactive Mathematics)

Statistics, linear algebra, combinatorics, and calculus power most of the book's applications, but computer science shapes them.

Computer Science

In modern society, computer science includes the study of computers and computational systems. It impacts our society's function through research, innovation, and product cycles. As you may recall from Figure 1-3, the study of 3D data science is at the intersection of mathematics, computer science, and 3D data expertise. When we think of this through the lens of computer science, there are three families that are essential to grasp: core computer science concepts, the computer science concepts used in 3D machine learning, and the computer science concepts used heavily in 3D data processing (Figure 2-2).

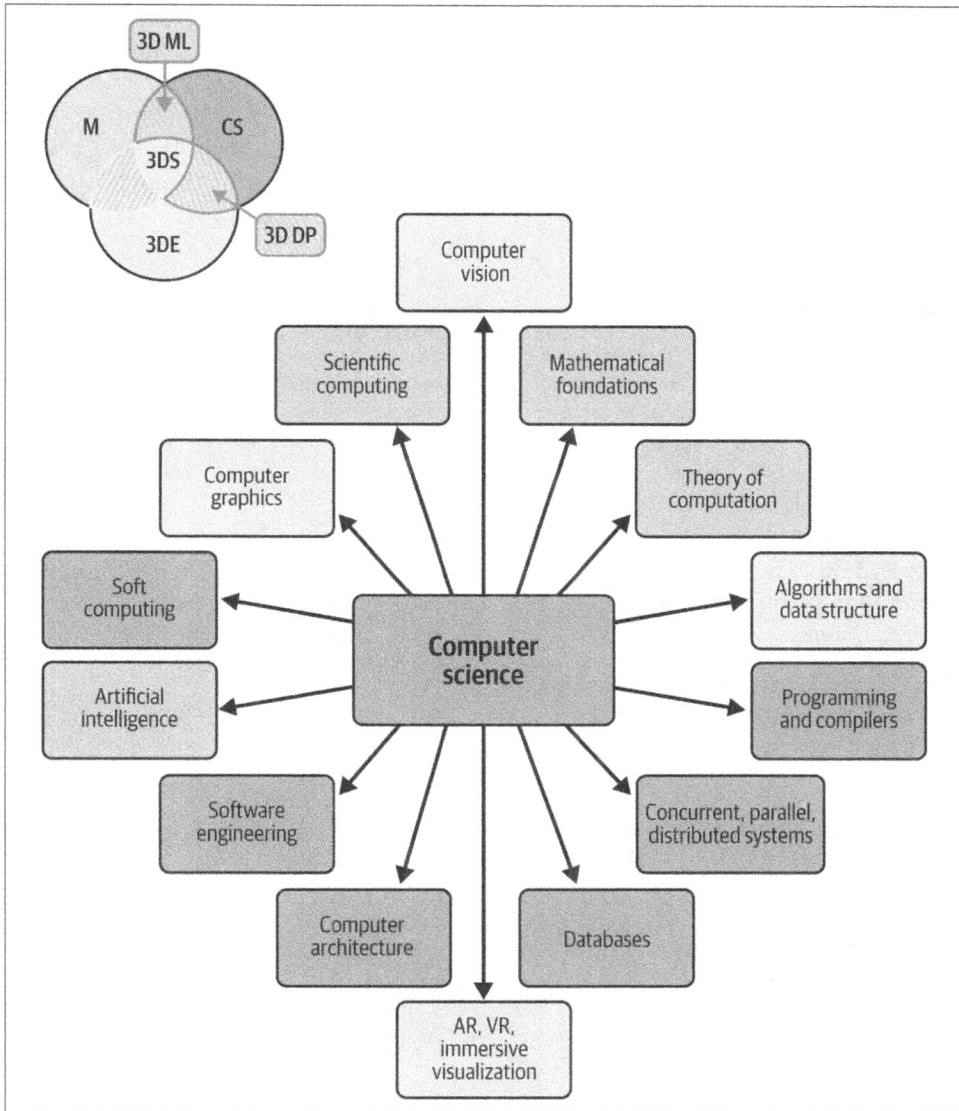

Figure 2-2. The various branches of computer science that unlock the potential of 3D data science

The rest of this section will focus on the core concepts of computer science, and I will share essential resources along the way. The remaining branches (scientific computing, computer graphics, artificial intelligence, mathematical foundations, theory of computation, computer vision, algorithms, data structure, AR, VR, and immersive visualization) are also part of this book's primary learning outcomes, but they will be detailed in subsequent chapters. Table 2-2 contains some helpful computer science resources.

Table 2-2. Core computer science concepts

Medium	Resource
Books	*Computer Science Illuminated*, 7th Edition (O'Reilly). Authored by the award-winning team Nell Dale and John Lewis, this is a great 101-level book on computer science that boasts accessibility and in-depth coverage while incorporating all-new material on cutting-edge issues in computer science.
Videos/ courses	"Computer Science 101" (*https://oreil.ly/WjO22*) (StanfordOnline). An introductory course in computer science for a zero-prior-experience audience.

Since this chapter will leverage Python, let's start with some basics of programming with Python. There are several key concepts that are fundamental for building a strong foundation in programming, especially in Python. Here are some of the most important ones:

Variables and data types
Understanding how to declare, assign, and work with different types of data (integers, floats, strings) is fundamental to programming.

Control structures
This includes loops (e.g., for, while) and conditional statements (e.g., if, else, elif). These allow you to control the flow of execution in your program.

Functions
Functions allow you to group blocks of code that can be reused. Learning to define, call, and work with functions is crucial for writing modular and maintainable code.

Data structures
It is essential to understand data structures like lists, dictionaries, tuples, and sets and how to manipulate and access data within them.

Algorithms
Learning fundamental algorithms like sorting and searching is crucial for solving problems efficiently.

Object-oriented programming (OOP)

In OOP, you organize code around objects and their interactions. This involves classes, objects, inheritance, polymorphism, and encapsulation.

Error handling

For robust and reliable code, it is important to know how to handle exceptions and errors gracefully.

File handling

Many applications commonly involve working with files and filesystems, including reading and writing data to files.

Libraries and modules

It is essential to understand how to use libraries and modules to extend Python's capabilities.

We use Python in this book because it has emerged as a first-hand language for numerous compelling reasons that are difficult to overlook. Python's popularity stems from its breadth, readability, and extensive collection of libraries and tools. Libraries like NumPy (*https://numpy.org*), SciPy (*https://scipy.org*), pandas (*https://pandas.pydata.org*), and Matplotlib (*https://matplotlib.org*) provide robust support for numerical operations, data manipulation, and visualization—all core tasks in 3D data science. Moreover, Python's simplicity and clean syntax make it accessible to both beginners and experienced programmers, facilitating rapid development and experimentation. However, the concepts covered in this book can be leveraged without too much refactoring using Matlab (*https://oreil.ly/TfWmS*) or R (*https://www.r-project.org*). If you want to push performance, I recommend investigating Java (*https://www.java.com*), C++ (*https://oreil.ly/woKZ8*), or Rust (*https://oreil.ly/o4cGK*) at the cost of a steeper learning curve and verbosity (less ideal for rapid prototyping and exploratory data analysis, which are crucial in 3D data science).

3D Data Expertise

3D data science requires expertise in 3D data. This includes knowledge of the different types of 3D data and how to acquire and represent it. Compared to mathematics and computer science, there is a lack of resources on this subject. I distinguish three families of data expertise that are heavily used in 3D data science, as shown in Figure 2-3.

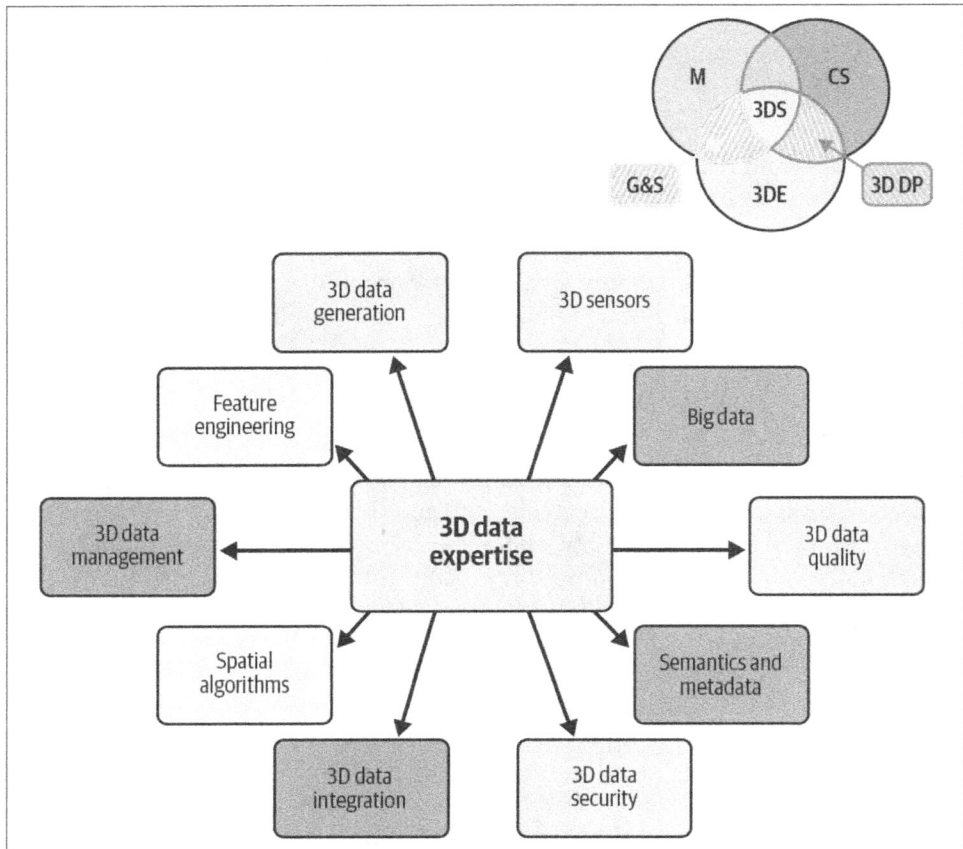

Figure 2-3. 3D data expertise and 3D data science concepts

First, we consider the areas heavily used at the intersection with mathematics, i.e., geometry and space concepts (located at the left intersection of the Venn diagram in Figure 2-3). The 3D data expertise that is required here includes geometric feature engineering and spatial algorithms. These concepts are used throughout the lifecycle of the 3D data science project and are covered in depth throughout this book.

If we move to the right side of the Venn diagram, we find elements that help 3D data processing workflows, such as 3D data management, big data, semantics/metadata, and 3D data integration. Again, the idea is to be able to handle the specificities of 3D datasets within the scope of a data processing workflow that leverages several computer science concepts (e.g., computer graphics, computer vision, algorithms and data structures, AR, VR, and immersive visualization).

Finally, we are left with the core concepts of 3D data expertise, i.e., laying the foundation for 3D data from its generation to its quality, security, and management. The last branch we need to clarify is at the intersection of mathematics and computer science: artificial intelligence for 3D.

Artificial Intelligence for 3D

The field of AI covers a broad spectrum of concepts and techniques. It is a highly active and exciting field, which may cause "fear of missing out" (FOMO) syndrome. Indeed, every week, we see new architectures or AI solutions as part of a research endeavor or startup.

My goal here is to ensure that you have a high-level overview of where each tech plays so that when we leverage AI in this book, you know which specialization to learn from. The circles in Figure 2-4 showcase the three main AI specializations (machine learning, deep learning, and generative AI) with some examples of known solutions.

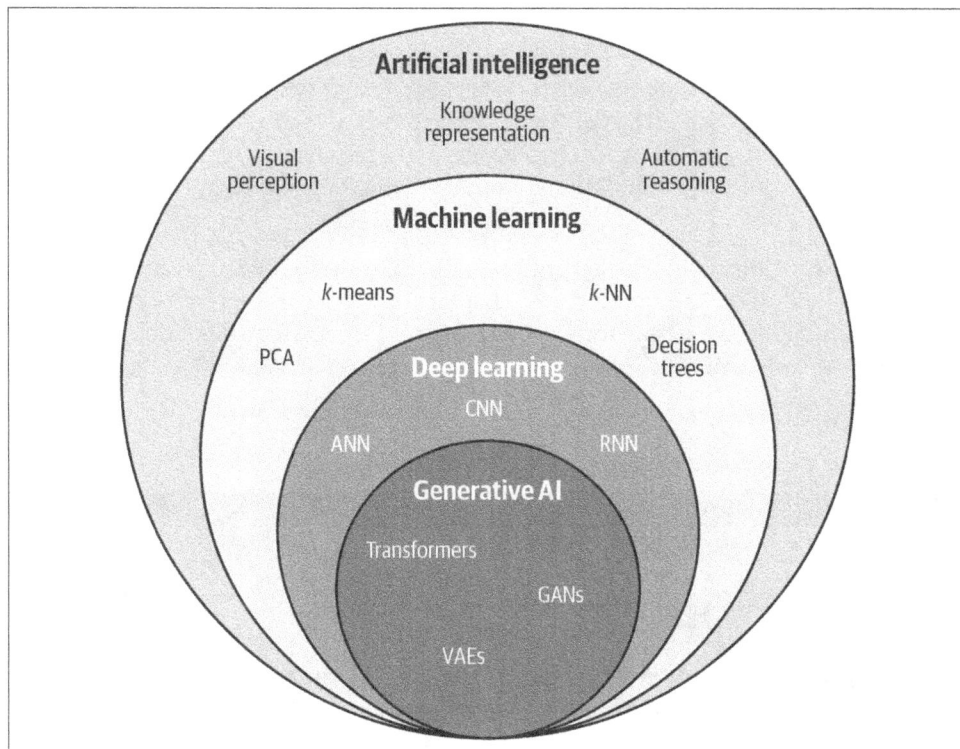

Figure 2-4. The scope and level of specialization of artificial intelligence

To ensure we're all on the same page about what each of these areas of AI are, let's review some brief definitions:

Artificial intelligence (AI)
    Any technique that enables machines to solve a task in a way like humans do

Machine learning (ML)
    Algorithms that allow computers to learn from examples without being explicitly programmed

Deep learning (DL)
    A subset of ML that uses deep artificial neural networks as models and automatically builds a hierarchy of data representations

Generative AI (GenAI)
    Leverages deep learning and neural networks to generate outputs that resemble real-world examples

To tie that with your 3D data science journey, I have meticulously compiled a list of 18 concepts with definitions available on the 3D Resource Hub (*https://learngeo data.eu/3d-data-science-with-python*) (supervised/unsupervised learning, regression, clustering, ANN, CNN, RNN, NLP, etc.). While this is a good refresher for anyone, if you haven't worked with AI or ML much yet, take a moment to read through this list to ensure that you have a foundational understanding of these terms and can get the most out of this book.

In the context of this book, we are interested in how these concepts can be used for 3D datasets and 3D data science. Later chapters, especially Chapters 12 through 16, cover this in depth. Since this topic will be covered in detail in this book, my list of recommended machine learning and deep learning resources is fairly short (Table 2-3).

Table 2-3. Resources for learning AI and machine learning

Medium	Resource
Books	*Hands-on Machine Learning With Scikit-Learn, Keras, and TensorFlow*, 2nd Edition by Aurélien Géron (O'Reilly)
Videos/courses	"Machine Learning Specialization" (*https://oreil.ly/U8uci*) by Andrew Ng
Articles	"Machine Learning Algorithms" (*https://oreil.ly/T86KJ*) (Khan Academy)

At this stage, we have reviewed all the essential concepts and resources that we use for 3D data science. Now, we can investigate some hardware recommendations for 3D.

Hardware Recommendations for 3D

Relying solely on Python is feasible for 3D data science. However, when we create working solutions, especially in production, we need to integrate our new developments into a broader ecosystem of software and tools (see Figure 2-5). If you recall the 3D Data Science Starter Pack referenced in Figure 2-1, we're now shifting to step 2.

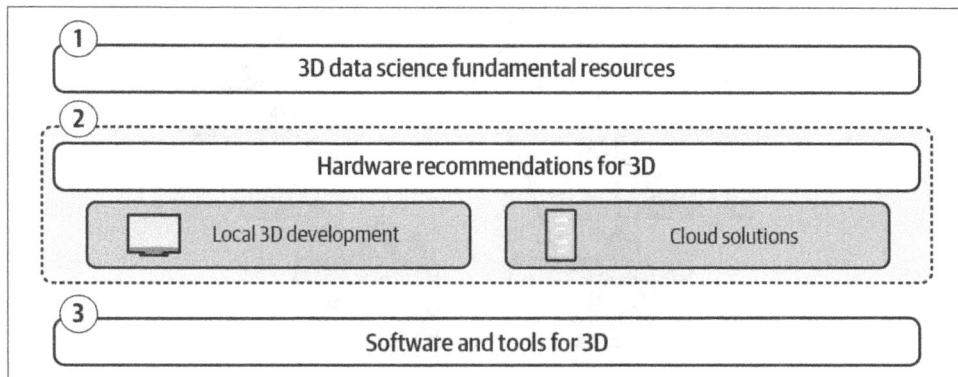

Figure 2-5. Hardware recommendations for 3D include local 3D development and cloud solutions

Additionally, during the design phase, it is usually a great idea to leverage existing software functions to speed up the time it takes to analyze and assess the effectiveness of our approach. Luckily, we can use open source software developed with some primary goals in mind. But first, we need to identify a computing system's key components to ensure that we are set for the next five years.

Local 3D Development

Hardware configuration is the bedrock upon which we develop our 3D data science projects. A judiciously chosen hardware setup expedites computations and unlocks the full potential of handling large datasets. Let's take a closer look at the key components and considerations for an optimal hardware configuration that is tailored to the demands of 3D data science (see Figure 2-6).

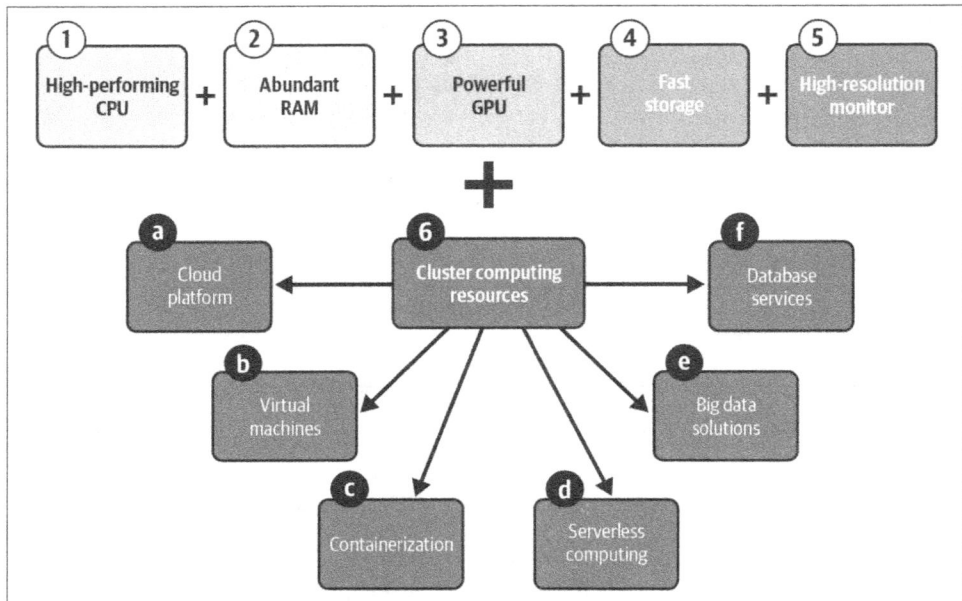

Figure 2-6. Hardware recommendations for 3D data science

Multicore central processing unit

A robust central processing unit (CPU) is fundamental for overall system performance, aiding tasks such as data preprocessing, feature extraction, and mesh processing. I recommend you opt for a multicore processor that can achieve high clock speeds. CPUs like Intel's Core i9 or AMD's Ryzen 9 series provide a balance of multiple cores and high frequencies.

Random access memory

As 3D datasets are often substantial in size, ample random access memory (RAM) is essential to ensure smooth loading and manipulation of volumetric data during analysis and modeling. I recommend having a minimum of 16 GB of RAM, which is advisable for handling moderate-sized 3D datasets, while more extensive datasets may necessitate 64 GB or higher.

Graphics processing unit

A graphics processing unit (GPU) plays a pivotal role in accelerating complex computations inherent in 3D data science tasks such as rendering, simulations, and deep learning. I recommend a high-end GPU with parallel processing capabilities, like the GeForce RTX 4060+ series.

Storage configuration

Due to the magnitude of 3D information, storage velocity is essential for effective data retrieval and processing. Solid-state drives (SSDs) significantly enhance read and write speeds. Thus, a combination of a fast NVMe SSD (e.g., 2 TB) for the operating system (OS) and frequently accessed data, coupled with a larger-capacity HDD or SSD for long-term storage (e.g., 5 TB), provides an optimal balance.

High-resolution monitors

Given the visual nature of 3D data science, high-resolution monitors with color accuracy are crucial for precise visualization and analysis. Opt for a 4K monitor with good color reproduction capabilities or multiple monitors for enhanced workspace flexibility.

Cluster computing for large datasets

For handling extensive datasets, cluster computing using multiple interconnected machines can distribute computations, enabling scalability. Deploying a cluster with high-speed interconnects coupled with load-balancing mechanisms can handle large-scale 3D data processing efficiently. This cluster will either be part of your company's hardware or accessible via cloud solutions. These are my recommendations. Additionally, you can leverage cloud computing if you want to leverage resources outside your local environment.

Cloud Computing

In simple terms, the word *cloud* refers to the internet, and *cloud computing* targets a set of computing services over the internet. Users can rent physical data servers, databases, storage, and computing power from cloud providers with a pay-as-you-go payment model. In some cases, cloud providers also offer software and analytics services on the cloud. Cloud computing has three main characteristics, namely services, tasks, and characteristics, as illustrated in Figure 2-7.

There is a guide on the 3D Resource Hub (*https://learngeodata.eu/3d-data-science-with-python*) to help beginners better understand how to get started with cloud computing. This guide is useful if you want to make sense of technical terms (XaaS, SaaS, PaaS, IaaS) that can be overwhelming. These shape how cloud computing combines computing, storage, and databases to form the foundational infrastructure.

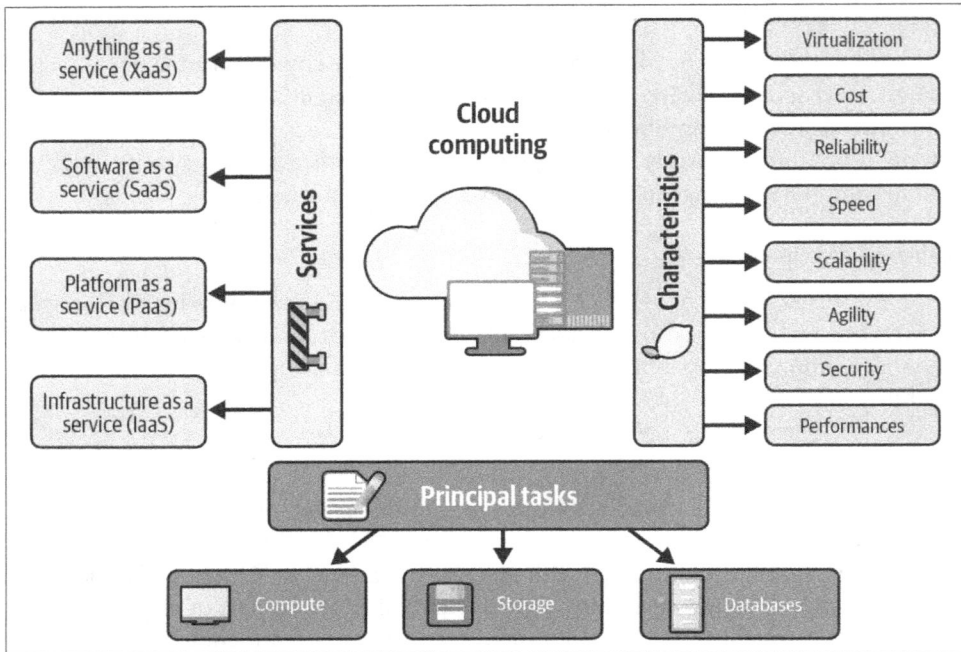

Figure 2-7. The main components of cloud computing

Industry Platforms

Did you know that 3D data science and cloud computing are not just for big corporations with deep pockets? Smaller companies can also benefit from these technologies, which help them store and analyze large amounts of data without breaking the bank. Finding and accessing the data can be a real challenge, too. Without cloud computing, smaller companies would have to store data on local servers and transfer it manually every time a data scientist needs to analyze it. Luckily, cloud computing makes it easy for small companies to access big data, democratizing data and making it available to everyone. There are three main providers of such infrastructures. The first one is Amazon Web Services. They offer cloud computing platforms for a variety of usage types. It is one of the most popular platforms and provides several products that could be useful for either computing, storage, or database needs. Google Cloud, a cloud computing platform provided by Google, offers similar solutions. Finally, we have Microsoft Azure, another popular cloud computing platform created by Microsoft.

Now that we better understand what we need to invest in to get started, let's look at some open source solutions to complement your work in 3D data science.

Essential Software and Tools for 3D

Choosing the right software stack and tools is an accelerator for 3D innovations. While cloud-based solutions offer convenience, a local setup often provides greater control, performance, and privacy (see Figure 2-8). If you recall the 3D Data Science Starter Pack from Figure 2-1, we are now at step 3.

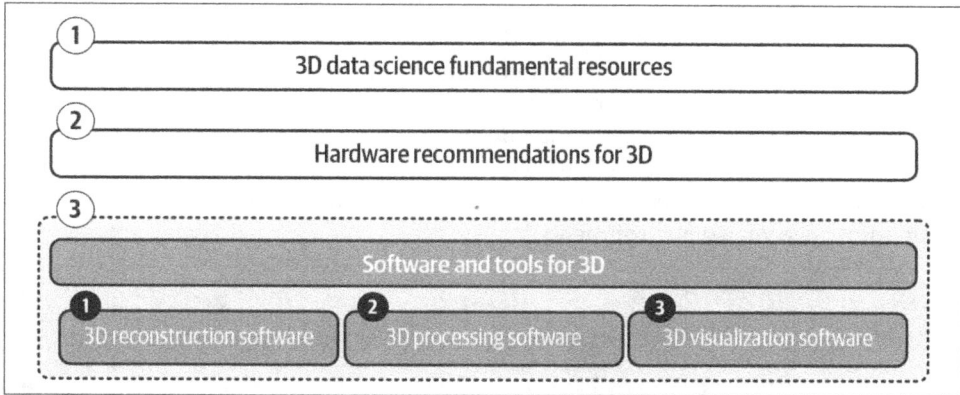

Figure 2-8. Software and tools for 3D

I'll walk you through how to set up a completely local experience that makes it possible to work offline with little to no dependency on large infrastructures. By setting up your 3D software environment on your machine, you gain the flexibility to customize your workflow, ensure data security, and optimize performance based on your specific needs.

> While Python is a powerful language for 3D data science, I recommend starting with specialized 3D modeling and visualization tools to gain a solid understanding of the fundamentals. Once you have a strong foundation, you can gradually incorporate Python and its libraries to enhance your workflow and tackle more complex tasks.

A robust 3D data science toolkit typically involves a combination of open source software and specialized tools. This approach fosters interoperability and simplicity, allowing you to tailor your 3D data science workflow to the specific tasks at hand. I decided to cover free-to-use software that can be categorized into three families: 3D reconstruction software, 3D processing software, and 3D visualization software.

3D Reconstruction Software

Now that we have explored the fundamentals of 3D data, let's delve into the world of 3D reconstruction software. These powerful tools can transform seemingly ordinary photos or scans into stunningly detailed 3D models.

RealityCapture

RealityCapture is a software solution that specializes in creating 3D models from photographs and laser scans. Unlike traditional modeling techniques, RealityCapture doesn't require users to be experts in 3D sculpting or design. Instead, it utilizes photogrammetry, automatically analyzing and piecing together pictures or scans to build a realistic digital replica (Figure 2-9). This renders it an invaluable instrument across many domains, from architects conserving historical edifices to game developers creating immersive virtual environments.

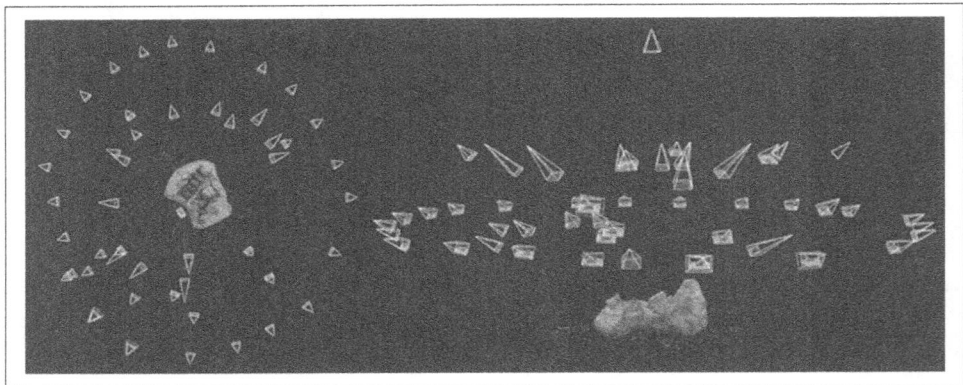

Figure 2-9. Point cloud and camera positions: top view (left), side view (right)

At its core, photogrammetry is a sophisticated 3D reconstruction technique. It leverages the triangulation principle to infer 3D spatial information from multiple 2D images of a scene or object taken from different viewpoints. Think of it like this: our brains effortlessly perceive depth and three-dimensionality by processing the slightly different perspectives our two eyes capture. Photogrammetry extends this concept by using many images, meticulously analyzing corresponding features across them, and mathematically determining their 3D coordinates in space. This process generates a dense point cloud, which can be further processed into a mesh, a surface representation composed of interconnected triangles, effectively creating a digital model of the physical object or scene (these principles will be covered in more detail in Chapters 3, 10, and 18).

However, RealityCapture does have limitations. While it automates much of the process, users still need to capture high-quality images or scans for optimal results.

Additionally, the software's processing power scales directly with the complexity of the project. So, creating intricate models with a vast amount of detail can take significant time and computing resources.

RealityCapture's website (*https://capturingreality.com*) makes it easy to download and install this software. Once the program is installed, you can follow online tutorials and discussions to help you get started. RealityCapture and photogrammetry can be quite efficient and surprisingly straightforward, even for nonexperts in the field.

> I am cautious with product placement and genuinely believe in objective testing. I do not wish to endorse solutions but rather provide explicit information and knowledge to guide users to make informed decisions. Therefore, I want to state that out of all the photogrammetry solutions, RealityCapture is excellent for some things, but other solutions are better for others. Still at the time of the writing, RealityCapture currently offers a free license for individuals and companies with an annual revenue under 1 million dollars.

Meshroom

Meshroom is an open source 3D reconstruction software that can be downloaded from their website (*https://alicevision.org*). This makes it accessible to hobbyists and educators, while its underlying framework, AliceVision, offers a powerful toolkit for researchers and developers. Meshroom's interface allows users to build a workflow by connecting different processing nodes. This provides flexibility for customizing the reconstruction process to specific needs (see Figure 2-10).

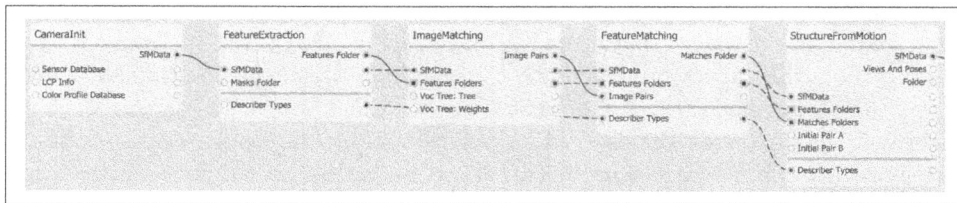

Figure 2-10. Node workflows in Meshroom

However, this freedom can be overwhelming for novices, and achieving high-quality results often requires some experimentation and understanding of the underlying 3D reconstruction techniques. Additionally, while Meshroom can handle large datasets, its reliance on powerful graphics cards can limit its accessibility for users with less robust computer systems.

Open Source Photogrammetry Solutions

Meshroom can be complemented by other open source resources that you might find helpful:

MicMac (https://oreil.ly/ADDmR)
An open source photogrammetry suite developed by IGN (French National Geographic Institute). It's particularly robust for challenging datasets and offers a wide range of functionalities.

OpenMVG (Open Multiple View Geometry) (https://oreil.ly/Hs8Cz)
A library specializing in the geometric aspects of multiple view reconstruction, providing a solid foundation for photogrammetric processing.

Colmap (https://oreil.ly/V8rzZ)
A general-purpose structure-from-motion (SfM) and multiview stereo (MVS) pipeline with a user-friendly graphical interface and robust performance. It's excellent for both ordered and unordered image sets.

Postshot

Postshot (*https://www.jawset.com*) is cutting-edge software that empowers users to create radiance fields from a set of input images. Radiance fields (and neural radiance fields, aka NeRF) are representations of a scene that capture color, depth, and lighting information, providing a rich and immersive experience. Imagine you're looking at a scene, any scene—a bustling street corner, a serene forest, or even your living room. From your perspective, light rays travel from every point in that scene to your eye, carrying color and intensity information. A radiance field is essentially a mathematical function that models this process. Given a specific 3D point in space and a viewing direction, this function predicts the color and density of the light radiating from that point along that particular direction. So, instead of explicitly representing the scene with polygons or voxels, a radiance field implicitly captures its appearance based on how light interacts with it. This allows for realistic novel view synthesis and other exciting applications detailed in Chapter 18.

One of Postshot's key advantages is its user-friendly interface, which simplifies creating radiance fields, making it accessible to users of all skill levels. The software leverages advanced algorithms like NeRF and Gaussian splatting, ensuring fast and memory-efficient training, even with large datasets. Additionally, Postshot prioritizes user privacy by processing all data locally on the user's device, eliminating the need for cloud-based solutions (see Figure 2-11).

Figure 2-11. An example of 3D Gaussian splatting datasets generated by Postshot

Postshot integrates with popular digital content creation tools like Adobe After Effects, allowing users to easily incorporate radiance fields into their projects. This integration expands the possibilities for creating immersive and interactive experiences.

Thus, Postshot is a valuable tool for researchers, artists, and professionals working with 3D data. Its ease of use, efficient algorithms, and local processing capabilities make it a powerful choice for creating high-quality radiance fields.

3D Data Processing Software

3D data processing software is a specialized suite of tools designed to handle, analyze, and manipulate three-dimensional data (point clouds, meshes, and volumetric data), which are commonly generated by 3D scanners, cameras, simulations, and 3D reconstruction software.

CloudCompare (point cloud processing)

CloudCompare (*https://cloudcompare.org*) stands out as an open source software that specifically processes, edits, and analyzes 3D point cloud data. Unlike some reconstruction software that focuses on creating final models, CloudCompare empowers users to delve deeper into 3D data processing for various tasks like registration (aligning multiple point clouds), segmentation (isolating specific features), and intricate analysis of spatial properties (see Figure 2-12).

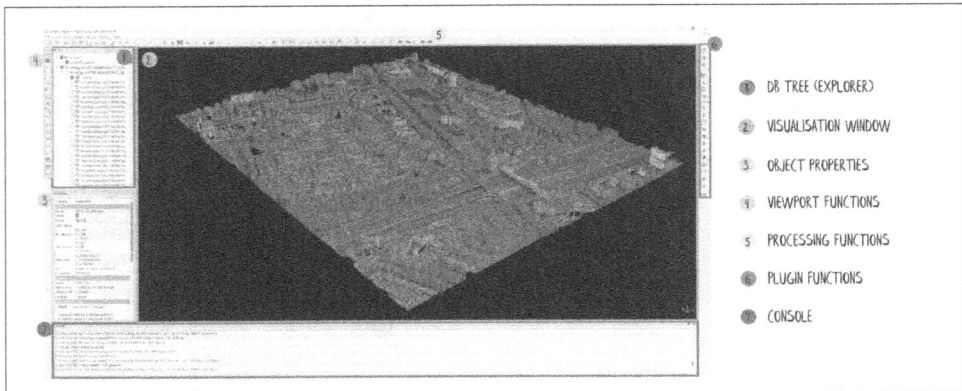

Figure 2-12. CloudCompare GUI with a LiDAR point cloud in it

This makes CloudCompare invaluable for researchers in fields like geology, survey-ing, and robotics, where precise manipulation and understanding of 3D point cloud data is crucial. However, CloudCompare's extensive feature set can present a steeper learning curve compared to more user-friendly reconstruction software. While there are tutorials and a supportive community, navigating its functionalities might require some experience with 3D data processing concepts. We leverage CloudCompare in this book for manipulating 3D datasets.

QGIS (geospatial solution)

QGIS (*https://qgis.org*) is a bit more geospatial oriented. Again, it's an open source initiative, and you can download the one that is best suited to your OS. After instal-ling everything, you have the ability to import your data, including 3D point clouds. What is interesting is you have the spatial context. Wherever your point cloud is to your reference, you will see that it's aligned with an Open Street Map layer (see Figure 2-13).

As you can see, it creates a nice overlay between the layers. QGIS is my go-to recom-mendation whenever you are dealing with a geospatial application where you need to make some queries or handle different types of modalities like spatial raster imagery, vector datasets, and point cloud/CityGML models.

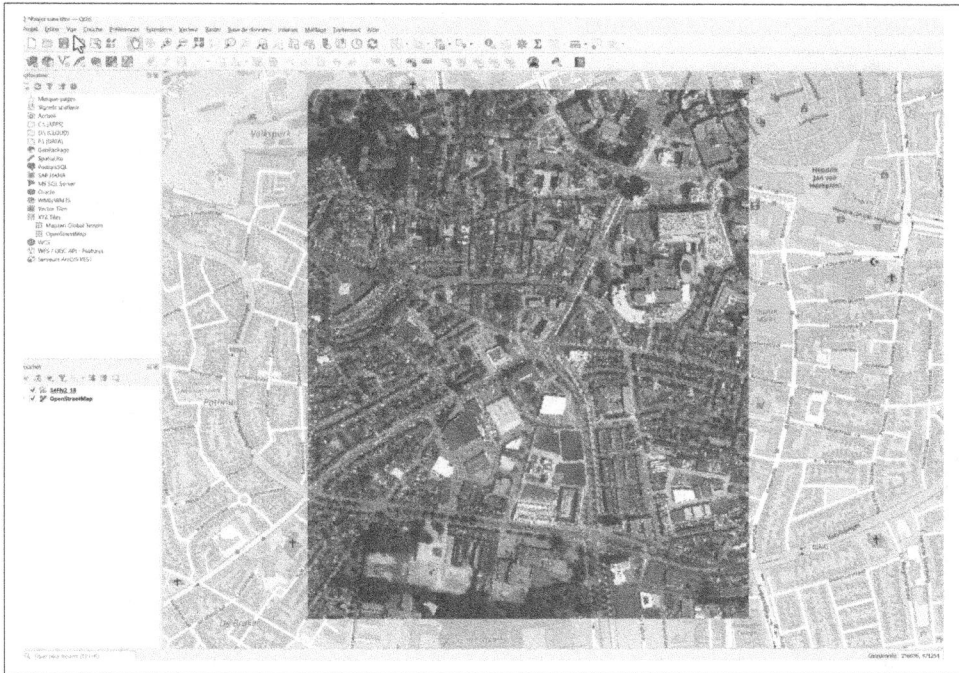

Figure 2-13. QGIS GUI with the point cloud and Open Street Map layers overlaid

Blender (geometric modeling and rendering)

Blender is very well known in the computer graphics community, especially for geometric modeling and rendering. More than that, Blender offers the possibility of working with meshes and 3D models on various tasks: rendering, rigging, VFX, sculpting, simulations, animation, modeling, motion tracking, and more. You will often see Blender used for creating animated films, 3D applications, visual effects, and more (see Figure 2-14).

Suppose we tie Blender to 3D data science. Now, we can use scripting to tie any of our workflows to Blender's abilities to improve the data visualization and modeling stages.

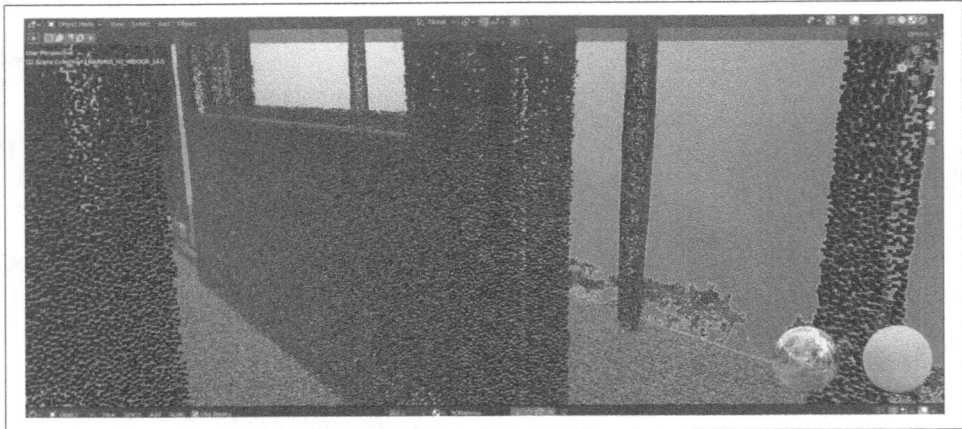

Figure 2-14. A 3D point cloud inside Blender

MeshLab (geometric modeling)

MeshLab (*https://www.meshlab.net*) is another open source initiative (see Figure 2-15). This software is mainly geared at processing and editing 3D triangular mesh modalities, but you can actually use it for point clouds as well. While the GUI may look a bit outdated, it hosts some of the most advanced features like aligning datasets, color processing, 3D reconstruction, 3D change detection, model conversion, simplification, refinement, and more. MeshLab stands as the open source reference for 3D mesh processing, and CloudCompare stands for 3D point cloud processing. Therefore, having it as part of your starter pack is a major benefit to quickly prototyping solutions.

Figure 2-15. The same 3D point cloud inside MeshLab

MagicaVoxel (voxel editing)

MagicaVoxel (*https://ephtracy.github.io*) is another open source software that is geared at visualizing, editing, and creating wonderful renderings using 3D voxels. Once downloaded, you can directly execute it and start visualizing and editing your 3D data. A nice feature lets you import meshes directly, which are then displayed as voxel assembly on the fly. It is also a very intuitive way to understand a data structure or the relationships between the constituents of your scene (see Figure 2-16).

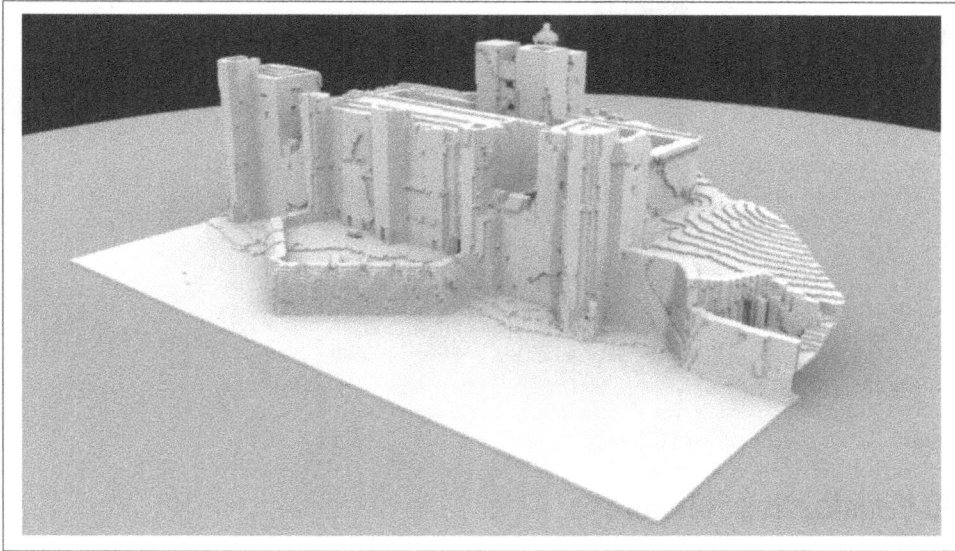

Figure 2-16. A voxel assembly of a castle

If you would like to visualize only your datasets, there is a standalone called Magica-Voxel Viewer (*https://oreil.ly/CSXOh*). With it, you can add your mesh files (*.obj*) or depth maps to a folder and visualize them as voxel assemblies on the fly. MagicaVoxel is a useful additional tool that will allow you to have outbounds from your 3D data science projects.

FreeCAD (CAD)

When it comes to computer-aided design (CAD), the largest reference that you may be aware of is AutoCAD (Autodesk). Unfortunately, it does not fall under the open source initiative; therefore, here is an alternative: FreeCAD. FreeCAD is an open source 3D modeler made to design real-life 3D objects of any size. It is also largely based on the parametric-modeling approach, which permits the editing of the size of various components and seeing the changes in their relations. Also, FreeCAD has some interesting finite element analysis (FEA) tools and dedicated building information modeling (BIM) tools. As such, FreeCAD is very useful to explore 3D datasets related to engineering applications (see Figure 2-17).

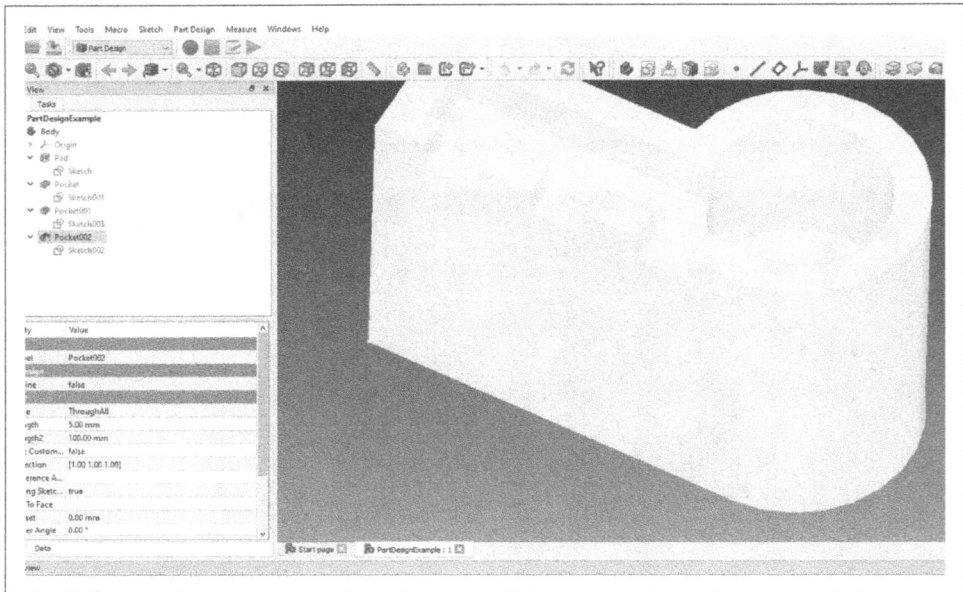

Figure 2-17. The FreeCAD GUI

3D Visualization Software

3D data visualization software is designed to create visual representations of 3D data. By transforming complex datasets into interactive and understandable experiences, they largely aid in the analysis, interpretation, and communication of 3D information. In particular, they enable interactive exploration for users to interact with immersive scenes.

Unity (game engine)

Unity (*https://unity.com*) helps you transform complex 3D data into interactive, immersive experiences that captivate and inform. As a platform renowned for its game development capabilities, Unity also excels in 3D data visualization. One of Unity's standout features is its real-time rendering engine. This allows you to visualize 3D data with stunning visuals and smooth animations, bringing your datasets to life in a way that's both engaging and informative (see Figure 2-18).

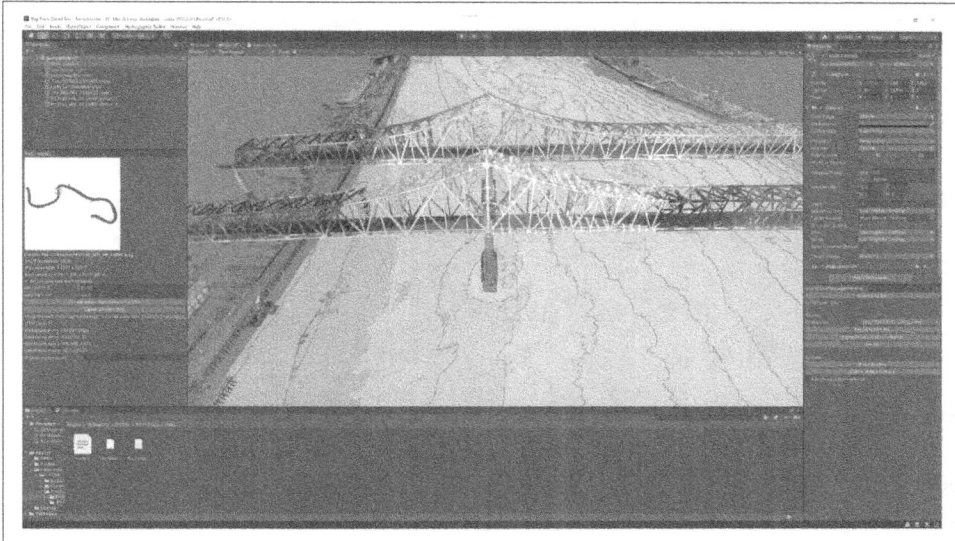

Figure 2-18. The GUI of a 3D dataset within the Unity game engine

Unity's interactive capabilities enable users to explore data from all angles, manipulate objects, and interact with simulations in real time. This fosters a more immersive and engaging experience, enabling users to acquire deeper insights from their data. Another advantage of Unity is its cross-platform compatibility. Whether you're targeting desktop computers, mobile devices, or virtual reality headsets, Unity allows you to create seamless experiences across various platforms.

Unity offers a vast asset store filled with prebuilt assets and tools to streamline the development process. This means you can quickly find and incorporate elements like models, textures, and scripts into your visualizations, saving you time and effort.

Unreal Engine (game engine)

While Unity has established itself as a powerhouse for 3D data visualization, another formidable contender is Unreal Engine (*https://www.unrealengine.com*). As someone who has explored both platforms, one of Unreal Engine's most striking features is its visual fidelity. Its real-time rendering engine, meticulously crafted for the gaming industry, pushes the boundaries of graphical detail. You can showcase complex 3D datasets with breathtaking lighting effects, photorealistic textures, and stunning particle systems. This level of visual polish can truly captivate your audience and immerse them in the data itself (see Figure 2-19).

Figure 2-19. Unreal Engine video game example

Similar to Unity, Unreal Engine has interactive capabilities. Users can explore data from all angles, manipulate objects in real time, and trigger simulations, creating deeper engagement and an intuitive understanding of the presented information. Unreal Engine has many use cases, including visualizing a scientific model, architectural design, or engineering projects. Both Unity and Unreal Engine can help you unlock the potential of creating an immersive 3D data experience through virtual reality and augmented reality.

Potree (point cloud web visualization)

My third recommendation is a solution for sharing your projects on the web. This "software" is Potree. Marcus Shutz mainly developed it based on work at TU Wien called Scanopy (see Figure 2-20).

With Potree, you can basically push your 3D point cloud onto the web by leveraging a data structure—a modified nested octree—that will handle massive datasets and allow you to interact with your point cloud in real time. To get started, you could try out the desktop version, which was built with Electron. To get your hands on it, just go to this GitHub repository (*https://oreil.ly/SFC41*) and download the latest release.

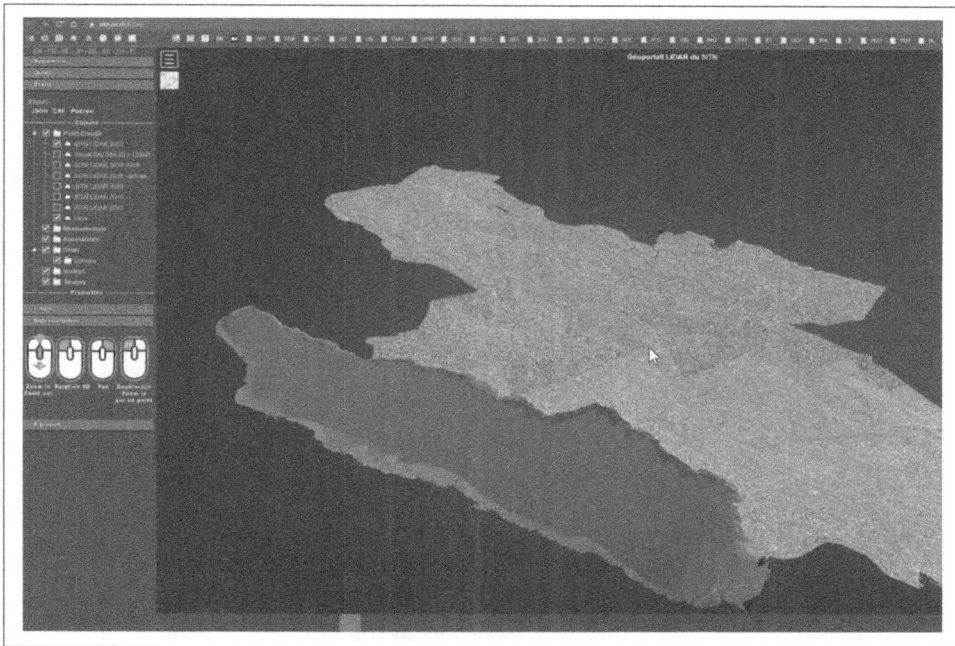

Figure 2-20. Potree desktop point cloud visualization

From there, you can unzip the archive, launch Potree, and drag and drop a 3D point cloud in the *.las* file format within the GUI. That will trigger an indexing process (e.g., Potree 2.0 version), which, upon completion, results in your point cloud being usable through an out-of-core mechanism in real time. You can use Potree to share your creations with other people very quickly and give them access to tools like measurements, taking some surfaces, extracting information, and more. It is great to offer the possibility to visualize and interact slightly with the point cloud in full 3D on the web without having to depend on a specific hardware configuration.

ParaView (3D data visualization and analysis)

I have found ParaView (*https://oreil.ly/ThZIN*) to be an invaluable tool in my arsenal. It's not a flashy game engine, but it excels in what it does best: providing a powerful and versatile platform for scientific data visualization and analysis.

ParaView's strength lies in its ability to handle various data formats, from point clouds and meshes to complex volumetric data. ParaView can ingest everything, whether you're working with intricate simulations, medical scans, or geological surveys (see Figure 2-21).

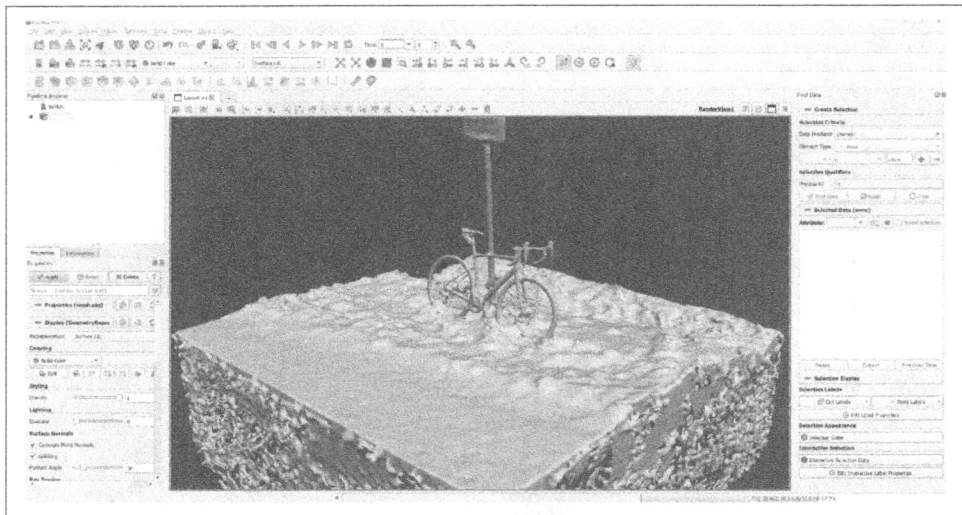

Figure 2-21. ParaView software, which is used to handle 3D datasets

Upon loading your data, you are presented with a wide range of visualization tools. From basic rendering techniques to advanced filtering and volume rendering, ParaView helps you to create clear, informative, and aesthetically pleasing visualizations of your data.

ParaView has a focus on interactivity, so you can explore your data from all angles, slice and dice it in real time, and apply different visualization techniques on the fly. This enables you to acquire a more profound comprehension of the fundamental linkages and patterns within the data.

Additionally, ParaView integrates seamlessly with Python, allowing us to automate tasks and create custom workflows that suit specific needs. Here's where ParaView truly shines for me:

Scientific data exploration
Dive into complex scientific datasets like climate models, fluid simulations, or protein structures, uncovering hidden trends and patterns through interactive visualization.

Engineering analysis
Visualize the results of engineering simulations, such as stress distribution in a bridge or airflow around a vehicle, gaining valuable insights for design optimization.

Medical data visualization
> Examine medical scans such as MRI or CT pictures, altering and viewing them meticulously to assist in diagnosis and therapy planning.

If you're looking for a robust and user-friendly platform for scientific data visualization and analysis, ParaView is worth exploring.

To ensure smooth curation, I compiled a complete list (updated) of all the 3D software (open source and paid) through the repository accessible on the 3D Resource Hub (*https://learngeodata.eu/3d-data-science-with-python*). While these are very convenient, to grasp the full might of 3D data science, it is paramount to leverage Python, which is the goal of our next chapter.

Summary

In this chapter, I covered the essential components of a successful project: the resources to fall back on, a sound hardware setup, and the local software solutions. You can think of this chapter and the topics covered within it as your 3D Data Science Starter Pack.

Here are the key components:

Knowledge and resources
> A solid foundation in mathematics, computer science, and 3D data makes things easy to grasp. We have explored areas like linear algebra, geometry, and AI, which are super useful as we delve into deeper layers.

Hardware setup
> The hardware recommendations for setting up your local 3D development environment and the possibilities if you choose cloud computing.

Software and tools
> The right tools can make all the difference. We've touched on essential software categories and their applications in 3D data science.

With these building blocks in place, you are well prepared to tackle the most complex projects, but naturally. We'll explore how to set up your programming environment and how to source, create, and manage 3D data next.

Hands-on Project

To succeed in 3D data science, you need the right resources. In this chapter, I shared many resources on mathematics, computer science, and 3D data. This exercise will help you deepen your understanding and build a customized resource guide.

Think about the 3D data science project idea you want to build. Write down the core concepts you need to master, with this list in mind:

1. Go back to "Fundamental Resources" on page 45 in this chapter.

2. Carefully review the mathematical concepts. Identify areas where you need to improve. Select relevant resources from the curated lists for each branch of mathematics.

3. Examine the computer science concepts. Note areas where you need additional knowledge. Choose appropriate resources from the curated list for each branch.

4. Focus on "3D Data Expertise" on page 49. Identify concepts that are essential for your project. Refer back to specific sections of this book for in-depth coverage.

5. Compile the selected resources (books, articles, and videos) into a structured guide. Organize them by subject or area of focus.

3D Python and 3D Data Setup

Transforming ideas into innovative products is a life-changing experience. It creates a strong sense of accomplishment when you build something you invented. I still get a rush of ecstatic emotion just thinking about my first execution: turning my 3D modeling ideas into an automated 3D Python app. I managed to do this with my little hands, my computer, and a curated set of methods and tools.

By the same token, nothing is more frustrating than having a life-changing idea, and no clue how to make it a reality. Or perhaps worse, getting stuck when it comes to laying down the necessary components and tools to make it happen. I have been there, and it is painful. If I learned anything, it is that first-principles thinking is key, especially in the early stages of 3D product development.

In this chapter, you will build your very own 3D data science lab using Python. Think of it as setting up a workshop, but instead of hammers and saws, we wield Python and a tiny selection of robust yet powerful libraries. You will learn how to load, manipulate, analyze, visualize, and export point cloud data, setting the stage for developing innovative 3D products and solutions. This process will include four steps, as illustrated in Figure 3-1.

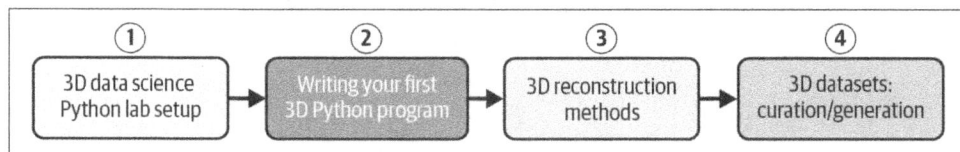

Figure 3-1. The full workflow to establish a coherent 3D data science solution

Our first stage is to create a solid foundation for building digital environments. This means setting up our 3D data science Python lab so that we can most efficiently develop 3D programs.

Once our lab is set up, we will dive into the heart of the matter: writing your first 3D Python program. It's like taking your first steps on a new planet, and with a minimal set of libraries, you'll have the tools to explore this exciting new world.

You can find all the materials for this chapter on the 3D Data Science Resource Hub (*https://learngeodata.eu/3d-data-science-with-python*). You can use the base link and add *"/chapter-X"* to access the files (code, datasets, articles) and resources; you may need to share an email address, a personal password, or proof that you are the book's owner.

But before we can start building this lab, we need to gather our materials. That's where 3D data curation comes in. We'll discuss different methods to obtain 3D datasets, from real-world scanning to leveraging open source repositories. One particularly intriguing method is photogrammetry, where we can use a series of images to reconstruct a 3D model.

3D Python Setup and Libraries

The time has come to set up our first environment for 3D data science (refer back to Figure 3-1 for the full 3D data science solution workflow). Let's walk through the major components necessary to create a 3D data science setup with Python.

As illustrated in Figure 3-2, we are going to cover the choice of OS (1), environment setup (2), the libraries to install (3, 4), and integrated development environment (IDE) setup (5). We then test that everything works properly with a 3D dataset (6) that we are going to visualize in Python, and export to handle it in CloudCompare (7).

> In this example, I will cover a setup for noninitiates that will enable you to get started coding locally in under ten minutes. Therefore, some choices privilege simplicity over efficiency. We will cover how to deepen the setup in future chapters (e.g., other IDEs, cloud-based setup, GPU and CUDA management, virtual environment management, etc.).

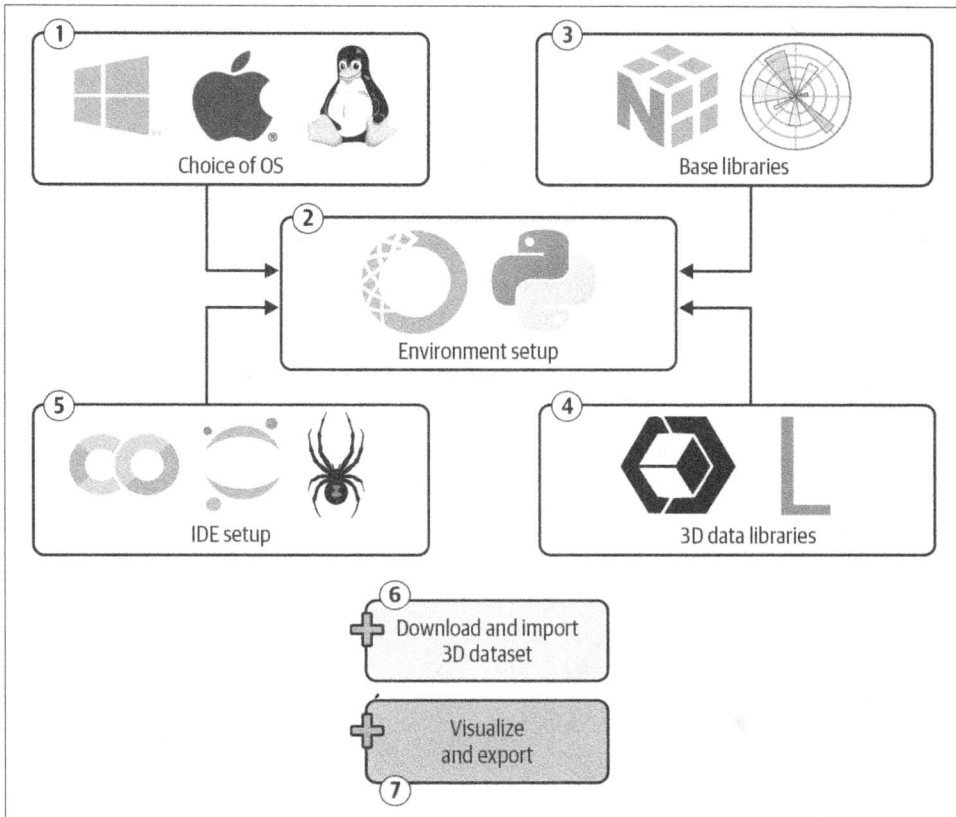

Figure 3-2. The major components of a 3D data science Python project

Choice of OS

When it comes to choosing an OS, we have the luxury of being pretty free. Indeed, the choice of the OS will have a minor influence on what you will extract from your project. Therefore, let me make a bold move: I will follow a democratic approach, where we first source data points on commonly used OSs worldwide. If we look at computers (laptops and desktops), Microsoft's Windows is the most popular OS, accounting for 69% of all users at the time of this writing. Apple's macOS comes in second at 17%, followed by Google's ChromeOS at 3.2% (in the US, it's up to 8.0%), and desktop Linux at 2.9%. Additionally, 5% of users have an "unknown" OS, likely a form of BSD or an obscure variety of Linux. Therefore, I will describe a setup that meets the needs of the majority of users and will cover both Windows and macOS.

Environment Setup

We want to ensure our environment is stable. Therefore, we are not going to directly download and install Python from the Python website (*https://oreil.ly/s21s0*). First, let's explore Anaconda (*https://oreil.ly/_oYxk*) and Miniconda (*https://oreil.ly/C_rI-*), software packages designed to manage Python environments and packages. They provide a convenient way to install and manage multiple Python versions, as well as various libraries and packages that we are going to leverage.

One key advantage of using Anaconda or Miniconda is its ability to create isolated environments, each with its own set of Python versions and packages. This functionality is especially beneficial when managing numerous projects with conflicting requirements, as it guarantees that each project has its own dedicated environment.

> The choice between Anaconda and Miniconda depends on your specific needs. Anaconda offers a larger collection of prebuilt packages and a more user-friendly interface, while Miniconda provides a smaller, more lightweight installation and allows for greater customization. I encourage you to check out the documentation (*https://oreil.ly/HBBDx*) for clear step-by-step guidance.

To create your own 3D data science Python environment:

1. Download Anaconda or Miniconda: go to the relevant website (Anaconda (*https://oreil.ly/_oYxk*) or Miniconda (*https://oreil.ly/C_rI-*)), and choose the installer for your OS.

2. Install Miniconda:

    a. For Windows: Execute the *.exe* file and adhere to the installation instructions.

    b. For macOS/Linux: Open the terminal and run `bash Miniconda3-latest-MacOSX-x86_64.sh` (replace with your downloaded filename).

3. Verify the installation:

    a. Open a new Anaconda (or Miniconda) prompt terminal window.

    b. Run `conda -version` to confirm the installation and get your current version.

4. Create a new environment: in the terminal window, you are by default in the "base" environment. It is best to create a new isolated environment with Python 3.9. For this, you can type `conda create --name DATAS3D python=3.9`.

5. Activate the environment by typing `conda activate myenv`.

6. Install a package manager named "pip" to handle the installation of all future libraries. You can just type `conda install pip`.

Going through these steps will create a new environment called "DATAS3D," in which we can organize our little 3D lab. Also, you can find some handy commands to move in and out of your environment and other conda commands (Table 3-1) that are extracted from this complete cheat sheet (*https://oreil.ly/LBqtY*).

Table 3-1. Anaconda command summary

Objective	Terminal command
Creating environment with Python version	conda create -n DATAS3D python=3.10
Activating environment (do this before installing packages)	conda activate DATAS3D
Cloning environment	conda create --clone DATAS3D -n DATAS4D
Deleting environment by name	conda remove -n DATAS4D --all
Exporting cross-platform compatible environments	conda export --from-history>DATAS3D.yml
Importing environments from a *.yml* file	conda env create -n D3D --file DATAS3D.yml
Getting help with any command	conda COMMAND --help

For the sake of completeness, know that once you have installed the Anaconda software that matches your OS, you can use the provided GUI (Anaconda Navigator) to do the same Anaconda-related setup without using the terminal.

Using the GUI called "Anaconda Navigator," go to the "Environments" tab on the left. Create a brand-new isolated Python environment by clicking on "Create." Once you create an environment, click on it. You will see a "play" icon next to its name. This opens the possibility of launching an Anaconda terminal directly acting in the selected environment, which loops back to the first terminal-based way of managing Anaconda environments.

At this point, we are ready to proceed to the next stage: populating our environment with valuable libraries.

I spent countless hours testing all the code on various machines with various configurations and different operating systems. My aim is to ensure maximum flexibility and robustness of all the code given to you in this book. Nevertheless, ensuring 100% replication without minor adjustments is hard, especially as the versions are not static and evolve with time. If you want to get as close to my Python setup as possible, all experiments were run with conda 23.3.1 (*https://oreil.ly/oayBf*) and Python version 3.9.19 (*https://oreil.ly/sSiI7*) (on Windows 11 (*https://microsoft.com*), macOS (*https://apple.com/macos*), and Ubuntu (*https://ubuntu.com*)).

Base Python Libraries

Let us first move on to the base libraries that provide the building blocks for handling, manipulating, and visualizing data efficiently. There are three cornerstone libraries: NumPy (*https://numpy.org*), pandas (*https://oreil.ly/hG3k5*), and Matplotlib (*https://matplotlib.org*). These versatile tools will become your go-to companions as you explore the intricacies of 3D data. Inside our environment (DATAS3D), we can install the three libraries from the Anaconda terminal with the package manager pip, one line at a time (for starters):

```
pip install numpy
pip install matplotlib
pip install pandas
```

> If you are asked to agree to install some components at some point, just press Y and the installation will continue. The installation retrieves the necessary components from the internet; this is the only time you will need access to the internet.

Now that they are installed, let us test their usage. For that, in the terminal, you can type:

```
python
```

This will trigger Python (*https://oreil.ly/s21s0*), in which we can write instructions. Let's start with NumPy.

This library is utilized for manipulating arrays and matrices. NumPy is a high-performance library that offers rapid and effective operations on extensive arrays with multiple dimensions, rendering it a potent instrument for scientific computation and data processing. To use it in our current Python session, we first have to import the library (and use the alias "np" to be quicker thereafter):

```
import numpy as np
```

From there, we can move onto a hands-on example of how to use NumPy to create a point cloud as a set of data points in the 3D Euclidean space (*https://oreil.ly/4ri6p*). To do this, you can create a NumPy array with three columns, where each row represents a single point in the point cloud:

```
point_cloud = np.array([[1, 2, 3], [4, 5, 6], [7, 8, 9]])
```

This example features a point cloud with three points with coordinates (1, 2, 3), (4, 5, 6), and (7, 8, 9). To print the `point_cloud` variable, you can type:

```
print(point_cloud)
```

Now, what about another library? Let's try an example with pandas. This library is more geared toward data manipulation and analysis. It provides robust data structures for handling structured data, such as CSV files, spreadsheets, and databases.

While it's not specifically designed for 3D data processing, it can still be used to write and access point clouds in a tabular format. First, we import pandas in our current Python session:

```
import pandas as pd
```

Subsequently, we construct a DataFrame object with three columns. They are respectively holding the X, Y, and Z coordinates of each point:

```
points_df = pd.DataFrame({ 'X': [1, 4, 7], 'Y': [2, 5, 8], 'Z': [3, 6, 9]})
```

Finally, we could save our DataFrame as a CSV file in a result folder, which we give as an absolute path:

```
points_df.to_csv('absolutepath/point_cloud.csv', index=False)
```

This example features a point cloud with three points with coordinates, namely (1, 2, 3), (4, 5, 6), and (7, 8, 9), as derived using NumPy.

> pandas can be extended with another Python module: GeoPandas (*https://oreil.ly/Goa0b*). This library makes it possible to work directly with spatial data stored, e.g., in shapefiles or PostGIS databases. This extends the scope of the current tutorial, but it is good to know because we will use it in other cases.

OK, that makes two interesting libraries for handling our data. But what about visualization? Well, it would not be complete without Matplotlib. This library is well known for its data visualization support. It provides many tools for creating high-quality charts, graphs, and other visualizations. First, let us import the needed modules and functions from Matplotlib, alongside NumPy:

```
import matplotlib.pyplot as plt
from mpl_toolkits.mplot3d import Axes3D
import numpy as np
```

While it's not specifically designed for 3D data visualization, it can still create 3D scatter plots of point clouds. To do this, let us first randomize a point cloud with 1,000 points leveraging NumPy:

```
points = np.random.rand(1000, 3)
```

Then, we use the `Axes3D` class to create a 3D plot and the `scatter` function to populate the plot with the point cloud:

```
fig = plt.figure()
ax = fig.add_subplot(111, projection='3d')
ax.scatter(points[:,0], points[:,1], points[:,2])
```

All that is left is to show the plot:

```
plt.show()
```

This returns the point cloud with 1,000 random points, as shown in Figure 3-3.

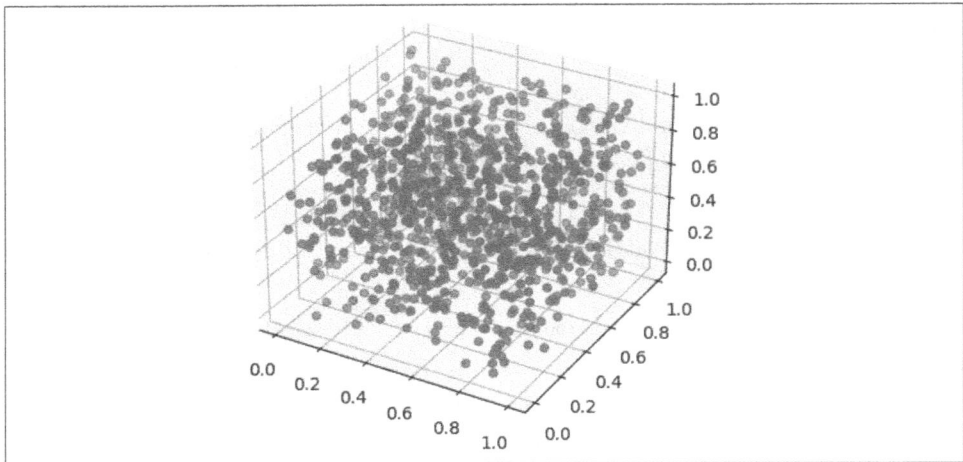

Figure 3-3. A point cloud visualized in Matplotlib

> When used as discussed here, Matplotlib opens a new contextual window, which you must close before you can use the terminal again. This is also the case with some other functions and tools that we will see later.

These libraries, combined with your understanding of 3D data, form the bedrock of your 3D projects. Having installed and tested their functionality, we can now log out of Python and get back to the packages in our environment. To do so, you can type this line:

```
exit()
```

This will exit Python and automatically revert to your current Anaconda environment. Let's now move onto the fun stuff: 3D Python libraries.

3D Python Libraries

By now, you should know that it is best to stay simple. When it comes to 3D Python libraries, let us find one with the ability to handle the unique challenges of working with three-dimensional information. Open3D (*https://www.open3d.org*) is a robust and adaptable library for the processing and visualization of 3D data. This is an open source library developed in C++ with Python bindings, rendering it available to both C++ and Python developers. Let's install Open3D:

```
pip install open3d
```

Once it is installed, let's put it to the test by entering Python again. For this, you know the drill; type the following in your terminal:

```
python
```

Open3D is a specialized library designed to process and visualize 3D data. It offers functions to manipulate point clouds, meshes, and other types of 3D data, including voxels. First, we import the library:

```
import open3d as o3d
```

Now, we create an empty point cloud object using built-in Open3D functions:

```
point_cloud = o3d.geometry.PointCloud()
```

From there, we can add some points to the point cloud:

```
point_cloud.points = o3d.utility.Vector3dVector([[1,9,3], [8,5,9], [8,8,9]])
```

For this particular illustration, the point cloud consists of three points with coordinates: (1, 9, 3), (8, 5, 9), and (8, 8, 9), respectively.

In the preceding code block, we created an Open3D `PointCloud` object and then passed a list of `points` attributes through the `o3d.utility.Vector3dVector` function that converts the list of points into a format that can be added to the point cloud object.

To visualize, we can also leverage Open3D and pass a list to the `draw_geometries` function:

```
o3d.visualization.draw_geometries([point_cloud])
```

This opens a window, and we can see three tiny squares, which are our floating points. With a minimalist mindset, we are fully equipped to move on to the next stage: upgrading our 3D data science lab.

The Python IDE

Now, it is time to facilitate the way we code by choosing a Python IDE (*https://oreil.ly/iCr4L*). This stands for integrated development environment, which provides a comprehensive set of useful tools for developing, testing, and debugging Python code. They offer an all-in-one environment where we can write, edit, and execute code; manage project files; track changes; and collaborate with other developers. Some popular desktop Python IDEs include PyCharm (*https://oreil.ly/IjTf4*), Visual Studio Code (*https://oreil.ly/VDzxV*), Cursor AI (*https://oreil.ly/ogltf*), and Spyder (*https://oreil.ly/5j3Q0*), each offering unique features and capabilities to suit different programming needs.

Desktop IDE: Spyder

In the Anaconda GUI (*https://oreil.ly/OKUEZ*) or using the Anaconda prompt, you can install Spyder. With the package manager, this is as simple as typing:

```
pip install spyder
```

Once everything is installed in your virtual environment, you can launch Spyder by typing:

```
spyder
```

Spyder is a fantastic option, especially for beginners, because of its user-friendly interface and functionalities that are specifically suited for scientific computing. It offers a clean and clutter-free layout that makes it easy to navigate your code, data, and variables. Additionally, Spyder boasts powerful features like code introspection, variable exploration, and inline plotting, all essential tools for understanding, debugging, and visualizing your 3D data manipulations. With Spyder, you can seamlessly switch between writing code, examining data, and visualizing results, making it a good fit for 3D data science projects.

Another impressive tool is JupyterLab, known for its ability to present interactive code in a visually appealing manner, making it ideal for showcasing to higher management.

Web-based IDE: JupyterLab (Jupyter Notebook)

One of the main benefits of JupyterLab (*https://oreil.ly/MnURe*) is its notebook interface, which provides a visual, interactive environment for working with code. It makes it easy to create, edit, and run code cells in real time without switching between different windows or applications. Additionally, JupyterLab offers a diverse array of data visualization tools, which makes it a useful option for tackling various data science or 3D machine learning endeavors.

JupyterLab IDE also offers support for multiple programming languages, including Python, R, and Julia. This allows one to work with various tools and libraries within a single environment, which is massively cool, as R and Julia are lovely languages.

Before we can use JupyterLab, we need to install it in our current Anaconda environment. Similarly to Spyder or our libraries, in the Anaconda terminal, write the following line:

```
pip install jupyterlab
```

Note that it may ask for your approval to install some needed library, which you need to accept by typing "y" followed by pressing the Enter key.

Once the process is done, you have a JupyterLab IDE installed in your conda environment. Do not close the console; we will test that everything works smoothly in four simple steps:

1. To launch JupyterLab, enter the command **jupyter lab** in the console. Upon execution, the JupyterLab interface will seamlessly open in your default web browser.

2. Let's create a new notebook. To initiate a new notebook and commence coding, select the File menu. You will find the menu in the upper-left corner of the JupyterLab interface. From there, select New Notebook to get started.

3. Begin coding. It is now time to commence writing code in the notebook. To add a new code cell, click the "+" button (from the toolbar), or alternatively, you can use the shortcut Ctrl + Shift + Enter. At this stage, you can write your Python code in the "cell" component, and run it by pressing Shift + Enter.

4. Ensure the safety of your work. It is crucial to regularly save your work by utilizing the Save button in the toolbar or the convenient Ctrl + S keyboard shortcut.

These are the basic steps to get started with JupyterLab. As you become more familiar with the interface, you will likely be happy about the ability to use more advanced features, such as adding markdown text, importing and exporting data, and working with JupyterLab extensions.

When you launch JupyterLab, it starts a server process on your local machine. This server manages the communication between your web browser and the Python kernel that executes your code. Press Ctrl + C in the console where you launched JupyterLab if you want to interrupt the server process and shut down JupyterLab.

Creating a 3D Python Program

Now that you've set up your 3D data science environment, it's time to write your first Python program (step 2 in the workflow shown in Figure 3-1). With the right tools and a little bit of guidance, you'll be exploring the exciting world of 3D data in no time.

This section will walk you through a simple yet informative example. We are going to create a Python script that loads a 3D point cloud, visualizes it, and performs basic operations like filtering and downsampling. Let's get started!

For our experiments, let's use a sampled point cloud that you can retrieve from the chapter's repository. Figure 3-4 shows a three-dimensional point cloud of a statue from the Glyptotek Museum in Copenhagen, created through photogrammetry. Do not worry; I will cover how you can make such models yourself, using just your phone's camera.

Figure 3-4. An example point cloud of a jaguar

From the repo, use the chapter's folder structure as it stands; you should have something like this in the *CHAPTER_3* folder:

```
CHAPTER_3
-- CODE
---- first_steps.py
-- DATA
---- sample.xyz
```

This means that if you use the *first_steps.py* script, your working directory is the *CODE* folder. As you move along in this book, you will see that retrieving datasets is done by using relative paths, i.e., the path to the data starts with the folder that holds the script.

> Working with relative paths from the *CODE* folder means that we need to exit the folder before entering the *DATA* folder. To go one level up, you can use this: ../.

Importing 3D Data in Python

First, we will utilize the NumPy library. You can begin writing your initial shortcode:

```
import numpy as np
file_data_path='../DATA/sample.xyz'
point_cloud= np.loadtxt(file_data_path, skiprows=1, max_rows=1000000)
```

This code snippet imports the NumPy library and assigns it the alias np. It then creates a variable that stores the filepath of a file containing points and imports the point cloud data into a variable named `point_cloud`. The first row of the file, which typically contains information such as the number of points, is skipped. Additionally, a maximum number of rows is set to prevent memory shortages during testing.

To access the first point of the entity that contains your data (`point_cloud`), you can now input the following command in the console:

```
point_cloud[0]
```

Subsequently, you will receive an array that encompasses the content of the initial point, which includes the X, Y, and Z coordinates. The output is an array with the values 0.480, 1.636, and 1.085.

This was your initial introduction to using Python for 3D data analysis. Now that you are familiar with the method of loading point data, let's examine some efficient first-hand procedures.

Extracting Specific Attributes

We possess a point cloud containing six attributes: X, Y, Z, R, G, and B. It's worth mentioning that when working with NumPy arrays, the indexes always begin at 0. So, if the data is loaded without field names, obtaining the second point can be achieved by typing in the console:

```
point_cloud[1]
```

This returns an array:

```
array([0.480, 1.636, 1.085, 25, 44, 68])
```

To obtain the red (R) attribute (with a NumPy "column" index of 3), we can do the following:

```
point_cloud[1][3]
```

This returns 25. To obtain the Z attribute for every point in the point cloud, we can proceed with the following steps:

```
point_cloud[:,1]
```

This returns array values: [2.703, 2.716, 2.712, ..., 2.759, 2.741, 2.767]. If we want to exclusively isolate the X, Y, and Z properties for each point:

```
point_cloud[:,:3]
```

This returns an array as follows:

```
array([
[4.933, 2.703, 2.194],
...,
[5.191, 2.767, 0.279]
])
```

You have just experimented with multidimensional indexing. In the given example, the third column (R) is not included in the selection. If a result is intended to be utilized multiple times, it can be kept in variables. We can put the geometry in an xyz variable:

```
xyz = point_cloud[:,:3]
```

And the colors in an rgb variable:

```
rgb = point_cloud[:,3:]
```

Conducting Attribute-based Data Analysis

Now, let's get into some helpful analysis. To calculate the mean height of your point cloud, you can follow these steps:

```
np.mean(point_cloud,axis=0)[2]
```

This will return 2.6785763 (meters).

By setting the axis to 0, we are instructing the analysis of each "column" separately. If the value is disregarded, the mean will be calculated for all the values. If the value is set to 1, each row's average will be calculated. If you now desire to extract points that fall within a 1-meter buffer zone from the average height (assuming the value is kept in mean_Z):

```
point_cloud[abs( point_cloud[:,2] - mean_Z) < 1]
```

Python, like programming in general, provides various methodologies for solving problems. The method provided is succinct and efficient; however, it may not be the most straightforward. Using a for loop to resolve it is a commendable approach. The goal is to attain a balanced state of clarity and efficiency, as specified in the PEP-8 guidelines.

Now, you possess the knowledge required to configure your working environment and utilize Python, JupyterLab, and NumPy for your coding pursuits. You can import point clouds and manipulate their attributes, as well as explore other possibilities, such as color filtering and analyzing point proximities.

3D Data Visualization and Export

To begin our second example with plotting 3D point clouds, we will use Matplotlib. We include packages in the import part of the original script, enabling us to utilize their functionalities:

```
import matplotlib.pyplot as plt
from mpl_toolkits import mplot3d
```

`matplotlib.pyplot` manages the plotting, whereas `mpl_toolkits.mplot3d` facilitates the creation of a three-dimensional projection. Subsequently, we can incorporate the previously examined code:

```
file_data_path="../DATA/sample.xyz"
point_cloud= np.loadtxt(file_data_path,skiprows=1)

mean_Z = np.mean(point_cloud,axis=0)[2]
spatial_query = point_cloud[abs( point_cloud[:,2]-mean_Z)<1]

xyz=spatial_query[:,:3]
rgb=spatial_query[:,3:]
```

The result of the spatial query on the initial point cloud yields two variables: xyz and rgb. Now, let's explore the process of graphically representing the outcomes. First, let's generate a three-dimensional axis object. To do this, we specify `parameter projection='3d'` when calling `plt.axes`, which will result in the return of an `Axes3D Subplot` object:

```
ax = plt.axes(projection='3d')
```

This is the blank canvas (Figure 3-5) on which we will create our artwork. Next, we will create a scatter plot and aim to assign a color to each and every point. This process can be fairly laborious, but it is made easier by the following line of code:

```
ax.scatter(xyz[:,0], xyz[:,1], xyz[:,2], c = rgb/255, s=0.01)
```

The function is passed the coordinates x, y, and z using xyz [:, 0], xyz [:, 1], and xyz [:, 2]. The colors for each point are passed using c = rgb/255, which is normalized to a [0:1] interval by dividing by the highest value. The on-screen size of each point is set at s = 0.01.

Ultimately, we generate the graph by using the command provided and appreciate the visual representation:

```
plt.show()
```

Wonderful! We have now successfully visualized our point cloud in Python with some lines of code and processed it (Figure 3-6).

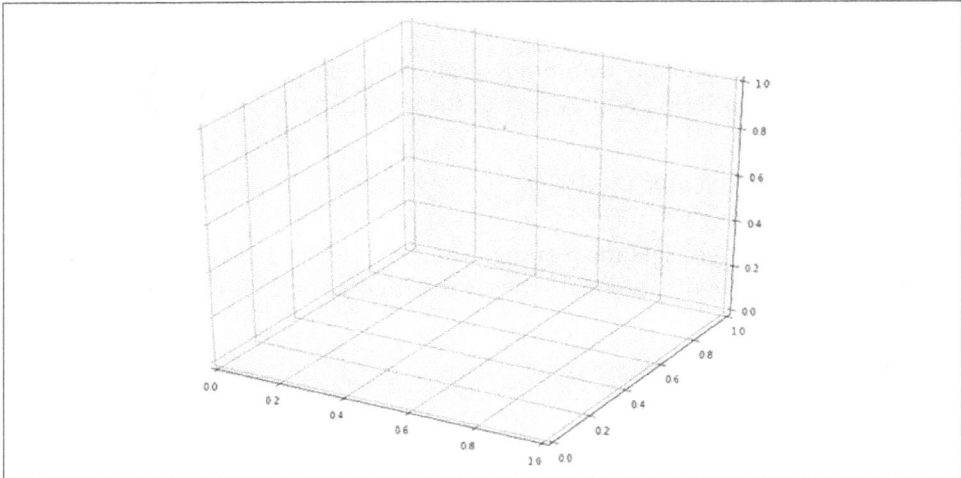

Figure 3-5. The blank canvas of Matplotlib

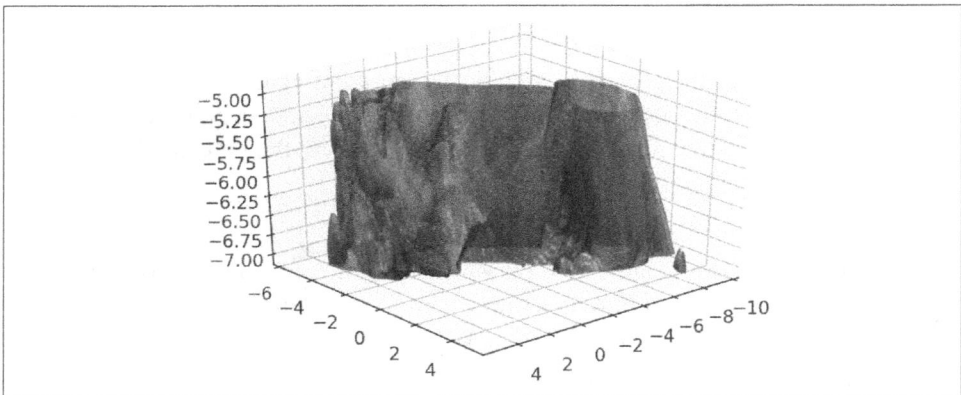

Figure 3-6. The result of slicing the point cloud

Let us move on to the next stage: how do we gather datasets, and especially, how do we gather datasets with 3D reconstruction methods?

3D Reconstruction Methods

3D data science often begins with creating a 3D model from scratch or reconstructing it from various sources. This relates to 3D reconstruction methods, the third stage of the 3D data science workflow (Figure 3-1). Let me guide you through some of my favorite techniques to transform sensory outputs or generative content into a 3D digital environment.

Before we use 3D datasets, it's essential to understand where they come from, the context associated with each dataset, and the intrinsic nature of the data we play with. To illustrate this, I delineate six ways we obtain 3D datasets (Figure 3-7).

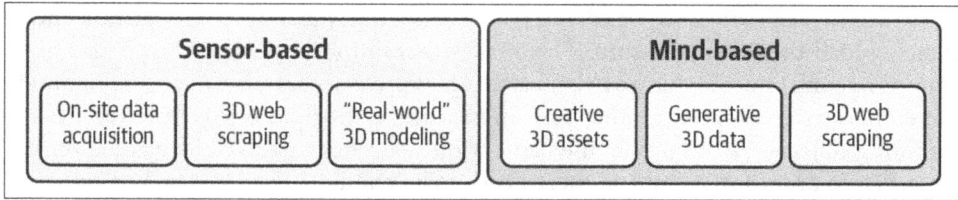

Figure 3-7. *How to reconstruct 3D data with methods from sensor-based and mind-based approaches*

To contextualize, the dataset we used previously (the jaguar) was generated from a set of pictures and leveraging photogrammetry while on a trip to Norway. You will find in all the coming chapters of the book that I provide 3D datasets from my own collection for you to play around with. These are usually derived from 3D data acquisition missions I managed. This means most of them are sensor-based and relate to on-site data acquisition with real-world reconstruction techniques. Throughout the book, I will use various technologies (ToF, stereo vision, LiDAR, etc.). Therefore, I believe it is necessary to define these terms and explain the concepts behind them. Let me do that right now.

Real-World 3D Reconstruction (Sensor-Based)

Today, we have the capability to describe objects on Earth at various scales, which is incredibly interesting. By using 3D reality capture, we can obtain datasets that describe an entire country, such as the A4 highway in the Netherlands. We can even go further and describe a whole city, like what you would find on most open data portals, and even deeper to describe a building or an object. This object can be anything from a statue to a molecule or even a cup of coffee. To accomplish this, we can utilize various platforms and sensors, and it's essential to note that there is a significant distinction between them.

It's important to understand that the sensor must be attached to a platform, which will act as its host. For instance, you could have a LiDAR sensor that is attached to your backpack, resulting in a backpack-mounted LiDAR scan. Alternatively, you could attach the same LiDAR sensor to a drone, producing a UAV-based LiDAR map of an area. This means that you can always choose which platform to use with which sensor, depending on the mission you want to accomplish when it comes to collecting data for a specific application. Most platforms you encounter will be static, meaning that they won't move from a single position. However, you may encounter mobile mapping platforms that can move across the ground at various altitudes, such as low altitude, high altitude, and satellite level. There's a lot to say about the different types of platforms available, so it's important to carefully consider which one is best suited for your needs.

There are two primary categories of sensors: active and passive. Active sensors emit a signal, which they then measure. This makes them more expensive, but it also gives you more control over what you're measuring. Passive sensors, on the other hand, do not emit any signals. Instead, they rely on measuring the light that is reflected by surrounding objects. For example, the camera on your smartphone is a passive sensor. However, in order for it to work, there needs to be enough light for it to measure.

So, I'll explain what's happening. When you want to take a picture in the dark, you usually end up with nothing unless you have a camera with a very low-pressure timer or something similar. In general, to get a nice picture with low light quality, you need to be in good lighting conditions. Otherwise, you may have to deal with a lot of noise or other issues. This highlights the major difference between platforms and sensors.

On-site data acquisition

On-site data acquisition is the most prevalent method of obtaining 3D datasets (Figure 3-8). Imagine having a camera or a laser scanner at your disposal as you survey a specific region, capturing data points in the form of point clouds that beautifully represent the geometric details of the area. This method allows for precise control over the sensors used and the accuracy of the data collected. So that's one way to get data. But if we get back even more with history, we do not have to wait to generate massive datasets.

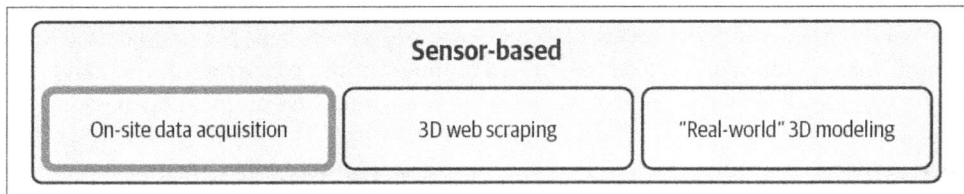

Figure 3-8. Sensor-based methods, including on-site data acquisition

Say we are going on-site, for example, with a total station or a theodolite to select points of interest directly. We are shooting or aiming at one point, describing, for example, the corner of the roof of a house. We then shoot some laser that records, thanks to a ToF technology, the angle from this initial seed point and the orientation so that we can compute a distance from the center of our system. And then, we can create 3D points that describe this semantic attached to the geometry we want to describe.

The first distinction to make is the complementarity between sensors, platforms, and methods. Platforms host sensors, and methods permit easy adjustment to a specific context. You may encounter active sensors (LiDAR systems, single measurement system, InSAR, radar) and passive sensors (image-based reconstruction, photogrammetry), which are ways to obtain 3D point clouds from on-site acquisition. If you want to explore this further, I encourage you to read the relevant tutorial offered within the deep dive section of the 3D Data Science Resources Hub (*https://learngeo data.eu/3d-data-science-with-python*).

Most of the datasets that you will have will fall under this category.

Sensor-dependent web scraping

The second family of datasets, still sensor dependent, is gathered through 3D web scraping. In this setting, you will source data that is related to a sensor through the web. For example, you have a lot of open data catalogs and open data platforms that allow you to get your hands on some dataset, but you also have commercially available data that you can buy or get with a license.

This approach is incredibly powerful today, but it wasn't five years ago. With the emergence of various quick ways to generate data, we now find a lot of it online. However, on-site data acquisition and 3D web scraping have advantages and disadvantages.

With 3D web scraping, we depend on existing databases and can't immediately access what we need if it hasn't been created yet. On the other hand, we have almost instantaneous access to data. We don't need to travel to a specific location to get a fancy reconstruction of a building; we can scrape the web and get the data ourselves. The downside, however, is that we don't have control over the precision and accuracy of the data we obtain.

3D web scraping for sensor-based data is getting better each year. Indeed, we see a lot of national (as well as international) programs opening up their nationwide height maps, LiDAR point clouds, and 3D city models as open data initiatives. We saw how to source these datasets in Chapter 2. If you want to learn more and get a good look at the most relevant open 3D datasets available, I encourage you to explore the relevant tutorial offered within the deep dive section of the 3D Data Science Resources Hub (*https://learngeo data.eu/3d-data-science-with-python*).

Sensor-related 3D modeling

The third category is what I call a "real-world 3D model," which is related to sensors but focused on 3D modeling. In this category, we work with interpreted data that is not obtained directly from the sensors themselves but through modeling. We start with a representation of something that exists in the real world and then use sensor data to build a 3D model. Essentially, we're still representing reality, but in a modified or extended way.

The base layer of this type of modeling is real and obtained through sensors that gather information. This sensor-related 3D modeling is widely used and can be applied to 3D data acquisition or 3D web scraping. It extends the technology by allowing us to modify or transform existing elements into new standards, formats, and ways to use datasets.

To provide a clear example of this, let's consider a famous application in the construction industry: a cantilever beam. When working on a construction site, we might obtain a point cloud that describes the building. However, what we actually want is an extended model called a building information model (BIM), which includes 3D models, semantics, attributes, and relationships. This model is useful for monitoring a building's lifecycle over time, starting from the beginning of the project.

With sensor-related 3D modeling, we start with a base layer that is related to reality, but we need to create something for a specific application. This concept is powerful and is present in digital twin workflows, which aim to create a digital representation of the real world. In these workflows, we almost always go through a modeling phase to fit the specific application. Going through this extra step is the third way to obtain a dataset.

Creative 3D Reconstruction

While we see how to turn sensor outputs into 3D data representation, we can also consider techniques that help generate entirely new 3D forms.

Creative 3D assets

There is another possibility apart from the virtual versus reality paradigm. It involves creating 3D assets or datasets without copying anything (Figure 3-9). This is pure creativity, like an artist starting with a blank canvas. This approach is widely used in video game development.

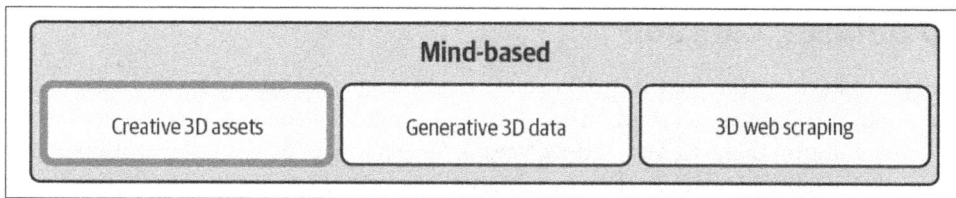

Figure 3-9. Mind-based approaches, including creative 3D assets

If you need to create something based on a real-life object, like a forest, then you need to use sensor-related 3D modeling. But if you're making a video game set in space or any other imaginary world, you'll need to create everything from scratch with your imagination. This is where 3D datasets come in handy.

In addition, you can use 3D web scraping or sensor-related 3D data sourcing to inspire or generate new ideas. These techniques are very powerful and can help you create amazing things.

Generative 3D

Another alternative to get your hands on a 3D dataset is what we call generative 3D. This includes generative 3D methods that rely on AI or non-AI algorithms. The choice is usually first triggered by the scope and the scale at which you are generating data. This can be a powerful method if you want to create various versions of something specific or if you wish to augment your dataset. We will use a specific tag for datasets generated with a generative algorithm or generative method.

> With the rise of generative AI, new solutions for 3D data science workflows have emerged. We can generate 3D models with various levels of detail and fidelity/creativity, especially when reconstructing 3D datasets from text or image prompts. Due to the large impact this has on a lot of applications, we'll explore a hands-on solution with generative AI for 3D datasets in Chapter 18.

3D creative web scraping

The last remaining way to access a 3D dataset is to scrape the web. But here, we scrape for creative 3D assets instead of sensor-based ones. With these categories, you understand the full scope of gathering 3D datasets. Next, let me show you how to curate or generate your 3D dataset from real-world scenes with just your camera.

3D Dataset: Curation

Before embarking on your 3D data science adventures, solid methods for sourcing high-quality 3D datasets are essential. This section guides you through the process of 3D data curation, helping you source, create, and manage the data that will fuel your projects. Imagine 3D data as the raw materials of your 3D universe. Just as a sculptor needs the suitable clay to craft a masterpiece, you need the correct 3D data to build your data science projects. In this section, we explore various methods for acquiring 3D datasets, from real-world scanning to leveraging open source repositories.

 Remember that you can leverage the software stack shared in the previous chapter to visualize and interact with the datasets gathered in this chapter. I recommend this approach, particularly at an early stage of development. Indeed, it is usually counterproductive to start developing a solution without at least initiating the formulation and hypothesis test. Using software with implemented tools is usually an excellent way to bypass this friction.

3D Data from Image-based Reconstruction

As a young expert in 3D data science, generating your datasets from real-world objects gives you a great edge. However, it is important to understand the fundamental principles amid the emergence of advanced techniques like neural radiation fields (NeRF) and 3D Gaussian splatting. This enables you to navigate through the hype and address various challenges.

First, let's consider reconstructing 3D objects from images, which has been the subject of extensive research. It has a fascinating history, which can be divided into three main research areas: computer vision, robotics, and geomatics. Computer vision focuses on techniques like stereo vision and structure from motion (SfM), while robotics explores different iterations of simultaneous localization and mapping (SLAM). Geomatics, on the other hand, approaches the problem from a photogrammetry perspective. These terms focus on a sensor's ability to determine its position and orientation in relation to its surroundings while also creating a 3D map of those surroundings.

The metric approach has the ability to transform a collection of unstructured overlapping images into intricate 3D objects. And what better way to gain understanding than by actively engaging in the process?

Let me quickly provide a workflow as defined in Figure 3-10, which leverages Reality-Capture and Meshroom. It consists of five main steps: subject definition, scene parameter settings, sensor parameter settings, data preparation, and 3D reconstruction with photogrammetry.

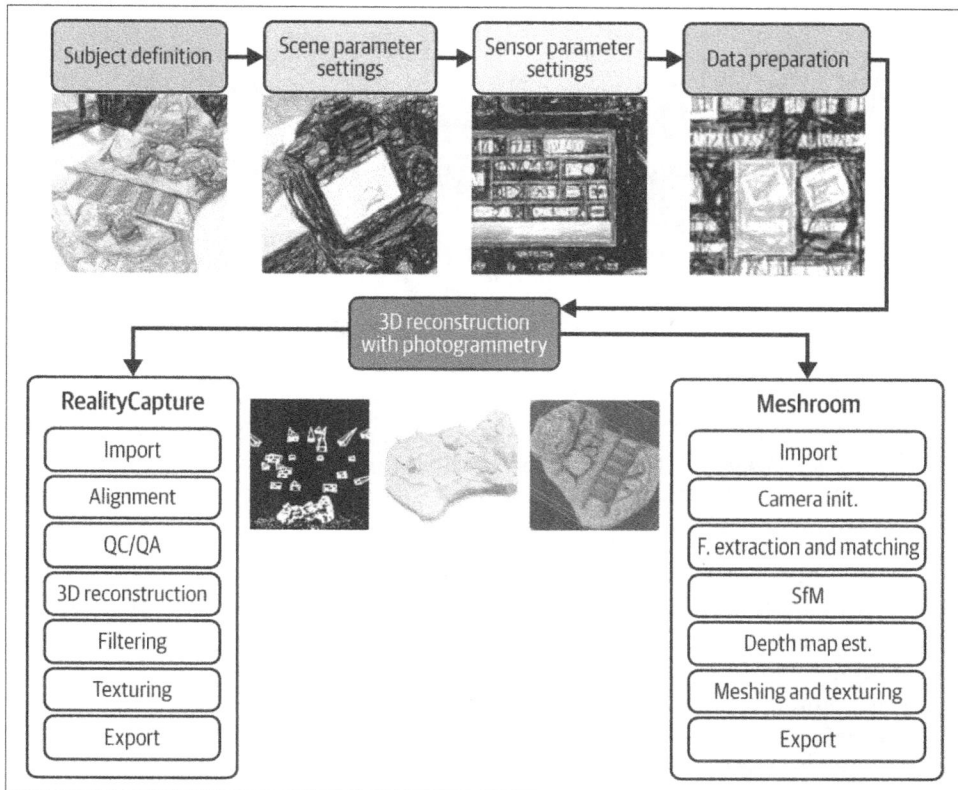

Figure 3-10. The 3D photogrammetry workflow comprises a straightforward usage of the RealityCapture and Meshroom software solutions

This process is what allows you to take a few pictures of a scene of interest and obtain a 3D digital reconstruction of your subject. The 3D dataset for the Glyptotek statue used earlier in this chapter comes from this process.

The necessary equipment is fairly accessible. Only a camera is needed (your smartphone works), and a tripod can be used if more stability and control are needed, but it is possible to work without it. In experiments, we set the camera parameters as follows:

Focal length
> A fixed focal length of 28 mm, giving a good balance between distance to the subject and workable area.

ISO
> 400, which is a good value to start with, but if your scene appears too dark, you can push this value up to 1600 (at your own risk).

Diaphragm opening

F/3.5, which allows for enough light to be drawn in at the cost of a limited depth of field. In our case, as we focus only on a tiny object, we are good here.

Shutter speed

125 ms (1/8 s), which our tripod allows thanks to more stability. Without a tripod, below 1/200s is super tricky.

White balance

4000 K. This adapts to our scene's ambient lighting.

The acquisition strategy is straightforward. We place the camera on the tripod and adjust the height and the angle so that the object completely fits in the field of view. We then move the tripod along a circular trajectory all around the object. We need to remain at a constant distance from the object so that the focus stays right. We check the photos as we go. Once we have achieved the first circle around the object, we diminish the tripod's legs and lower the downward angle of the camera. We take a few more photos all around the subject.

Finally, we remove the camera from the tripod and take a vertical photo of the complete object. To do this, we adjust the focus again. Then, we take detailed photos to ensure that the texture of all parts of the object is fine. The idea is basically to take enough images to allow for a complete and detailed reconstruction while keeping in mind that an unnecessarily large number of inputs will only slow down and complicate the process. Ideally, for such an object, you have 25 pictures (eight horizontal positions multiplied by three layers of heights, plus one on top; see Figure 3-11).

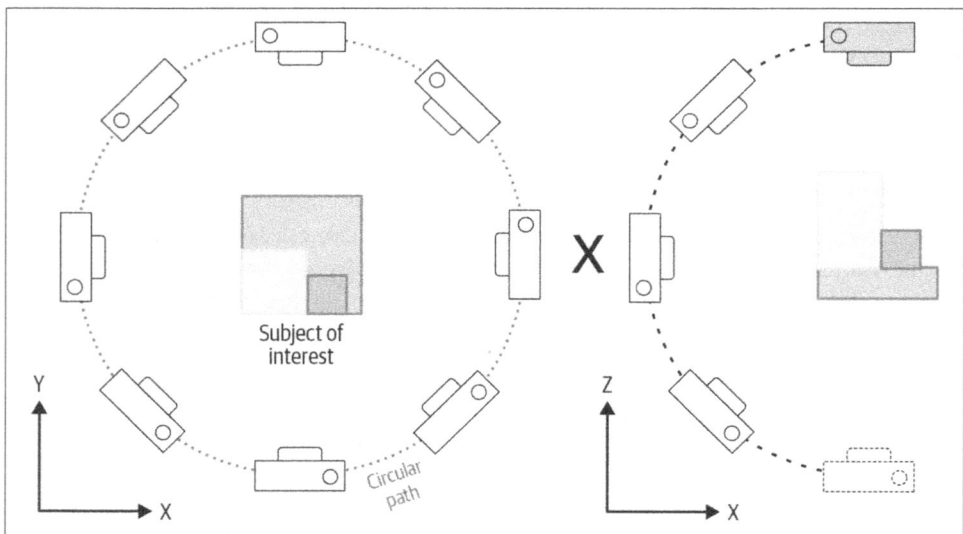

Figure 3-11. The 3D acquisition setup for photogrammetry of unitary objects

After this stage, you can leverage the 3D reconstruction software to turn images into 3D models. This video tutorial will be helpful: "2D Images to 3D Models: Quick Guide (End-to-End 3D Reconstruction (*https://oreil.ly/5rUfa*))." This approach makes me think about another solution: how about quick ways to access various multimodal datasets through web scraping?

Multimodal Web Scraping

To kick things off, what is this word: multimodal? It refers to data that spans different types and contexts, such as images, point clouds, text, and sound.

> At this stage, I will not discuss the specificities of each modality but rather help you constitute a base layer of datasets that we can use as we deepen our 3D expertise and 3D data science skills.

My goal is to identify a zone of interest and gather as much data as possible to allow us to build strong 3D data science solutions as we move through this book. The zone of interest selected is a part of Enschede, a city in Eastern Netherlands in the province of Overijssel (Twente region).

> This location is near and dear to me because I had the chance to work at the University of Twente and educate students from all nationalities in geospatial data processing. Of course, you are free to choose any other region and replicate the same approach.

We are going to get our hands on eight modalities as illustrated in Figure 3-12. We use various 2D/2.5D/3D modalities, including 3D point clouds, 3D meshes, city models, voxels, spatial rasters, 360° imagery, tabular data, and vector data.

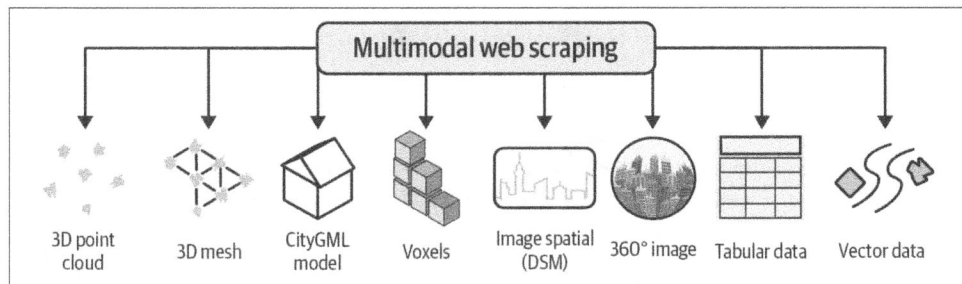

Figure 3-12. Multimodal web scraping modalities

We are now ready to source different datasets. For clarity concerns, I organized this sourcing into four categories by searching for 3D datasets, spatial rasters, vector datasets, and finally other sources, as shown in Figure 3-13.

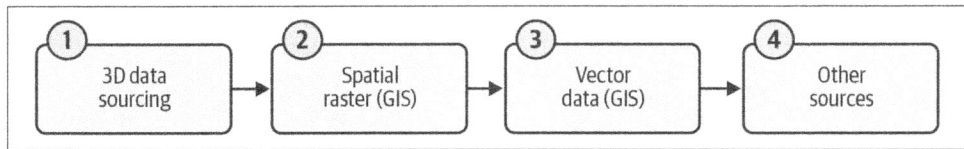

Figure 3-13. How to source datasets in four steps

3D data sourcing

The first step is to source datasets from an open data repository with a data license that allows us to do some experiments. On that front, for the Netherlands, there is the possibility of obtaining LiDAR data, elevation data models, and raster imagery from one place: GeoTiles.nl (*https://oreil.ly/GnASB*). You can zoom in on the original tiles and get access to the various datasets download links, as shown in Figure 3-14.

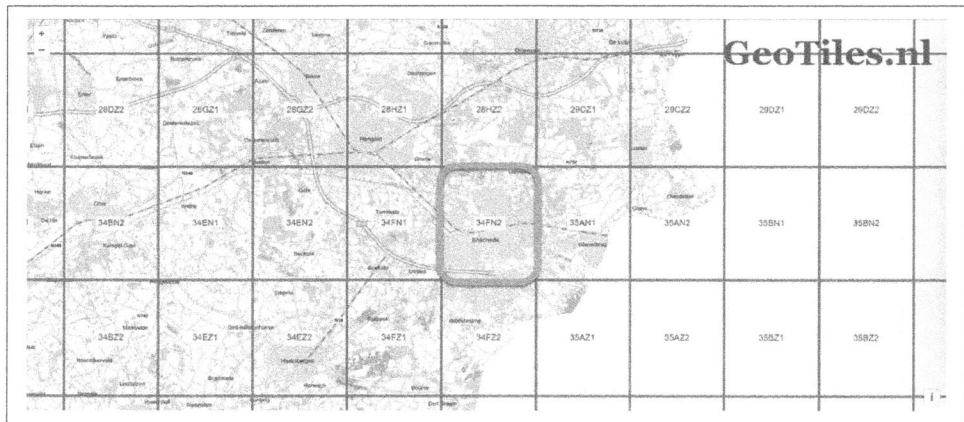

Figure 3-14. Extracting the 34FN2 Tile of the AHN4 dataset through GeoTiles.nl (https://oreil.ly/GnASB)

The second place you can explore for city models is 3DBAG, which stands for 3D Register of Buildings and Addresses; it is the most detailed, openly available dataset on buildings and addresses in the Netherlands. It contains 3D models at numerous levels of detail, generated through the combination of two open datasets: the building data that stems from the BAG website (*https://oreil.ly/rAdHd*) and the height data from AHN (National Height Model of the Netherlands) website (*https://oreil.ly/McIdU*). Using the 3DBAG viewer with tile queries, you can explore and access buildings (Figure 3-15).

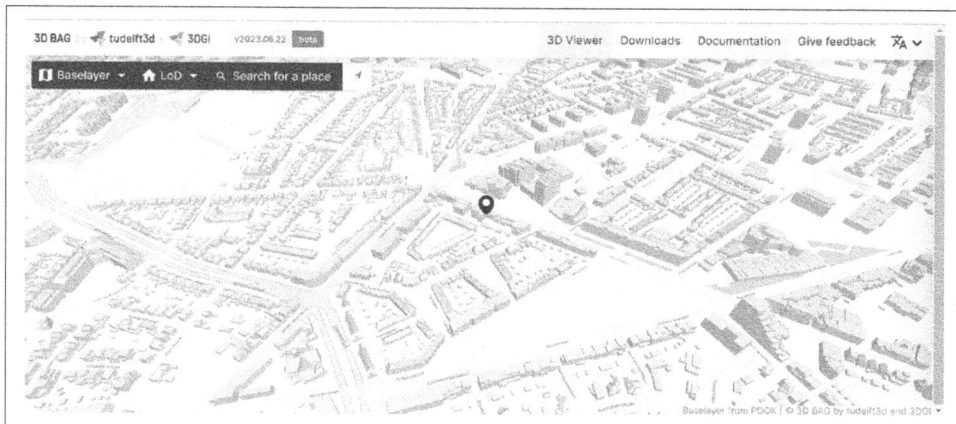

Figure 3-15. The BAG viewer (this data is licensed under CC BY 4.0, 3DBAG by tudelft3d and 3DGI (https://oreil.ly/xSazG))

The tile extent differs from what we got from GeoTiles.nl (*https://oreil.ly/GnASB*), making it interesting when we attack the integration phase. At this stage, we already have exciting datasets available to us. We can now move on to exploring spatial raster datasets.

> Several open-data portals help you gather aerial LiDAR datasets from national campaigns. You can gather datasets from OpenTopography (*https://oreil.ly/HfCoZ*), LiDAR HD (*https://oreil.ly/xyIHS*), the National LiDAR Program (*https://oreil.ly/kHT1o*), etc. The list goes on, and search engines can be the best source for the most relevant open data for your goals.

Spatial raster data sources

If we want to get our hands on satellite imagery, the USGS EarthExplorer (*https://oreil.ly/f8VlR*) is one of the largest free data sources. It is available worldwide, with a friendly user interface that makes accessing remote sensing data simple. For some cases where we need to get our hands on more than one data tile, there is also the ability to buy a download (Figure 3-16). If this is your first time accessing this data source, you will need to execute one additional step: creating an account, which is free and quick.

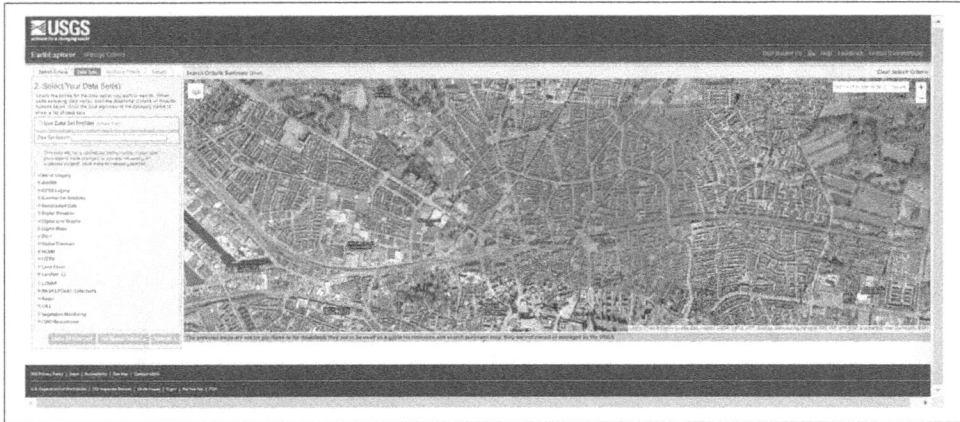

Figure 3-16. The USGS EarthExplorer (the open data is part of the US public domain, credit USGS (https://oreil.ly/BvZ0v))

I drew a polygon from the web UI and then asked to get the *Landsat > Landsat Collection 2—Level 1* group (the most recent Landsat imagery is L8–9 OLI/TIRS and L7 ETM+). The main differences between the two collections are linked to the data quality and also the level of processing involved. USGS has classified images into tiers (*https://oreil.ly/ndck1*) based on quality and processing level (see this tutorial on downloading Landsat data). Once in the Results section, you can check the footprint before deciding which would best fit your needs, as shown in Figure 3-17.

Figure 3-17. The polygons used to extract Landsat images

For digital elevation models, I suggest sticking with the GeoTiles.nl (*https://oreil.ly/GnASB*) data service as it is already on point with the most up-to-date and precise elevation models from AHN.

Vector data curation

If you are in the GIS community, I hope presenting the power of OpenStreetMap (OSM) (*https://oreil.ly/I7Zoj*) will not insult your knowledge base (Figure 3-18). OSM provides different maps and layers with a crowd-sourcing initiative that makes it highly exhaustive, including a precision flag. Indeed, OSM is open to the public and created by a general audience.

This means that accuracy can vary based on the creator and their mapping expertise level. However, OSM is a gold mine for openly licensed street-level GIS data. We have several ways to download OSM data to get our hands on some of this gold. Conveniently, there is even an OSM data Wikipedia page (*https://oreil.ly/KaaIg*) with all the available OSM extracts.

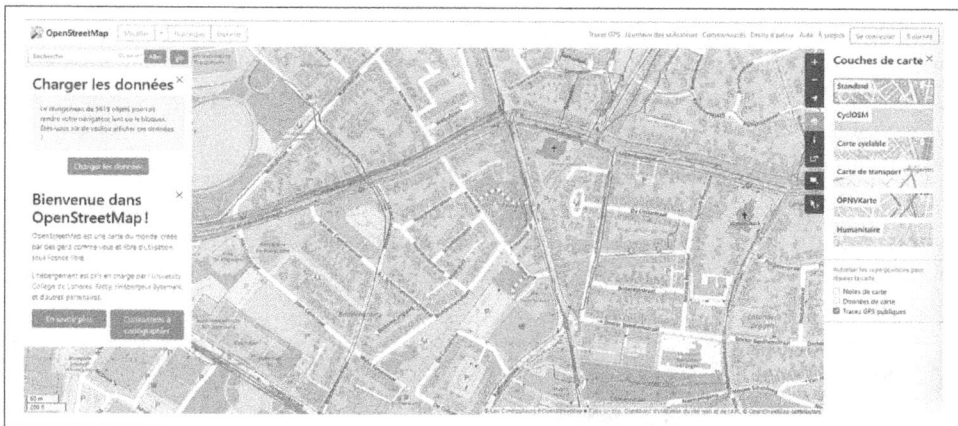

Figure 3-18. OpenStreetMap data curation portal (data available under the open database license [ODbL], credit OpenStreetMap (https://oreil.ly/jRRSb))

My recommendation is to use the tool Geofabrik (*https://oreil.ly/J6cqK*). You can then leverage data organization by semantic spatial extent (e.g., country, state, continent, etc.). You can quickly choose a geographic location and then download OSM data, as shown in Figure 3-19.

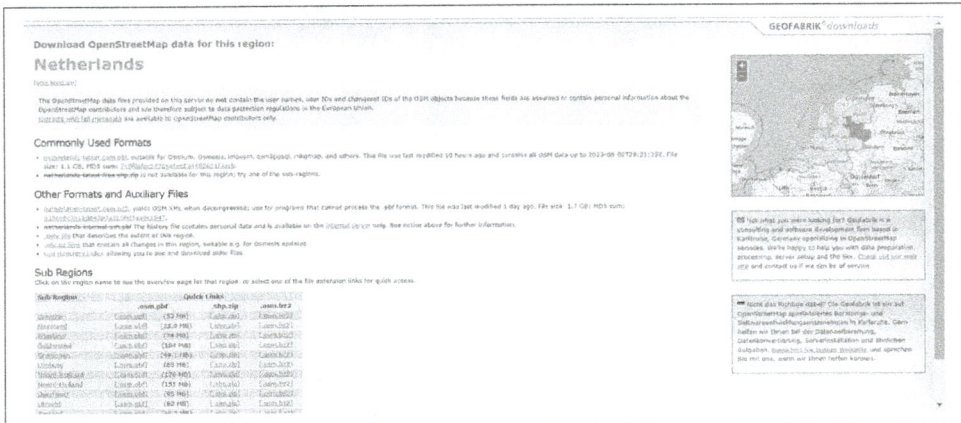

Figure 3-19. The Geofabrik portal, where you can gather vector datasets

Other sources

When it comes to dataset curation, the web is your ally. You can find anything you want to tie to your analyses, from web pages to news to sounds to real-time data feeds. The sky (or your bandwidth) is the limit! But let me be very pragmatic and also guide you toward one platform: Mapillary (*https://oreil.ly/XJnbW*).

Mapillary is a platform that makes street-level images and map data available to scale. Using the Map Explorer, you can explore it and download 360° imagery and points of interest (Figure 3-20). You can also filter out elements by their class if you want to select only some aspects of interest or cross-validate information with OSM data, for example.

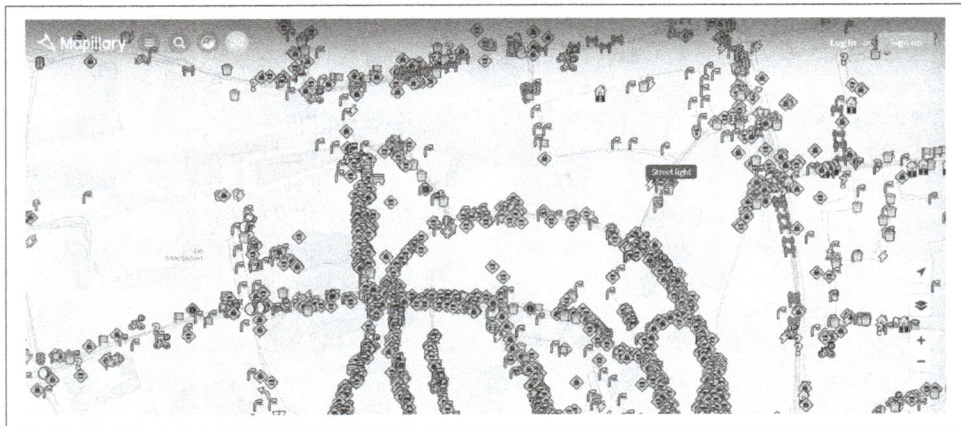

Figure 3-20. The Mapillary database portal (if you download images, they are licensed under CC-BY-SA (https://oreil.ly/UzYJU) by Mapillary)

In summary, both on-site data acquisition and 3D web scraping are fundamental pillars that have a significant impact on the strategy we choose for data sourcing. Depending on your needs, you can choose the most suitable approach.

Summary

In this chapter, we covered the 3D data science lab pack, which is essential for creating a successful 3D data science project. It complements the resources and tools from the previous chapter, constituting a starter pack with four main components: knowledge and resources for 3D data science, software with tools for 3D, programming/environment setup with libraries, and 3D datasets. By this point, you have a clear guide for setting up a 3D data science environment in Python and leveraging various methods for obtaining and utilizing 3D datasets.

By the end of this chapter, you should understand how to handle the following tasks:

- Setting up and managing a 3D Python project
- Using key Python libraries for 3D data manipulation and visualization, especially for basic operations with NumPy, pandas, Matplotlib, and Open3D
- Writing a Python script to load data from a file, extract specific attributes from the point cloud, perform attribute-based analysis, and visualize the results using Matplotlib
- The origins and nature of 3D datasets, classifying them into five categories: real-world 3D reconstruction, sensor-dependent web scraping, sensor-related 3D modeling, creative 3D assets, and generative 3D
- Leveraging image-based 3D reconstruction techniques for creating 3D models from images using photogrammetry (including aspects like equipment, camera settings, acquisition strategy, and software utilization)
- The value of working with multimodal datasets, encompassing 3D point clouds, meshes, CityGML models, voxels, spatial rasters, 360° images, tabular data, and vector data

Now that you have everything set up and tested as shown in this chapter, you are ready to tackle the entire book. From this foundation, we will improve some aspects as we move along or add new libraries, but the foundations are never going to change. Next, we will look at building 3D data tools to handle representations and data structures.

Hands-on Project

I have a quick hands-on exercise for you: testing your setup for point cloud manipulation and visualization:

1. Download the Chapter 3 resources (*https://learngeodata.eu/3d-data-science-with-python*). This includes a sample point cloud (*sample.xyz*) and a Python script (*first_steps.py*).

2. Examine the structure of the *sample.xyz* file. Identify the attributes associated with each point (e.g., X, Y, Z, R, G, B), and create another sample file to generalize on.

3. Open the provided Python script in your preferred IDE (Spyder or JupyterLab).

4. Utilize NumPy (`np.loadtxt()`) to import the point cloud data into a variable (e.g., `point_cloud`). Remember to skip the header row (if applicable) and set a maximum number of rows to prevent memory issues.

5. Extract the attributes of all points and store them in variables.

6. Conduct some basic 3D data analysis with spatial queries:

   a. Calculate point cloud statistics.

   b. Calculate the mean height (z-coordinate) of the point cloud.

   c. Calculate the standard deviation of the height values to understand the height distribution.

   d. Perform spatial queries: extract all points within a one-meter radius from a point of your choice (e.g., the point with the highest z-coordinate). You can achieve this using NumPy's array operations to calculate distances between points.

7. Visualize the point cloud using Matplotlib to create a 2D and 3D scatter plot of the point cloud.

   This exercise follows the logic of your first 3D Python program, but it should leave enough room for making small mistakes that help you grasp the intricacies of 3D Python code.

## CHAPTER 4

3D Data Representation and Structuration

3D data representations, such as point clouds, meshes, parametric models, voxels, and depth maps, offer various methods to describe and handle 3D geometric entities. Knowing their differences and how to manage their uniqueness is essential. It helps us pick the right format for a task, improve workflows, and ensure different applications work well together.

Moreover, with the rise in 3D data volume, we need efficient ways to handle increasingly complex 3D datasets that often appear in different shapes and formats. Indeed, this can cause significant problems when attempting to integrate and establish robust 3D analytical workflows. Consider the time you spend changing formats, handling compatibility problems, and trying to find common ground for your analyses. These are not just minor issues; they affect your productivity and restrict the potential of your advanced work.

A unified way to represent 3D data is essential for opening up new opportunities and overcoming the challenges of different formats. Picture a process where you can easily switch between 3D meshes, point clouds, and volumetric models.

This chapter examines point clouds as a key data link and shows how they can connect different representations. We'll also cover how to integrate various 3D data representations to establish core data structures to speed up computation (see Figure 4-1).

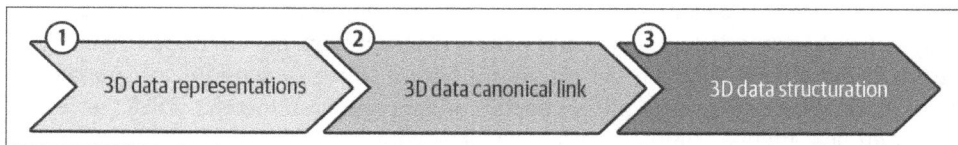

Figure 4-1. High-level workflow for establishing core data structures

This workflow covers (1) how we can represent 3D data, (2) how we can link it to a base 3D data layer, and (3) how to structure it to gain efficiency in computational tasks. Only then can we attack the development of a multimodal 3D viewer directly within Python, which is the purpose of Chapter 5.

You can find all the materials for this chapter on the 3D Data Science Resource Hub (*https://learngeodata.eu/3d-data-science-with-python*). You can use the base link and add "*/chapter-X*" to access the files (code, datasets, articles) and resources. Note that you may need to share an email address, a personal password, or proof that you are the book's owner.

3D Data Representations

As you grasped from the short introduction, there are various ways to represent 3D geometric entities. They are summarized and organized in Figure 4-2.

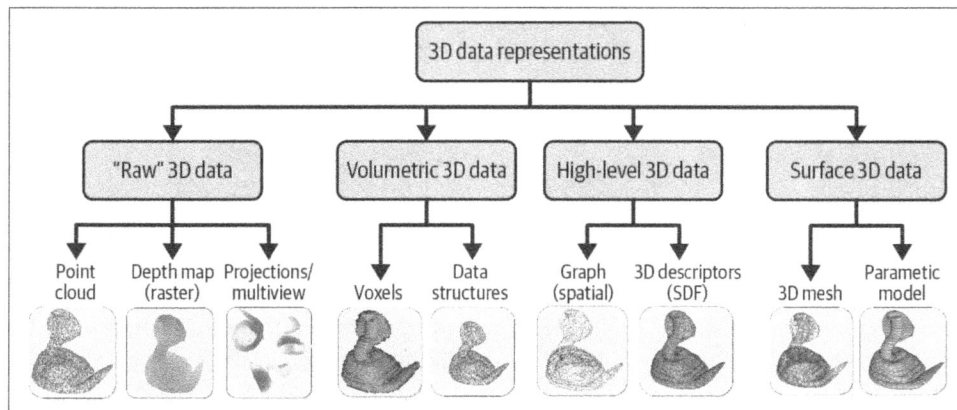

Figure 4-2. 3D data representations

On the left you'll see raw data, i.e., the closest from the sensory output, and toward the right is surface data, usually denoted as a 3D model with Python.

3D Point Clouds

A point cloud comprises a set of data points, usually expressed in a three-dimensional coordinate system (see Figure 4-3). In a Cartesian coordinate system, these points are spatially defined by x, y, and z coordinates and often represent the envelope of an object. Reality capture devices obtain the external surface in three dimensions to generate the point cloud. These are commonly obtained through photogrammetry, LiDAR (terrestrial laser scanning, mobile mapping, aerial LiDAR), depth sensing, and, more recently, deep learning through generative adversarial networks (see Chapter 18).

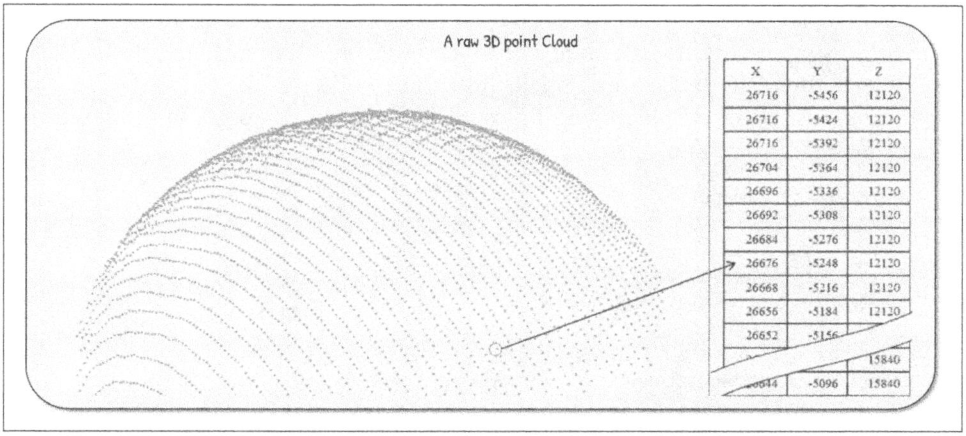

Figure 4-3. A 3D point cloud "representing" a sphere surface

Point clouds provide simple yet efficient 3D data representations. I've summarized the primary operations, benefits, and disadvantages that come with them in Table 4-1.

Table 4-1. Overview of point cloud operations, benefits, and disadvantages

Main operations	Main benefits	Main disadvantages
• Transformations: You can multiply the points in the point list with linear transformation matrices. • Combinations: Objects can be combined by merging the point list. • Rendering: Projects and draws the points onto an image plane.	• Fast rendering • Exact representation • Fast transformations	• Numerous points (obj. curve, exact representation) • High memory consumption • Limited combination operations

It's easy to manipulate point clouds due to fast rendering and easy transformations. Recent developments are trending toward even better support within pure mesh-based rendering platforms, with a recent example within the Unreal 5 game engine.

3D point clouds in Python

Now, what does this look like with Python? We have a point cloud loader function for which usage depends on the need to support a specific file format and leverage a set of libraries. The most direct route within Python is to use NumPy (*https:// numpy.org*), which works perfectly for ASCII files and allows handling any extra attribute as a tabular file. To do this, we can use the following lines:

```
import numpy as np
pcd_np = np.loadtxt('../DATA/bike_florent.txt', delimiter=';', skiprows=1)
pcd_np[0:5]
```

This script loads a point cloud from the file *bike_florent.txt* located in the *../DATA/* directory using `np.loadtxt`. It skips the first row (containing headers) and specifies that semicolons delimit the data. Finally, to print a subset of the loaded point cloud data, we can use the variable name followed by filtering based on the index of the point (the index starts at 0), e.g.:

```
#Print the first point
print('First point of the data\n',pcd_np[0])

#Print the first 5 points
print('First 5 points\n',pcd_np[0:5])

#Print the first three columns (i.e., X, Y, Z) for all points
print('First three columns X,Y,Z\n',pcd_np[:,0:3])

#Print the last three columns (i.e., X, Y, Z) for the first 4 points
print('R,G,B information of the first four-point \n', pcd_np[0:4,3:6])
```

Now, there is another way to handle this kind of data. We could use a DataFrame, where pandas (*https://oreil.ly/dWnTv*) may be a more suited choice to handle the data in a structured way:

```
import pandas as pd

# Load the point cloud data from the specified CSV file using pd.read_csv
pcd_pd = pd.read_csv('../DATA/bike_florent.txt', delimiter=';')

# Display the subset of the loaded point cloud data (rows 1 to 5, index 0 to 4)
pcd_pd.iloc[0:5]
```

This results in the data in Table 4-2.

Table 4-2. Example of point cloud data loaded as a pandas DataFrame

	X	Y	Z	R	G	B
0	1.171487	1.247825	2.320436	205	201	190
1	1.459993	1.341689	2.305161	197	197	189
2	1.329194	1.299692	2.297502	162	161	157
3	1.515316	1.356915	2.300854	198	199	193
4	1.350535	1.296824	2.292063	122	121	115

For large point clouds, memory mapping is recommended (see "3D Data Structures: k-d Trees, Octrees, BVH" on page 139).

We also have a second possibility for point clouds in file formats such as *.ply*: leverage Open3D (*https://www.open3d.org*). Here is what the import looks like:

```
import open3d as o3d
pcd_o3d = o3d.io.read_point_cloud("../DATA/Seychelles-beach.ply")
o3d.visualization.draw_geometries([pcd_o3d])
```

In this example, the function `o3d.visualization.draw_geometries([pcd])` invokes the `draw_geometries()` function from the visualization module in Open3D. The function accepts a collection of geometries as a parameter and presents them in a visualization window. Here, the list consists of only one geometry, which is represented by the `pcd` variable and represents the point cloud. The `draw_geometries()` function generates a three-dimensional viewing window and displays the point cloud. Manipulate the visualization window to rotate, zoom, and examine the point cloud from various viewpoints.

On top is a bidirectional system to go from or to NumPy point clouds. Therefore, we are in a nonclosed system:

```
# Create an Open3D PointCloud object
pcd_np_o3d = o3d.geometry.PointCloud()

# Set the points of the PointCloud using the x, y, z coordinates from pcd_np
pcd_np_o3d.points = o3d.utility.Vector3dVector(pcd_np[:, 0:3])

# Set the scaled colors of the PointCloud using the RGB values from pcd_np
pcd_np_o3d.colors = o3d.utility.Vector3dVector(pcd_np[:, 3:6] / 255)

# Visualize the PointCloud using Open3D's visualization capabilities
o3d.visualization.draw_geometries([pcd_np_o3d])
```

A mask is captured as a 3D point cloud and placed on a beach (see Figure 4-4).

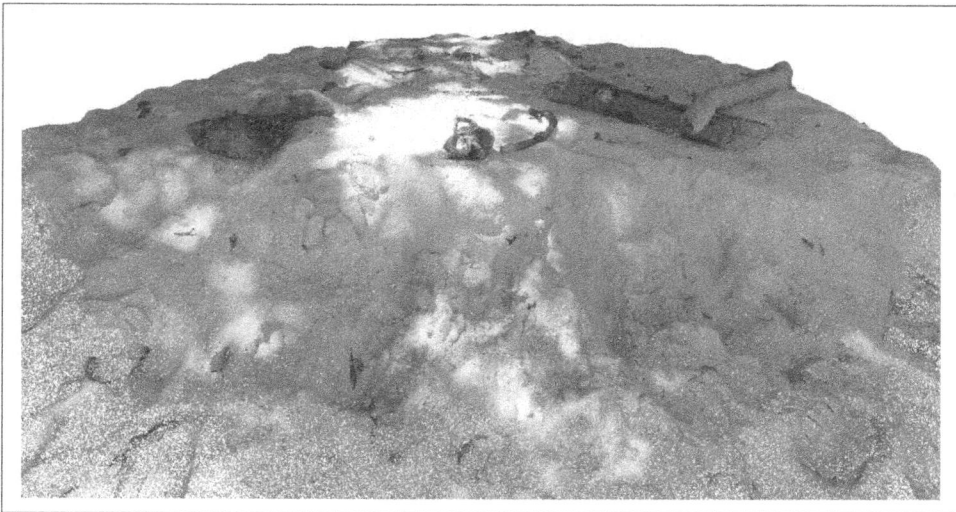

Figure 4-4. A 3D point cloud of a mask on the beach

Let's consider the same data using pandas:

```
# Create an Open3D PointCloud object
pcd_pd_o3d = o3d.geometry.PointCloud()

# Set the points of the PointCloud using the x, y, z coordinates from pcd_pd
pcd_pd_o3d.points = o3d.utility.Vector3dVector(np.array(pcd_pd[['X', 'Y','Z']]))

# Set the colors of the PointCloud using the RGB values from pcd_pd
pcd_pd_o3d.colors = o3d.utility.Vector3dVector(np.array(pcd_pd[['R',   'G',
'B']]/255))

# Visualize the PointCloud using Open3D's visualization capabilities
o3d.visualization.draw_geometries([pcd_pd_o3d])
```

Finally, the last solution is to use a specific loader for the massive point cloud that we usually find in the *.las* or *.laz* file format. For this, we use the library laspy (*https://oreil.ly/t7g9d*), which is a bit more complex:

```
import laspy as lp

# Load LAS file using laspy
las = lp.read('../DATA/beach-5M.las')

# Extract x, y, and z coordinates and RGB colors from the LAS file
points = np.stack([las. x, las. y, las. z]).T
colors = np.stack([las.red, las.green, las.blue]).T / 65535

# Create an Open3D PointCloud object and set points and colors
pcd_las_o3d = o3d.geometry.PointCloud()
pcd_las_o3d.points = o3d.utility.Vector3dVector(points)
pcd_las_o3d.colors = o3d.utility.Vector3dVector(colors)

# Visualize the PointCloud using Open3D
o3d.visualization.draw_geometries([pcd_las_o3d])
```

A 3D point cloud of a bike is enriched with RGB color information, resulting in a more visually appealing representation (see Figure 4-5).

These are essentially the three ways we load point clouds. As you can see, once within Python, it is easy to push your various datasets toward Open3D point cloud objects if visualization or access to specific Open3D functions is needed.

Figure 4-5. A 3D point cloud of a bike, with RGB color information

Common point cloud file formats

As you get deeply involved in 3D data science, you are likely to navigate various point cloud file formats. Table 4-3 shows common point cloud formats with Python in mind.

Table 4-3. Common point cloud file formats and their Python libraries

File format	Definition and link	Type	Python library of choice	Notes
.ply	Polygon File Format (*https://oreil.ly/adUSy*)	ASCII/Binary	open3d, pyvista	Simple, versatile. Great for quick prototyping or when human readability is a plus. Can handle various data attributes.
.las/.laz	LAS/LAZ formats (*https://oreil.ly/cJWTU*)	Binary (*.laz* is compressed)	laspy	Industry standard for LiDAR. *.laz* offers significant compression benefits for large datasets—essential for efficient storage and processing.

File format	Definition and link	Type	Python library of choice	Notes
.xyz	Simple text file (https://oreil.ly/PztlA)	ASCII	pandas, numpy	A barebones format often used for basic point coordinates (x, y, z). Simplicity is its virtue but it lacks metadata capabilities.
.pts	Leica point cloud format (https://oreil.ly/PNClj)	ASCII/Binary	pdal, open3d	Commonly used in surveying and CAD. While versatile, it sometimes requires tailored parsing due to its proprietary nature.
.e57	ASTM E57 3D Imaging Data Exchange Format (https://oreil.ly/LGHbx)	Binary	pye57, e57, pdal	An emerging standard designed for interoperability. Handles rich metadata and point attributes, crucial for complex 3D projects.
.obj	Wavefront OBJ (https://oreil.ly/Xjwy9)	ASCII	open3d, pywavefront	More focused on mesh data but can represent point clouds. Commonly used in computer graphics and 3D modeling. Often paired with .mtl files for material definitions.
.pcd	Point Cloud Data file format (https://oreil.ly/5Uwqw)	Binary	open3d, pcl, pyntcloud	The preferred format for the Point Cloud Library (PCL), commonly used in robotics and computer vision research.

For less common or proprietary formats like *.pts*, you might need to roll up your sleeves and write some Python parsing logic. Some libraries can be beneficial in these situations, but working with NumPy is usually the best way to ensure open I/O operations. Remember, choosing the proper format depends heavily on the specific application and the tools involved. Now, let us detail raster-based representations, such as depth maps and projections.

Image-based Representations

Working with images is very appealing. We have manipulated images since we were very young and were viewing pictures of family members and baby books. In the digital era, images are one of the most used datasets, especially when we speak about data science. This means that we can easily find some libraries tailored to handle images quickly, and nowadays, many deep learning models are tailored to images. Knowing this, it becomes evident that adding another arrow to our data representation quiver may become helpful for future applications. But how can we tie 3D representations to 2D images? Let's first detail depth maps.

Depth maps

Depth maps (https://oreil.ly/jhDvI) are raster images, where the main image channel provides depth-based separation between the pixels that make up a scene, as seen from one particular perspective. Even though we're accustomed to working with RGB images, color-coding intensity values on one channel are the simplest way to convey depth (see Figure 4-6).

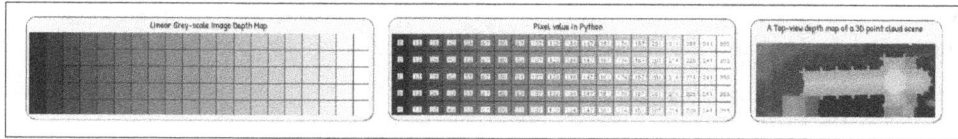

Figure 4-6. Single-channel image and its representation as a matrix, where each pixel value ranges between 0 and 255

In a depth map, bright pixels have the highest value, and dark pixels have the lowest value. A depth image presents values according to objects' distance, whereas pixel color gives the distance from the camera. The depth map is related to the z-buffer, where the "z" relates to the direction of a camera's central axis of view, not the absolute z scene coordinate.

This works well if you require surface data connected to a preexisting viewpoint, which is the case for autonomous vehicles, where a 360-projected depth map allows you to map the surroundings at every location rapidly. However, the big caveat is that you are not working with complete 3D data but rather 2.5D, as you cannot represent two different values for one line of sight. In Table 4-4 you can find the operations, benefits, and disadvantages of depth maps.

Table 4-4. The operations, benefits, and disadvantages of depth map modeling

Operations	Benefits	Disadvantages
• Transformations: Multiply the pixels in the image with linear transformation matrices • Combinations: Objects can be combined by merging the point lists • Rendering: Draws pixels on the image plane	• Low memory requirements • Very well-known raster format • Transformations are quick and easy	• Essentially, a 2.5D representation • Cannot describe a whole 3D scene on its own • Weak topology

In addition, we can extend depth maps to a particular case called RGB-D. The widespread use of RGB-D sensors has led to the popularity of representing 3D data as RGB-D images in recent years. RGB-D data provides 2.5D information about the captured 3D object by attaching the depth map and 2D color information (RGB).

To start working with depth images that already have a pose, Open3D provides a loader function that permits the leveraging of these files. To import the image with the code:

```
depth_raw = o3d.io.read_image(
"../DATA/depth.png")
```

RGB-D images

As just mentioned, there is an exceptional case called RGB-D: an RGB image with an additional depth channel. To create an RGB-D image object, we first must load both the depth channel and the RGB channels from the RGB image:

```
color_raw = o3d.io.read_image("../DATA/rgb.jpg")
depth_raw = o3d.io.read_image("../DATA/depth.png")
```

After being transformed into grayscale, the color image is saved in float ranges [0, 1]. The depth image represents the depth value in meters and is saved in float:

```
# Create an RGB image object
rgbd_image = o3d.geometry.RGBDImage.create_from_color_and_depth(
    color_raw, depth_raw, convert_rgb_to_intensity=False)
```

We can render the transformed photos as NumPy arrays with Matplotlib:

```
plt.subplot(1, 2, 1)
plt.title('Olive branch grayscale image')
plt.imshow(rgbd_image.color)
plt.subplot(1, 2, 2)
plt.title('Olive Branch depth image')
plt.imshow(rgbd_image.depth)
plt.show()
```

This allows us to visualize the grayscale image and a depth image of a leaf side by side to illustrate the different types of information captured (see Figure 4-7).

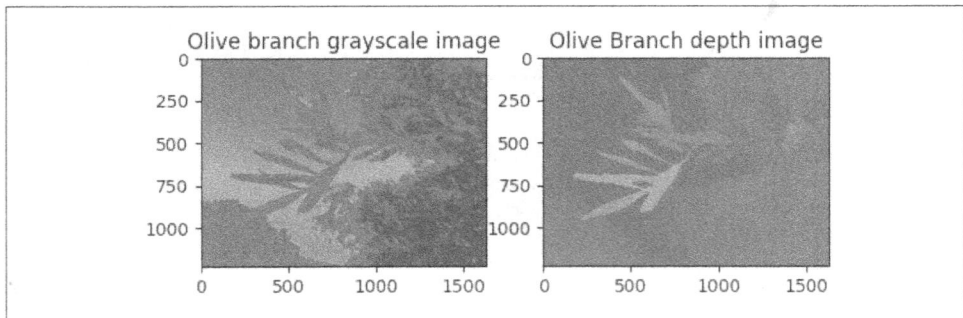

Figure 4-7. Grayscale and depth image of a leaf

If your image is coded with 16 bits, you may have to convert it to pass it to the Open3D constructor object. To do this, you have to go through NumPy and back to the Open3D Image object with:

```
depth_8 = (np.array(depth_raw)/256).astype('uint8')
depth_raw_8 = o3d.geometry.Image((depth_8).astype(np.uint8))
rgbd_image = o3d.geometry.RGBDImage.create_from_color_and_depth(color_raw,
depth_raw_8)
```

In addition to being affordable, RGB-D data is a straightforward but efficient representation of three-dimensional objects for various applications, including correspondence, posture regression, and identity identification. The quantity of RGB-D datasets available is tremendous compared to other 3D datasets, like point clouds or 3D meshes. As a result, it is the recommended method for training deep learning models using extensive training datasets.

Projection

Another way to display raw 3D data is to project it onto a different 2D space, capturing some of the essential characteristics of the original 3D shape (see Figure 4-8).

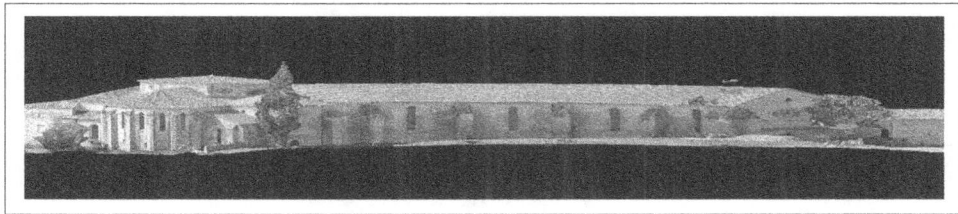

Figure 4-8. Example of a widely deformed cylindrical projection of the point cloud

Several projections exist, each of which turns the 3D object into a 2D grid that contains specific data. One popular way to express 3D data in such a format is to project it onto the spherical and cylindrical domains. Because of their Euclidean grid structure, these projections facilitate the processing of 3D data and aid in the projected data's invariance to rotations around the primary axis.

However, because projections lose information, they are not ideal for complex 3D computer vision applications like dense correspondence (*https://oreil.ly/Zjwwp*). Handling projected images is not directly included within core libraries; thus, let me give you some ways to get to 3D representations from projections. Once mastered, we can iterate projections based on a certain number of poses to create multiview datasets.

One central concept is understanding the relationship between a camera and the real-world scene. Projections often play on placing a virtual camera inside a 3D scene (world space) to render what this camera captures (every object, each having a local space) in a view space (see Figure 4-9).

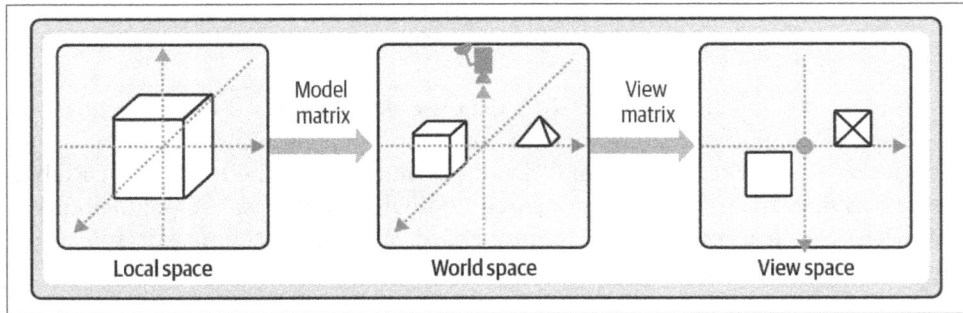

Figure 4-9. Relationships between a local space, a world space, and a view space

The view space is influenced by the camera's parameters: its intrinsic properties (focal length, distortion parameters) and extrinsic properties (position and orientation), as shown in Figure 4-10.

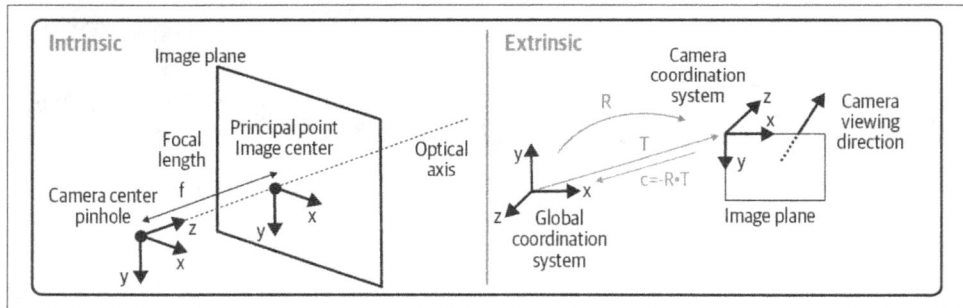

Figure 4-10. Intrinsic versus extrinsic parameters

Let me clarify what that means with Python. First, we load a point cloud and get the color vector:

```
x,y,z,r,g,b = np.loadtxt('../DATA/bike_florent.txt', delimiter=';', skiprows=1,
unpack=True)
colors = np.transpose(np.stack([r,g,b]))
```

If we have unlimited viewpoints, we can plot an orthographic projection, which is a view of the 3D scene (see Figure 4-11A):

```
def orthographic_projection(x, y, z):
    fig = plt.figure()
    ax = fig.add_subplot(111, projection='3d')
    ax.scatter(x, y, z, s =0.001, c = colors/255)
    plt.show()

orthographic_projection(x, y, z)
```

The next stage is to define a camera with a set of parameters and place it in the scene:

```
def plot_3d_scene(x, y, z, colors, camera_params):
    fig = plt.figure(figsize=(10, 8))
    ax = fig.add_subplot(111, projection='3d')

    # Plot the point cloud
    ax.scatter(x, y, z, c=colors/255, s = 0.005)

    # Plot the camera position
    camera_position = np.array(camera_params['position'])
    ax.scatter(camera_position[0], camera_position[1], camera_position[2], c='r',
    marker='x', s=100, label='Camera')

    ax.set_xlabel('X')
    ax.set_ylabel('Y')
    ax.set_zlabel('Z')
    ax.set_title('3D Scene with Point Cloud and Camera Position')
    ax.legend()

    plt.show()

# Camera parameters
camera_params = {
    'focal_length': 5,
    'position': [0, 0, 5],
    'orientation': (np.pi/4, np.pi/4, np.pi/4)  # (roll, pitch, yaw) in radians
}

# Plot the 3D scene
plot_3d_scene(x, y, z, colors, camera_params)
```

We see where our camera is positioned (the X in the upper left corner of Figure 4-11B). This makes it easy to understand the various projections.

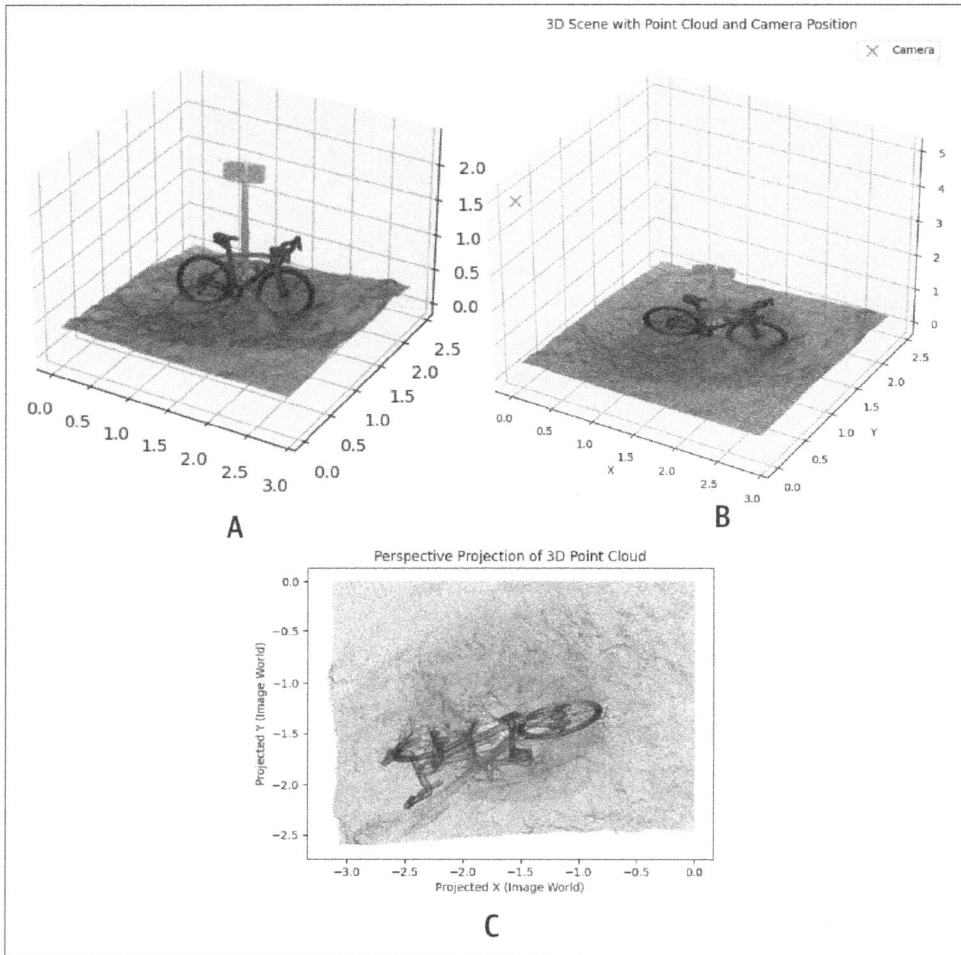

Figure 4-11. (A) A 3D point cloud local space; (B) 3D perspective projection, world space; (C) perspective projection, view space

The last thing is to define a projection function. We first define a transformation matrix, composed of a rotation and a translation matrix:

```
def perspective_projection_3d(x, y, z, colors, camera_params):
    # Extract camera parameters
    focal_length = camera_params['focal_length']
    camera_position = camera_params['position']
    camera_orientation = camera_params['orientation']

    # Transformation matrix to account for camera position and orientation
    rotation_matrix = np.array([
        [np.cos(camera_orientation[1]) * np.cos(camera_orientation[2]),
```

```
            -np.cos(camera_orientation[0]) * np.sin(camera_orientation[2]) +
            np.sin(camera_orientation[0]) * np.sin(camera_orientation[1]) *
            np.cos(camera_orientation[2]),
            np.sin(camera_orientation[0]) * np.sin(camera_orientation[2]) +
            np.cos(camera_orientation[0]) * np.sin(camera_orientation[1]) *
            np.cos(camera_orientation[2]),
            0],

        [np.cos(camera_orientation[1]) * np.sin(camera_orientation[2]),
            np.cos(camera_orientation[0]) * np.cos(camera_orientation[2]) +
            np.sin(camera_orientation[0]) * np.sin(camera_orientation[1]) *
            np.sin(camera_orientation[2]),
            -np.sin(camera_orientation[0]) * np.cos(camera_orientation[2]) +
            np.cos(camera_orientation[0]) * np.sin(camera_orientation[1]) *
            np.sin(camera_orientation[2]),
            0],

        [-np.sin(camera_orientation[1]),
            np.sin(camera_orientation[0]) * np.cos(camera_orientation[1]),
            np.cos(camera_orientation[0]) * np.cos(camera_orientation[1]),
            0],

        [0, 0, 0, 1]])

translation_matrix = np.array([
    [1, 0, 0, -camera_position[0]],
    [0, 1, 0, -camera_position[1]],
    [0, 0, 1, -camera_position[2]],
    [0, 0, 0, 1]])

transformation_matrix = np.dot(translation_matrix, rotation_matrix)
```

We then apply and plot the transformation:

```
# Apply transformation and perspective projection
points_3d = np.stack((x, y, z, np.ones_like(x)))
transformed_points = np.dot(transformation_matrix, points_3d)
projected_x = focal_length * transformed_points[0] / transformed_points[2]
projected_y = focal_length * transformed_points[1] / transformed_points[2]

# Plot the 2D projection
plt.scatter(projected_x, projected_y, s =0.001, c = colors/255)

plt.xlabel('Projected X (Image World)')
plt.ylabel('Projected Y (Image World)')
plt.title('Perspective Projection of 3D Point Cloud')
plt.grid(False)
plt.show()
```

From there, you can test by putting up some parameters and checking the results:

```
# Camera parameters
camera_params = {
    'focal_length': 5,
```

```
        'position': [0, 0, 5],
        'orientation': (0, 0, 0)  # (roll, pitch, yaw) in radians}
    perspective_projection_3d(x, y, z, colors, camera_params)
```

A perspective projection of a point cloud from a different viewpoint highlights the impact of camera position on the resulting 2D representation (see Figure 4-11C). If we iterate (with specific rotations, such as yaw, pitch, and roll), we get the result in Figure 4-12.

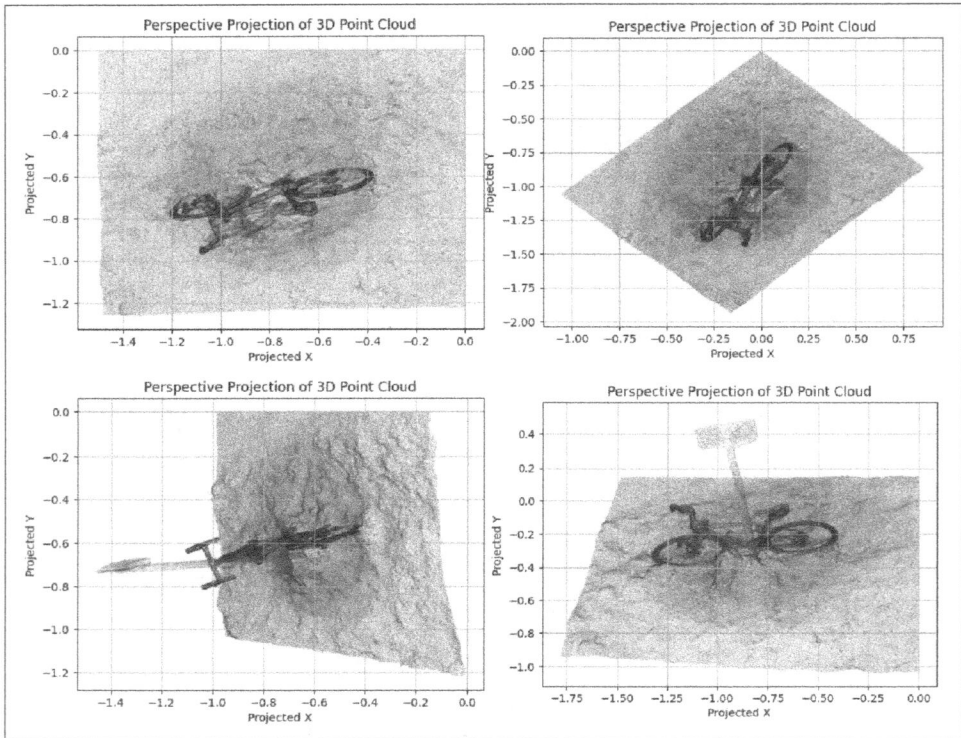

Figure 4-12. Perspective projection of a 3D point cloud

As you can see, by playing with the camera's orientation, we can generate various 2D projections of the original point cloud. Therefore, going multiview is only one stretch.

Multiview

We can access 3D information from a multiview image, a 2D-based 3D representation in which we access information by matching several 2D images of the same object from different points of view (see Figure 4-13).

Figure 4-13. Multiview example of a 3D dataset

Using this representation for 3D data can help in learning various feature sets to lessen the impact of noise, incompleteness, occlusion, and lighting issues on the acquired data. The acquisition process for photogrammetric reconstructions is linked to the unresolved topic of how many views are needed to model the 3D shape: an object with too few views may not capture the attributes of the entire 3D shape (particularly for 3D scenery) and may result in an overfitting issue. Multiview datasets are more suitable for analyzing rigid data with minimal deformations. This is complemented by volumetric models.

Volumetric (Voxel) Models

If you enjoy Lego (or nowadays Minecraft), using voxels may be very attractive. Why? Simply put, voxels are the 3D analog of 2D pixels. It is a simple way to initially structure a 3D dataset that is unordered (like point clouds). You get an assembly of primitive blocks that can easily be linked to Lego assemblies. Apart from the physical versus digital property, the primary difference is that voxels only play with one type of base element (a cube). A 3D voxel model is shown in Figure 4-14, representing a 3D object using discrete volumetric elements known as voxels.

Figure 4-14. A 3D voxel model

In contrast, Lego lets you play with various pieces of different dimensions. But if we adopt a multiscale vision and theoretical reasoning, "Voxels are the perfect modeling technique for replicating reality and can go far beyond what we take them for."[1]

In fact, many of the materials in our three-dimensional universe may be approximated by extremely small voxels. Consequently, if you combine enough density with suitable rendering methods, *you can use voxels to replicate real-world objects that would be impossible to differentiate from the real thing—in appearance and behavior.*

This is an interesting benefit compared to raw point clouds: you get a behavioral possibility to simulate real-world physics that would be either impossible or hair-loss techniques with other modeling methods (see Table 4-5). Voxels can store a true "matter" volumetry. This makes them convenient for unlocking new simulation techniques that would be tricky with other methods.

Table 4-5. Advantages and disadvantages of voxel-based modeling

Advantages	Disadvantages
Voxels can depict geometries with a nice balance between structure and accuracy.	Building objects using voxels is not straightforward without using 3D scanning techniques.
Other modeling techniques would not be feasible without voxels, opening up new simulation methodologies.	Voxel modeling lacks the mathematical precision of other modeling methods, like B-representation.
Voxels are one of the quickest ways to model and visualize volumetric data.	We lack specialized technology to generate high-resolution voxels efficiently, and our current computer hardware is optimized for generating polygons.

I will cover the theory and implementation for voxels in depth in Chapter 10. Voxels works well for rendering and fluid visualization but introduce aliasing problems when approximating the original geometry. The highly ordered grid structure of voxel models is helpful for processing tasks like classification using 3D convolutional neural networks (see Chapter 16).

To experiment with voxels, we can use Open3D, which has a loader for voxel objects:

```
vox_read = o3d.io.read_voxel_grid("../DATA/voxels.ply", format='auto')
o3d.visualization.draw_geometries([vox_read])
```

You can then decide how to express your voxel (center + length, width and height, etc.).

---

1 This quote is drawn from the blog post "The Main Benefits and Disadvantages of Voxel Modeling" (*https://oreil.ly/SgOzg*), where you can learn more.

High-level 3D Data Representation

Let me quickly discuss some 3D representation alternatives: spatial graphs and 3D descriptors such as signed distance functions (SDFs) (*https://oreil.ly/2Ykmn*). They both provide distinct benefits for certain applications.

We can treat 3D data as a graph, with nodes as points and edges connecting nearby points. I dedicate a complete chapter to using graphs for 3D data science in Chapter 13, including graph theory (*https://oreil.ly/uBjER*) fundamentals. These representations are interesting as they allow the use of practical graph-based algorithms for tasks such as segmentation, classification, and shape analysis. Graph-based representations are helpful for showing complex relationships between points, especially when dealing with noisy or irregular data. You can find the complete Python code to generate 3D graphs such as the one in Figure 4-15 on the 3D Resource Hub (*https://learn geodata.eu/3d-data-science-with-python*). Here's a snippet:

```python
import networkx as nx
import matplotlib.pyplot as plt
from mpl_toolkits.mplot3d import Axes3D
import numpy as np

def create_3d_graph():
    # Create a random graph
    G = nx.erdos_renyi_graph(10, 0.3)

    # Generate 3D positions
    pos = {node: (np.random.uniform(0, 10), np.random.uniform(0, 10),
    np.random.uniform(0, 10)) for node in G.nodes()}

    # Create 3D plot
    fig = plt.figure(figsize=(10, 8))
    ax = fig.add_subplot(111, projection='3d')

    # Draw edges
    for edge in G.edges():
        x = [pos[edge[0]][0], pos[edge[1]][0]]
        y = [pos[edge[0]][1], pos[edge[1]][1]]
        z = [pos[edge[0]][2], pos[edge[1]][2]]
        ax.plot(x, y, z, color='gray', alpha=0.6)

    # Draw nodes
    xs, ys, zs = zip(*[pos[node] for node in G.nodes()])
    ax.scatter(xs, ys, zs, c='red', s=100)

    ax.set_title('3D Network Graph')

    plt.show()

# Run the 3D graph visualization
create_3d_graph()
```

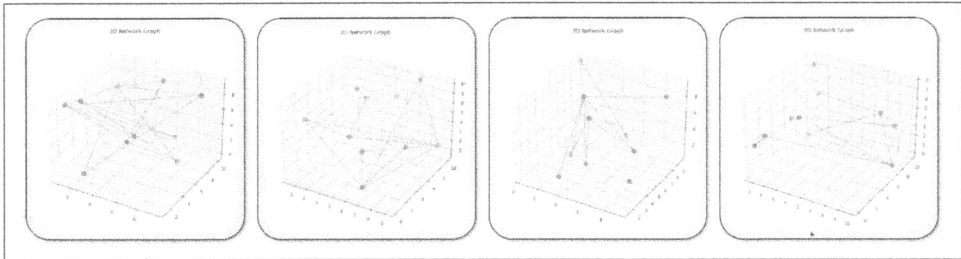

Figure 4-15. Randomly generated 3D graph

Aside from graphs, SDFs define 3D shapes by indicating the signed distance from any point in space to the surface. This representation is very adaptable and can capture detailed and complex shapes. SDFs are often used in procedural modeling, level design, and fluid simulation because they allow for precise control over creating and manipulating shapes. Now, let us detail how we can handle 3D surface models to grasp the range of possibilities better.

3D Surface Models

These models depict the object's surface or border rather than its volume. Boundary representations (*https://oreil.ly/9U-Uy*) (B-reps) are found in almost all visual models used in games, movies, and reality capture workflows. We would not be complete if we did not bring up a comparison to solid 3D models (which we will study in Chapter 10). Indeed, these usually define the volume of an object they represent. These solid models are often used for simulations (e.g., engineering and medical applications) and are built with constructive solid geometry (*https://oreil.ly/X8KsM*) or voxel assemblies (*https://oreil.ly/BUVik*).

Solid and surface modeling can produce objects with the same functionality. The main distinctions between them are the methods used to produce and modify them. This naturally orients some standards in different domains and approximations between the model and reality.

Three main strategies permit geometry description through 3D modeling: constructive solid geometry (*https://oreil.ly/X8KsM*), parametric modeling (including B-reps), and 3D meshes. Constructive solid geometry is exciting, but we'll wait until Chapter 10 to cover it in depth. First, let's focus on the most common 3D models: 3D meshes.

3D mesh

A mesh is a geometric data structure that enables a collection of polygons to represent surface subdivisions. In modeling, meshes are used to discretize a continuous or implicit surface, or they are used to represent surfaces in computer graphics. A mesh consists of faces (also known as facets) of a polygonal shape, which are vertices (also known as vertex) joined by edges. Triangle meshing is the term used when all of the faces are triangles. These are the most common in 3D data science workflows (see Figure 4-16).

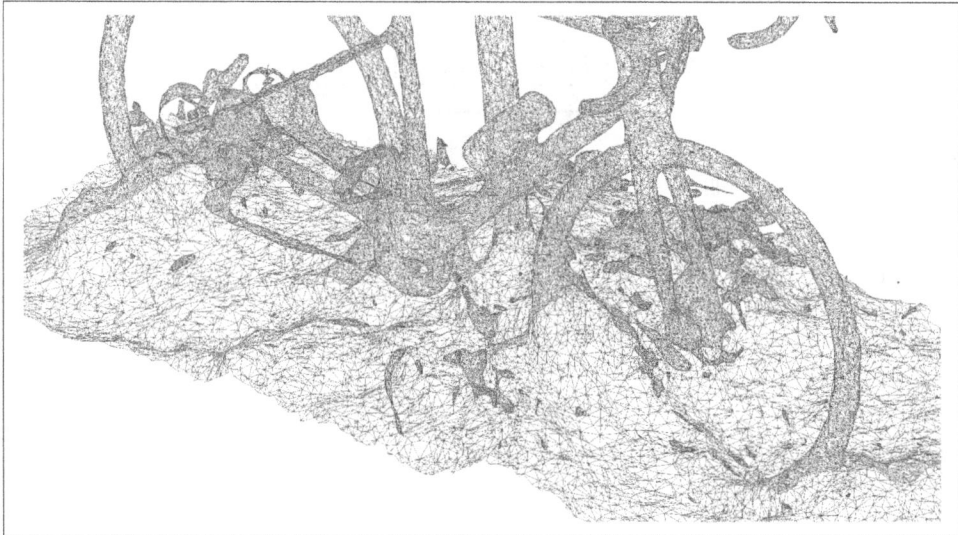

Figure 4-16. The mesh's vertices, edges connecting them, and faces (mostly triangular)

Quadrilateral meshes are also very interesting but often obtained through mesh optimization techniques to get more compact representations. It is also possible to use volumetric meshes connecting the vertices by tetrahedrons, hexahedrons (cuboids), and prisms. Meshes are based on the boundary representation, which is dependent upon the wireframe model. 3D lines simplify the object, and a line in the model represents each edge of the object. Let's expand on the theory.

The topology, elemental arrangement, and geometry make up most of the boundary representation of 3D objects (surfaces, curves, and points). Faces, edges, and vertices are the primary topological elements. I have schematized a basic B-rep for a cube, as shown in Figure 4-17. For an overview of B-rep operations, benefits, and disadvantages, see Table 4-6.

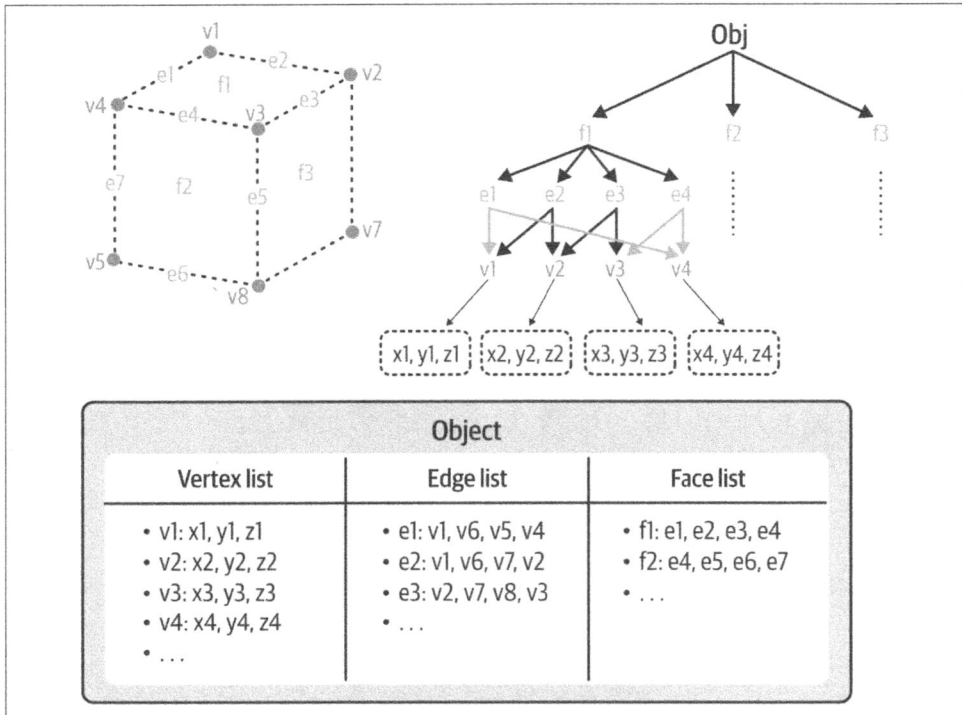

Figure 4-17. The schematization of the boundary representation underlying the structure

Table 4-6. Overview of boundary representation (B-rep) operations, benefits, and disadvantages

Operations	Benefits	Disadvantages
• Transformations: All points are transformed with the wireframe model (multiply the points in the point list with linear matrices). In addition, the surface equations or normal vectors can be transformed. • Combinations: Objects can be combined by grouping point lists and edges; operations on polygons (divide based on intersections, remove the redundant polygons, and combine them). • Rendering: Hidden surface or line algorithms can be used because the surfaces of the objects are known so that visibility can be calculated.	• Well-adopted representation • Model generation via "new-gen scanning" • Transformations are quick and easy	• High memory requirements • Expensive combinations • Curved objects are approximated

Meshes are an excellent tool for expressing a point cloud's geometry and frequently allow for a significant reduction in the number of points that must be used as vertices. On top of that, it permits us to get a sense of the relationship between objects through facial connectivities.

However, because of the mesh's intricacy, meshing is frequently an approximation of the original point-based geometry and can only partially reflect the data. So, how can we get to work with these ensembles within Python? Well, we can use the Open3D library, which comes with a specific set of loaders for various data formats:

```
# Import the open3d library and alias it as 'o3d'
import open3d as o3d

# Read a triangle mesh from a PLY file at '../DATA/mesh_terrain.ply.'
mesh = o3d.io.read_triangle_mesh('../DATA/mesh_terrain.ply')

# Compute vertex normals for the mesh. This step is essential for
# lighting calculations and visualization.
mesh.compute_vertex_normals()

# Visualize the mesh using open3d's built-in visualization tool.
# The draw_geometries function takes a list of geometry objects as input and
# displays them.
o3d.visualization.draw_geometries([mesh])
```

3D mesh common file formats

At this point, burdening ourselves with every possible way to load 3D mesh is unnecessary. Other libraries, file formats, and ways to use them exist, and they are showcased in this book as part of projects. For the sake of completeness, Table 4-7 presents a breakdown that I find useful, tailored for Python users.

Table 4-7. Common 3D mesh file formats for Python users

File format	Definition and link	Type	Python library of choice	Notes
.stl	Stereolithography (*https://oreil.ly/A2Bdw*)	ASCII/Binary	meshio, open3d	Widely used in 3D printing. Simple, represents surfaces as triangles. Binary version is more compact. ASCII is easier for debugging.
.obj	Wavefront OBJ (*https://oreil.ly/hNK1c*)	ASCII	pywavefront, trimesh, open3d	Common in computer graphics. Supports materials, textures, and vertex normals. Often accompanied by *.mtl* (material library) files.
.ply	Polygon File Format (*https://oreil.ly/lcp7q*)	ASCII/Binary	plyfile, trimesh, open3d	Versatile format that can also represent point clouds. Handles various data attributes.
.glb/.gltf	glTF (GL Transmission Format) (*https://oreil.ly/eYY-Q*)	Binary (.glb), JSON (.gltf)	pygltf, trimesh, open3d	Modern format optimized for web and real-time applications. *.glb* is a single binary file, while *.gltf* uses separate JSON and binary files.
.fbx	Autodesk FBX (*https://oreil.ly/dXFg6*)	Binary	Blender (Python scripting possible within Blender)	Used in many 3D modeling and animation software packages. Handles complex scenes, animations, and rigging information. Often requires dealing with proprietary SDKs.

File format	Definition and link	Type	Python library of choice	Notes
.off	Object File Format (*https://oreil.ly/iz85y*)	ASCII	trimesh, open3d	A simple format primarily used in academic and research settings. Represents meshes as lists of vertices, faces, and edges.
.dae	COLLADA (COLLAborative Design Activity) (*https://oreil.ly/pcHoK*)	XML-based	pycollada, cloudcompare	Designed for data exchange between different 3D applications. Supports complex scenes and animations.

Choosing the best format depends entirely on your project requirements, and this table should provide a good starting point for making informed decisions.

Parametric models (e.g., B-reps)

The term "parametric" refers to a shape's capacity to alter its underlying geometry by adjusting a parameter to a desired value. This is helpful if you want to represent walls by just setting up their orientation, length, breadth, and height.

Parametric modeling is suited to using computing capabilities that can model component attributes with the aim of real-world behavior. Parametric models use a composition of feature-based (parametric), solid, and surface modeling to allow the manipulation of the model's attributes. One of the most essential features of parametric modeling is that interlinked attributes can automatically change values. In other words, parametric modeling allows defining entire classes of shapes, not just specific instances. However, breaking down the model object into smaller entities (such as segments as illustrated in Figure 4-18) combined into classes necessitates a very smart structuration of the underlying point cloud geometry.

These parametric models can often be combined or extracted using 2D+Z CAD files. However, working with parametric models in Python is limited, so we often resort to external software.

> B-rep parametric models are a great asset in computational workflows. However, when we need to display them, we resort to approximating the geometry with 3D meshes. Nevertheless, they are largely used in manufacturing and for computer-aided design (CAD) workflows (*https://oreil.ly/80aWV*), and you can find additional resources in the 3D Resource Hub (*https://learngeodata.eu/3d-data-science-with-python*).

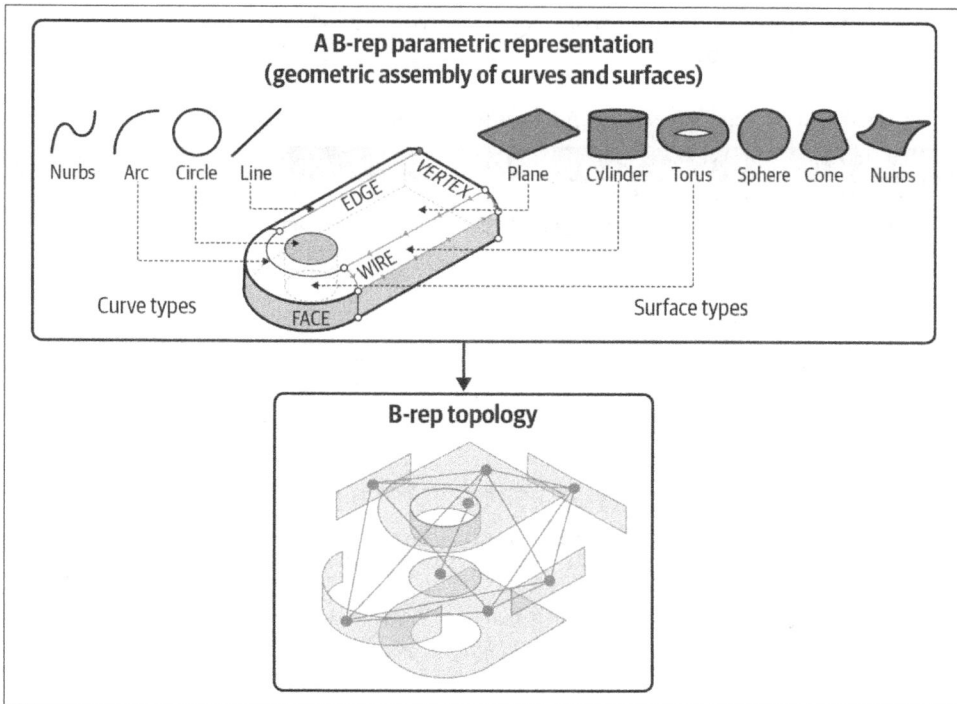

Figure 4-18. The functioning and constitution of a parametric model following a B-rep strategy

Advanced parametric models with Python will be covered in Chapter 10. For now, let's make a parametric cube and test it using NumPy and Open3D. First, we define the vertices of the cube's base layer (thinking in indexes along each axis, x, y, and z) and move up to the cube's upside (see Figure 4-19).

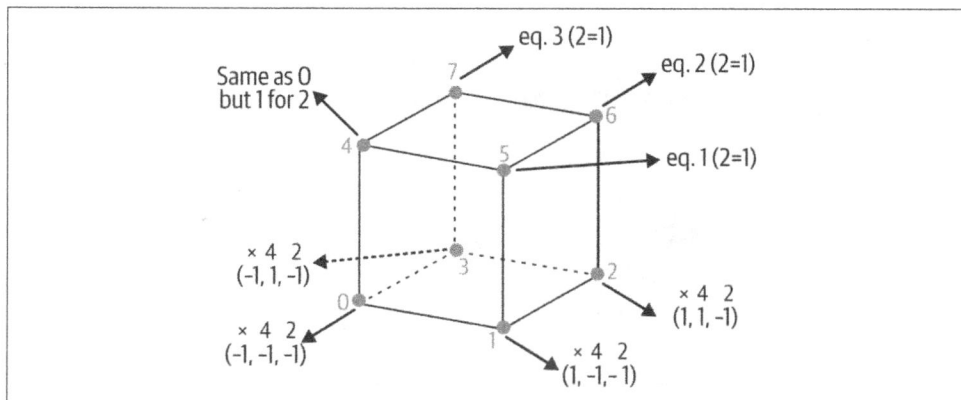

Figure 4-19. Representation of a cube

All in all, we create a function from where, as you can see in Table 4-8, we parametrize the size of the cube.

Table 4-8. The definition of vertices and faces with Python

Python code for the vertices	Python code for the faces
``# Define the vertices of the cube
vertices = size / 2 * np.array([
[-1, -1, -1],
[1, -1, -1],
[1, 1, -1],
[-1, 1, -1],
[-1, -1, 1],
[1, -1, 1],
[1, 1, 1],
[-1, 1, 1],
])```	```# Define the triangles (faces) of the
# cube using vertex indices
triangles = np.array([
[0, 1, 2], [0, 2, 3],
[4, 5, 6], [4, 6, 7],
[0, 4, 5], [0, 5, 1],
[2, 6, 7], [2, 7, 3],
[0, 4, 7], [0, 7, 3],
[1, 5, 6], [1, 6, 2],
])```

From there, we can create a triangle mesh object using vertices and triangles and compute normals:

```
mesh = o3d.geometry.TriangleMesh(o3d.utility.Vector3dVector(vertices),
o3d.utility.Vector3iVector(triangles))
mesh.compute_vertex_normals()
```

Next, we can pull all the preceding into a function (`create_parametric_cube(size)`) and create two cubes with different sizes; as we visualize them, do not forget to paint and translate them, with `.paint_uniform_color()` and `.translate()`:

```
# Visualize the cube
o3d.visualization.draw_geometries([parametric_cube1,parametric_cube2],
mesh_show_back_face=True)
```

The generation of parametric cubes demonstrates the flexibility of parametric modeling in creating different sizes and shapes (see Figure 4-20).

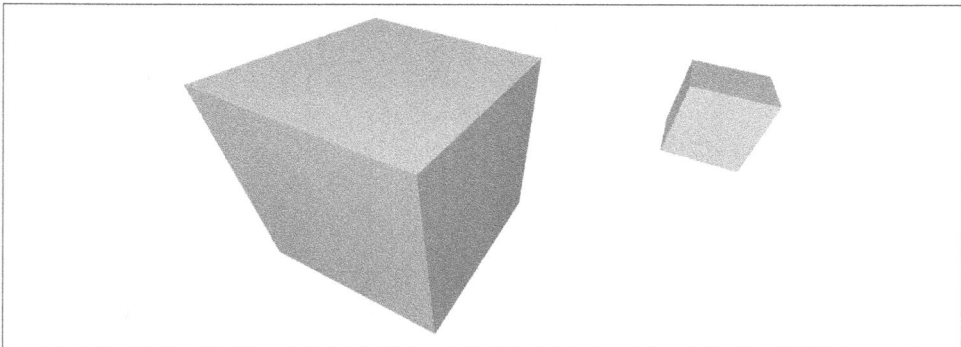

Figure 4-20. Parametric cube generation

Wonderful! This allows us to then proceed with advanced parametrization over 3D datasets!

3D parametric model common file formats

These formats can be modified by adjusting parameters, making them fundamental to CAD, engineering, and product design. Table 4-9 shows a breakdown of common formats and their interaction with Python.

Table 4-9. Overview of 3D parametric model file formats

File format	Definition and link	Type	Python library of choice	Notes
.step/.stp	STEP file format (*https://oreil.ly/Ic-7p*)	ASCII	FreeCAD, OCPNet	ISO standard for exchanging product model data. Excellent for interoperability between CAD systems.
.iges/.igs	IGES file format (*https://oreil.ly/pOupV*)	ASCII	FreeCAD, pythonOCC	An older but still widely used standard. Can be complex to parse due to its verbose nature.
.stl	STL file format (*https://oreil.ly/pBfYs*)	Binary/ ASCII	open3d, meshio	Primarily used for 3D printing and rapid prototyping. Represents surfaces as a collection of triangles.
.obj	Wavefront OBJ (*https://oreil.ly/IJxEd*)	ASCII	pywavefront, trimesh, open3d	Common in computer graphics. Can represent both mesh and parametric surfaces.

Additionally, formats like *.stl* represent surfaces as triangle meshes, while *.step* uses parametric representation, which stores the underlying mathematical definitions of surfaces. This distinction is essential when choosing a format based on your analysis needs. Mesh-based formats are simpler but might lose precision for complex shapes, whereas parametric formats allows for precise geometric operations. However, *.step* and *.iges* files can become very large and complex, especially for detailed models. This can impact loading and processing times. Consider simplifying models or using more compact formats like *.glb/.gltf* where appropriate.

Now that you have a firm understanding of the various ways to represent 3D data, let's consider one way to focus processes on a central representation: 3D point clouds.

3D Data Canonical Link

Interestingly, we can successfully translate all the 3D representations we saw into point clouds because point clouds are canonical (see Figure 4-21). Indeed, the simplicity we explicated previously permits us to capture the most essential aspect of geometric representations directly within 3D point clouds. But how can we do that? This section gives you the main methods to choose from if you want to work in a coherent point cloud–based workflow that you can then link to your modality of choice.

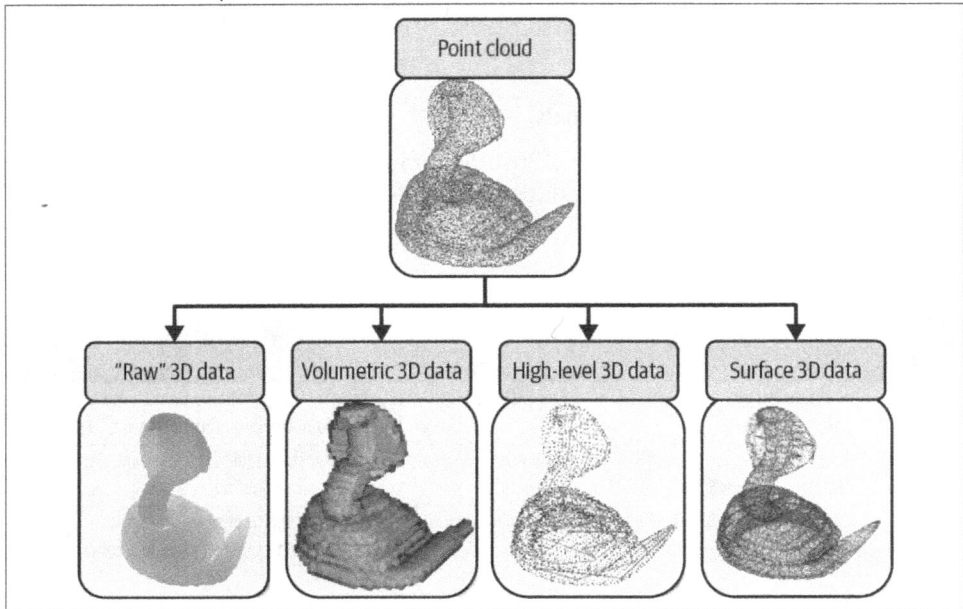

Figure 4-21. Point cloud as canonical data link

Mesh to Point Cloud

The mesh already retains vertices we could hold as a point cloud. However, if we want to keep a better transcription of the geometry, we may want to also get some points out of the edges and faces. This means sampling the surface into a point cloud by defining a parameter. Here is how to go from a 3D mesh to a 3D point cloud using Python:

```
bunny = o3d.data.BunnyMesh()
mesh = o3d.io.read_triangle_mesh(bunny.path)
```

Or you can import it after downloading it from this link:

```
mesh = o3d.io.read_triangle_mesh("data/bunny.ply")
```

From there, we can display the 3D model to see what it looks like. You can move the mouse to view from different viewpoints:

```
# Visualize:
mesh.compute_vertex_normals() # compute normals for vertices or faces
o3d.visualization.draw_geometries([mesh])
```

Now, to sample a point cloud, several methods exist. In this example, we sample 1,000 points uniformly from the imported mesh, and we visualize it:

```
# Sample 1000 points:
pcd = mesh.sample_points_uniformly(number_of_points=1000)

o3d.visualization.draw_geometries([pcd])
```

We can save the created point cloud in *.ply* format as follows:

```
o3d.io.write_point_cloud("output/bunny_pcd.ply", pad)
```

There you have it: a direct path from a 3D mesh to a 3D point cloud.

Voxel to Point Cloud

The voxel entity can be described in various ways. You usually encounter the voxel described as a point with a cubic extent (width = length = height) relating to the point position. This point position describes the voxel, which is often the down left point or the center point.

With voxel entities, going to a point cloud can follow the same principle used previously: sampling. However, here the sampling is done on the volume, or you retain representative points for each entity. This is the easiest way to turn a voxel entity into a 3D point cloud:

```
pcd_vox_np = np.as array([voxels_dataset.origin +
pt.grid_index*voxels_dataset.voxel_size for pt in voxels_dataset.get_voxels()])
```

From this little trick, we can then transform it into an Open3D object and get the visualization:

```
pcd_vox_o3d = o3d.geometry.PointCloud()
pcd_vox_o3d.points = o3d.utility.Vector3dVector(pcd_vox_np)
o3d.visualization.draw_geometries([pcd_vox_o3d])
```

Figure 4-22 shows a point cloud of a neighborhood loaded for further processing.

Figure 4-22. Initial loaded point cloud of a neighborhood

Finally, if you need color, you can input the following:

```
color_voxels=[v.color for v in voxels_dataset.get_voxels()]
pcd_vox_o3d.colors = o3d.utility.Vector3dVector(color_voxels)
o3d.visualization.draw_geometries([pcd_vox_o3d])
```

Beautiful! Now, let's study how to generate 3D point clouds from depth maps.

Raster to Point Cloud

As we saw in the previous section, finding a relationship between a raster projection and Cartesian coordinates opens up many possibilities. Figure 4-23 illustrates three main cases: perspective projection, orthographic projection, and spherical projection. Note how the perspective view is linked to a single point of view that skews dimensions.

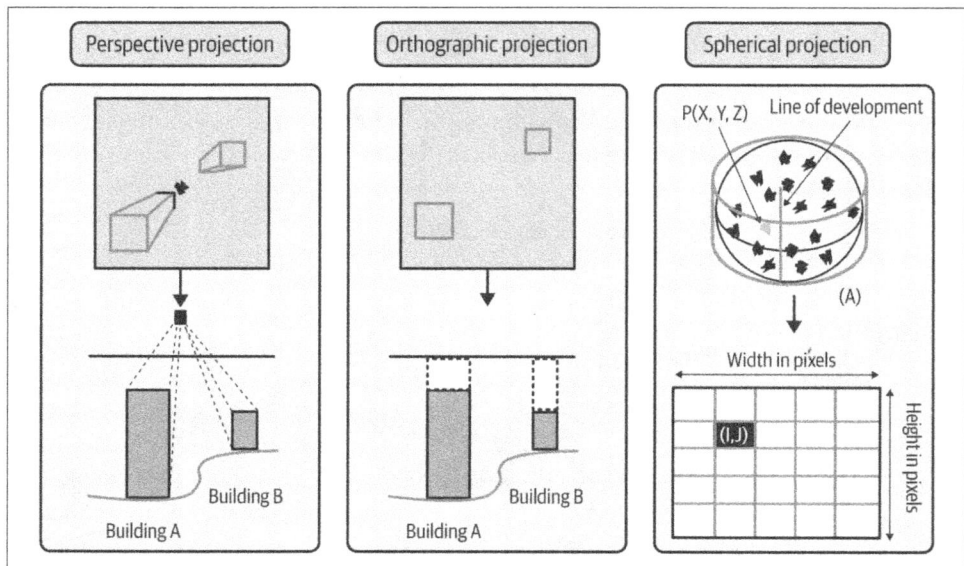

Figure 4-23. The difference between orthographic, perspective, and spherical views for 3D projections

We used perspective projection in the previous section when we used a camera model with intrinsic and extrinsic parameters and generated a point cloud from it. This means that, with our previous depth map, we can interpret the depth information from the image as the z component along the lines of sight to generate our 3D points:

```
pcd = o3d.geometry.PointCloud.create_
from_rgbd_image(rgbd_image, o3d.
camera.PinholeCameraIntrinsic(
o3d.camera.PinholeCameraIntrinsicParameters.PrimeSenseDefault))
o3d.visualization.draw_geometries([pcd])
```

In the preceding snippet, we use a standard pinhole camera model (*https://oreil.ly/ 81wOT*). You can explore the 3D Resource Hub (*https://learngeodata.eu/3d-data-science-with-python*) for more models and perspective projection cases. Now, let us consider orthographic projection.

Orthographic projection

Orthographic, or ortho, projection (*https://oreil.ly/a8wue*) is an excellent bridge between the multidimensional complexities of 3D point clouds and the comprehensible world of 2D images. Through ortho projection, we flatten dimensions and unveil a direct way to turn the point cloud into a raster data representation while keeping dimensions. The idea is to generate a top-down view plane and an image not constrained by a single perspective. You could think of ortho projection as pushing visible points from the point cloud (the highest ones) onto the plane that holds the empty image to fill all the necessary pixels just above those points.

To work out this process, we can define a 3D-to-2D projection function that takes the points of a point cloud alongside its color and a desired resolution to compute the ortho projection and return an orthoimage from the point cloud. This would translate into the following:

```
def cloud_to_image(pcd_np, resolution):
    minx = np.min(pcd_np[:, 0])
    maxx = np.max(pcd_np[:, 0])
    miny = np.min(pcd_np[:, 1])
    maxy = np.max(pcd_np[:, 1])
    width = int((maxx - minx) / resolution) + 1
    height = int((maxy - miny) / resolution) + 1
    image = np.zeros((height, width, 3), dtype=np.uint8)
    For a point in pcd_np:
        x, y, *_ = point
        r, g, b = point[-3:]
        pixel_x = int((x - minx) / resolution)
        pixel_y = int((maxy - y) / resolution)
        image[pixel_y, pixel_x] = [r, g, b]
    return image
```

This effectively projects the 3D point cloud (as a NumPy array) onto a 2D ortho-image, where each point's position is translated into a pixel location, and its color is preserved, creating a top-down view of the point cloud data. To do so, let us load a point cloud dataset, transform it to a NumPy array, apply the function, and export an image of this point cloud:

```
# Reading the point cloud with laspy
pcd = laspy.read("../DATA/34FN2_18.las")

# Transforming the point cloud to NumPy
pcd_np = np.vstack((pcd.x, pcd.y, pcd.z, (pcd.red/65535*255).astype(int),
(pcd.green/65535*255).astype(int), (pcd.blue/65535*255).astype(int))).transpose()

# Ortho projection
orthoimage = cloud_to_image(pcd_np, 1.5)

# Plotting and exporting
fig = plt.figure(figsize=(np.shape(orthoimage)[1]/72, np.shape(orthoimage)[0]/72))
fig.add_axes([0,0,1,1])
plt.imshow(orthoimage)
plt.axis('off')
plt.savefig("../DATA/34FN2_18_orthoimage.jpg")
```

3D point cloud spherical projection

Spherical projections (*https://oreil.ly/9qSck*) offer a unique perspective, enabling us to visualize the data by simulating a virtual 360° scan station. To do just this, we proceed in four steps as illustrated in Figure 4-24:

1. Load a 3D point cloud.
2. Project each point onto a sphere.
3. Define a geometry that will retrieve the pixels.
4. "Flatten" this geometry to produce an image.

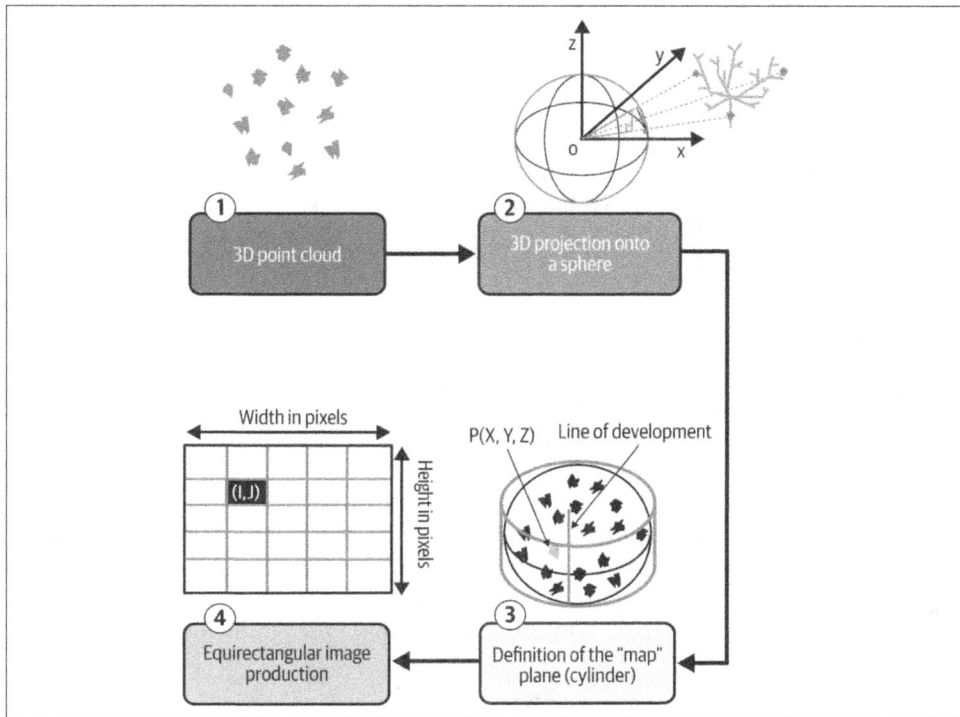

Figure 4-24. The 3D point cloud spherical projection workflow

This means we design a mechanism to project 3D points onto a sphere and unroll them as pixels onto a cylinder. Let me now detail the function that allows just this:

```
def generate_spherical_image(center_coordinates, point_cloud, colors,
resolution_y=500):
    # Translate the point cloud by the negation of the center coordinates
    translated_points = point_cloud - center_coordinates

    # Convert 3D point cloud to spherical coordinates
    theta = np.arctan2(translated_points[:, 1], translated_points[:, 0])
    phi = np.arccos(translated_points[:, 2] / np.linalg.norm(translated_points,
    axis=1))

    # Map spherical coordinates to pixel coordinates
    x = (theta + np.pi) / (2 * np.pi) * (2 * resolution_y)
    y = phi / np.pi * resolution_y

    # Create the spherical image with RGB channels
    resolution_x = 2 * resolution_y
    image = np.zeros((resolution_y, resolution_x, 3), dtype=np.uint8)

    # Create the mapping between the point cloud and image coordinates
```

```
mapping = np.full((resolution_y, resolution_x), -1, dtype=int)

# Assign points to the image pixels
for i in range(len(translated_points)):
    ix = np.clip(int(x[i]), 0, resolution_x - 1)
    iy = np.clip(int(y[i]), 0, resolution_y - 1)
    if mapping[iy, ix] == -1 or np.linalg.norm(translated_points[i]) <
    np.linalg.norm(translated_points[mapping[iy, ix]]):
        mapping[iy, ix] = i
        image[iy, ix] = colors[i]
return image, mapping
```

It is essential to digest this function. It looks pretty straightforward, but there are nice tricks at several stages. For example, what do you think about the 3D point cloud to spherical coordinates step? What does the mapping do? What is the point of using the mapping as a conditional statement while assigning points to pixels?

To make sense of the information that appears to be dispersed, we convert unstructured point data into structured raster representations. The compass that directs us when processing 3D point clouds via 2D projection is called point cloud mapping. Now, to use this handy function, let us load and prepare the point cloud first:

```
# Loading the last file from the disk
las = laspy.read("../DATA/NAAVIS_EXTERIOR.las")

# Transforming to a numpy array
coords = np.vstack((las.x, las.y, las.z))
point_cloud = coords.transpose()

# Gathering the colors
r=(las.red/65535*255).astype(int)
g=(las.green/65535*255).astype(int)
b=(las.blue/65535*255).astype(int)
colors = np.vstack((r,g,b)).transpose()
```

Once prepared, we can define the necessary parameters for projection. These are the center of projection (basically the position from which we want a virtual scan station) and the resolution of the final image (expressed in pixels, as the minor side of the image):

```
resolution = 500

#Defining the position in the point cloud to generate a panorama
center_coordinates = [189, 60, 2]
```

Finally, we can call the new function, plot, and export the results as an image:

```
# Function execution
spherical_image, mapping = generate_spherical_image(center_coordinates,
point_cloud, colors, resolution)
```

```
# Plotting with matplotlib
fig = plt.figure(figsize=(np.shape(spherical_image)[1]/72,
np.shape(spherical_image)[0]/72))
fig.add_axes([0,0,1,1])
plt.imshow(spherical_image)
plt.axis('off')

# Saving to the disk
plt.savefig("../DATA/ITC_BUILDING_spherical_projection.jpg")
```

This process results in Figure 4-25. You can experiment with the various parameters, such as the resolution or the center of projection, to ensure that you balance no-data pixels with relevant panoramas.

Figure 4-25. The 3D point cloud transformed into an equirectangular image from the projection

With the generation of 3D point cloud to equirectangular images, you have just unlocked a formidable new ability. It lets you create virtual scans anywhere you think it makes sense, which opens up the prospect of analyzing photos using deep learning and image processing techniques. We will leverage these outcomes in Chapter 13.

3D Data Structures: k-d Trees, Octrees, BVH

As we delve deeper into the world of 3D data, efficient organization becomes paramount. While 3D representations describe and store the geometric information of a three-dimensional object (or scene), 3D data structures, on the other hand, are ways to organize and manage 3D data for efficient processing and analysis. These structures are not concerned with how the 3D data is represented (e.g., as a point cloud or a mesh) but rather with how to store and access this data efficiently. This section introduces you to robust data structures that organize and manage 3D datasets such as point clouds (see Figure 4-26).

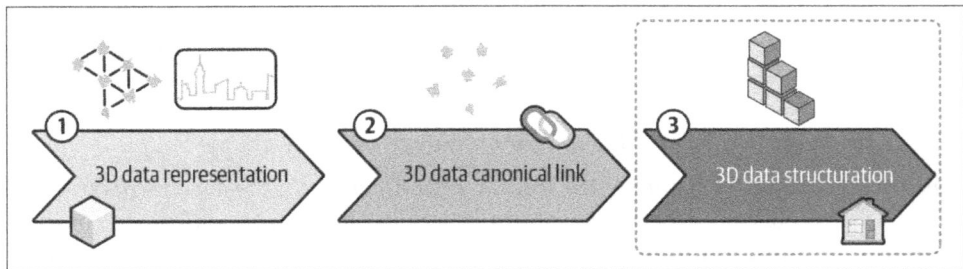

Figure 4-26. 3D data structures for efficient organization

These are essential for tasks like nearest neighbor searches and efficient retrieval of specific data points within your 3D world. Understanding their strengths and functionalities will unlock a new level of performance and control when working with 3D datasets.

k-d Trees

A *k*-d tree (*https://oreil.ly/zQm-X*) is a data structure that separates space with hyperplanes. They are heavily used and primarily studied in industry. If you look at their name, there are three main components: "k," "d," and "tree." Let us move from the end to the beginning.

A tree is a structure with branches, precisely like the ones you are used to seeing daily (hopefully) that participate so heavily in our well-being and functioning as a society. So digitally, it means we have some refined process that can drive the data it holds. How do we conceive them then? Well, you first have roots, which is where the tree starts and grows. The root usually pinpoints the entire dataset. Then the tree grows and creates branches, and these branches have other branches until, at some point, the leaves appear!

A *k*-d tree has the same structure but with a high level of control. It provides a binary partitioning of the initial root into two branches, which will also get split into two, etc. (see Figure 4-27). This is the "d" of dimension.

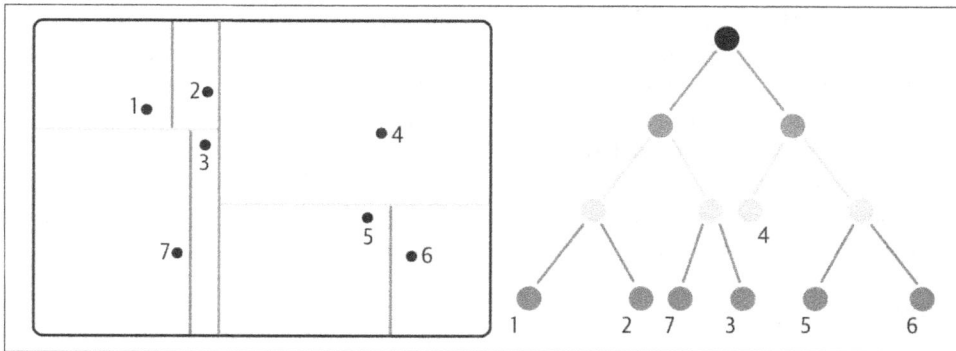

Figure 4-27. Representation of a k-d tree data structure from the top view of a 2D dataset

So, down the line, we have a splitting approach to how the data should be distributed in our structure (or linked to). The *k* means that we can work with 2D data, 3D data, and *k*-d data while keeping the binary split. But how do we split then? With the *k*-d tree, we try to get the same number of "data points" on each branch that a "hyperplane" defines. Why a hyperplane? Well, in 2D, it is a line; in 3D, it is a plane; and in *k*-D, it is a hyperplane.

We can use Open3D or SciPy to construct a *k*-d tree:

```
# Calculate the k-d tree from the point cloud
pcd_tree = o3d.geometry.KDTreeFlann(point_cloud)

# k-d tree (using scipy)
points = np.array(point_cloud.points)
kdtree = cKDTree(points)
```

k-d trees play a crucial role in 3D data science because they efficiently handle spatial data structures. They are particularly valuable for their ability to perform rapid nearest-neighbor searches and range queries in multidimensional spaces. One of the key advantages of *k*-d trees is their logarithmic time complexity for searches in average cases, allowing them to maintain high performance even when dealing with large datasets.

These tree structures are well suited for low- to medium-dimensional data, typically up to about 20 dimensions, encompassing most 3D applications. Due to their versatility, they are well suited for various tasks involving the processing and interpretation of 3D data. They find extensive use in point cloud processing, facilitating operations such as finding nearest neighbors for surface reconstruction, normal estimation, and feature matching.

In future chapters, you will see how they are instrumental in matching features between images or 3D scenes. The robotics field benefits from *k*-d trees in applications like collision detection and path planning, where spatial awareness is critical. Another significant application area is 3D object recognition, where *k*-d trees assist in matching local descriptors or key points. They also prove invaluable in spatial databases, enabling efficient querying of 3D spatial data. In scientific simulations, particularly those involving particle simulations or *n*-body problems, *k*-d trees offer a performant solution for spatial organization and query.

Octrees

Octrees (*https://oreil.ly/A3n1O*) are also a way to structure space with another hierarchy: cubes (or voxels). They are a subdivision of space in octants, i.e., eight cubes recursively subdivided into eight children until a specific criterion is achieved. It is a divide-and-conquer approach where these cubes (nodes) have eight children representing eight parts of space called octants (see Figure 4-28).

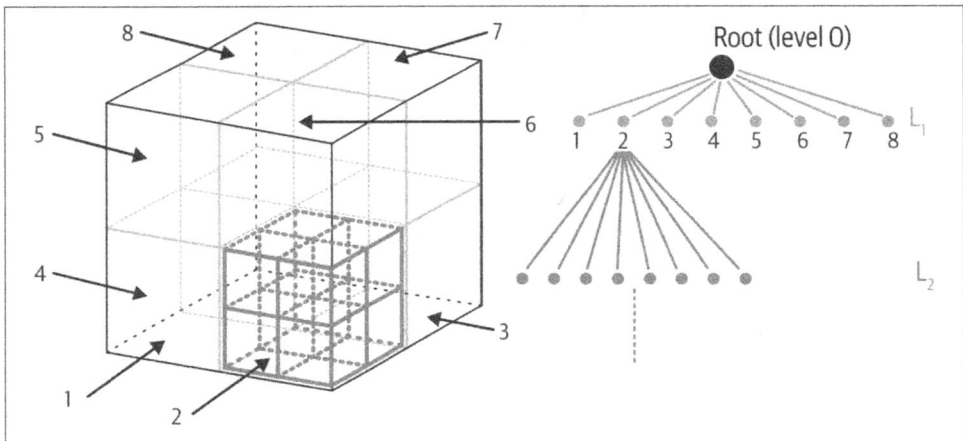

Figure 4-28. Representation of an octree for a 3D point cloud dataset

Why the term "octree"? It's a combination of *oct* (meaning eight) and *tree* (well, because it is structured like one). Let's dive into Python code and explore how octants can break down data structures within our 3D dataset. Use the following code to load a new point cloud and compute an octree with a maximum depth of 8:

```
curr_max_depth = 8

point_cloud = o3d.io.read_point_cloud('../DATA/structure_verviers.ply')

octree = o3d.geometry.Octree(max_depth = curr_max_depth)
octree.convert_from_point_cloud(point_cloud)
o3d.visualization.draw_geometries([octree])
```

You can see the octree, representing the hierarchical partitioning of a 3D dataset for efficient spatial data management, in Figure 4-29.

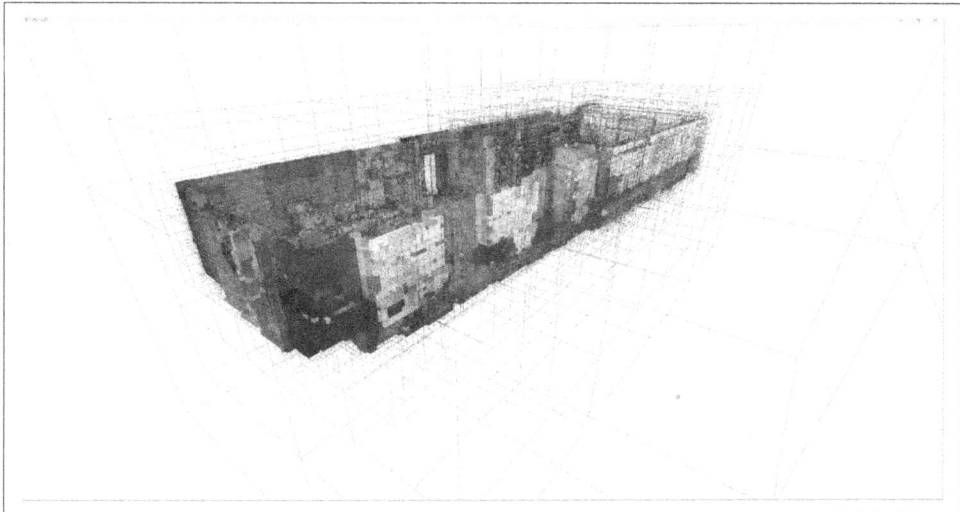

Figure 4-29. Visualization of an octree for a 3D dataset

From there, you can change the depth to see the impact of the data structure on the way the data is stored visually. Now that we've divided our 3D space, it's time to navigate through the maze of octants. Here's a snippet to search for a point in the octree:

```
# Check if the target point is within the bounds of the current octant
octree.locate_leaf_node(point_cloud.points[0])
```

There you have it—a compass for navigating your way through the intricate branches of your octree. Now, let me highlight some performance differences between octrees and *k*-d trees.

Let's import four libraries:

```
import numpy as np
from scipy.spatial import cKDTree
import open3d as o3d
import time
```

Then, we generate a random point cloud with 1 million points:

```
num_points = 1000000
points = np.random.rand(num_points, 3)
```

If you were to check for timings (see the 3D Resource Hub (*https://learngeodata.eu/3d-data-science-with-python*) for the code), you would find that *k*-d trees are quicker to build than octrees. For 10 million points:

```
KD-Tree build time: 5.211558 seconds
Octree build time: 23.283945 seconds
```

While this is largely due to the implementation of both, there is still a discrepancy in how efficient *k*-d tree constructions are compared to octrees. Nevertheless, octrees are important in 3D data science because they efficiently partition and organize three-dimensional space. Compared to *k*-d trees, octrees have some distinct characteristics. While *k*-d trees partition space using axis-aligned splitting planes at each level, octrees always divide space into eight equal-sized octants. This makes octrees particularly well suited for uniform spatial subdivision and situations where the data distribution is roughly uniform in all three dimensions.

They are often preferred over *k*-d trees when dealing with 3D data that benefits from regular spatial partitioning. They excel in voxel grid scenarios, where the natural octant structure aligns well with the data representation. Octrees also tend to perform better for range queries in 3D space, as their structure allows for efficient pruning of large spatial regions. *k*-d trees, on the other hand, are more flexible in terms of partitioning strategy and can adapt better to nonuniform data distributions. They are particularly advantageous in scenarios where memory efficiency is crucial, as they can represent large empty regions of space very compactly.

> Other 3D data structures exist but covering them all would extend the book's scope. You can find these in the 3D Data Science Resource Hub (*https://learngeodata.eu/3d-data-science-with-python*), and I recommend specifically focusing on bounding volume hierarchies (BVHs) (*https://oreil.ly/uvRpQ*). These decompose the overall space or dataset into bounding volumes, which are usually bounding boxes with a notion of scale and hierarchy. They are handy in ray tracing and collision detection for complex 3D scenes, especially those with dynamic or deformable objects.

File Organization

3D data science deals with big data problematics. Thus, it is essential to understand that how you structure your dataset may impact how you address your computation. Let me try to describe what we are dealing with here in a bit more detail. Your computer's memory is like a cozy library, and your 3D dataset is a collection of weighty tomes. The difficulty? Not all the books fit on the shelves at once. Memory versus out-of-core becomes the librarian's choice—do we squeeze all the books onto the shelves and risk an avalanche, or do we keep some in the basement, ready to be fetched as needed (see Figure 4-30)?

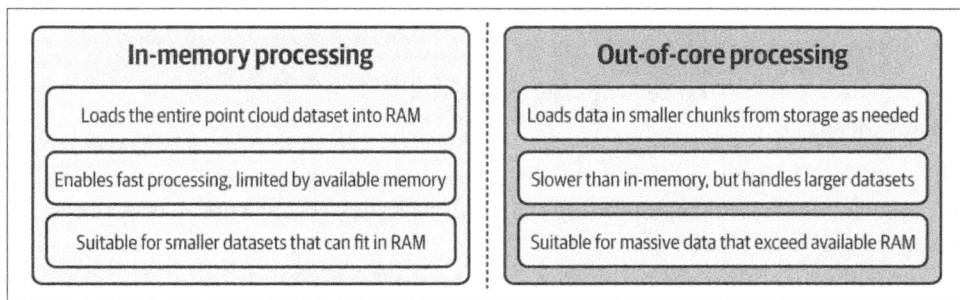

Figure 4-30. File organization in 3D data science (memory versus out-of-core)

In-memory processing loads the entire dataset into RAM, like having all the books open on a giant table—quick and snappy, but watch out for that table space! When the dataset dwarfs your memory, out-of-core (*https://oreil.ly/FASV1*) processing becomes the unsung hero.

It reads chunks of data from storage, allowing you to wrangle datasets larger than your memory's capacity. It's like reading chapters of a book, swapping them out as needed—slower, perhaps, but with more scaling potential. So, let us define the functions with Python:

```python
# In-Memory
def process_in_memory(dataset):
    # Load the entire dataset into memory
    entire_dataset = np.loadtxt(dataset, delimiter=';', skiprows=1)

    # Now you can perform computations on the entire dataset
    result = np.mean(entire_dataset, axis=0)

    return result

# Out-of-Core
def process_out_of_core(dataset, chunk_size=1000):
    # Read the dataset in chunks
    reader = pd.read_csv(dataset, delimiter=';', chunksize=chunk_size)

    # Initialize an empty array to accumulate results
    result_accumulator = np.zeros((1, 6))

    counter=1
    for chunk in reader:
        # Perform computations on each chunk
        result_chunk = np.mean(chunk.to_numpy(), axis=0)

        # Accumulate results
        result_accumulator += result_chunk
        counter+=1
```

```
# return result_accumulator / len(reader)
return result_accumulator/counter
```

You can test the difference with any point cloud, such as `dataset = '../DATA/bike_florent.txt'`.

The `process_in_memory` function loads the entire dataset into memory using `np.loadtxt` and calculates the mean. The out-of-core processing function processes the data in chunks using pandas.

Summary

This chapter provided methods to capture the intricacies and nuances of three-dimensional datasets. More specifically, you can now leverage a robust and adaptable framework that allows you to seamlessly transition between different 3D data modalities while preserving geometric fidelity. This chapter will refresh your in-depth exploration of 3D data representations as you continue reading (point clouds, meshes, parametric models, voxel models, and image-based representations, such as depth maps and multiview images).

It's important to understand each representation's unique properties, advantages, and limitations. This will enable you to make informed decisions based on your application requirements. I covered the pivotal role of point clouds as a canonical representation, demonstrating how other modalities can be effectively translated into point cloud format, providing a unified foundation for your 3D data science workflows. You are now well equipped to leverage these representations and structures to develop sophisticated algorithms and applications. The insights gained from this chapter will empower you to:

- Improve the accuracy and efficiency of your 3D data processing pipelines.
- Design more robust and adaptable algorithms for 3D object recognition, segmentation, and reconstruction.
- Create innovative visualizations and interactive experiences.

In Chapter 5, I will guide you through creating a 3D multimodal viewer experience directly within Python that leverages the data structures and representations of this chapter. But before moving there, I have designed a hands-on project to help you ensure you master the techniques.

Hands-on Project

In this chapter, we explored the topic of 3D data representation and structure. You now have a very clear understanding of the different forms a 3D dataset can take and how it can be structured to gain efficiency in computational tasks and other tasks. As 3D data scientists, we often face the challenge of extracting meaningful insights from complex point cloud data. One practical approach is to project the point cloud onto a 2D image plane, creating a more manageable representation for analysis. Let's explore how to enhance 3D scene understanding through ortho projection and spherical projection. Here is a workflow you can put in place with Python:

1. *Load the point cloud data*: Begin by loading your point cloud data using a suitable library, such as laspy.

2. *Extract point coordinates and colors*: Extract the x, y, and z coordinates of each point, along with any associated color information (R, G, B). Store these values in a NumPy array for easy manipulation.

3. *Ortho projection*: To create an orthoimage, project the point cloud onto a 2D plane from a top-down perspective. The code snippet provided in the chapter demonstrates how to implement this projection. I encourage you to experiment with different resolutions to find the optimal balance between detail and image size.

4. *Spherical projection*: You can project the point cloud onto a sphere for a more immersive perspective. This technique involves transforming the point cloud into spherical coordinates and then mapping these coordinates onto an equirectangular image. You can experiment with different center coordinates to simulate various viewpoints within the scene.

5. *Visualize and analyze*: Once you've generated the ortho and spherical projections, visualize the resulting images. These projections provide valuable insights into the spatial distribution of the point cloud data. As you'll see in Chapter 12 and onward, they can be used for object detection, classification, and segmentation tasks.

Mastering these projection techniques is a powerful tool for extracting meaningful information from complex 3D point clouds.

Developing a Multimodal 3D Viewer with Python

I take little risk saying that, as you explore 3D data science, you will deal with more than one 3D modality. However, we have an unfair advantage: we can represent 3D data as point clouds, meshes, parametric models, or voxel assemblies (see Chapter 4). Each modality offers unique perspectives and insights, yet effectively integrating and visualizing these diverse formats remains challenging.

Current 3D viewers often struggle to handle multimodal data, limiting our ability to analyze and interpret complex 3D scenes comprehensively. It would help if we had a solution that seamlessly combines these modalities, enabling us to visualize and interact with them in a unified environment.

Imagine a scenario where you need to analyze a flood-prone area. You have access to a 3D point cloud of the terrain, a mesh model of buildings, and other relevant geospatial data. A conventional 3D viewer might only display these datasets separately, making it difficult to assess the spatial relationships and understand the overall situation. It would help if you had a powerful tool that combines these datasets, allowing you to visualize the building heights, built coverage, and other crucial information in a single, interactive environment.

But is it possible to combine all of these in a simple solution? Could that be the base for multimodal digital environments?

It is time to take what was learned in Chapter 4 one step further and integrate the modalities. In this chapter, I'll present a step-by-step guide to developing a multimodal 3D viewer using Python. You will learn to combine point clouds, meshes, and other data formats, creating an interactive visualization that empowers you to explore and analyze 3D scenes efficiently.

The rest of this chapter will use the following scenario as an example. You are part of a cutting-edge 3D data science team in the Netherlands. You were recently warned by the crisis department in the Netherlands that a flood will likely happen within the next 48 hours. After initial investigations, they localized a zone at a very high risk that they want you to examine.

In this zone, they want to know the height of the residential buildings and the built overage so that they can plan protective actions. Armed with a will to best serve and protect your community, you develop a 3D viewer that can combine available datasets and provide a base for answers. You establish a sound workflow composed of four main stages as shown in Figure 5-1.

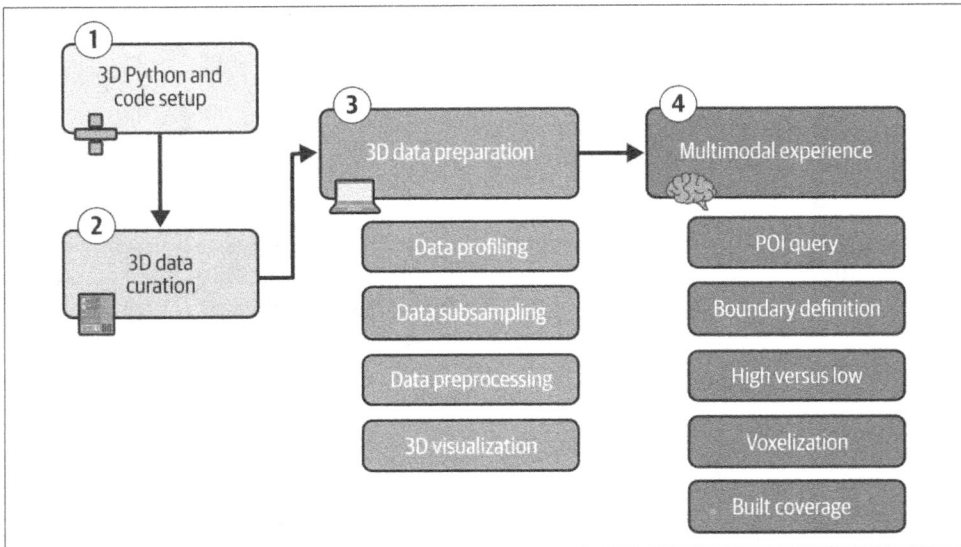

Figure 5-1. The workflow to create a multimodal 3D viewer

> This chapter is based on a real-world scenario and provides new solutions to 3D data preprocessing and modality-specific solutions. This is a great way to combine previous hands-on knowledge in an integrated project to ensure you are properly mastering the learning to this point in the book. You can find all the materials for this chapter on the 3D Data Science Resource Hub (*https://learngeo data.eu/3d-data-science-with-python*). You can use the base link and add "*/chapter-X*" to access the files (code, datasets, articles) and resources. Note that you may need to share an email address, a personal password, or proof that you are the book's owner.

3D Python and Code Setup

The first step in creating a multimodal 3D viewer is to launch our Python virtual environment manager (refer back to Chapter 3 if you need a refresher). Using Anaconda is a great way to refresh knowledge and integrate it into your skill set. You can use the GUI or go with the command-line solution. We'll use five critical libraries in this chapter: NumPy (*https://numpy.org*), pandas (*https://oreil.ly/-vbZ4*), Open3D (*https://www.open3d.org*), Matplotlib (*https://matplotlib.org*), and Shapely (*https://oreil.ly/3R-NK*).

> If you want to use the aforementioned libraries, we must ensure they are installed and available in your Python environment. As you remember from Chapter 3, you must first create this environment. Assuming you have Anaconda installed, launch an Anaconda terminal session and create your environment: `conda create-n NEWENV python=3.10`. Once your environment is created, activate it: `conda activate NEWENV`. Now, you are ready to install libraries.

We'll start by exploring a library setup with pip. First, make sure your Anaconda environment is active. Then, you can install the libraries:

```
pip install numpy pandas matplotlib open3d shapely
```

As defined in Chapter 3, we stick to a minimal number of libraries to deepen our expertise in their usage. If you do not have an IDE, one way to get started is to use JupyterLab to work with notebooks (`pip install jupyterlab`). Once the installation is finished, just type in the terminal:

```
jupyter lab
```

We are now ready to write some lines of code in our Python notebook. First off, we import the libraries mentioned earlier with the following code snippet:

```
import numpy as np
import pandas as pd
import open3d as o3d
import matplotlib.pyplot as plt
from shapely.geometry import Polygon
```

Now, we are all set up and can move on to loading the datasets.

3D Data Curation

We now move on to the second step: 3D data curation. We aim to gather datasets before making them usable for our analysis with Python. Therefore, this step also acts as a 3D data visualization phase, where we can qualitatively assess what we are dealing with. If you are ready, let's get started. To achieve a robust 3D city neighborhood analysis, we first gather some excellent datasets from open data sources. I illustrate a specific tile of interest in the Netherlands, but I encourage you to study your house or any point of interest.

The point cloud is gathered using the GeoTiles.nl (*https://oreil.ly/4A8LM*) portal (see Chapter 3), which provides excellent datasets under the CC-BY 4.0 license. The mesh dataset is collected from the same spot as the LiDAR area we downloaded. We thus use the 3DBAG (*https://oreil.ly/f3Ooq*) platform created by TUDelft, which allows us to retrieve 10M buildings in the Netherlands in LoD1.2, LoD1.3, and LoD2.2 from the CityGML specification (*https://oreil.ly/_DEk7*). At this step, we are mainly interested in the geometry.

> After downloading the datasets, you should have one point cloud in the *.laz* file format and one or more *.obj* datasets (with their accompanying *.mtl* files) that describe approximately the same extent with different levels of detail. You can download the selection from the chapter's resources for convenience to avoid going from *.laz* to Open3D objects.

Let's define our paths:

```
data_folder="../DATA/"
results_folder="../DATA/RESULTS/"

pc_dataset = "30HZ1_18_sampled.xyz"
mesh_dataset = "NL.IMBAG.Pand.0637100000139735.obj"
```

We can load the mesh, in *.obj*, with any ASCII reader. Nevertheless, we will wait for the next step to leverage Open3D's ability to load and visualize the mesh. We can now move on to data preparation.

3D Data Preparation

3D data preparation is a critical step in the 3D data processing pipeline, as it ensures the quality and consistency of the data. This involves cleaning, filtering, and normalizing the data to remove noise, outliers, and inconsistencies. By preparing the data, we can improve the accuracy and reduce the computational costs of our downstream processes. While we cover this in depth in Chapter 6, let me give you a quick primer here.

Initial Profiling

First, we'll prepare the point cloud withs a pandas DataFrame object called `pcd_df`, which hosts the point cloud data:

```
pcd_df = pd.read_csv(data_folder + pc_dataset, delimiter=";")
print(pcd_df.columns)
```

This returns the name of the columns in this `pcd_df` DataFrame: `['X', 'Y', 'Z', 'R', 'G', 'B', 'Classification']`. This is handy for selecting only the explicit "columns" without blurry indexes. At this stage, it would be useful to quickly profile our pandas DataFrame. For this, I recommend printing out the following commands:

```
print(pcd_df.shape)
print(pcd_df.describe())
```

The first line returns the shape, i.e., the size of your DataFrame, with the number of rows (7,103,848 rows, each being a point), and columns (7, as we saw before, one column per feature). The second line returns the main distribution of each of our features and can help us adjust subsequent processes.

We see that we have large coordinates that we may need to adjust and have more than 7 million points that we could try to bring quickly below five million for our experiments. This paves the way to 3D data subsampling, a feature engineering step covered in Chapter 6.

3D Data Downsampling

Let's initiate some first-hand processing. First, we'll do some light subsampling to speed up subsequent steps. 3D data subsampling is an essential method for handling the complexity and memory needs of large 3D datasets. Subsampling speeds up processing, analysis, and visualization by reducing the number of data points. It is beneficial when iterating experiments or real-time applications that need to manage large datasets effectively. Subsampling can reduce noise and improve the signal-to-noise ratio, resulting in more accurate and reliable results.

By choosing the proper subsampling methods, we can balance reducing data and keeping important information, ensuring the subsampled data still reflects the original dataset. While this is detailed in Chapter 6, let's get a feeling of its impact in this chapter. We select the DataFrame and apply a decimation to keep one point every two points. This allows us to go to less than five million points, in a random manner:

```
pcd_subsampled = pcd_df.iloc[::2, :]
```

Next, we prepare our data so that we can use it with Open3D. We will create an Open3D `PointCloud` object, and select only the `['X','Y','Z']` coordinates for the points, and `['R','G','B']` for our colors:

```
pcd_o3d=o3d.geometry.PointCloud()
pcd_o3d.points=o3d.utility.Vector3dVector(np.array(pcd_subsampled[['X','Y',
'Z']]))
pcd_o3d.colors=o3d.utility.Vector3dVector(np.array(pcd_subsampled[['R','G',
'B']]))/255)
```

This is an excellent way to understand that with a different library, we cope with different mechanisms to transform the dataset in Python objects. Here, we go from a pandas DataFrame to an Open3D `PointCloud`, which is necessary to use the functions implemented in the Open3D library. As you can see, we had to be mindful of transforming the R, G, and B values to float values between [0,1] and make sure that we pass a `Vector3dVector` object to the `colors` attribute. Now, we can load the 3D mesh in a mesh variable using the `read_triangle_mesh()` method from Open3D. We can also paint the mesh with the `paint_uniform_color()` method:

```
mesh=o3d.io.read_triangle_mesh(data_folder+mesh_dataset)
mesh.paint_uniform_color([0.1, 0.4, 0.8])
```

You can visualize these independently with Open3D (results in Figure 5-2):

```
# When finished, close the rendering window, which blocks the Python flow.
o3d.visualization.draw_geometries([pcd_o3d])
o3d.visualization.draw_geometries([mesh])
```

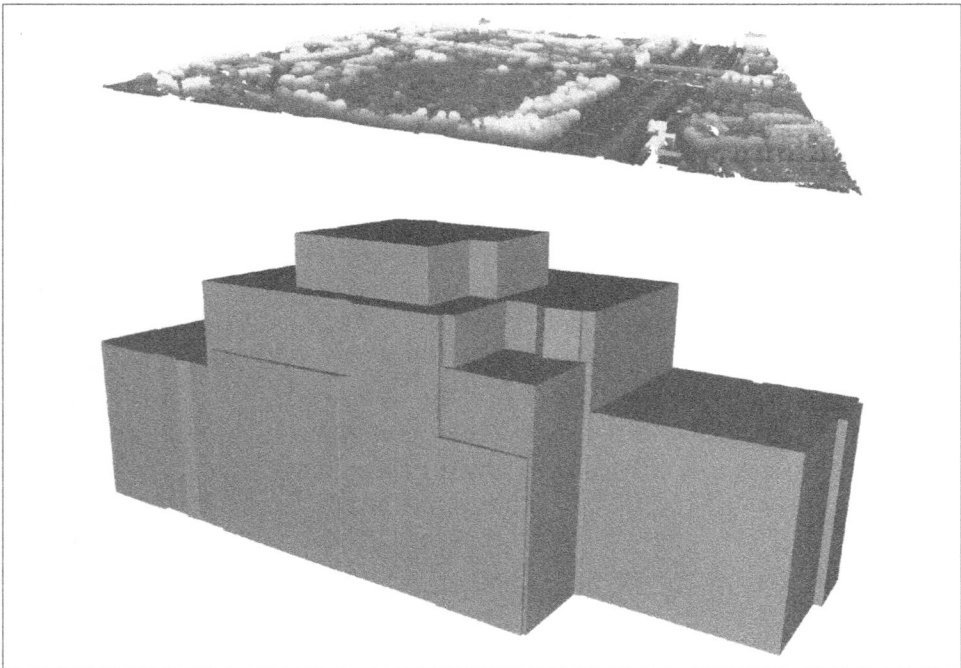

Figure 5-2. The 3D point cloud and the 3D mesh loaded with Open3D

Once the subsampling step is complete, we get a resulting subsampled point cloud that we can use for downstream processes. We now have two Open3D objects: A PointCloud with 3,551,924 points and a TriangleMesh with 674 points and 488 triangles. Let's now ensure we have an easy time working with the data in a local context.

Data Preprocessing

The real world is messy, and so is our point cloud. Indeed, we have a few stray data points due to sensor glitches and reflections. These outliers can skew our analysis. We can tackle this challenge by exploring a statistical analysis technique to identify and remove these disruptive points. However, let us leave this for Chapter 6 so that we can see the impact of such noise on our analysis. Next, we'll center our dataset to ensure we avoid issues due to large coordinates when viewing 3D datasets (this could cause rendering glitches):

```
pcd_center = pcd_o3d.get_center()
pcd_o3d.translate(-pcd_center)
mesh.translate(-pcd_center)
```

3D Data Visualization

To visualize different 3D objects in Open3D, we have to pass a Python list holding these Open3D objects, as seen earlier. Our list is thus composed of one Open3D PointCloud and one Open3D TriangleMesh, which gives [pcd_o3d, mesh]. To visualize this combination in a standalone window, use the following:

```
o3d.visualization.draw_geometries([pcd_o3d,mesh])
```

To improve the perception of how light propagates, we can use normals, a very useful feature for a lot of processes (we'll cover these in Chapter 6), which we can quickly compute with Open3D:

```
pcd_o3d.estimate_normals()
mesh.compute_vertex_normals()
```

There is one handy trick to play with the colors to display: using a color variable that we pass to the colors attribute of the PointCloud Open3D object. This variable should hold R, G, and B float values ranging from 0 to 1. Let's say that we want to visualize the point cloud colored based on the classification attribute. To do this, we'll follow the four-stage process outlined in Figure 5-3.

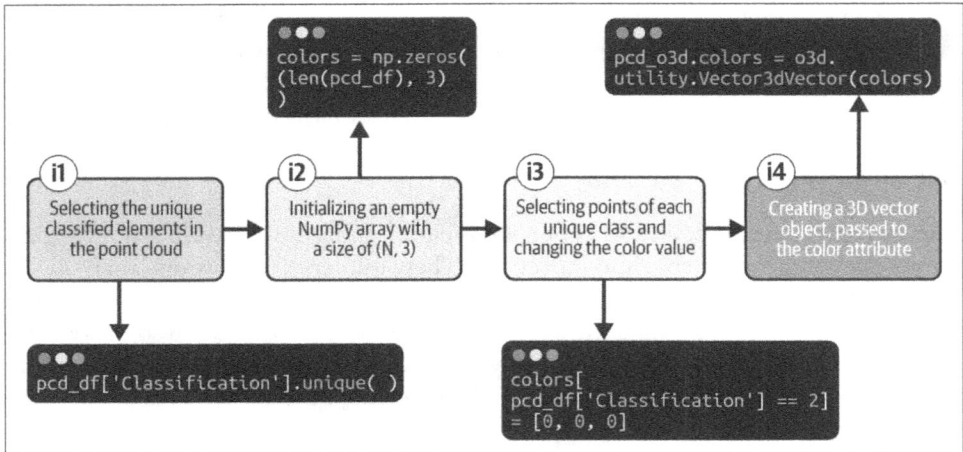

Figure 5-3. The code workflow uses classification as a color for Open3D

This translates to the following Python code and results in Figure 5-4:

```python
pcd_subsampled['Classification'].unique()
colors=np.zeros((len(pcd_subsampled), 3))
colors[pcd_subsampled['Classification'] == 1] = [0.611, 0.8, 0.521]
colors[pcd_subsampled['Classification'] == 2] = [0.8, 0.670, 0.521]
colors[pcd_subsampled['Classification'] == 6] = [0.901, 0.419, 0.431]
colors[pcd_subsampled['Classification'] == 9] = [0.564, 0.850, 0.913]
colors[pcd_subsampled['Classification'] == 26] = [0.694, 0.662, 0.698]
pcd_o3d.colors = o3d.utility.Vector3dVector(colors)
o3d.visualization.draw_geometries([pcd_o3d,mesh])
```

Figure 5-4. The point cloud colored with the class information

Because we want to have a hand in the color of each class, we can adapt the preceding code to the number of unique classes. Then, the correspondence is done based on the LAS (1.4) Specification (*https://oreil.ly/zt5Pv*). We have variables loaded, we can see both the point cloud and the 3D mesh, and everything looks like it is working smoothly (Figure 5-4)! It's time to define the various 3D city analytical tasks we want to do. For this, we relate to our overall high-level workflow and adapt it to our application. This means we have inputs, a processing pipeline, and an output for each scenario. The conceptual workflow from input to output in the context of point cloud data is illustrated in Figure 5-5.

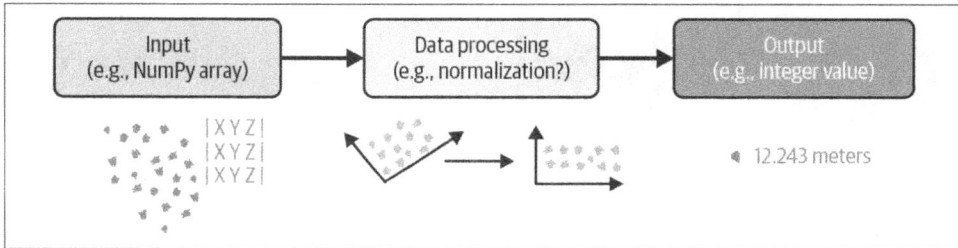

Figure 5-5. Conceptual workflow for 3D data processing

The input can vary, but in most cases, it is a NumPy array that holds the spatial information. The output can be an integer, a list, another NumPy array, or anything else you need.

Multimodal 3D Experience

We are going to complete this mission. This means that we want to find a solution to four challenges by utilizing multimodal 3D data representation: point of interest query, boundary definition, high point extraction, and built coverage. Let's start with the point of interest query.

Point of Interest Query

The objective of this challenge is to keep only the data points that fall within a certain distance from your point of interest (the building). Our inputs are the point cloud and the mesh (Open3D objects), and our output is the filtered point cloud, which answers the distance to the point of interest (POI) criterion (see Figure 5-6). To get there, we'll follow a six-stage process as illustrated in Figure 5-7.

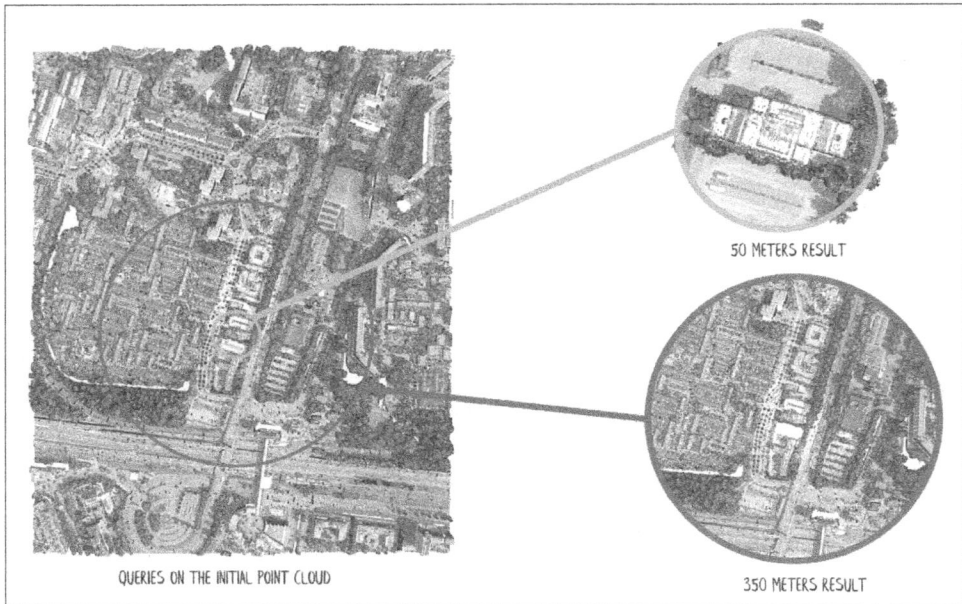

Figure 5-6. *The close-up view of the radius query*

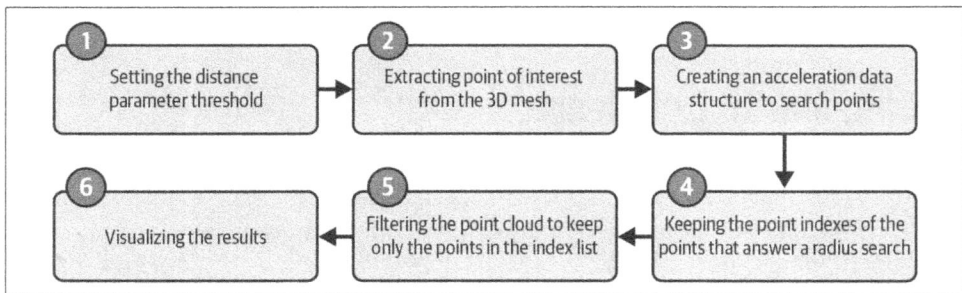

Figure 5-7. *The conceptual workflow for a radius search in a 3D point cloud*

To set the distance threshold, we pass the radius value we want to use as a threshold (e.g., 50) to a new variable dist_POI (1). Then, we get the POI from the BAG dataset mesh using the get_center() Open3D method of the mesh object (2). This gets us the center of the mesh as a POI:

```
dist_POI=50
POI=mesh.get_center()
```

After that, we can create a *k*-d tree (3) to organize points in space and allow for fast nearest-neighbor searches and range queries. Using a *k*-d tree makes it possible to find the points closest to a given point quickly, or all the points within a certain distance (our POI), without having to search through all the points in the point cloud. You can use the Open3D or SciPy version:

```
# Creating a k-d tree data structure
pcd_tree = o3d.geometry.KDTreeFlann(pcd_o3d)
```

As a reminder, the *k*-d tree works by recursively partitioning the space into smaller regions, dividing it along the median of one of the dimensions at each level (x is a dimension, y another, z another). As a result, a "tree" structure is created, with each leaf node representing a single point and each node representing a space split. From there, we can use the `search_radius_vector_3d()` Open3D method and select the points from the output index to finally visualize our result (4 + 5 + 6):

```
# Selecting points by distance from POI (your house) using the k-d tree
[k, idx, _] = pcd_tree.search_radius_vector_3d(POI, dist_POI)
pcd_selection=pcd_o3d.select_by_index(idx)
o3d.visualization.draw_geometries([pcd_selection,mesh])
```

The results of performing a radius search within a 3D point cloud with this Python code are shown in Figure 5-8. Now that you have a working solution let's extract the parcel area from that selection.

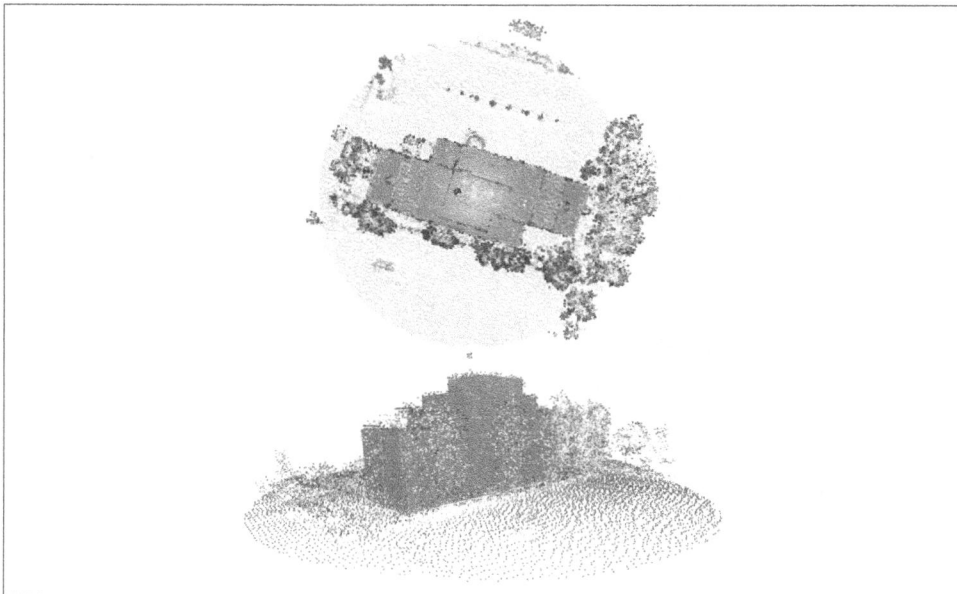

Figure 5-8. The 3D point cloud and mesh underlay of the radius search

Manual Boundary Selection

To extract the unofficial boundaries, we need to move to a semiautomated and interactive approach. The good news is that we can do that directly within Python with Open3D.

The first step is to create an interactive Open3D window with the following `draw_geometries_with_vertex_selection()` method:

```
o3d.visualization.draw_geometries_with_vertex_selection([pcd_selection])
```

You can then select the points that define the corners of your parcel. The manual boundary selection process uses the Open3D GUI; you hold Shift + Mouse to select points of interest. The results will then appear under your cell in your notebook (or your REPL) upon closing the window.

> Using R, G, and B coloring may be easier at this step. You will need to change this before the selection if you want to work with the correct indexes. If you want to extract the cadastral boundary, you can import the official 2D vector shape and cutting based on this data constraint.

From your REPL, you can copy and paste the different indexes (e.g., `34335,979,21544,19666,5924,21816,38008`) of the selected points into the `select_by_index()` method to define an `o3d_parcel_corners` variable:

```
o3d_parcel_corners=pcd_selection.select_by_index([34335 ,979 ,21544 ,19666 ,
5924 ,21816 ,38008 ])
```

We must still prepare the corners further to avoid considering the z value. Therefore, we will filter out the coordinates to drop the z value, but beware: doing this means that we assume that we are in a flat area (which is the case in the Netherlands):

```
o3d_parcel_corners=np.array(o3d_parcel_corners.points)[:,:2]
```

From there, it is time to compute the area of the parcel with the Shapely library. For this, we can directly use the `Polygon` function to create a polygon out of the set of provided corners (see Figure 5-9):

```
Polygon(o3d_parcel_corners)
```

This looks somewhat wrong. If you do not believe me, we can compute the area to check:

```
pgon = Polygon(o3d_parcel_corners)
print(f"This is the obtained parcel area: {pgon.area} m²")
```

This outputs 43.583 m². That sounds weird for a big building. Ha, a simple problem becomes a bit more complicated! Indeed, the problem here is that computing the area requires a polygon that is constituted so that we obtain a closed shape. This is not always the case and depends on the "order" in which corners have been added to the data structure. Now, the problem becomes sorting out the coordinates to avoid edge intersections. One way of dealing with the problem is to perceive all coordinates from the perspective of the center point. We can then compute the angle between each corner in the list and our center point. We do this to know how wide the individual angles are and provide us with the means to sort out the coordinates based on their values. Translating this into code allows defining a sorting function as follows:

```
def sort_coordinates(XY):
    cx, cy = XY.mean(0)
    x, y = XY.T
    angles = np.arctan2(x-cx, y-cy)
    indices = np.argsort(-angles)
    return XY[indices]
```

I use the NumPy arctan2() method to estimate the angle for every coordinate. This will return the array of angles in radians. All that's left is to sort out the angles in ascending order to receive a list of indices correctly. The list can then fix the original list indices with the argsort() method. At this step, it will not work with U-shape buildings.

From there, we can apply our sorting function to the corners, create a new sorted variable, and compute the polygon and the associated area:

```
np_sorted_2D_corners=sort_coordinates(o3d_parcel_corners)
pgon = Polygon(np_sorted_2D_corners)
Polygon(np_sorted_2D_corners)
print(f"This is the parcel area: {pgon.area} m²")
```

This returns a polygon with an area of 3408.9 m², a much more plausible number (see Figure 5-10).

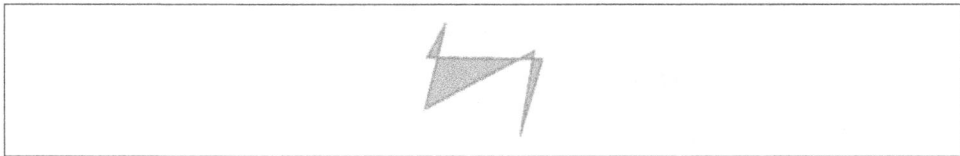

Figure 5-9. Incorrect geometry

Figure 5-10. Correct geometry

We now have a good idea of the area of our parcel, by leveraging 3D point clouds, 3D meshes, and 2D polygonal vectors. Now, let's find the high and low points in the zone.

Find High and Low Points

To get the low and high points of the area, one option is to use the `get_max_bound()` and `get_min_bound()` Open3D methods:

```
pcd_selection.get_max_bound()
pcd_selection.get_min_bound()
```

However, doing this will not hint at which point is the highest and which is the lowest. What we need are the point indexes to retrieve the coordinates.

First, create a NumPy array object, `np_pcd_selection`, from the Open3D `PointCloud` object that only holds x, y, and z coordinates. Then, gather the indexes of the min and max values over the z dimension with the `argmax()` method and store the results in `lowest_point_index` and `highest_point_index`:

```
np_pcd_selection=np.array(pcd_selection.points)
lowest_point_index=np.argmin(np_pcd_selection[:,2])
highest_point_index=np.argmax(np_pcd_selection[:,2])
```

We can check the results by generating parametric models. For this, we select the points with the indexes, create spheres, translate them to the selected points' positions, and visualize:

```
# We create 3D spheres to add them to our visual scene
lp=o3d.geometry.TriangleMesh.create_sphere()
hp=o3d.geometry.TriangleMesh.create_sphere()

# We translate the 3D spheres to the correct position
lp.translate(np_pcd_selection[lowest_point_index])
hp.translate(np_pcd_selection[highest_point_index]))

# We compute some normals and give color to each 3D sphere
lp.compute_vertex_normals()
lp.paint_uniform_color([0.8,0.1,0.1])
hp.compute_vertex_normals()
hp.paint_uniform_color([0.1,0.1,0.8])

# We generate the scene
o3d.visualization.draw_geometries([pcd_selection,lp,hp])
```

We now have our 3D scene with the high and low points (see Figure 5-11).

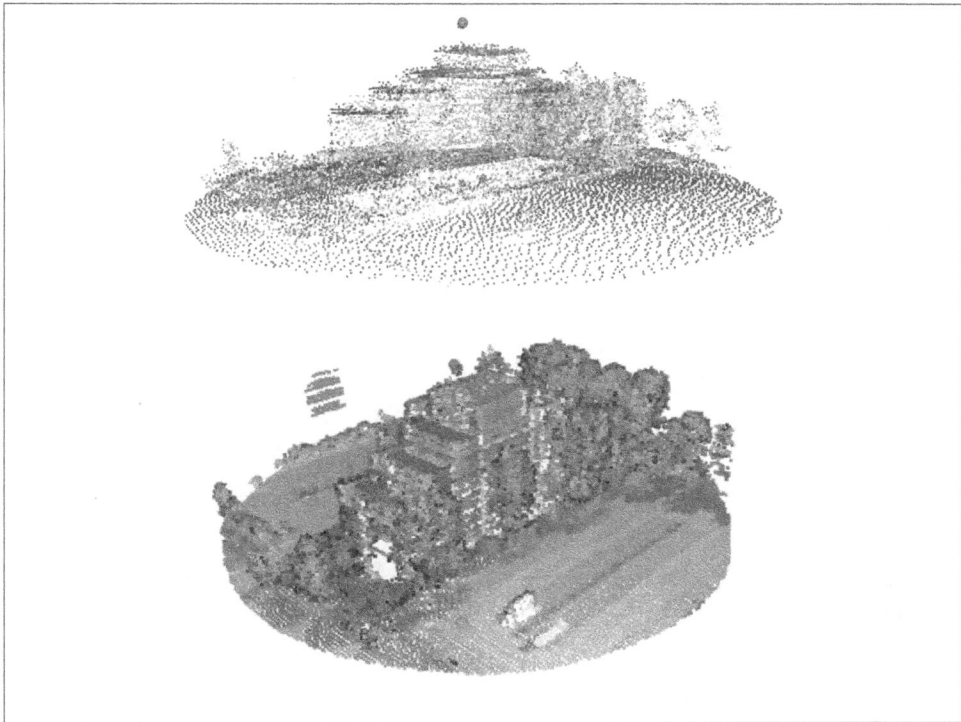

Figure 5-11. The 3D scene with high and low points detected (top: without noise filtering; bottom: with noise filtering)

As you can see, noise impacts our results: if we do not filter outliers, the results can be skewed. This hints at 3D data engineering (we'll cover this in greater detail in Chapter 6). To finish up, let's study the building coverage over 350 meters. For this, we re-execute the code with a POI = 150 m.

Point Cloud Voxelization

We want to extract the built coverage. For this, we take an intuitive approach by first transforming the point cloud modality to a filled analog of the 2D pixels: 3D voxels. This allows us to pursue the methodology outlined in Figure 5-12.

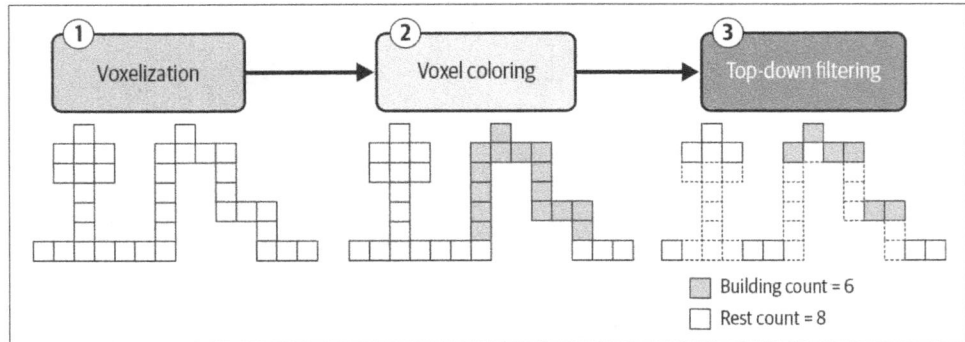

Figure 5-12. The point cloud voxelization process

We are extracting high points by constructing a 3D voxel data structure. First, we voxelized the point cloud, then colored each voxel from a binary perspective, and then filtered by top-down indexes.

Before going there, let me show you how to select a subset of points and voxelize it. To voxelize the ground, enter the following:

```
df_vox = pcd_subsampled[pcd_subsampled['Classification'] == 2]
color_vox = np.tile([0.8, 0.670, 0.521], (len(df_vox), 1))
pcd_vox=o3d.geometry.PointCloud(o3d.utility.Vector3dVector(np.array(df_vox[['X',
'Y','Z']])))
pcd_vox.colors=o3d.utility.Vector3dVector(color_vox)
pcd_vox.translate(-pcd_center)
o3d.visualization.draw_geometries([pcd_vox])
```

We essentially filter out the points classified as ground and create a point cloud that we then translate with a specific coloring. The next stage is to voxelize the geometry. This allows us to combine all the various modalities into one (Figure 5-13):

```
voxel_grid = o3d.geometry.VoxelGrid.create_from_point_cloud(pcd_selection_vox,
voxel_size=2)
o3d.visualization.draw_geometries([pcd_selection,lp,hp, mesh, voxel_grid])
```

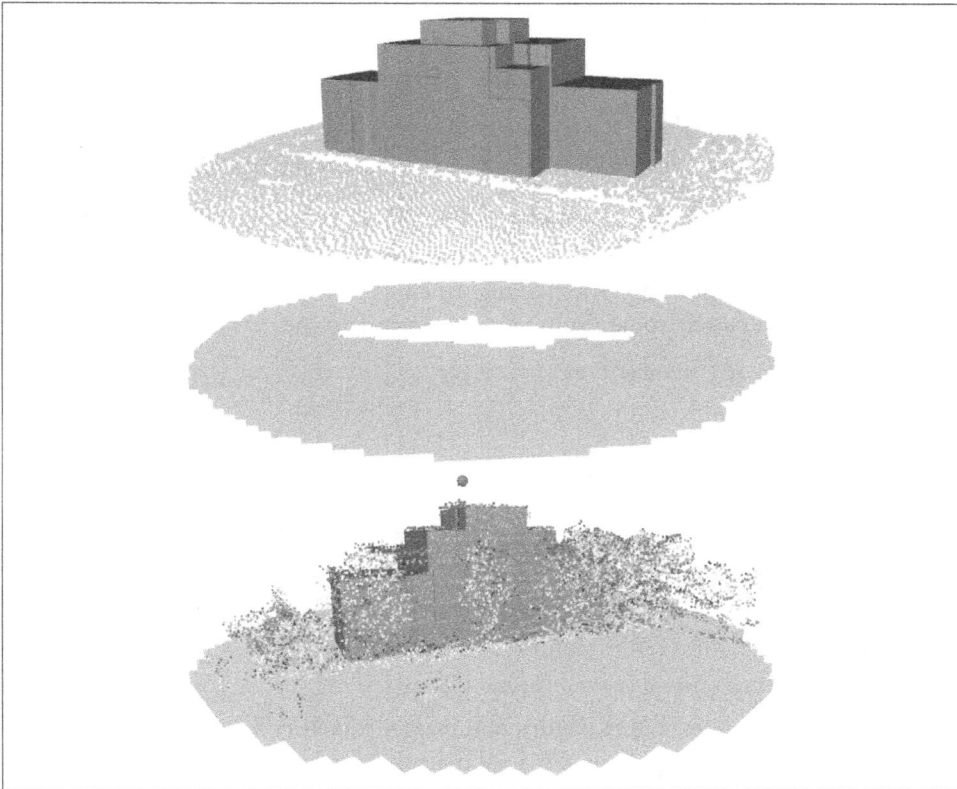

Figure 5-13. The voxelization process for the ground, from the mesh with point cloud to the voxel ground to the multimodal view

Using this particular coloring trick, let's (1) generate voxels where points exist, then (2) color them based on the classification of the points they hold. Finally, we filter the voxels to keep and count only the highest voxel per x and y coordinate. This way, we can avoid counting elements that bias the built coverage. The first step is to create a voxel grid. We fit a voxel grid to the point cloud, where each voxel is a cube of 20 cm:

```
voxel_grid = o3d.geometry.VoxelGrid.create_from_point_cloud(pcd_selection,
voxel_size=2)
```

With Open3D, we have to change the color of the voxel's points. Then, the `VoxelGrid` method will average the colors and retain the result as a color for the voxel. Now that we know how to generate 3D voxels from point clouds, we use their color scheme to bypass the coloring averaging problem.

For this, a simple method would be to handle the color in a binary way: either it is black ([0,0,0]) or not (all the rest). This means that we can initialize the colors variable to 0, and then select all the points that are classified as a building and give them another color, e.g., red ([1,0,0]):

```
colors=np.zeros((len(pcd_df), 3))
colors[pcd_df['Classification'] == 6] = [1, 0, 0]
pcd_o3d.colors = o3d.utility.Vector3dVector(colors)
```

Now, because we played with the original point cloud, we must redefine the POI and the selection to our choosing. For the sake of efficiency, you will find here the code block that you can run to update your voxel rendering to the new color scheme:

```
# Defining the POI and the center of study
dist_POI=150
POI=mesh.get_center()

# Querying all the points that fall within based on a k-d tree
pcd_tree = o3d.geometry.KDTreeFlann(pcd_o3d)
[k, idx, _] = pcd_tree.search_radius_vector_3d(POI, dist_POI)
pcd_selection=pcd_o3d.select_by_index(idx)

# Computing the voxel grid and visualizing the results
voxel_grid = o3d.geometry.VoxelGrid.create_from_point_cloud(pcd_selection,
voxel_size=2)
o3d.visualization.draw_geometries([voxel_grid])
```

Awesome! Now, we need to get the discrete integer indexes of each voxel in a list, as well as the colors and the bounds. This will permit us to loop over each voxel and its color later to check if a voxel is the highest in a "voxel column." To accomplish this, an efficient method is a list comprehension to vectorize our computation and avoid unnecessary loops:

```
idx_voxels=[v.grid_index for v in voxel_grid.get_voxels()]
color_voxels=[v.color for v in voxel_grid.get_voxels()]
bounds_voxels=[np.min(idx_voxels, axis=0),np.max(idx_voxels, axis=0)]
```

We have a grid of $150 \times 150 \times 16$ voxels on a selection of 150 meters, which amounts to 360,000 filled voxels. Next, we'll extract the built coverage.

Built Coverage Extraction

Now that we have a voxelized point cloud with binary colors, we can focus on the third stage from Figure 5-12. As we see, selecting only the top voxels is a bit more complex than it seems. Therefore, let's design a robust pipeline (Figure 5-14).

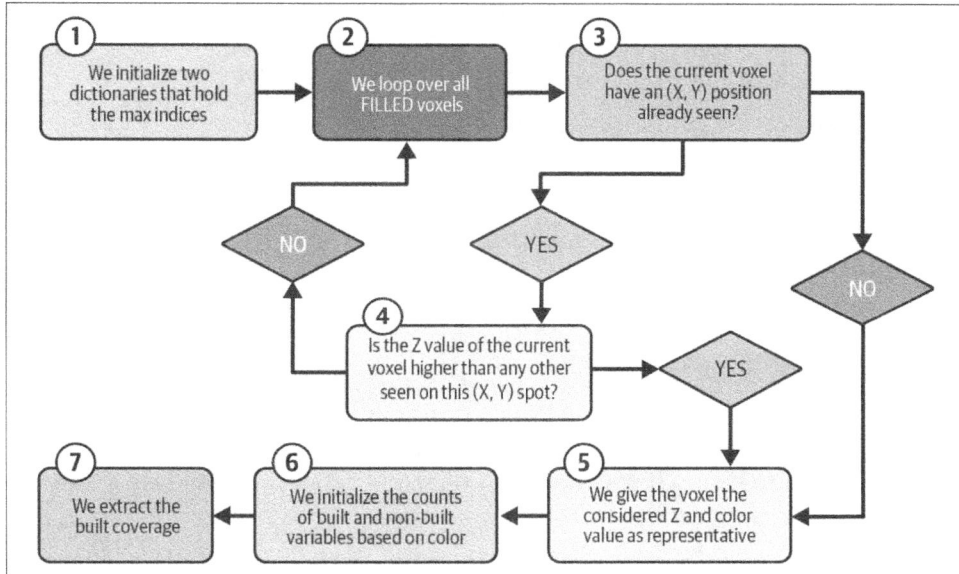

Figure 5-14. The voxel selection workflow to extract the built coverage in seven substeps

Let's walk through this step by step:

Step 1: First, we initialize two dictionaries that hold the max indices:

```
max_voxel={}
max_color={}
```

Steps 2 to 5: We loop over all filled voxels to check if they are the highest in a voxel column or should be dropped, which gives the following for loop:

```
for idx, v in enumerate(idx_voxels):
    if (v[0],v[1]) in max_voxel.keys():
        if v[2]>max_voxel[(v[0],v[1])]:
            max_voxel[(v[0],v[1])]=v[2]
            max_color[(v[0],v[1])]=color_voxels[idx]
    else:
        max_voxel[(v[0],v[1])]=v[2]
        max_color[(v[0],v[1])]=color_voxels[idx]
```

The first thing to notice is the `enumerate()` method in the loop definition. This permits looping over each value of the variable `idx_voxels` while keeping track of the index of the list. The second thing is that we are using the "tuples" data type as (x, y), which gives the integer position of our voxel. This ensures that we are continuously checking on the identical x and y grid positions. Finally, the "if" statement permits testing the expressed condition and will execute if the condition returns true. If it does not, it will execute the else statement (in the case there is one) or pass and exit the condition check.

Step 6: We initialize the counts of voxels tagged as built or nonbuilt, and we check if the color of the top voxel retained is black or not, in which case we update the voxel count of the respective category:

```
count_building_coverage,count_non_building=0,0
for col in list(max_color.values()):
    if np.all(col==0):
        count_non_building+=1
    else:
        count_building_coverage+=1
```

`np.all()` is also a Boolean check and will return true only if all values within are true. In our case, the black color is [0,0,0], which will thus return true because all R, G, and B are set to 0 in this case. If one of them is not zero, not all are 0, and `np.all()` will return false, triggering the else statement.

Step 7: We can extract the area covered for each type (built and nonbuilt) without forgetting to multiply the count variable by the actual 2D voxel area (4 m^2 in our case) and get the ratio:

```
print(f"Coverage of Buildings: {count_building_coverage*4} m²")
print(f"Coverage of the Rest: {count_non_building*4} m²")
print(f"Built Ratio: {(count_building_coverage*4)/(count_building_coverage*2+
count_non_building*2)} m²")
```

For 150 meters, this gives us coverage of 17.9%, which amounts to 12,812 m^2 of built area and 58,372 m^2 belonging to the rest. So, we have a nice balance of building occupation on the selected point of interest compared to the rest. Our 3D pipeline is fully functional, so we can save the results to one or more files to work on them outside Python. For this, we will use two valuable possibilities. For complex outputs, where you want to save any attributes, I recommend using NumPy with the following line of code:

```
np.savetxt(result_folder+pc_dataset.split(".")[0]+"_selection.xyz",
np.asarray(o3d_parcel_corners),delimiter=';', fmt='%1.9f')
```

If you only want to use spatial attributes with color and normals, Open3D is useful:

```
o3d.io.write_point_cloud(result_folder+pc_dataset.split(".")[0]+
"_result_filtered_o3d.ply", pcd_selection, write_ascii=False, compressed=False,
print_progress=False)
```

> split(".") allows us to split the pc_dataset string object into a list of two strings, before and after the ., and we keep only the first element with [0]. NumPy exports a variable o3d_parcel_corners with a delimiter in a *.xyz* ASCII file. Open3D will write a *.ply* file from the Open3D object pcd_selection.

The bulk of the Python automation structure is now up and running! We have imported the libraries, stored the datasets in different explicit variables, and ensured we can handle various 3D data representations in a single multimodal viewer experience, which allows interactivity.

Summary

You should be proud of successfully developing a robust and versatile 3D viewer using Python. By leveraging various libraries and techniques, we've created a powerful tool capable of combining diverse datasets, visualizing complex 3D models, and extracting valuable insights.

To relate to our scenario, this 3D viewer is key in the context of the impending flood in the Netherlands. By integrating elevation data, building footprints, and other relevant information, your program can:

- Visualize the affected area in 3D to comprehensively understand the terrain, building heights, and potential flood zones.
- Identify vulnerable areas where building heights and coverage can be analyzed to pinpoint locations most susceptible to flooding.
- Support decision making to aid in planning protective measures, such as evacuations and sandbagging efforts through the 3D viewer.

Moving forward, this 3D viewer can be further enhanced by incorporating additional features, such as flood simulations, integration with IoT sensors, and advanced analytics capabilities. But before continuing to develop this tool, let's explore the critical aspects of 3D data engineering needed to extract relevant insights from 3D datasets and move one step closer to creating digital environments for spatial AI systems.

Hands-on Project

This chapter's exercise will combine point cloud data with a mesh model. We will visualize both data types in Open3D, which is useful for tasks like flood analysis, where buildings need to be seen in context with the terrain:

1. *Data acquisition*: First, gather a point cloud dataset of the terrain and a mesh model of buildings in the same area. You can find sample datasets online or use data from your own projects.

2. *Library setup*: Ensure that Open3D and NumPy are installed in your Python environment. Then, import these libraries into your script.

3. *Load and prepare datasets*: Load the point cloud and the mesh using Open3D's `read_point_cloud` and `read_triangle_mesh` functions. You can add color to the point cloud using its `colors` attribute.

4. *Create visualization*: Combine the point cloud and the mesh in a list. Use the `draw_geometries` function to visualize both in a standalone Open3D window.

5. *Add interactivity*: Use the capacity of Open3D to interact with the viewer, to allow users to make measurements, extract coordinates, and preview results of computation based on a custom coloring of your 3D point cloud.

Make sure the user can explore the scene by rotating, panning, and zooming. Experiment with different color schemes to highlight specific features of the data.

CHAPTER 6

Point Cloud Data Engineering

It is time to reveal the intricacies of 3D—specifically point cloud—data engineering. While point clouds offer a rich representation of 3D geometry, extracting meaningful information requires specialized techniques to cope with noise and redundancy and to help discern relevant patterns. We need a way to transform raw point cloud data into a refined form, extracting essential features that enhance the performance of downstream tasks such as 3D shape detection (Chapter 9), classification (Chapter 15), and object detection (Chapter 16). The base unit is our data. We want to refine it to make more sense for our end application.

Let's consider a short example. Imagine that you want to analyze an MRI scan of the brain (a selection of 2D slices of the brain) to identify if specific zones are subject to the potential threat of a tumor. We could pass the set of images that constitute our dataset directly to a skilled neurosurgeon. It is up to his expertise to interpret the first image, repeat this prediction on the various slices, and then push out his diagnosis and analysis. This puts a high level of liability on the quality of the data. If you have a lot of noise (for example, if the patient is moving while capturing data) or you have an uncalibrated range of colors, the quality of the interpretation can suffer drastically.

But suppose we inject a bit of data engineering. In that case, we may be able to work to eliminate the noise and highlight specific edges and parts of the data without going into the interpretation step by just playing with the initial channels we are given. We aim to do this with our data engineering step: we want to ensure we put our downstream processes in the best possible conditions. However, without any context, and especially if we do not clarify the 3D geometry to which we apply data engineering, it may be hard to grasp the full extent of what data engineering can help us achieve.

In this chapter, we target 3D point cloud data engineering. We previously saw that point clouds have magnificent canonical properties that make them a good candidate for centralized processing and understanding. We can then push out the middleman result to other data modalities within our workflow. This is why starting with 3D point clouds will make it easy to understand and build a 3D data engineering pipeline to propel our further downward processes with higher robustness.

I will guide you through point cloud data engineering fundamentals, focusing on the crucial feature extraction step. You will learn how to derive meaningful, nonredundant features that capture the essential geometric properties of point clouds. This also permits the highlighting of essential 3D preprocessing stages, including principal component analysis (PCA) and 3D data registration, as illustrated in Figure 6-1.

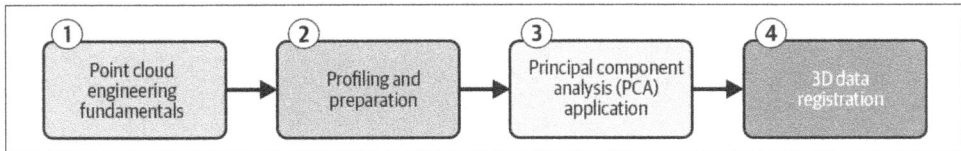

Figure 6-1. The four core steps of the point cloud data engineering workflow

> You can find all the materials for this chapter on the 3D Data Science Resource Hub (*https://learngeodata.eu/3d-data-science-with-python*). You can use the base link and add "*/chapter-X*" to access the files (code, datasets, articles) and resources. Note that you may need to share an email address, a personal password, or proof that you are the book's owner.

Fundamentals

The foundation of most 3D projects involves the raw data itself: point clouds (*https://oreil.ly/TwTD_*). This section dives into point cloud data engineering fundamentals, equipping you with the knowledge to tackle these intricate datasets (see Figure 6-2). We explore how to store, preprocess, and manage point clouds efficiently.

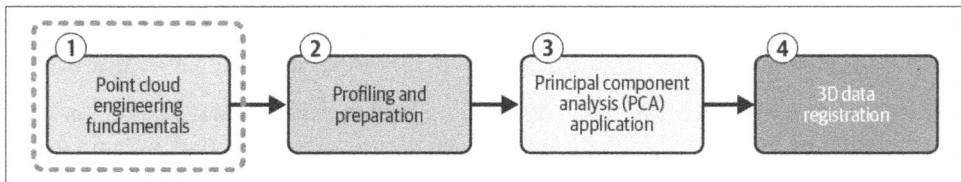

Figure 6-2. Fundamentals of point cloud data engineering

We use a simple Python setup, as shown in Chapters 3, 4, and 5. Then, we build a solution on top of an Anaconda environment (*https://oreil.ly/dkSiM*) with Python 3.10 (*https://oreil.ly/4713_*), leveraging Open3D (*https://www.open3d.org*), NumPy (*https://numpy.org*), pandas (*https://oreil.ly/G5pBG*), and Matplotlib (*https://matplotlib.org*).

Initial Preprocessing

Having previously examined this topic, we can immediately demonstrate the outcomes of specific preprocessing procedures commonly employed in a 3D point cloud. For this instance, we utilize a dataset acquired through a terrestrial laser scanner (see Figure 6-3).

Figure 6-3. The 3D dataset that we use in this chapter is composed of three main areas (indoor 1, indoor 2, and outdoor)

After acquiring a thorough understanding of the data on your local system, you can easily import the dataset into your Python execution runtime with Open3D (*https://www.open3d.org*) using just two lines of code:

```
import open3d as o3d
pcd = o3d.io.read_point_cloud("../DATA/techshop.ply")
```

The `pcd` variable now contains the point cloud data. Subsequently, diverse preprocessing methodologies must be employed to improve quality and extract significant data: transformation, downsampling, denoising, outlier removal, normalization, and feature extraction (Figure 6-4).

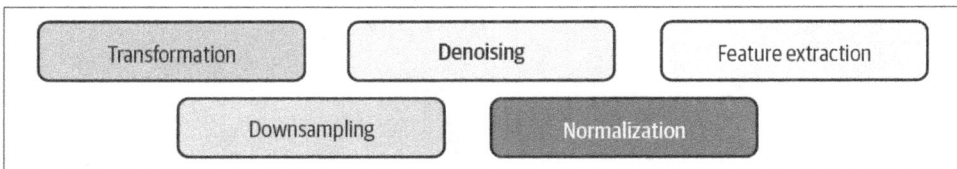

Figure 6-4. The various processing tasks useful for all applications

Let's cover each step with Python in detail.

Transformation: Initial shift

Transformation is recommended if you plan to visualize the point cloud using Open3D or if you want to prevent truncation or loss of precision during computations. Under such circumstances, it is advisable to relocate your point cloud to circumvent the approximation of huge coordinates. To apply a shift to your `pcd` point cloud, begin by obtaining the center of the point cloud. Next, a translation is performed by subtracting the center from the original variable (you can visualize this after with `o3d.visualization.draw_geometries([pcd])`):

```
pcd_center = pcd.get_center()
pcd.translate(-pcd_center)
```

Excellent. We have completed the necessary preparations to evaluate various sampling techniques to streamline our downstream procedures. Downsampling aims to decrease the number of samples or data points in a dataset.

Point cloud subsampling

Let's examine sampling techniques that can efficiently decrease the size of a point cloud while maintaining its overall structural integrity and representativeness. A point cloud can be defined as a matrix with dimensions $m \times n$. At the matrix level, decimation is performed by retaining points from every nth row based on the value of the n factor (see Figure 6-5). Naturally, this is derived from how the points are recorded in the file. The function `random_down_sample()` from +Open3D is used to randomly downsample the original point cloud `pcd` with a specified `retained_ratio`:

```
retained_ratio = 0.2
sampled_pcd = pcd.random_down_sample(retained_ratio)
```

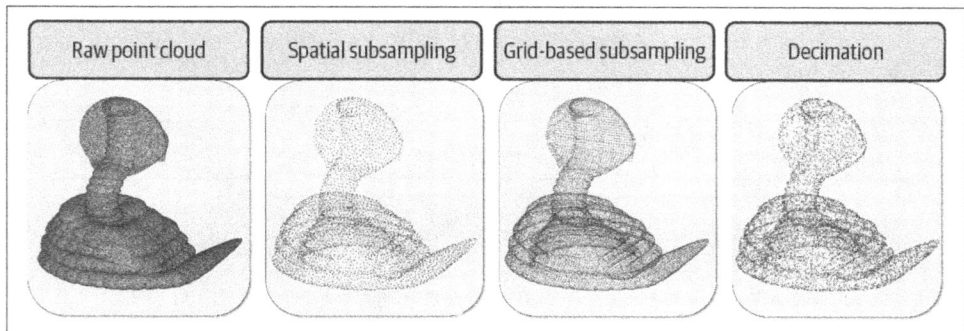

Figure 6-5. Point cloud downsampling strategies: spatial, grid, and decimation

The resulting downsampled point cloud is stored in the variable `sampled_pcd`. The `retained_ratio` argument specifies the percentage of points that will be kept after downsampling. For instance, when `retained_ratio` is set to 0.5, about 50% of the points will be randomly chosen and kept in the sampling point cloud.

Random sampling in the study of 3D point clouds is subject to restrictions that may lead to omitting crucial information and imprecise analysis. It fails to consider the spatial aspect or the connections between the points. As a result, it is interesting to employ alternative techniques to guarantee a more thorough examination, such as spatial or grid-based subsampling, as illustrated in Figure 6-5. We'll explore this further in Chapter 18, which covers hybrid methods. Next, let's deal with any outliers using statistical techniques to remove them. This guarantees the quality and dependability of the data for further analysis and processing.

Let's look at the grid subsampling approach, which involves partitioning the three-dimensional space into uniform voxel cells. Each cell in the grid is represented by a single point, and there are various ways to choose this representative point. During the subsampling process, we can select the closest point of each cell to its barycenter. Concretely, we define a `voxel_size` that we then use to filter our point cloud:

```
voxel_size = 0.05
pcd_downsampled = filtered_pcd.voxel_down_sample(voxel_size = voxel_size)
```

This function executes voxel downsampling on the filtered point cloud, `filtered_pcd`. The parameter `voxel_size` determines the dimensions of each voxel used to reduce the resolution of the point cloud. Increasing the size of the voxels leads to a greater decrease in the density of the point cloud. Examine the distribution after downsampling with `o3d.visualization.draw_geometries([pcd_downsampled])`. We have a set of outlier points that have yet to be modified for further processing, so let's address the issue.

Denoising (outlier removal)

Noise in 3D point clouds refers to unwanted or erroneous data points arising from various sources, such as sensor inaccuracies, environmental factors, or object movement. Removing noise is crucial as it improves the quality of 3D data. Applying an outlier filter to 3D point cloud data can effectively detect and exclude any data point that deviates considerably from the overall dataset. These outliers may arise due to measurement errors or other variables that can distort the findings.

Eliminating these extreme values may allow us to obtain a more accurate depiction of the data and make more precise adjustments to algorithms. Nevertheless, we must exercise caution to avoid erasing significant aspects. Let's now define a statistical outlier removal filter to remove all the points that are far off their neighbors by taking the average distance between points in the entire point cloud as a base.

To achieve this, we take two input parameters. The first, `nb_neighbors`, specifies how many neighbors are considered to compute the average distance for any given point. The second, `std_ratio`, permits us to set our threshold level. It is based on the standard deviation of the average distances across the entire point cloud. It is good to note that the lower this number is, the more aggressive the filter will be. This amounts to the following:

```
nn = 16
std_multiplier = 10
filtered_pcd, filtered_idx = pcd.remove_statistical_outlier(nn, std_multiplier)
```

We utilize a statistical outlier elimination on the point cloud by employing a defined number of nearest neighbors (nn) and a multiplier for the standard deviation (std_multiplier). The function yields two outputs: filtered_pcd and filtered_idx. The first holds the filtered point cloud, while filtered_idx contains the indices corresponding to the points in the original point cloud that were kept after removing outliers. To represent the outcomes of this filtering process, we assign outliers a red color and include them in the list of point cloud objects we intend to depict:

```
outliers = pcd.select_by_index(filtered_idx, invert=True)
outliers.paint_uniform_color([1, 0, 0])
o3d.visualization.draw_geometries([filtered_pcd, outliers])
```

The data points significantly deviating from the norm are identified and displayed in darker shades within the point cloud (see Figure 6-6). Once the statistical outliers have been eliminated from the point cloud, the last stage involves a normalization strategy.

Figure 6-6. The data points that deviate significantly from the norm are identified and displayed in darker shades within the point cloud.

3D data normalization

Let's discuss 3D data normalization. It's a crucial preprocessing step, especially when dealing with diverse 3D datasets originating from different sources or acquisition techniques. For example, a point cloud from a LiDAR scan of a building will have drastically different scales and orientations compared to a 3D mesh of a human hand generated from a structured light scanner. Without normalization, these discrepancies can severely impact the performance of downstream tasks like registration, segmentation, and especially deep learning models.

I've seen firsthand how neglecting normalization can lead to AI models that overfit specific dataset characteristics, resulting in poor generalization to unseen data. Imagine training a model to recognize different types of furniture from 3D scans. If the training set consists only of large objects like sofas and tables, the model might struggle to classify a small chair if it hasn't been exposed to that scale during training. So, what can we do about it?

Several normalization techniques are particularly relevant to 3D data. Let's break down a few key ones:

Centering

This involves translating the 3D object so that its centroid lies at the origin. We calculate the average coordinates across all points in the point cloud or vertices in the mesh and subtract this average from each point's coordinates. This ensures that the object is centered in the coordinate system.

Scaling

We essentially resize the object. A common approach is to scale the object to fit within a unit-bounding box or sphere. This can be achieved by dividing each coordinate by the maximum absolute coordinate value or by the object's diameter. This step helps to standardize the size of different objects.

Alignment

This rotates the object to align it with a specific axis or plane. PCA (*https://oreil.ly/8A6vp*) is often employed to determine the principal axes of the 3D object, allowing us to align the object along these axes. This is beneficial for aligning objects with different orientations.

Here's a simplified Python example demonstrating centering and scaling using NumPy:

```
def normalize_3d_data(data):

    centroid = np.mean(data, axis=0)
    data_centered = data - centroid

    max_abs_coord = np.max(np.abs(data_centered))
    data_normalized = data_centered / max_abs_coord

    return data_normalized
```

Remember, the choice of normalization technique should align with your specific application. For instance, alignment and scaling are crucial in shape retrieval tasks, whereas centering might suffice for object detection. Now, let's explore feature extraction in the context of preprocessing. For this, we will highlight the role of normals.

Feature extraction preview: Normal estimation

A point cloud normally denotes the orientation of a surface at a particular point within a 3D point cloud. One possible application of this technique is segmentation, which is done by splitting the point cloud into sections that have similar normals (Figure 6-7). In our scenario, normals will aid in the identification of objects and surfaces inside the point cloud, hence facilitating visualization.

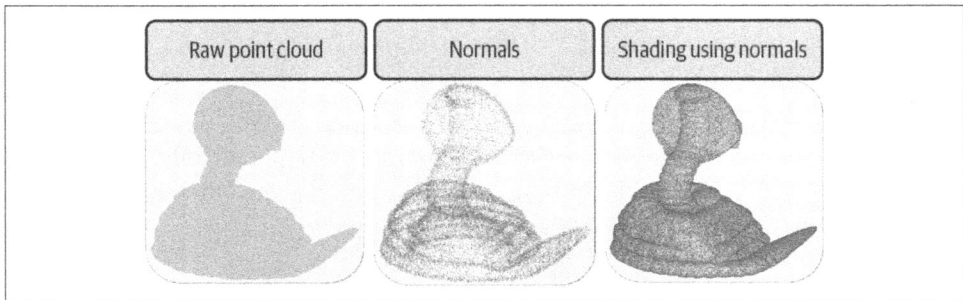

Figure 6-7. Normals computed on the raw point cloud and used for shading

Furthermore, this is a remarkable occasion to provide a method for computing these normals in a partially automated manner. Initially, we establish the mean distance between each point in the point cloud and its neighboring points:

```
nn_distance = 0.05
```

Subsequently, we utilize this data to obtain a restricted number of maximum nearest neighbor points within a specific vicinity:

```
# Compute a normal for each point in the 3D point cloud:
radius_normals=nn_distance*4
pcd_downsampled.estimate_normals(search_param=o3d.geometry
.KDTreeSearchParamHybrid(radius=radius_normals, max_nn=16),
fast_normal_computation=True)
```

The point cloud object, `pcd_downsampled`, now possesses normals and is prepared to showcase its most aesthetically pleasing aspect. You are familiar with the routine by now:

```
pcd_downsampled.paint_uniform_color([0.6, 0.6, 0.6])
o3d.visualization.draw_geometries([pcd_downsampled,outliers])
```

As you can see, already at this stage, we can grasp how features can help further downward processes. Let's refine these feature extraction fundamentals.

Feature Extraction Fundamentals

The objective here is clear: we aim to extract meaningful information, known as features, from the data within point clouds. These features must adhere to three key criteria. First, they must be informative. Extracting data that doesn't provide additional insight is pointless. Secondly, they should not be redundant. It is essential to focus on data that offers significant, unique information. Lastly, they must facilitate subsequent processes, particularly in classification tasks. The features should be practical and enhance the performance of classification models.

The process of feature extraction from point cloud data requires a strategic approach. Direct extraction of these features isn't feasible. Instead, we leverage the data's inherent geometry, which is rich in information. However, identifying a specific neighborhood around each point is crucial to describing each point effectively.

Strategies for Point Cloud Feature Extraction

Unlocking the hidden insights within a point cloud requires extracting its key features. This is done by profiling and preparing our data, which is stage 2 in Figure 6-1. Think of it like identifying the building blocks that define the shape and properties of a 3D object. We have two main methods that analyze local geometric properties, capture the overall shape of the point cloud, and even identify surface textures: global feature extraction and local feature extraction (see Figure 6-8). Let's first cover global feature extraction.

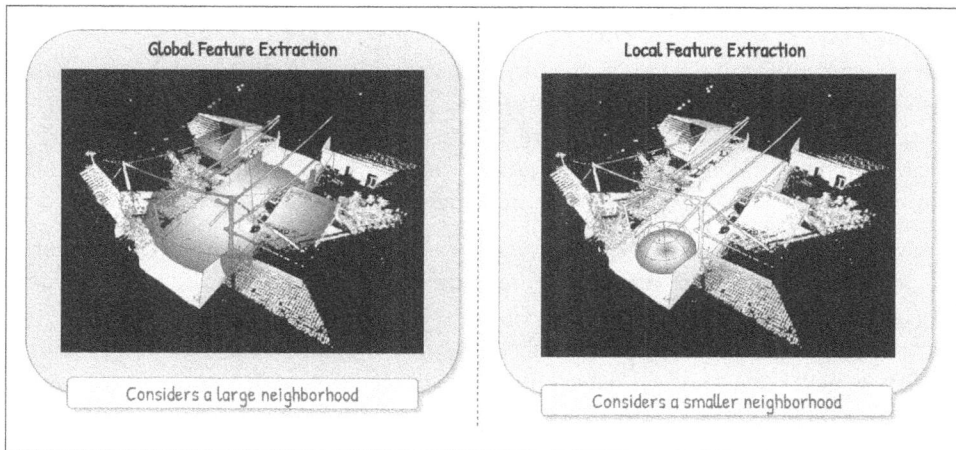

Figure 6-8. Methods for feature extraction in 3D point clouds

Global Feature Extraction

Global feature extraction is a pivotal method in point cloud processing. It focuses on delineating a given point cloud's overarching shape and structure. It usually entails analyzing the entire point cloud instead of isolating individual points or localized areas (see Figure 6-9).

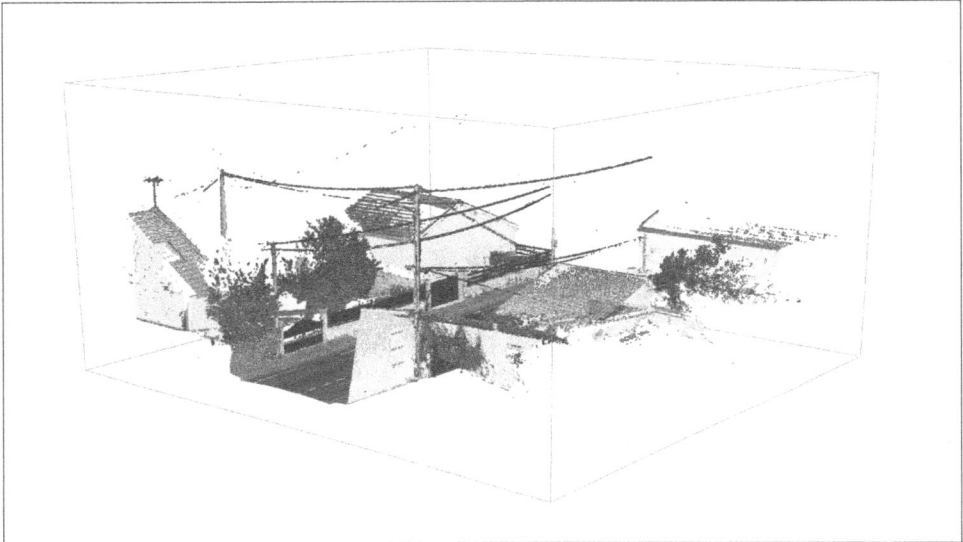

Figure 6-9. Global feature extraction in point clouds

The primary objective of global feature extraction is to distill crucial information from the point cloud, which proves invaluable for many uses, including object recognition, scene segmentation, and the registration process. Among the extracted features are statistical metrics—such as the mean, variance, and skewness—and geometric characteristics, like curvature and various shape descriptors.

A notable benefit of employing global feature extraction lies in its ability to condense the point cloud into a streamlined representation that encapsulates its fundamental traits. This condensed representation significantly simplifies the storage and processing of voluminous point clouds, which may comprise millions or billions of points. Moreover, global features facilitate comparing and classifying distinct point clouds by examining their geometric and structural attributes (see Chapter 8). In essence, global feature extraction stands as a crucial strategy within point cloud processing, enabling the extraction of valuable insights from expansive and intricate datasets.

Local Feature Extraction

To effectively extract per-point features, it is critical to employ either a localized or a more encompassing global approach to describing the surrounding area of a point. This neighborhood context is critical, as different strategies for defining it can lead to varied feature values. The significance of this cannot be overstated; for instance, a narrowly defined neighborhood, focusing on a point and its immediate neighbor, severely restricts the scope of information obtainable (see Figure 6-10).

Figure 6-10. Local feature extraction in point clouds

Conversely, a broader neighborhood might miss out on finer details, such as the presence of an edge or the surface's planar or spherical nature, which are vital for understanding the geometry at play. Features derived from these observations necessitate an appropriately sized neighborhood (see Figure 6-11).

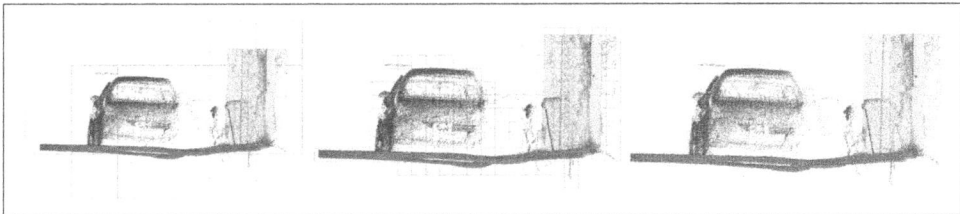

Figure 6-11. Balancing neighborhood size in local feature extraction

One standard method involves the *k*-nearest neighbors (*k*-NN) approach (*https:// oreil.ly/6mCpZ*). The strategy might include gathering five surrounding points as the neighborhood for a focus point. This is the essence of the *k*-NN technique, with "*k*" set to 5 in this scenario.

Alternatively, a strategy might revolve around a defined radius, establishing a spherical or cylindrical neighborhood. For instance, identifying all points within a one-meter radius of a focal point might only include three points, offering a different perspective on the neighborhood based on the scale parameter of the radius.

Beyond these, there are more neighborhood-defining techniques, such as using a grid to include all points within a voxel or adopting a multiscale approach. The latter varies the scale parameter across different levels to extract a range of features, underscoring the critical nature of neighborhood selection before proceeding to feature extraction.

Once a connected 3D neighborhood is established around a point of interest, the next step is extracting relevant features for the specific application. Here, we primarily focus on geometric features derived from methods like covariance analysis. However, depending on the application's requirements, it is important to note that features can extend beyond geometry to include aspects like color (RGB), radiometry, semantics, and even topology (see Chapter 9 for a deeper explanation of these).

In the feature extraction phase, with our selected neighborhood—whether defined by radius or another method—we aim to derive meaningful features from the point and its XYZ coordinates. A point of emphasis is the significant role of specific extraction methods, particularly those highly effective for various applications, including indoor mapping and aerial classification. While these methods are precious, they come with limitations, and they remain efficient tools in our processing arsenal. It is time to explore one of the most fundamental feature extraction methods: PCA (*https:// oreil.ly/rMEUP*).

Principal Component Analysis

We discussed principal component analysis (PCA) earlier, but now we'll explore it further. PCA is an instrumental technique in handling three-dimensional point clouds and is step 3 of our point cloud data engineering workflow (shown in Figure 6-1).

However, here we consider a two-dimensional point cloud involving the x- and y-axes for simplicity. This method is centered on reorienting the data to align with axes that reflect its most significant variance. PCA serves primarily in feature extraction, significantly aiding in reducing the dimensionality of data from potentially numerous dimensions down to one, two, or three principal components. These components are remarkably revealing, often encapsulating most of the data's variance. In this context, variance refers to the spread of the data points around the mean. For instance, considering the x-axis, we can observe the data's spread from a minimum to a maximum value, with variance measuring the dispersion around the dataset's mean.

PCA aims to align the first principal component with the maximum variance direction, followed by identifying subsequent orthogonal axes that capture the remaining variance in order of significance (see Figure 6-12). In a three-dimensional scenario, this would result in three principal axes. The process intuitively involves identifying a line (or, in higher dimensions, a plane or hyperplane) along which the data's variance is maximized. This line represents the first principal component, with subsequent orthogonal components capturing variance in other directions.

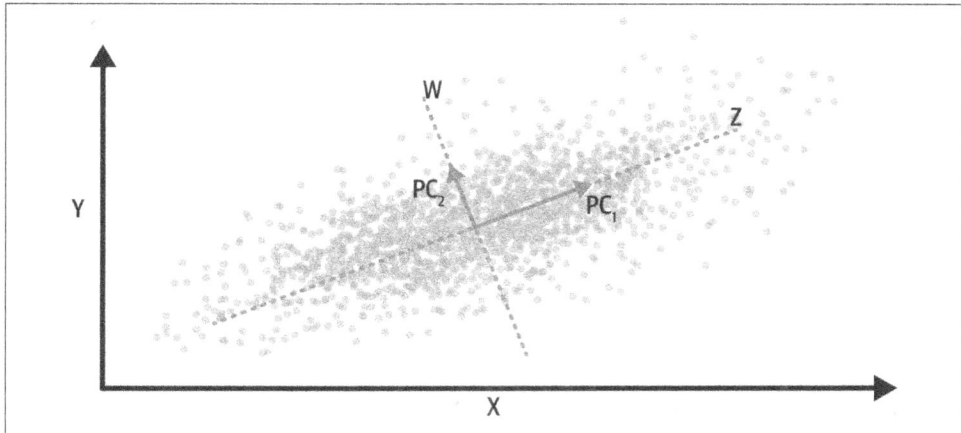

Figure 6-12. The principal component analysis process for 3D data

Crucially, each principal component is associated with an eigenvector, indicating the direction of maximum variance and an eigenvalue, quantifying the magnitude of this variance. Through PCA, these eigenvalues can be ingeniously leveraged to derive more nuanced features like omnivariance, planarity, and others, enriching the analysis by quantitatively describing the data's geometric properties.

Let's study the practical applications of PCA to extract meaningful features and visually demonstrate the process efficacy in revealing the inherent structure of point clouds. Let's begin anew, focusing on a 3D point cloud treated as a singular entity. Imagine working with a point cloud comprising various segments (extracted from a process shown in Chapter 12); for this exercise, we concentrate on just one of these segments, treating it as a standalone entity (see Figure 6-13). Our objective is to find the three principal directions within the point distribution of this segment. We aim to identify the direction with the maximum variance within our point sample, followed by the second and third highest variances, ensuring each direction is orthogonal. Our PCA approach is crucial for setting the stage for subsequent processing steps.

So, are we ready to dive in? The first significant step we tackle is data preparation.

Figure 6-13. Example of the detection of the various objects in a 3D scene

Python and Data Preparation

Let's use three essential libraries: NumPy, pandas, and Matplotlib. Open3D will be excluded from this setup to work with first principles logic (low abstraction level). The inclusion of the time library is strictly for performance analysis purposes:

```
# Base libraries
import numpy as np
import pandas as pd
#Viz' libraries
import matplotlib.pyplot as plt
# To monitor timings
import time
```

To improve the resolution of your figures, adjust a specific setting by running:

```
plt.rcParams['figure.dpi'] = 600
```

Setting the figure DPI to 600 ensures high-resolution visuals, facilitating detailed visualization. Now, let's load the point cloud data with pandas.

pandas was one of the various ways we covered for handling point clouds in Chapter 4. Let's use this approach to load the point cloud data from an ASCII format file (e.g., *.xyz*) generated by a process like DBSCAN (*https://oreil.ly/ZFafO*) (Chapter 12). The following sample code shows how to use the pd.read_csv method to load such a file and assign column names for simple data manipulation:

```
# Load a point cloud .xyz from the previous step with pandas
data_folder="../DATA/"
dataset="velodyne_pca.XYZ"
pcd = pd.read_csv(data_folder+dataset, delimiter=";", names=['X', 'Y', 'Z',
'LABEL'], header=None)
pcd['LABEL'] = pcd['LABEL'].astype(int)
```

This method simplifies loading a point cloud file and preparing the data for analysis by assigning meaningful column names and ensuring the data types are appropriate for subsequent operations. When handling point cloud data with pandas, directly specifying the data structure by naming each column proves helpful, especially for files without headers. This approach allows for intuitive and straightforward data manipulation and analysis. Now, let's prepare our cluster for analytics.

Cluster Identification with pandas

Moving forward, our focus shifts to preparing clusters for in-depth analysis. This begins with examining the labels assigned to each point within our dataset. The goal is to aggregate points sharing identical labels into distinct clusters, a step that enhances the coherence and structure of our analysis. To accomplish this task, we use the groupby function available in the pandas library. This feature allows us to partition the dataset into clusters based on the labels:

```
segments=pcd.groupby(['LABEL'])
```

The groupby functionality in pandas occasionally presents a quirky syntax, yet its utility in organizing data is clear. By specifying the column name (in this case, 'LABEL') as the criterion for grouping, we initiate the creation of clusters. This yields a groupby object, a structure from which specific clusters can be extracted as needed. For instance, to explore the contents of a particular cluster—say, the second or third one—a simple method chain involving .get_group() is enough. Moreover, suppose one's interest lies in the indices of the points within a cluster. In that case, pandas offer a direct pathway to this information through the groups attribute, followed by the cluster label of interest.

We can apply specific operations to neatly organized clusters. Initially, we select a single cluster to ensure the efficacy of our intended operation, typically involving data transformation or analysis techniques such as PCA. In our example, we opt for the third cluster for preliminary testing. A crucial aspect of this operation is selecting relevant data dimensions, the x, y, and z coordinates, for the analysis. This selection process is facilitated by specifying the desired columns within brackets:

```
cluster=segments.get_group(3)[['X',' Y',' Z']]
```

This systematic approach ensures that each cluster is prepared with precision, allowing for applying complex analytical techniques to a well-defined subset of the dataset. The result is a streamlined workflow that enhances the clarity and effectiveness of data analysis within the broader project framework.

3D Data Normalization

Let's begin by normalizing our dataset, as seen in the previous section. This involves adjusting our data to center around a common origin, a fundamental step for PCA's effectiveness in identifying patterns. The initial stage in normalization consists in calculating the mean of our dataset:

```
#Compute the mean and center it
m = np.mean(cluster, axis=0)
```

This is straightforward with NumPy's `np.mean` function. We apply this function to our previously determined cluster, specifying `axis=0`. This computes the mean column-wise across the dataset. Following this, we normalize our cluster by subtracting the calculated mean from each point:

```
cluster_norm = cluster-m
```

This step effectively centers our dataset around the origin, a necessary condition for PCA.

Extracting the Principal Components

With our data normalized, we proceed to the covariance matrix. This matrix is pivotal in PCA, serving as the foundation from which we derive the eigenvalues and eigenvectors (see Figure 6-14). We are starting at step 3: extracting the covariance matrix. Then we will move onto steps 4 and 5.

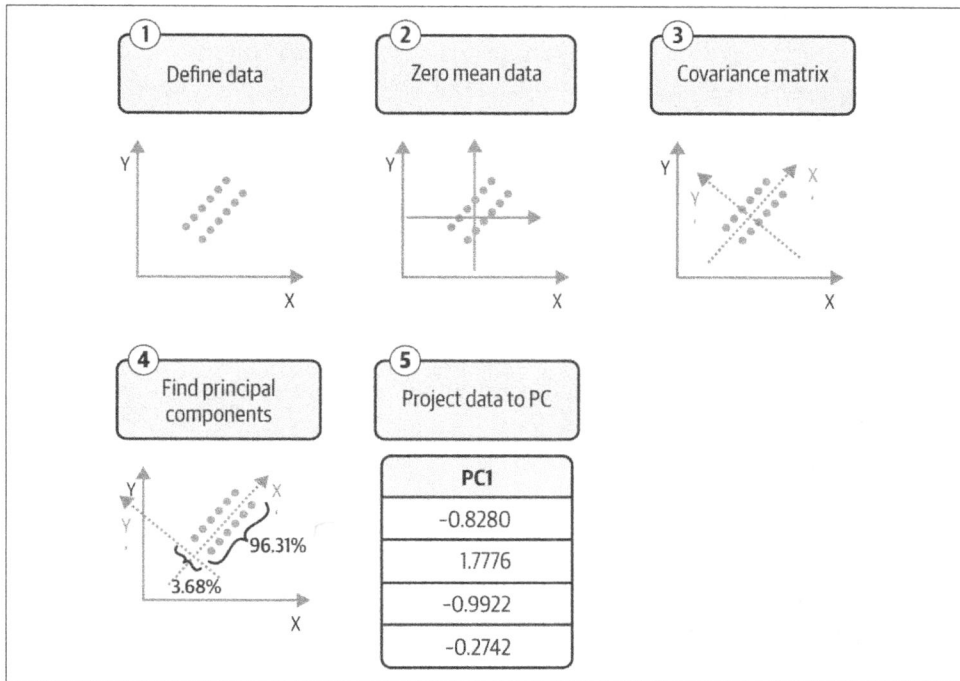

Figure 6-14. The five main steps of our PCA implementation.

NumPy's `np.cov` function facilitates the construction of the covariance matrix (*https://oreil.ly/X-O99*) (see the 3D Resource Hub (*https://learngeodata.eu/3d-data-science-with-python*) for more information), which is applied to the transposed normalized cluster. This transposition aligns the data correctly for the covariance calculation. Subsequently, we extract the eigenvalues and eigenvectors, the core components in PCA. This extraction is achieved using the linear algebra module in NumPy, `np.linalg.eig`, applied to the covariance matrix. These eigenvalues and their corresponding eigenvectors unveil the variances across our dataset's dimensions and the directions of these variances, respectively:

```
cov = np.cov(cluster_norm.T)
eig_val, eig_vec = np.linalg.eig(cov.T)
```

Upon execution, we observe the eigenvalues: an array showcasing variances such as 5.5, 0.02, and 0.77, among others. This gives us a snapshot of the variance captured by each principal component. The eigenvectors, on the other hand, provide the directions of these variances.

Let me explain the significance and magnitude of these eigenvalues. The eigenvalues rank the importance of each corresponding eigenvector based on the variance in our data. We can then sort these eigenvalues (and their corresponding eigenvectors) to prioritize our analysis. The goal is to arrange them from the most significant to the least, essentially from the highest variance to the lowest. We use the `np.argsort` function to achieve this, which ingeniously returns the indices that would sort an array. However, there's a twist—we seek to sort the eigenvalues in descending order, ensuring the principal component with the highest variance comes first:

```
sorted_indexes = np.argsort(eig_val)[::-1]
eig_val = eig_val[sorted_indexes]
eig_vec = eig_vec[:,sorted_indexes]
```

By ordering our eigenvalues and their corresponding eigenvectors, we can now define ways to visualize our PCA.

3D Visualization of PCA

When we sort the eigenvalues in descending order, the corresponding eigenvectors also align, forming principal components. This simplification leads us to the first and second principal axes. Let's define a function that allows us to plot the data points, the PCA axes for each cluster in our dataset. The `DrawPCA` function takes points and a cluster ID, computes the PCA, and visualizes it by plotting the first three principal components as vectors in a 3D space, highlighting the cluster's spatial variance:

```
def DrawPCA(points,id_cluster):
    m = np.mean(points, axis=0)
    cluster_norm = points-m

    cov = np.cov(cluster_norm.T)
    eig_val, eig_vec = np.linalg.eig(cov.T)

    sorted_indexes = np.argsort(eig_val)[::-1]
    eig_val = eig_val[sorted_indexes]
    eig_vec = eig_vec[:,sorted_indexes]

    u = eig_vec[0]
    v = eig_vec[1]
    w = eig_vec[2]

    fig= plt.figure()
    ax = plt.axes(projection='3d')
    ax.set_xlim([min(cluster['X'])-1,max(cluster['X']+1)])
    ax.set_ylim([min(cluster['Y'])-1,max(cluster['Y']+1)])
    ax.set_zlim([min(cluster['Z'])-1,max(cluster['Z']+1)])
    ax.set_xticklabels([])
    ax.set_yticklabels([])
    ax.set_zticklabels([])
    # ax.set_xlabel('X')
```

```
# ax.set_ylabel('Y')
# ax.set_zlabel('Z')
ax.set_title('PCA of Cluster '+str(id_cluster),fontsize = 10)


ax.scatter(cluster['X'], cluster['Y'], cluster['Z'], color='steelblue',
alpha=0.2)
ax.quiver(m[0],m[1],m[2],u[0],u[1],u[2], color ='dark turquoise')
ax.quiver(m[0],m[1],m[2],v[0],v[1],v[2], color ='royal blue')
ax.quiver(m[0],m[1],m[2],w[0],w[1],w[2], color ='salmon')
return
```

You can then loop over clusters. Here, we specify our image folder, which has been prepared in advance. Then, we initiate the loop for, let's say, 10 images:

```
image_folder="../RESULTS/"
for i in range(10):
    cluster=segments.get_group(i)[['X','Y','Z']]
    DrawPCA(cluster,i)
    plt.savefig(image_folder+'PCA of Cluster '+str(i)+'.png', dpi=600)
```

This results in Figure 6-15.

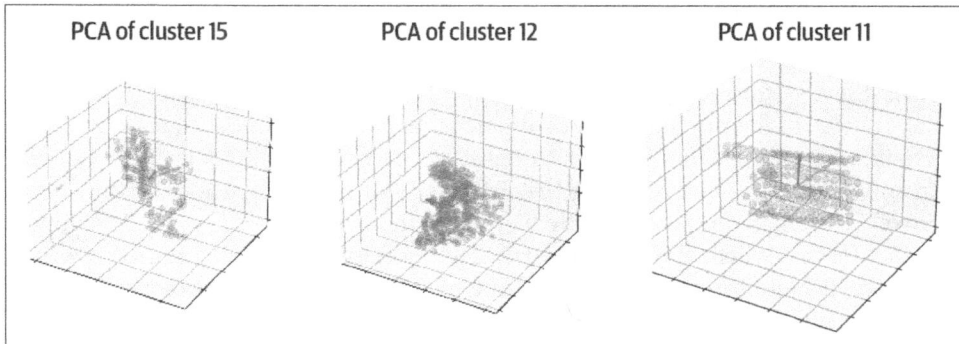

Figure 6-15. Visualizing PCA for 3D point cloud clusters

The three principal axes of the PCA for each cluster are shown. Now, let's analyze the results quantitatively. This allows us to understand the main directions for each cluster, which is very useful if you want to conduct advanced analytics (see Chapter 7). Let's define a custom PCA function before proceeding to its testing phase. This custom PCA function accepts points as input but does not generate a plot:

```
def CustomPCA(points):
    m = np.mean(points, axis=0)
    cluster_norm = points-m

    cov = np.cov(cluster_norm.T)
    eig_val, eig_vec = np.linalg.eig(cov.T)

    sorted_indexes = np.argsort(eig_val)[::-1]
```

```
eig_val = eig_val[sorted_indexes]
eig_vec = eig_vec[:,sorted_indexes]
return eig_val, eig_vec
```

From there, we can initialize empty columns, and then make sure to fill them with the eigenvalues and eigenvectors:

```
pcd['eig_1'],pcd['eig_2'],pcd['eig_3']=0,0,0
pcd['nx'],pcd['ny'],pcd['nz']=0,0,0

t1= time.time()
for i in range(len(segments)):
    cluster=segments.get_group(i)[['X','Y','Z']]
    eig_val, eig_vec = CustomPCA(cluster)
    pcd.loc[cluster.index.values,['eig_1','eig_2','eig_3']] = eig_val[0],
    eig_val[1],eig_val[2]
    pcd.loc[cluster.index.values,['nx','ny','nz']] = eig_vec[2][0],eig_vec[2][1],
    eig_vec[2][2]
t2=time.time()
print("Time to attribute features to segments ", t2-t1, " seconds")
```

Once the process concludes, the dataset is ready for export (we exclude the index columns for a cleaner dataset presentation):

```
pcd.to_csv("result_pca.xyz",float_format='%1.9f',index=False)
```

PCA is effective for reducing dimensions and is important for extracting useful patterns. It helps to identify the main components that show the most significant differences in a dataset, making it easier to represent and analyze the data. PCA can also help simplify 3D data by reducing the complexity of high-dimensional point clouds or voxel grids, making them easier to work with in later processing steps. It is important to note that whether you use it for extracting features, reducing noise, or visualizing data, PCA improves the quality of the processes. It also supports the success of 3D object recognition, scene understanding, and advanced AI solutions, as presented in Chapter 18. Before heading there, let's detail the process of 3D data registration.

3D Data Registration: Unifying Perspectives

Now that we have examined PCA-based feature extraction and normalization, let's look at the next step in the general workflow: 3D data registration (refer back to Figure 6-1 for the workflow). I am going to detail the registration fundamentals, the role of the iterative closest point (ICP) algorithm, and other central concepts as illustrated in Figure 6-16.

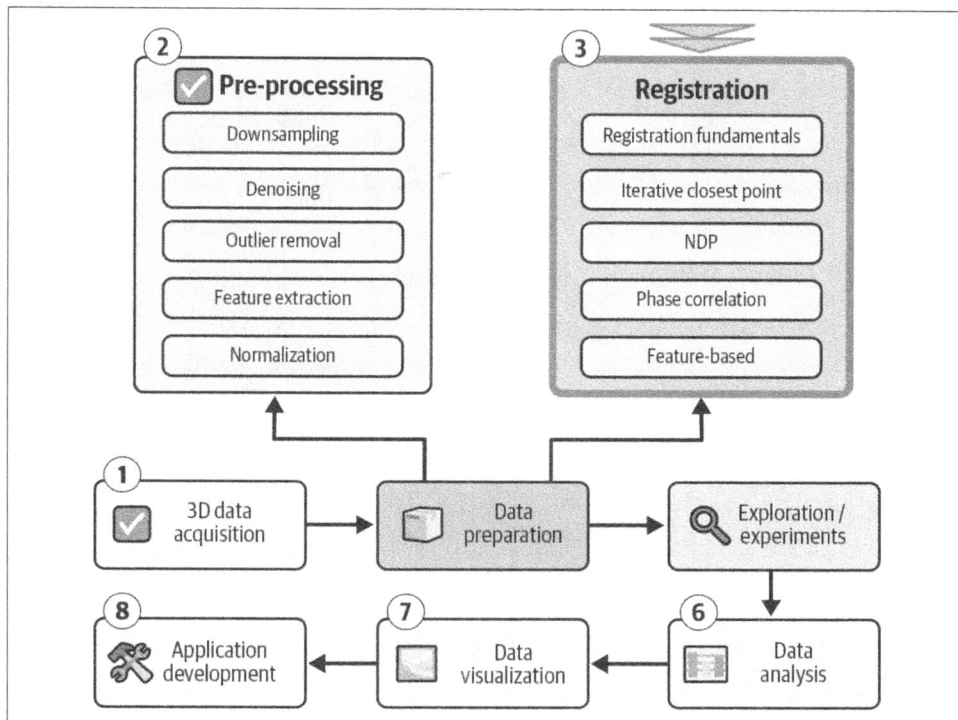

Figure 6-16. The 3D data registration workflow

Let's envision another use case. When a robot with a LiDAR scanner moves in a room, it captures "data flashes" from different viewpoints. So, how can we use these point clouds to build a "map of the environment" and figure out the relative motion between them? This permits the introduction of the challenge of data registration.

3D Data Registration Fundamentals

It has become apparent that even in the best-case scenario of acquired data, the data is not complete, and it is necessary to align it to a standard coordinate system. Applications such as geo-referencing or data fusion are relevant here; you can find detailed information in the 3D Resource Hub (*https://learngeodata.eu/3d-data-science-with-python*). In this section, we want to understand how to merge 3D datasets from different perspectives in a common coordinate system. In the context of 3D data science, there are three main registration strategies: localization from GPS (*https://oreil.ly/ziDcz*) or GNSS (*https://oreil.ly/S3Opg*) information, using targets as reference data, and coarse-to-fine registration without additional information.

In the first approach (sensor fusion), we have some location data at our disposal, such as WGS84 (*https://oreil.ly/Hi4R0*) localization information (latitude, longitude, altitude) from GNSS (*https://oreil.ly/S3Opg*) systems calibrated to be linked to the generating sensor. In the second approach, we can align multiple scans using targets or shared points of interest within the various scans. These could be flat targets or spherical markers that would appear in both scans as sources of information to align the separate scans.

The last way to align two or more scans is to use no extra measured information. In this situation, you use a method to get the scans almost aligned cyclically, such as using low-parameter coarse registration and then fine registration, where a rough alignment could be made using the geometric knowledge of the match algorithms and different shapes found in the data.

Let's delve deeper into these processes up to the coarse registration stage, based on identifying common points between the scans to get the initial transformation. Different methods are used to get the common points, mainly the point-to-point, line-to-line, and surface-to-surface methods. Each method has advantages and disadvantages based on the data density and noise. Let me bring you a solution to one of the best strategies: coarse-to-fine registration.

A coarse-to-fine registration strategy is a powerful technique for aligning 3D datasets, especially when dealing with large datasets or significant initial misalignments. This involves progressively refining the alignment from a coarse level to a fine level. Initially, a coarse alignment is established using robust feature-matching techniques to identify corresponding points between the source and target point clouds. This coarse alignment provides an initial estimate of the transformation parameters, such as rotation and translation. Then, a finer-grained alignment is achieved through optimization techniques, which iteratively refine the transformation parameters by minimizing the distance between corresponding points. This multiscale approach is practical because it efficiently handles large-scale transformations while ensuring accurate alignment at the fine level. Next, let's develop a Python solution for this idea.

Registration Initialization

First, let's import some useful libraries:

```
# Base library
import numpy as np

# 3D library
import open3d as o3d

import copy
```

Now, we define the source and target point clouds we will be working with:

```
#%% Load point clouds
source = o3d.io.read_point_cloud("../DATA/registration_source.ply")
target = o3d.io.read_point_cloud("../DATA/global_todo_registration_target.ply")
```

When visualizing both point clouds, we can see them as follows:

```
o3d.visualization.draw_geometries([source, target])
```

It is obvious that both the source and target are point clouds of the same environment but at different angles (see Figure 6-17).

Figure 6-17. Right: source point cloud; left: target point cloud

Now, let's try to align these point clouds by first going into a global registration step.

Coarse Registration

During the registration between the point clouds, common algorithms such as ICP (*https://oreil.ly/zhfsl*) do not converge to the global error minimum but to a local minimum. Accordingly, ICP registration is referred to as a local registration. It relies on an approximate alignment as an initial step. Another category of registration methods is global registration. These algorithms do not depend on alignment for initialization. Typically, they produce less precise alignment outcomes but aid in converging to the global minimum and are employed as the initial step for local methods. Let's look at an approach to this. First, we define a function that prepares a point cloud for registration:

```
def preprocess_point_cloud(pcd, voxel_size):
    pcd_down = pcd.voxel_down_sample(voxel_size)
    pcd_down.estimate_normals(o3d.geometry.KDTreeSearchParamHybrid(radius=
    voxel_size * 2, max_nn=30))
    return pcd_down
```

As you can see, we are performing two main operations:

Voxel downsampling
> This reduces the number of points in the cloud by creating a voxel grid and replacing all points within each voxel with their centroid. This helps to reduce computational complexity and noise.

Normal estimation
> This calculates surface normals for each point used in the registration process.

Now, let's create another obscure function to prepare our point clouds for global registration:

```
def prepare_dataset(source, target, voxel_size):
    source_down = preprocess_point_cloud(source, voxel_size)
    target_down = preprocess_point_cloud(target, voxel_size)

    source_fpfh = o3d.pipelines.registration.compute_fpfh_feature(
        source_down, o3d.geometry.KDTreeSearchParamHybrid(radius=voxel_size * 5,
        max_nn=100))
    target_fpfh = o3d.pipelines.registration.compute_fpfh_feature(
        target_down, o3d.geometry.KDTreeSearchParamHybrid(radius=voxel_size * 5,
        max_nn=100))
```

The function calls `preprocess_point_cloud()` on both the source and target clouds and computes a Fast Point Feature Histograms (FPFH) (*https://oreil.ly/685vL*) feature for both downsampled point clouds. FPFH is a descriptor that captures the local geometry around each point, which is used in the global registration step to find initial correspondences between the two point clouds. Now, we are ready to express our global registration function:

```
def execute_global_registration(source_down, target_down, source_fpfh,
target_fpfh, voxel_size):
    distance_threshold = voxel_size * 1.5
    result = o3d.pipelines.registration.registration_ransac_based_on_feature_
    matching(
        source_down, target_down, source_fpfh, target_fpfh, True,
        distance_threshold,
        o3d.pipelines.registration.TransformationEstimationPointToPoint(False),
        3,
        [o3d.pipelines.registration.CorrespondenceCheckerBasedOnEdgeLength(0.9),
         o3d.pipelines.registration.CorrespondenceCheckerBasedOnDistance
         (distance_threshold)],
        o3d.pipelines.registration.RANSACConvergenceCriteria(100000, 0.999))
    return result
```

This function performs global registration using the RANSAC (random sample consensus) algorithm. It aims to find an initial alignment between the source and target point clouds. Here's what it does:

- Sets a distance threshold based on the voxel size
- Uses RANSAC (*https://oreil.ly/_fJ2N*) (see Chapter 9) to repeatedly sample point correspondences based on FPFH (*https://oreil.ly/_nH-X*) features and estimate transformations
- Applies correspondence checkers to filter out bad matches
- Returns the best transformation that aligns the source to the target

This step is crucial for finding a good initial alignment, especially when the point clouds are far apart (as they are in our case). Now, let's set a voxel size for downsampling, and prepare our dataset using the defined function:

```
# Set voxel size for downsampling
voxel_size = 1.5

# Prepare dataset
source_down, target_down, source_fpfh, target_fpfh = prepare_dataset(source,
target, voxel_size)
```

We are now at a stage where our source and target point cloud do not have a single overlapping point. Let's now apply our global registration:

```
result_ransac = execute_global_registration(source_down, target_down,
source_fpfh, target_fpfh, voxel_size)
```

Next, we evaluate the fitness with two metrics:

`fitness`
Measures the overlapping area (`# of inlier correspondences / # of points in target`). The higher, the better.

`inlier_rmse`
Measures the root mean square error of all inlier correspondences. The lower the value, the better the registration results are.

```
print("Global registration result:")
print(result_ransac)
```

We get a `fitness` score of 0.9984 and an `inlier_rmse` of 0.8278517 (for a `correspond ence_set` size of 4444 points). This sounds coherent for a first approach. Now, let's print our transformation matrix and our results:

```
print("Transformation matrix:")
print(result_ransac.transformation)

#%% Visualize
source_global = copy.deepcopy(source)
source_global.transform(result_ransac.transformation)

o3d.visualization.draw_geometries([source_global, target])
```

You can see the results in Figure 6-18.

Figure 6-18. Point cloud global registration

Now, it would be interesting to increase the fitness score and fine-tune the registration process. Let's use a local registration method, the ICP, to optimize our results.

Iterative Closest Point

The ICP (*https://oreil.ly/j1QkI*) approach is a fundamental technique widely used in optimizing the alignment of point clouds because iterative optimization puts all point pairs on their closest points. It is employed to minimize the difference between two clouds of points. Utilizing this strategy allows us to identify the closest points within two sets of point clouds and, using this information, establish an optimal transformation.

This is accomplished by determining the nearest corresponding points via the given point clouds and solving an optimization equation such that each point in the source, department, or point cloud is reflected as accurately as possible within the destination, or target, point cloud. This process is iteratively repeated until an acceptable level of registration optimization is achieved. The versatility and effectiveness of ICP puts its use at the forefront of registration techniques, as does its practicality: it can be used in almost every situation. As mentioned earlier, during ICP registration, one of the two point clouds is kept as a reference (source) while transforming the other (target) to roughly align the point cloud (source) to the target point cloud.

The output is a refined transformation that tightly aligns the two point clouds. There are two ICP variants: the point-to-point ICP and the point-to-plane ICP. The are similar in concept. The main change is in the cost function. Unlike point-to-point ICP, where we look for the closest point in the other point cloud for all points and then try to minimize the square distances between them, in point-to-plane ICP, we additionally take the normal information into account.

We do so by computing surface normals on the target point cloud, projecting the error vector, and taking the dot product with the Euclidean distance between the points on the source and target point clouds:

$$\text{Cost function} = \min \sum \|y_n - x_n\|^2$$
Point − to − point

$$\downarrow$$

$$\text{Cost function} = \min \sum \left(\|y_n - x_n\| \cdot n_y \right)^2$$
Point − to − plane

As shown in the second equation, the only modification to the cost function is the addition of the dot product of the normal target point in the point cloud data, denoted as n_y, where y_n is the target point in the target point cloud data, and x_n is the source point in the source point cloud data.

Fine Registration: ICP

We start at a stage where we have our two point clouds, and a transformation initialization that results in a pretty good fit. It is time to iteratively refine this through ICP optimization. Let's define a function for this goal:

```
def refine_registration(source, target, result_ransac, voxel_size):
    distance_threshold = voxel_size * 0.4
    result = o3d.pipelines.registration.registration_icp(
        source, target, distance_threshold, result_ransac.transformation,
        o3d.pipelines.registration.TransformationEstimationPointToPlane())
    return result
```

This function refines the initial alignment obtained from global registration using the ICP algorithm. It performs the following steps:

- Sets a finer distance threshold for point matching
- Initializes ICP with the transformation from the global registration

- Runs the ICP algorithm, which iteratively refines the alignment by minimizing the distance between corresponding points
- Uses the point-to-plane error metric, which generally provides better results than point-to-point for surfaces

This ICP method fine-tunes the alignment, providing a more precise registration between the source and target point clouds. Let me show you this in action, but first, I want to work on the full-resolution point cloud; it's more fun. However, we need normals for that case. Therefore, let's compute that and check our new inputs:

```
source.estimate_normals(
search_param=o3d.geometry.KDTreeSearchParamHybrid(radius=2*voxel_size,
max_nn=30))
target.estimate_normals(
search_param=o3d.geometry.KDTreeSearchParamHybrid(radius=2*voxel_size,
max_nn=30))
o3d.visualization.draw_geometries([source, target])
```

Fine registration using ICP is illustrated in Figure 6-19.

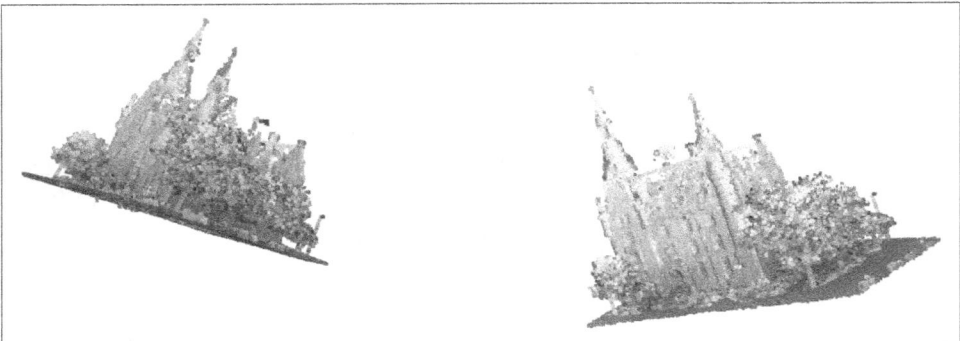

Figure 6-19. Fine registration using iterative closest point

Now, we can refine the registration:

```
result_icp = refine_registration(source, target, result_ransac, voxel_size)
print("Local registration (ICP) result:")
print(result_icp)
print("Refined transformation matrix:")
print(result_icp.transformation)
```

This results in the following: `Fitness: 0.9988617 %`, `Inlier RMSE: 0.07792996 m`, `Correspondance set: 264 132 points`. This looks much better. Now let's look at the qualitative results:

```
# Transform the source point cloud
source_local = copy.deepcopy(source)
source_local.transform(result_icp.transformation)
```

```
# Visualize the result
o3d.visualization.draw_geometries([source_local, target])
```

Now, we have obtained aligned 3D point clouds after ICP registration, as shown in Figure 6-20.

Figure 6-20. Perfectly aligned 3D point clouds after ICP registration

Indeed, we have two point clouds that are almost perfectly overlapping. Note that these are from different points of view, from real-world scans. You now have a way to bring multiple 3D perspectives into the same frame of reference for further analysis.

In this demonstration, we used additional 3D data engineering principles, such as feature extraction and data preprocessing. This is a perfect example of how intertwined the various steps are and that the proposed serialized logic is very flexible. In the next chapter, this is emphasized as concepts are shown more than once, within different contexts and paradigms.

Summary

We explored the fundamentals of 3D point cloud data engineering, specifically focusing on point cloud feature extraction. We saw how to optimize data to fulfill functional objectives, provide requirements, and make downstream processes more robust. I also presented a step-by-step, hands-on example of applying preprocessing techniques to a point cloud dataset followed by feature extraction. These preprocessing techniques include initial shift, downsampling, outlier removal, and normal estimation. By applying these techniques, we can enhance the quality of the 3D dataset while extracting meaningful information. You should now be comfortable with using 3D point cloud data engineering to improve the performance of downstream tasks such as classification, object detection, and scene understanding.

This chapter is crucial, as features are part of any subsequent workflow, and grasping their meaning provides a very high computing edge. Therefore, let's make sure you have a direct project to test yourself: creating a 3D analytical app that leverages 3D point cloud features.

Hands-on Project

Feature extraction from point clouds is crucial for many 3D data science tasks. We need a way to make the data more useful for downstream processing. This exercise will guide you through extracting geometric features using principal component analysis (PCA). We will then visualize the results to see how PCA helps us understand the structure of a point cloud:

1. *Data acquisition*: You will need a point cloud dataset for this exercise. Select a dataset that is representative of the type of data you typically work with. For example, if you are interested in urban analytics, you could use a point cloud of a city.

2. *Prepare environment*: Ensure you have the necessary libraries installed: NumPy, pandas, and Matplotlib. Import them at the beginning of your script.

3. *Load data*: Load your point cloud dataset using the pandas library. This will create a DataFrame that makes it easy to manipulate and analyze your data.

4. *Normalize data*: Normalize the x, y, and z coordinates of your point cloud. This ensures that all features contribute equally to the PCA calculation.

5. *Perform PCA*: Compute the eigenvectors and eigenvalues of the covariance matrix of your normalized point cloud data. Sort the eigenvectors by their corresponding eigenvalues in descending order.

6. *Visualize results*: Create a 3D scatter plot of your point cloud data. Plot the eigenvectors on the same plot, scaled by their eigenvalues. This will visually show the principal axes of your data.

Observe how the principal axes align with the main directions of variance in the point cloud. The first principal axis represents the direction of maximum variance, the second represents the next highest variance, and so on.

Building 3D Analytical Apps

We constantly seek ways to translate raw 3D data into actionable insights. While traditional analysis methods provide valuable information, they often lack the flexibility and interactivity needed for in-depth data exploration. Static visualizations and manual data manipulation hinder your ability to effectively segment, label, and extract knowledge from complex 3D datasets. It would help if you had dynamic tools to visualize and manipulate features in real time.

Imagine analyzing a point cloud of an old factory, aiming to isolate specific machinery and understand its spatial context. Conventional methods might involve tedious manual segmentation or scripting, limiting our ability to quickly explore different feature combinations and thresholds. We need an interactive application to visualize features like planarity, linearity, and elevation differences. This would enable the definition of thresholds that can be used to structure our dataset based on real-time feedback.

Our goal in this chapter is to design such an interactive thresholding method that enables us to selectively extract segments based on specified thresholds, directly within our application. This endeavor serves two main objectives: first, to establish a workflow for labeling 3D data, and second, to harness the power of 3D machine learning for developing AI models. Figure 7-1 demonstrates this workflow.

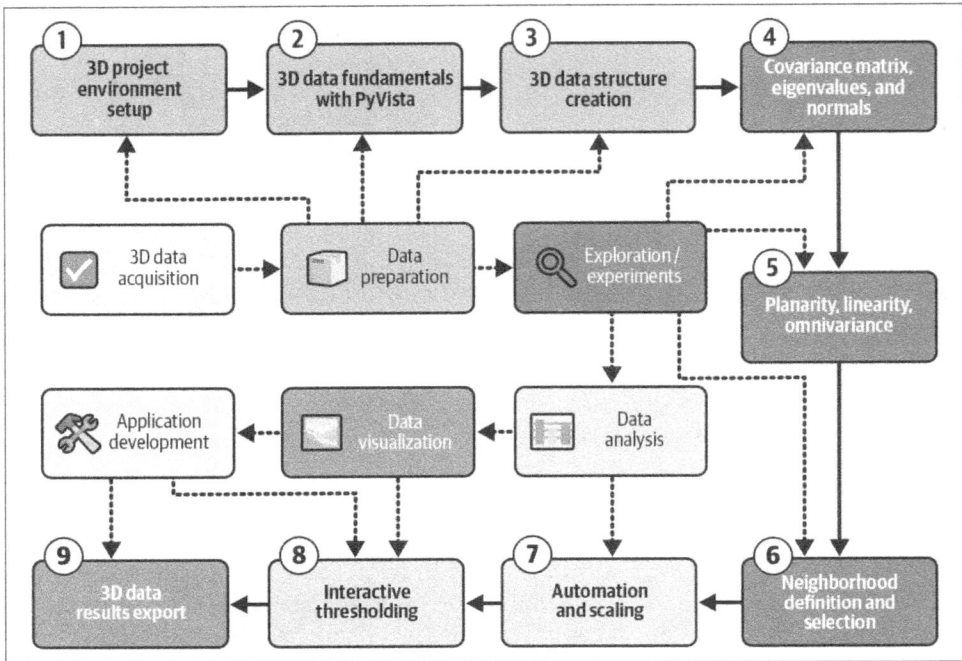

Figure 7-1. A nine-step workflow to build an app using the feature extraction process

We begin by collecting the dataset before setting up our Python environment. Next, we explore the basics of 3D data input and output, focusing mainly on the PyVista library (*https://docs.pyvista.org*). From there, we venture into the more experimental phase. This involves preprocessing and organizing the data, including defining 3D data structures used to compute features such as principal components. Our second stage is dedicated to automation, visualization, preparation, and segmentation.

This chapter presents a practical guide to building interactive 3D analytical applications for point cloud features. We will leverage the power of three Python libraries: PyVista (*https://docs.pyvista.org*) (used for 3D visualization and interaction for 3D point cloud and mesh analysis), SciPy (*https://scipy.org*) (utilized for scientific computing and technical computing; specifically, the KDTree module is used for efficient nearest neighbor searches in 3D space), and NumPy (*https://numpy.org*). You can find all the materials for this chapter on the 3D Data Science Resource Hub (*https://learn geodata.eu/3d-data-science-with-python*). You can use the base link and add "/chapter-X" to access the files (code, datasets, articles) and resources. Note that you may need to share an email address, a personal password, or proof that you are the book's owner.

3D Project Environment Preparation

Laying the groundwork is crucial for any successful 3D data science project. Let's step back and ensure we have all the necessary tools. I'll guide you through setting up a Python environment tailored explicitly for 3D data science tasks, selecting Spyder as the IDE to streamline your workflow, and sharing appropriate 3D datasets to practice your skills (see Figure 7-2).

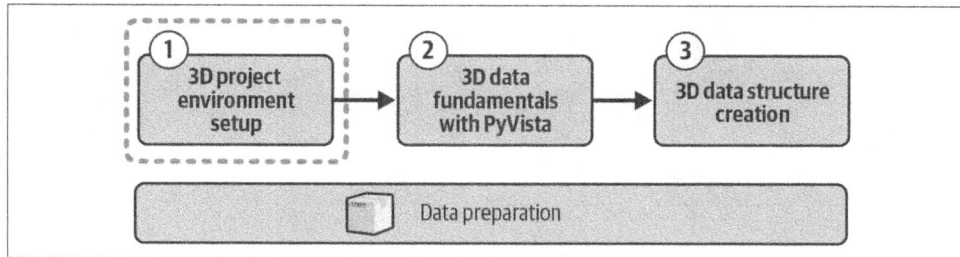

Figure 7-2. Preparing the 3D project environment

Gathering Datasets

Before we dive into the coding aspect, our first step involves data collection. Today, the process is quite simple. All you need to do is visit the 3D Data Science Resource Hub (*https://learngeodata.eu/3d-data-science-with-python*) and navigate to the Chapter 7 folder, where you find an intriguing dataset from an old wool factory in Belgium. Upon examining the dataset more closely, you will discover that it is a detailed point cloud showcasing a variety of ancient machinery once used in wool processing.

Our goal with this specific point cloud is ambitious; we aim to undertake numerous tasks. However, to effectively manipulate and analyze this data, it is imperative that we first identify and extract certain features. These features will be instrumental in distinguishing between the ground, walls, and machinery we wish to pinpoint within this dataset (see Figure 7-3). After acquiring the dataset, your next step is to organize it within your data folder.

> I adhere to a simple organizational structure for this project. Within the data folder, I have three subfolders, namely code, data, and results, all positioned at the same hierarchy level. For my setup, the dataset resides in the code folder.

Figure 7-3. An extract of the 3D point cloud dataset

Python and Environment Setup

Let's move on to the initial phase of setting up our Python environment. In this chapter, we will gloss over the detailed steps in setting up environments (detailed in Chapter 3). I recommend checking out previous chapters for those needing more in-depth explanation. Assuming your environment is ready, the next step involves installing a few essential libraries for our project. These include NumPy (*https://numpy.org*), SciPy (*https://scipy.org*), and PyVista (*https://docs.pyvista.org*), which are particularly useful for 3D visualization tasks. These three libraries alone are potent tools for performing sophisticated feature extraction. Installing these libraries is straightforward, utilizing pip commands:

```
pip install numpy scipy pyvista
```

After setting up your environment and installing these libraries, you can install and launch your IDE (for this chapter, I use Spyder: `pip install spyder`).

Now, let us import the libraries required for our project. We import NumPy (aliased as np for convenience), the time library to track the duration of our operations, and from SciPy, we specifically import the `KDTree` module for our spatial data structure needs. Lastly, we import PyVista (aliased as pv), a library that will be instrumental in handling 3D visualization:

```
#%% 1. Python Environment and Library setup

# Base libraries
import numpy as np
import time

# [OPTIONAL] Project Module
from scipy. spatial import KDTree


# 3D library
import pyvista as pv
```

These imports have laid the groundwork for our Python environment and library setup. Before executing these imports, ensure that the working directory is correctly set up. Having completed the setup, we are poised to advance to the next step: loading data and exploring the fundamentals of handling such data with PyVista.

3D Data Fundamentals with PyVista

Let's dive into using PyVista to handle and visualize point cloud data (see Figure 7-4). Initially, we need to load our data into a variable called PCD_PV using PyVista's read function. Here's how we do it:

```
pcd_pv = pv.read('../DATA/structure_verviers.ply')
```

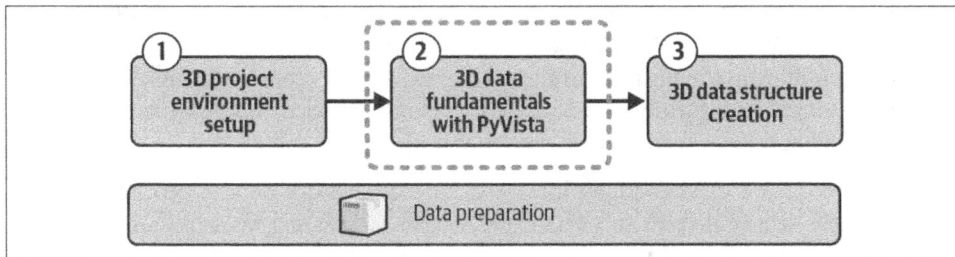

Figure 7-4. 3D data fundamentals with PyVista

After loading the data, it is beneficial to visualize it to understand its structure better. We can quickly plot the loaded point cloud with enhanced visualization options like eye dome lighting:

```
#Plot quickly with EDL on
pcd_pv.plot(eye_dome_lighting=True, rgb=True)
```

After executing this code, a window opens, providing us with an interactive 3D view of our point cloud, leveraging PyVista's rendering capabilities (see Figure 7-5).

Figure 7-5. The 3D point cloud dataset loaded in PyVista

Next, we delve into the structure of our `PCD_PV` variable, which is of the `PolyData` type. This format is particularly suited for handling 3D data like point clouds. To get more insights into the data and its structure, we can print or inspect `PCD_PV` to see its properties, including the number of points (vertices) and other metadata. `PolyData` is advantageous for its flexibility and the comprehensive set of operations it supports, making it an excellent choice for 3D data manipulation.

To further explore PyVista's capabilities, we can add custom data to our point cloud. For instance, we might want to visualize elevation or create a random variable based on the point coordinates. This is straightforward in PyVista:

```
# Storing in variable for pyvista
pcd_pv['elevation'] = pcd_pv.points[:,2]
pcd_pv['random'] = pcd_pv.points[:,0] * pcd_pv.points[:,1]
```

Here, `elevation` is derived directly from the z-axis coordinates of our points, and `random` is a custom variable created from a simple operation on the x and y coordinates.

We can render our points as spheres instead of the default point representation for a more engaging visualization. This gives the data a more tangible feel, especially when highlighting specific features or variables like our `random` variable. We achieve this by calling `pv.plot` again, with parameters to render points as spheres and to map colors based on the `random` variable, all while omitting the scalar bar for a cleaner view:

```
# Render as spheres
pv.plot(pcd_pv, scalars = pcd_pv['random'], render_points_as_spheres=True,
point_size=5, show_scalar_bar=False)
```

This command transforms our point cloud visualization, offering a fresh perspective by applying our custom color mapping and altering the point representation. It showcases the versatility of PyVista in customizing 3D data visualizations, making it an invaluable tool for data exploration and presentation (see Figure 7-6).

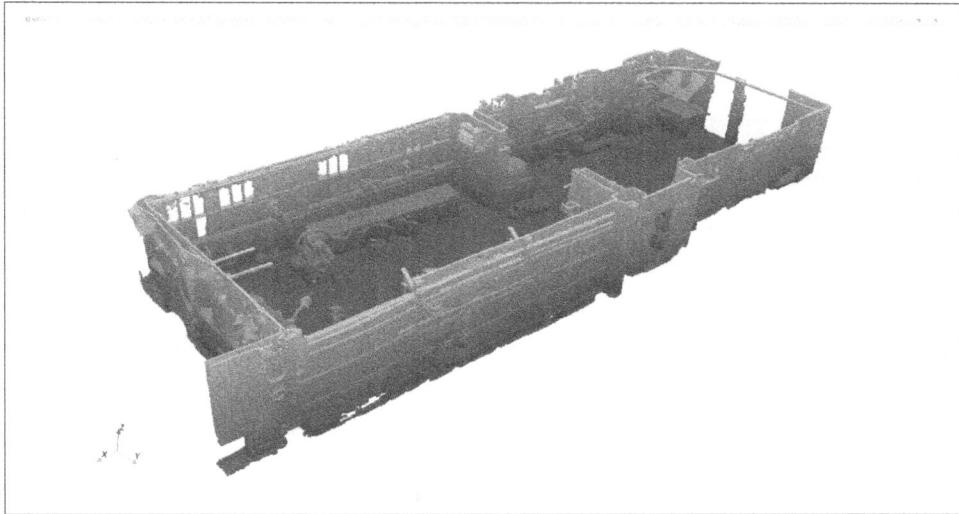

Figure 7-6. The 3D point cloud visualized with respect to the z-axis

Let's shift our focus to preprocessing and structuring our 3D data for further analysis. We explore the use of a KDTree for this purpose, highlighting PyVista's utility in preparing data for complex computational tasks or analyses.

3D Data Structure Creation (KDTree)

Next, we'll look at 3D data structure creation (see Figure 7-7). By now, you should have a pretty good understanding of *k*-d trees (see Chapter 4). If you happen to stumble upon these lines when opening the book, well, the *k*-d tree structure allows us to quickly navigate through its hierarchical structure, significantly reducing the search space and time. We already used both the Open3D and SciPy libraries to practically apply this concept in Python.

Let's consider another implementation using the PyVista library, which simplifies 3D data manipulation and visualization. An example code snippet to find the 20 nearest points to a given point using PyVista could look like this:

```
temp = pcd_pv.find_closest_point((1, 1, 0), n = 20)
print(temp)
```

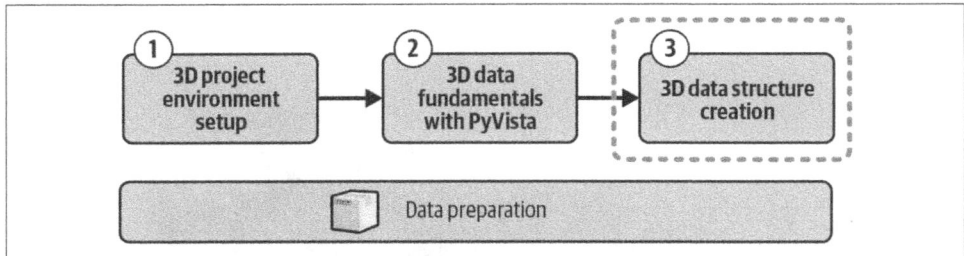

Figure 7-7. 3D data structure creation

A significant limitation of PyVista is its restriction to processing a single point at a time. To illustrate this, consider the scenario where I define a temporary variable to employ PyVista's `find_closest_point` function. By specifying coordinates (1, 1, 0), I request the 20 nearest points:

```
[1584544 1584611 1688219 1584564 1688248 1360375 1688402 1539639 1584601
 1447384 1446717 1584145 1688177 1360343 1539641 1688409 1688254 1584613
 1584259 1688374]
```

Upon executing this, the output is an array of point indices, providing a valuable means to identify and further utilize these points. However, this method's efficiency drops significantly when dealing with large datasets, such as a point cloud encompassing 500,000 points, due to the necessity of iterating through each point individually.

To overcome this, the recommended approach is to construct a k-d tree for the entire point cloud. This process begins with the creation of a tree object, humorously named `tree` for simplicity. This object is initialized with the KDTree function and applied to the dataset of interest, `pcd_pv.points`, effectively preparing the tree structure for use:

```
# Build a k-d tree for each point cloud
tree = KDTree(pcd_pv.points)
```

Following this, the objective is to query the tree for each point within the cloud:

```
t0 = time.time()

# Find nearest neighbors for each point in the point cloud
lists, indices = tree.query(pcd_pv.points, k=20)

# Get the neighbor points for each point
```

```
neighbors = pcd_pv.points[indices]

t1 = time.time()
print(f"Neihgbor Computation in {t1-t0} seconds")
```

The `tree.query` function identifies the 20 nearest neighbors for each point, yielding both distances and indices, which are then utilized to extract the neighboring points. The process is timed to evaluate efficiency, revealing a computation time of approximately 16 seconds. While not instant, this duration is deemed acceptable for offline processing, offering a structured data overlay for enhanced operations:

```
Neighbor Computation in 16.688420057296753 seconds
```

Inspecting the `neighbors` variable reveals a detailed list of the 20 closest neighbors for each point, complete with coordinates, enabling immediate application. With a total count of 500,000, this represents the coordinates of every point's `neighbors`, facilitating direct implementation and showcasing the utility of the k-d tree structure.

Having established this foundation, the focus can shift to the next phase: feature extraction from the point cloud data, specifically through principal component analysis (PCA), marking a progression to a more detailed analysis stage.

Covariance Matrix, Eigenvalues, and Eigenvectors

Let's now focus on the covariance matrix, eigenvalues, and normals as shown in Figure 7-8. For each point in our dataset, we have already determined its neighboring points. Let's proceed by selecting a specific neighbor for closer examination, arbitrarily choosing the selector to be one. We craft a function named `PCA_of_the_cloud` that encapsulates the PCA computation we delineated in Chapter 6. This function will accept a selector as an argument, enabling us to apply our analysis selectively.

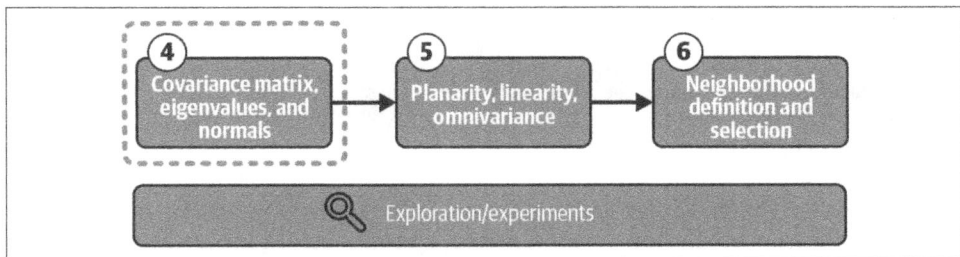

Figure 7-8. Covariance matrix, eigenvalues, and normals

I'm extracting the essential parts of the previous code related to sorting eigenvalues and eigenvectors for this purpose. Let's put this function to the test and verify its efficacy in yielding the expected outcomes. Here's the initial setup for computing PCA on our point cloud data:

```
X = pcd_pv.points

# Compute the mean of the data
mean = np.mean(X, axis=0)

# Center the data by subtracting the mean
centered_data = X - mean

# Compute the covariance matrix
cov_matrix = np.cov(centered_data, rowvar=False)

# Get the eigenvalues and eigenvectors
eigen_values, eigen_vectors = np.linalg.eig(cov_matrix)

# Sort the eigenvectors by decreasing eigenvalues
sorted_index = np.argsort(eigen_values)[::-1]
sorted_eigenvalue = eigen_values[sorted_index]
sorted_eigenvectors = eigen_vectors[:, sorted_index]

print(' sorted_index: ', sorted_index, '\n sorted_eigenvalue: ',
sorted_eigenvalue, '\n sorted_eigenvectors: \n', sorted_eigenvectors)
```

Upon execution, this segment confirms the successful identification of eigenvalues and eigenvectors, indicating the correct functionality for our selected neighbor.

Next, we aim further to distill distinctive features from our data through PCA, thereby calculating additional features. We proceed as follows:

```
sel = 2

def PCA(cloud):
    mean = np.mean(cloud, axis=0)
    centered_data = cloud - mean
    cov_matrix = np.cov(centered_data, rowvar=False)
    eigen_values, eigen_vectors = np.linalg.eig(cov_matrix)
    sorted_index = np.argsort(eigen_values)[::-1]
    sorted_eigenvalue = eigen_values[sorted_index]
    sorted_eigenvectors = eigen_vectors[:, sorted_index]
    return sorted_eigenvalue, sorted_eigenvectors
#Single unit test
f_val, f_vec = PCA(neighbors[sel])

print(' sorted_eigenvalue: ', f_val, '\n sorted_eigenvectors: \n', f_vec)
```

This step precisely applies our PCA function to a new selection, further illustrating the process of extracting nuanced features based on PCA and reinforcing our ability to discern and analyze specific characteristics within our point cloud data. We can now leverage PCA to extract more distinctive features.

Planarity, Linearity, Omnivariance, Verticality, Normals

For this step of our app building workflow, we will delve into four distinctive metrics, alongside the mean vector, to evaluate aspects of our dataset: planarity, linearity, omnivariance, and verticality (see Figure 7-9).

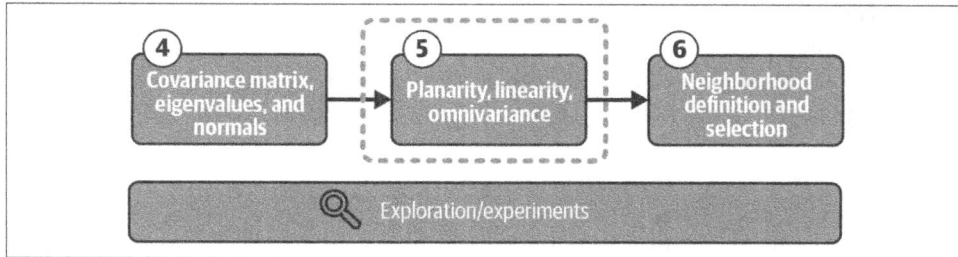

Figure 7-9. Planarity, linearity, omnivariance

These features offer insights into whether an area exhibits planar characteristics, linear formations such as poles, variance in a consistent direction, or vertical orientation. These concepts are derived from academic research, notably a paper by Martin Weinmann (*https://oreil.ly/Og9qA*) et al. that delineates these features comprehensively. Additionally, my work (*https://oreil.ly/CYuIz*), focused on voxel-based representations versus deep learning methodologies, further explores the distribution and significance of these features, underscoring their relevance at this juncture. For those keen on expanding their understanding, delving into additional features discussed in the literature is advisable, as it enriches comprehension and application of these analytical tools.

In this section, we introduce a function named `PCA_featuring`, which processes eigenvalues and eigenvectors to calculate the aforementioned metrics. The formulation of planarity, for instance, is derived directly from the academic literature, illustrating the methodological underpinning of our approach. Here's how these features are computed within our function:

```
def PCA_featuring(val, vec):
    planarity = (val[1] - val[2]) / val[0]
    linearity = (val[0] - val[1]) / val[0]
    omnivariance = (val[0] * val[1] * val[2]) ** (1/3)
    _, _, normal = vec
    verticality = 1 - normal[2]
    return planarity, linearity, omnivariance, verticality, normal[0],
    normal[1], normal[2]
```

To validate the function's efficacy, we conduct a test and inspect the output, which reveals the computed features in a structured manner:

```
#test
p, l, o, v, nx, ny, nz = PCA_featuring(f_val, f_vec)
print('', p, '\n', l, '\n', o, '\n', v, '\n', nx, '\n', ny, '\n', nz)
```

The successful execution of this test is a significant milestone, showcasing our ability to discern complex features within the dataset. However, the full implications of this achievement may not yet be fully apparent. Let's focus our understanding on feature-based neighborhood definition, aiming to refine our analytical capabilities further.

Neighborhood Definition and Selection

The concept of neighborhood is fundamental to 3D data processing, as it defines the spatial relationships between points within a dataset (see Figure 7-10). By considering the proximity of points to one another, we can perform various operations, such as noise reduction, feature extraction, and surface reconstruction. The choice of neighborhood strategy and scale significantly impacts the outcome of these operations. A smaller neighborhood size can capture fine-grained details, while a larger neighborhood can identify larger-scale structures. For instance, in point cloud denoising, a small neighborhood can remove isolated points, while a larger neighborhood can fill in missing data. By carefully selecting the appropriate neighborhood strategy and scale, we can effectively extract meaningful information from 3D data.

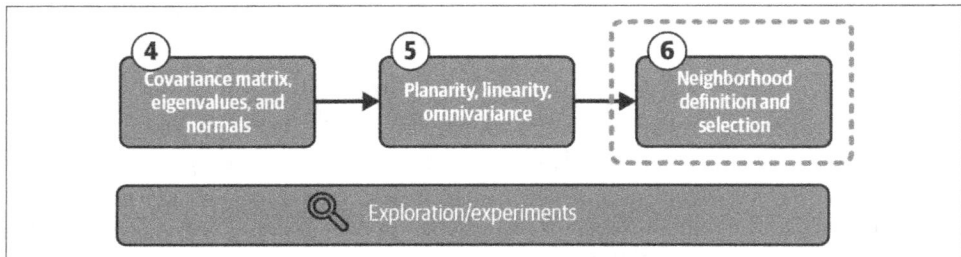

Figure 7-10. Neighborhood definition and selection

Let us first discuss the *k*-NN search. I emphasized the importance of employing a specific strategy for efficient search. For instance, try executing a *k*-NN search as shown in the following, and vary the number of neighbors:

```
tree_temp = KDTree(pcd_pv.points)
t0 = time.time()
_, idx_temp = tree_temp.query(pcd_pv.points, k=20)
t1 = time.time()
print(f"KNN Search Computation in {t1-t0} seconds")
```

Next, let's explore the radius search. This technique is similar to the previous one, but you define a search radius instead of specifying the number of nearest neighbors. Here, we are using a radius of 1 meter. Unlike the *k*-NN search, which returns distances, the radius search provides a list of indices:

```
t0 = time.time()
tree_temp = KDTree(pcd_pv.points)
idx_temp = tree_temp.query_ball_point(pcd_pv.points, 1)
t1 = time.time()
print(f"Radius Search Computation in {t1-t0} seconds")
```

The standard search took 15 seconds, slightly longer than the *k*-d tree search, but it is still manageable. Let us delve into a hybrid search, a custom, knowledge-driven approach. In this method, I have constructed a 2D tree by omitting the z-value from our dataset. This approach is particularly adept at handling top-down LiDAR data, allowing for more efficient detection of local maxima and minima. Be prepared for a longer wait time, approximately 2 minutes, for this process to complete:

```
t0 = time.time()
tree_2D = KDTree(pcd_pv.points[:,0:2])
idx_2D_rad = tree_2D.query_ball_point(pcd_pv.points[:,0:2], 1)
t1 = time.time()
print(f"Knowledge Driven Custom Search Computation in {t1-t0} seconds")
```

For the next part, we will focus on feature extraction relative to the points of interest:

```
t0 = time.time()
sel = 1
selection = pcd_pv.points[idx_2D_rad[sel]]

d_high = np.array(np.max(selection, axis = 0) - pcd_pv.points[sel])[2]
d_low = np.array(pcd_pv.points[sel] - np.min(selection, axis = 0))[2]
```

This code segment calculates the vertical distance from the selected point to its neighborhood's highest and lowest points. The resulting values indicate how close the point is to the ground and how far it is from the highest obstacle, providing valuable insights for applications such as terrain modeling or object detection.

Let's now address automation, scaling, visualization, interaction, preparation, exporting, and segmentation of results. At this juncture, it is crucial to recognize the minimal library dependencies required and the effectiveness of leveraging PyVista alongside *k*-d trees and PCA for feature extraction.

Automation and Scaling

To streamline the entire process, we will be functioning with automation in mind (see Figure 7-11). Let's set up a feature matrix so that we can streamline the entire dataset. Here's a savvy tip to minimize the need for numerous libraries. Though this could easily be achieved using pandas, I would like to demonstrate employing a structured array with NumPy:

```
dt = {'names':['planarity', 'linearity', 'omnivariance', 'verticality', 'nx',
'NY', 'nz', 'd_high', 'd_low'], 'formats':[float, float, float, float, float,
float, float, float, float]}

features = np.empty(len(pcd_pv.points), dtype=dt)
features[:] = np.nan
```

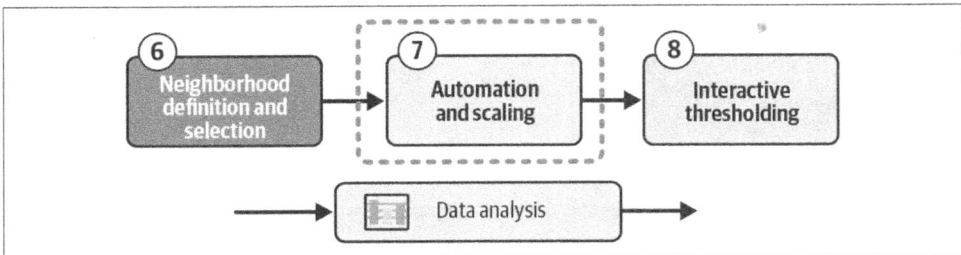

Figure 7-11. Automation and scaling

We construct a dictionary named dt, comprising names and formats. Each format is set to float, corresponding to the names of our features: planarity, linearity, omnivariance, verticality, x, y, and z, along with the highest and lowest points, amounting to nine features derived from the point cloud. Next, we initialize an empty NumPy array with the length of the points and the data type, dt, prepping all columns and initializing them. Then, all features are set to a NaN value, indicating "not a number." Inspecting the features reveals planarity and linearity, among others. A significant advantage here is the ability to reference features by name rather than indices, ensuring accurate feature selection. Now, let's loop and automate the feature computation:

```
pts_number = len(pcd_pv.points)

t0 = time.time()
for idx in range(pts_number):
    f_val, f_vec = PCA(neighbors[idx])
    features['planarity'][idx], features['linearity'][idx],
    features['omnivariance'][idx], features['verticality'][idx],
    features['next'][idx], features['ny'][idx],
    features['nz'][idx] = PCA_featuring(f_val, f_vec)
    if len(idx_2D_rad[idx])>2:
        selection = pcd_pv.points[idx_2D_rad[idx]]
        features['d_high'][idx] = np.array(np.max(selection, axis = 0) -
```

```
        pcd_pv.points[idx])[2]
        features['d_low'][idx] = np.array(pcd_pv.points[idx] -
        np.min(selection, axis = 0))[2]
    t1 = time.time()

    print(f"Full Point Cloud Feature Computation in {t1-t0} seconds")
```

Within our loop, we time the process. For each index up to the total point count, we utilize the previously defined PCA to obtain values and vectors. These are then fed into our feature computation function, ensuring the features array is populated accordingly. We calculate the highest and lowest feature values for points with more than two neighbors. The process will require approximately 2 minutes to execute. Upon completion, we use PyVista for quick visualization, and we can switch between features ('d_high', 'planarity', 'verticality', 'omnivariance'). Each feature sheds light on specific details, offering insights, for example, into distinguishing trees or evaluating the flatness of the terrain:

```
    pv.plot(pcd_pv, scalars = features['planarity'], render_points_as_spheres=True,
    point_size=5, show_scalar_bar=False)
```

Interestingly, the terrain is mostly flat. The appearance of walls can vary depending on the observation point, though some noise is present (see Figure 7-12).

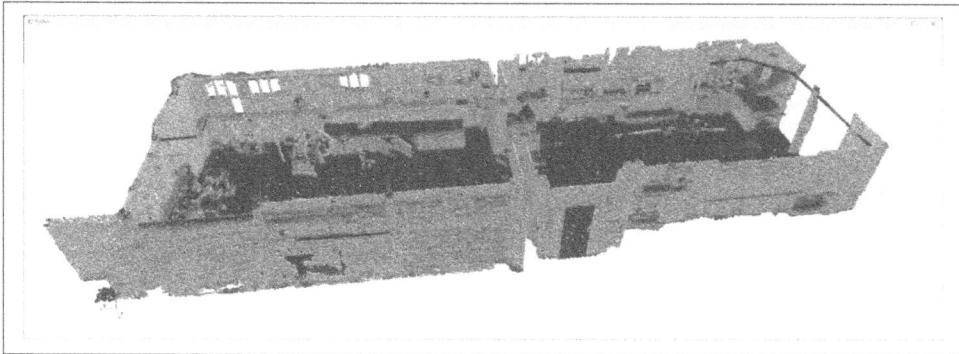

Figure 7-12. Results of the planarity or verticality-based visualization

This suggests a potential adjustment in our approach, such as the number of neighbors considered for PCA-based features, where using only 20 might not be optimal. A radius-based search could yield improved results. The discussion on verticality reveals it might overlap or offer slight distinctions from planarity. Yet, verticality uniquely identifies ground points, providing a valuable tool for isolating terrestrial data. A Boolean condition could refine the visualization, highlighting significant points amid the dataset. This hints at using the PyVista capability to apply thresholds.

Interactive Thresholding

Thresholding involves classifying features into foreground and background categories based on a predefined intensity threshold. It can be applied to segment objects or regions of interest within a 3D point cloud or voxel grid. By setting appropriate thresholds, we can isolate specific features or separate different materials (see Figure 7-13).

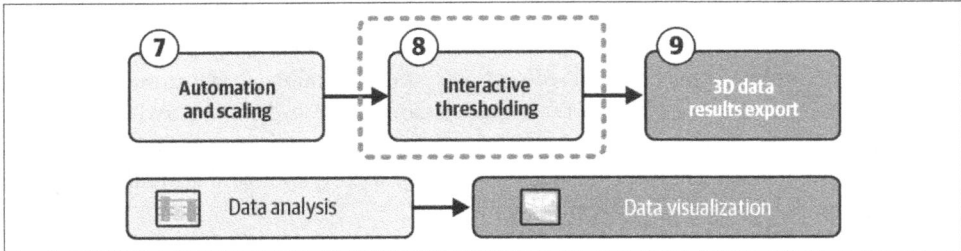

Figure 7-13. Interactive thresholding

This function enables the segmentation and extraction of features through histogram thresholding, offering a sophisticated means to analyze and interpret the computed features. To begin with, it is crucial to embed our features within the pcd_pv variable, assigning them descriptive names.

We need a unique approach to visualize and interact with features; we can configure a plotter object pv.Plotter(), followed by adding an interactive thresholding tool for comprehensive visualization. The feature to work with is called pcd_pv, our variable, followed by the scalars of interest, such as d_high or d_low; we then assign a title and execute the show command:

```
# Using the scalars
pcd_pv['d_high'] = features['d_high']
pcd_pv['d_low'] = features['d_low']
pcd_pv['verticality'] = features['verticality']

# Set up a plotter
p = pv.Plotter()

# Add the interactive thresholding tool and show everything
p.add_mesh_threshold(pcd_pv, scalars = 'verticality',
title = 'Verticality Pointer (%)', all_scalars = True,
render_points_as_spheres=True, point_size=7)
p.show()
```

This setup allows us to design an interactive 3D viewer capable of filtering data based on the distance from the highest point (see Figure 7-14).

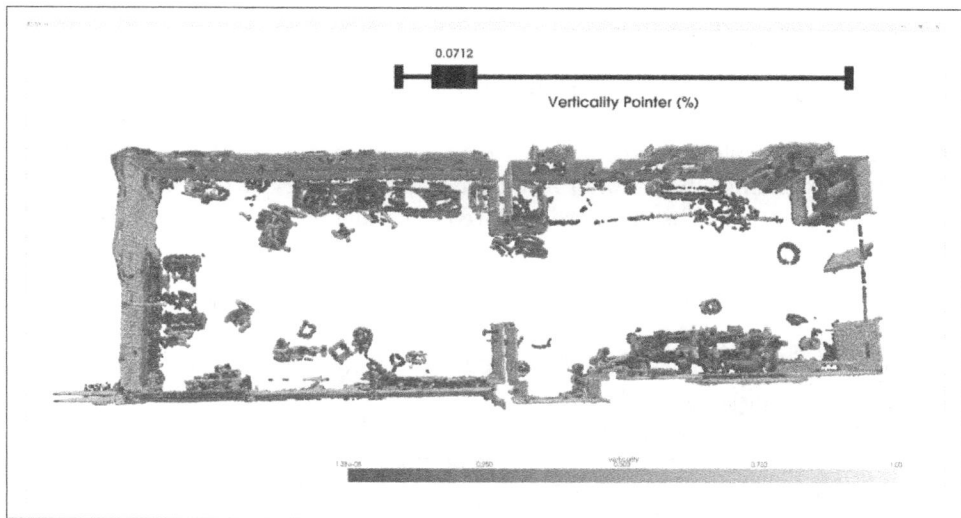

Figure 7-14. The interactive thresholding app

This methodology is advantageous for filtering desired points. For instance, applying this to low points can swiftly eliminate ground points. The potential applications, such as data labeling or segmentation, are direct and straightforward.

If you want another valuable trick, you can use interactive selection and persistence postthresholding for further interaction. This is achieved by replicating the initial steps but incorporating a critical line to enable scene picking. Following this action, the thresholding process can be executed to eliminate undesired points within a pipeline: eliminate points, then work with the remaining points until you reach a satisfactory result:

```
selection = p.picked_mesh
pcd_pv['planarity'] = features['planarity']

p2 = pv.Plotter()

# Add the interactive thresholding tool and show everything
p2.add_mesh_threshold(selection, scalars ='planarity', title = 'planarity (%)',
all_scalars = True, render_points_as_spheres=True, point_size=7)

# p2.enable_mesh_picking(selection)
p2.show()
```

After selection, pressing the P key captures the remaining points within an object, creating a remnant for future exploration. This final picked object is stored in the selection, allowing for further actions, such as exporting the selection for additional analysis or use.

3D Data Results Export

Let's retrieve the results through 3D data export (see Figure 7-15). Initially, we specified the path for our selection output as *segmentation_live.obj* using the variable `selection_path`. Following this, we use the previously picked mesh from the interactive session, initiating a new plotter instance to display and export our selection mesh to the designated path. This process is tailored to handle the unstructured grid format efficiently:

```
selection_path = '../RESULTS/segmentation_live.obj'

selection = p.picked_mesh
pl = pv.Plotter()
_ = pl.add_mesh(selection)
pl.export_obj(selection_path)
```

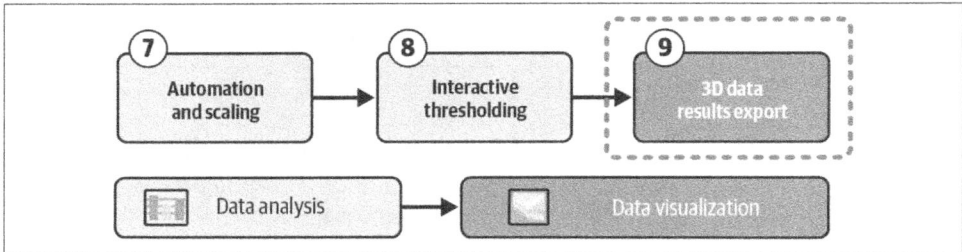

Figure 7-15. 3D data results export

Although the export format is *.obj*, not *.ply*, the critical takeaway is our successful isolation of the desired section without the ground or unnecessary parts, showcasing the efficiency of our method.

Scalar fields like planarity and linearity bring new insights, such as precise pole identification. Omnivariance requires slight adjustments to highlight its value, demonstrating its utility in distinguishing areas of uniformity from those with anomalies, such as trees. Verticality, with its distinctions between high and low, offers another analytical perspective. This feature aids in quickly labeling datasets, using planarity or verticality alongside elevation data to segregate ground points efficiently. This approach, combined with distinctions in high and low values, allows for rapid extraction of diverse elements that you can now use in your application.

Summary

We went through a comprehensive workflow using Python to extract features from 3D point cloud data. We focused on the open source library PyVista and emphasized the importance of understanding fundamental concepts like PCA and k-d trees for efficient data analysis. These are combined with an interactive thresholding method and derived features to better structure our dataset in functional entities, a first stage in 3D scene understanding. Now, what are the key takeaways here?

- When working with large point cloud datasets, efficient spatial searching is crucial. The use of k-d trees was particularly beneficial.

- Feature extraction with PCA plays a central role in extracting meaningful features from point cloud data. Characteristics such as planarity, linearity, and verticality can be quantified by analyzing the eigenvectors and eigenvalues of the covariance matrix.

- The choice of neighborhood definition significantly influences the extracted features.

- Visualizing the extracted features is essential for understanding their spatial distribution and relationships within the point cloud. PyVista is very helpful here.

- Interactive thresholding is a valuable technique for segmenting point cloud data based on specific feature values. This method allows for real-time exploration and isolation of points that meet defined criteria, facilitating data labeling or object extraction.

I do not think an exercise would complement this chapter, as it is largely hands-on. If you have been able to grasp and replicate the experiments, let's move on to the next stage: 3D data analysis.

3D Data Analysis

How often do you find yourself analyzing what you can see before you? If your answer is never, good news: you are one of the sane ones. But I cannot help but wonder: is the floor I stand on perfectly leveled? How big is the room in which I am working right now? Is there enough space to fit a new BESTÅ cupboard?

These are simple questions that are nearly impossible to answer accurately. But I have good news: we can leverage the ability to analyze what we see through 3D data science.

Well, 3D data analysis is the central stage in almost all aspects of 3D data science. What I mean is that this is not a processing module per se but more a global framework about how we should approach solving challenges and finding solutions. For that reason, it is tough to isolate the core components. To be fair, it would be like trying to tell a captivating story with only a few bullet points.

3D data analysis is heavily grounded in preprocessing approaches, segmentation workflow, and data modeling. Let's say you want to study the aerodynamics of a car. This objective is likely to take you through a 3D analytical workflow with several components. It could start with capturing the state of the vehicle (as a 3D point cloud), preprocessing to eliminate outliers, then segmenting the raw data to keep only study-relevant points, modeling its outer surface while refining the 3D "mesh" to fit the data points closely. We finally monitor deviations against an as-designed 3D car model. This would then allow executing airflow aerodynamics simulations to create a sound analytical base on which to take action (i.e., improving the car or designs).

This car example supports the idea that 3D data analysis results from combining 3D data science modules. It is a multifaceted discipline that combines statistical, geometric, and visual analytical techniques to facilitate communication and data understanding. It extends to many workflows. Whenever spatial information plays a role, you have an application for 3D data analysis.

In this chapter, I will provide the tools and inspired cases for manipulating 3D analytics. The purpose is to extract knowledge and insights from 3D models, such as point clouds, voxels, meshes, or any 3D data representation mentioned in Chapter 4. For this reason, it makes sense to understand and develop a set of analytical tools that we can leverage to serve these purposes: extracting insights, presenting visuals, and communicating efficiently. To do so, we first get an extensive view of 3D data analysis, then dive into using 3D analytical tools and finally 3D diagnostic tools (see Figure 8-1).

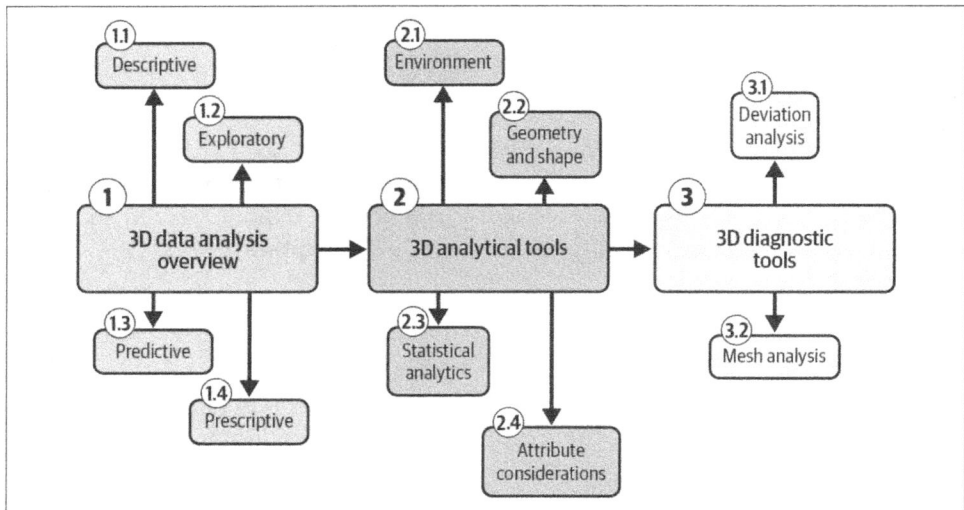

Figure 8-1. Complete overview of 3D data analysis categories

> You can find all the materials for this chapter, including a list of assets and libraries, on the 3D Data Science Resource Hub (*https://oreil.ly/Z_Sa2*). To access the files (code, datasets, articles) and resources, you may need to share an email address, a personal password, or proof that you are the book's owner.

Types of 3D Data Analysis

3D data provides a base layer of much richer information than 2D modalities. This is key, especially in engineering, urban planning, or other applications where decisions must be based on accurate data. Let's discuss how analytical tasks benefit from 3D datasets.

In scientific research, using 3D models of molecules permits a much better understanding of protein folding. Indeed, you unlock the ability to analyze patterns in their organization by directly working on their topology, geometry, and constituent relationships. We would be locked onto something much more abstract if we were using 2D classical insights.

Another example that I like is in sports analysis. If we take a soccer game, as an analyst, you are going to study the player's movement and patterns in a match and layout strategies. While working on 2D projections to explain the results of an analysis is efficient, working and analyzing the data on its based 3D reconstruction multiplies the capacity to extract relevant pointers. Indeed, insights, information, and patterns that could not be sequenced on 2D modalities or reduced to a single-shot context can now benefit from an extensive spatial contextual analysis, including the ability to study aerial strategies. Even at the individual player level, the ability to decompose a specific movement, like striking the ball with a 3D complementary view, permits the analysis of muscle activations, angle of shots, distance between arms, and so on.

You may have been intrigued by a key concept from the preceding examples: spatial cognition. 3D data analysis is critical to understanding spatial relationships and helps understand environments. With traditional 2D analysis, we often overlook the data's specific spatial location and orientation. By using 3D analysis, we breach this gap, which allows us to understand how a nearby chair is implicitly linked to a desk in an office context. So, before I become responsible for your next headache, let me explain how you can approach 3D data analysis. There are four main buckets where your analytics are used: 3D descriptive analysis, 3D exploratory analysis, 3D predictive analysis, and 3D prescriptive analysis (see Figure 8-2).

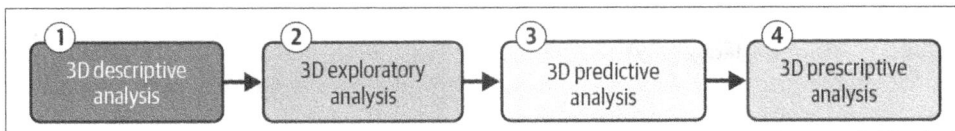

Figure 8-2. A linear workflow depicting the sequence of 3D data analysis types: descriptive, exploratory, predictive, and prescriptive

3D Descriptive Data Analysis

In the 3D descriptive data analysis (DDA) category, our goal is to study hidden patterns, relationships, and potential outliers. Its true aim is to assess the data bench and provide essential summaries to view the source of new knowledge/decisions clearly. It usually relies on quantitative and qualitative analytical steps to assess potential relationships between data points and features or possible anomalies that need to be pushed in preprocessing solutions.

DDA usually follows data collection and aims to prepare the data for some higher-end goals. We explore the structure, identify missing values, understand the distribution, determine correlations between variables, and produce key summary statistics.

We usually leverage DDA before data cleaning and preprocessing. Indeed, it can guide these steps to address inconsistencies, missing values, outliers, and anything else involving an imputation technique (filling in missing values, normalizing data by scaling to a joint range, and outlier removal). DDA also helps design feature engineering approaches to create and encode new feature variables.

> 3D descriptive data analysis is a central cog in the 3D data science system. In the "3D Data Analytical Tools" section, we develop a sound DDA that can be generalized to many applications. It is worth noting that DDA is helpful when automating workflows because it tunes algorithms to use data-driven parameters extracted with DDA.

It is very important to establish a sound workflow. Once you have framed the main statistical components of the dataset, we usually proceed to a more refined analysis: 3D exploratory analysis.

3D Exploratory Data Analysis

Exploratory data analysis (EDA) aims to diagnose a situation: why did it happen? It goes a step deeper into the initial qualitative and quantitative assessment of the dataset provided by sound DDA. This category of scenarios also includes detailed investigations of 3D geometric and spatial patterns.

Unlike traditional data analysis, which often deals with tabular data (rows and columns), geometric data analysis involves working with data representing shapes, objects, and spatial relationships. As we saw in previous chapters, this data type is becoming increasingly common due to advancements in sensor technologies, such as LiDAR, and imaging techniques like MRI and CT scans.

It is often linked to monitoring techniques like deformation analysis, which compares a specific state to a ground-truth element. This area falls within 3D monitoring and clash detection, where we want to study the deviation between two shapes or more. To diagnose a complete dataset, we can use several distance metrics with different approaches:

The Euclidean distance (https://oreil.ly/wQsZp)
> This is the most common distance metric. It calculates the straight-line distance between two points. It is appropriate for shapes representing points in a Euclidean space, such as 3D point clouds.

The Manhattan distance (https://oreil.ly/ClyIa)
This metric calculates the distance between two points as the sum of the absolute differences of their coordinates. It is also known as the L1 norm or taxicab distance, and it is appropriate for shapes defined by their grid coordinates, such as 3D voxels.

The Hausdorff distance (https://oreil.ly/IrHwE)
We calculate the distance between two sets of points as the maximum distance between any point in one set and any point in the other set. It is a valuable metric for comparing shapes that are not necessarily congruent.

The Fréchet distance (https://oreil.ly/1zSir)
This is the distance between two curves as the length of the shortest path connecting them while staying within a certain distance of both curves. It is useful for comparing shapes that are not necessarily rigid, such as curves or surfaces.

The choice of distance metric depends on the specific application and the properties of the studied shapes. It is important to choose a metric that is appropriate for the data and that captures the relevant geometric features.

> 3D monitoring and clash detection is usually dependent on 3D shape recognition tasks. This means we decompose our 3D object of interest into its constituents to better understand its state. I share some tools, examples, and processes to help you diagnose a 3D scene's verticality and planarity (if it is leveled), get the deviation to a reference, and extract an overall fine understanding of a 3D scene in Chapter 9.

Now that you understand 3D EDA (3D geometric analysis), let's orient our focus toward 3D predictive data analysis.

3D Predictive Data Analysis

3D predictive data analysis is concerned with what is likely going to happen. It is an advanced type of analysis that usually aims to build models that can predict future outcomes based on existing data patterns.

It usually allows us to forecast trends and make data review decisions. However, when we reach 3D predictive data analysis, we typically have a history of processing/analytical layers that provide extra information on which to base the predictive analytics. As seen earlier in the book, you likely approach this category with relevant features from a previous feature engineering step (see Chapter 6).

3D predictive data analysis is tied to concepts approached in later chapters: segmentation (Chapter 12), clustering (Chapter 13), and supervised AI (Chapter 14). Therefore, you are invited to refer to these future chapters to avoid overlap to see how we approach the 3D predictive analysis task in the context of these other topics.

Within scenarios tied to 3D predictive analysis, we usually need to intertwine 3D data representation and the creation and use of predictive models, including model training, evaluation, fine-tuning, and deployment. Topics such as this are not included in this chapter because they would be too lengthy. The fine details of the construction, architecture, and deployment of such models are given from Chapters 12 to 18.

3D Prescriptive Data Analysis

3D prescriptive data analysis aims to leverage insights from various types of analytics to recommend specific actions or strategies based on predicted outcomes. This means formulating optimization problems or creating decision support systems and experiences. In this case, the visual representation and possible human-machine interactions are useful. Therefore, how we communicate findings and information is paramount.

3D prescriptive data analysis is, therefore, highly application dependent. We are usually in a setting where we have identified that a decision has to be made. The goal is to identify, optimize, or support a series of actions or decisions.

I mentioned that I wanted to provide precise analytical tools and workarounds. Therefore, I will focus on DDA and EDA, which are leveraged in the later application-oriented buckets. They constitute a great compromise between learning rate and ease of extensibility for your applications. As always, your innovative spirit is the differentiator. However, to deepen your expertise feel free to explore the 3D Resource Hub (*https://learngeodata.eu/3d-data-science-with-python*).

Additional Considerations

Visualization and communication are crucial in 3D data analysis. They play a significant role in conveying complex data and unique relationships to all project stakeholders. They also enhance collaboration and improve stakeholders' understanding of complex phenomena, making them key steps in the analytical process.

Through the analytical process, we create an informative visualization to communicate our findings effectively afterward. We need to utilize appropriate plots, rendering approaches, and workflows depending on the analytics we want to convey. And that is only when we go on to the interpretation stage, based on clear presentations and communication. Analyzing the result and drawing conclusions based on evidence obtained is critical. We must ensure that this conclusion is statistically sound and supported by the data. Then, we must clearly and concisely communicate this finding to the audience (see Figure 8-3).

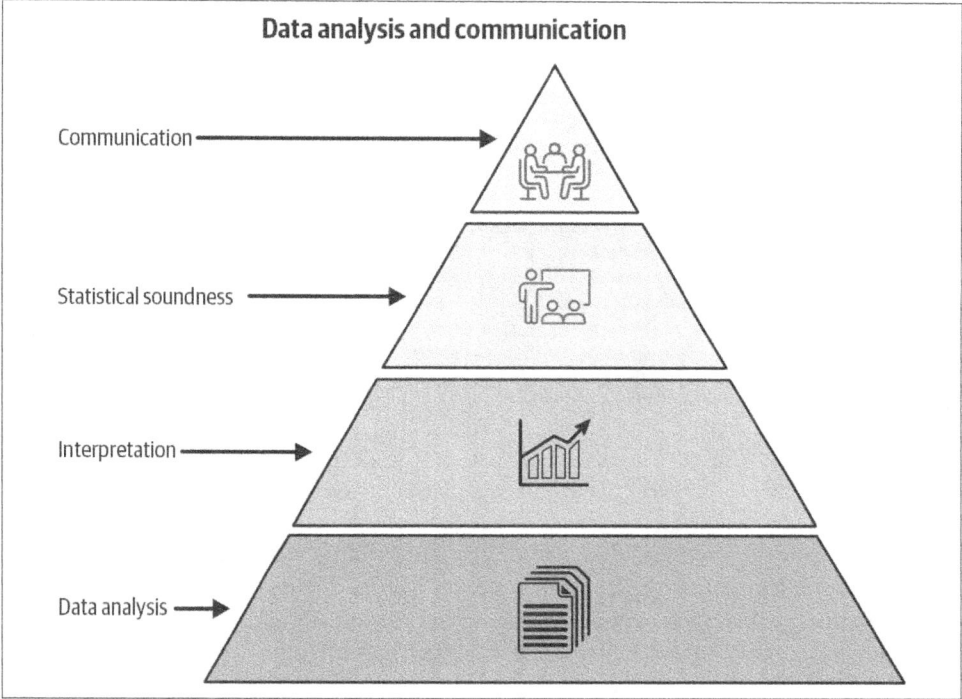

Figure 8-3. A hierarchical representation of the data analysis process, emphasizing effective communication of insights

That might mean reports, presentations, and interactive experiences to showcase our work. It is essential to understand and remember that these steps might not always be strictly linear and that we might need to revisit the previous step at some point. So, a structured approach is essential to systematically extracting knowledge and value from the data. Now that we have walked through this extensive introduction to 3D data analysis, let's dive into a concrete case that will tap into descriptive, exploratory, predictive, and prescriptive analytics.

3D Data Analytical Tools

3D descriptive analysis aims to answer the question: what happened (see Figure 8-4)? Therefore, we need to initialize our analysis by summarizing the critical characteristics of our data to get this high-level picture. You want to provide a foundational understanding of the data distribution, central tendencies, and variability.

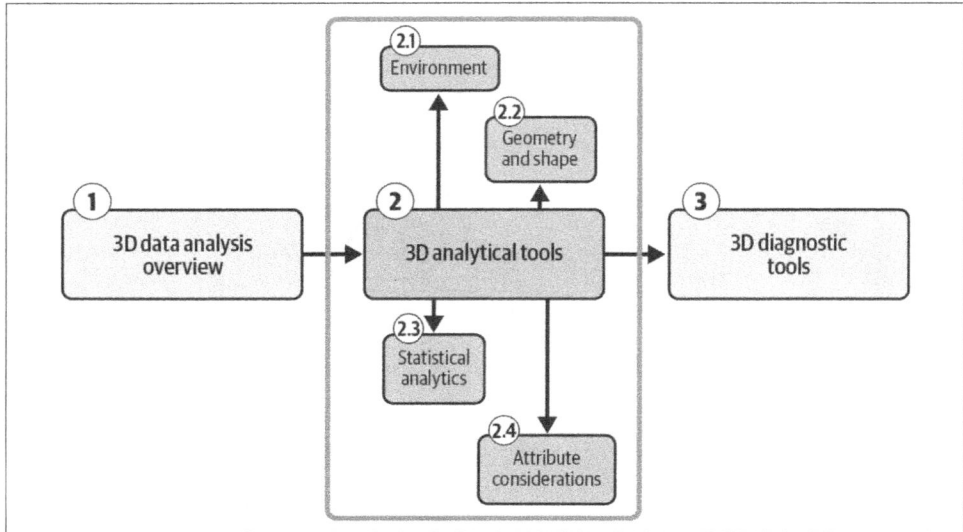

Figure 8-4. Overview of 3D data analysis tools, focusing on the analytical toolset

It is this high-level overview that will offer a snapshot of what we are going to play with. To best address this topic, let me decompose the 3D descriptive analysis workflow into five main steps: environment and data preparation, metadata analysis, geometry and shape analysis, statistical analysis, and attribute analysis, as shown in Figure 8-5).

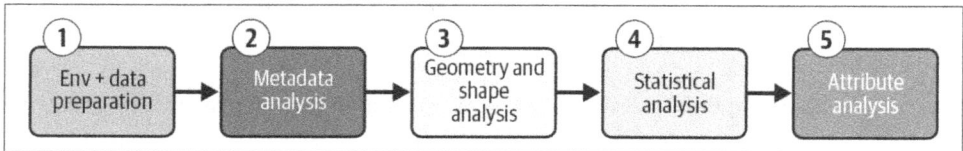

Figure 8-5. The complete 3D data analysis workflow, showing the progression from data preparation to attribute analysis

Environment and Data Preparation

The first step of the 3D data analysis workflow is ensuring we work on a sound dataset in a coherent environment. Let me detail both aspects to help replicate and extend the process from scratch.

Dataset

The dataset for this section represents a wool factory and was acquired by combining two techniques: terrestrial LiDAR data acquisition and 3D reconstruction with photogrammetry. The LiDAR data acquisition was made using a terrestrial laser scanner, which allowed me to gather 360° geometry representations of the surface for each station. Imagine that I placed the tripod in a specific location and launched my scanning process (stars in Figure 8-6). I gathered thousands of data points representing x, y, and z information. I dropped the intensity at this stage because RGB is predominant.

Once a scan location was done, I placed the terrestrial laser scanner in another spot between 5 and 15 meters from the initial one and launched another scan. I did that at various spots to capture each point from at least two stations.

The second process that I devised with the help of Roman Robroek (a fantastic photographer) involved taking many pictures with a super nice camera at various locations and fusing these camera poses with my scan poses (see the winding path in Figure 8-6). I also reconstructed 360° panoramas I took at the same spot as my laser scanning equipment (see the circles in Figure 8-6).

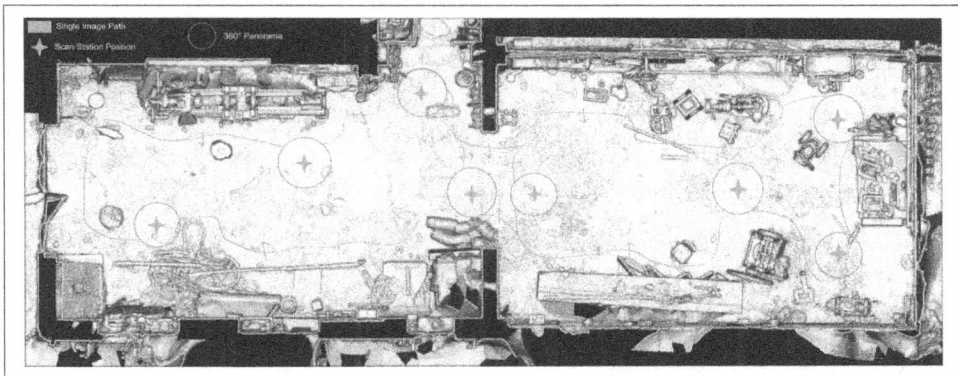

Figure 8-6. The 3D data acquisition setup (stars: the LiDAR scan locations; winding path: image path acquisition, the camera pointing inward [back to the wall]; circles: 360° camera acquisition, pointing outward)

These allowed us to generate the scene's geometric representation with high accuracy. The next stage is to go through the Chapter 4 method to extract additional features. Of course, at this stage, we are extracting features with a slight hint of what will be helpful, but we have no clear quantifiable or qualifiable information as to how they are instrumental. So, this is the process of creating the dataset and extracting the context around it.

To summarize the backlog and the starting point:

- First, I got on-site and carried out the 3D data acquisition using the described methodology.
- I then returned to my computer and processed the LiDAR data with the imagery to extract a 3D point cloud with RGB information.
- These 3D point clouds were injected into Chapter 4's code to compute many features.
- I added each feature to my data file and then exported it to a sampled version in ASCII file format (*.csv*) so that I could showcase the data analytics.

Let's now embark on the environment setup for this phase.

Environment setup

To be faithful to previous chapters, I set up my environment with Anaconda and Python 3.9 on a machine running the Windows 11 OS. In my environment, I install pip:

```
conda install pip
```

Then, using pip, I install all the libraries (NumPy (*https://numpy.org*), pandas (*https://oreil.ly/oE02A*), SciPy (*https://scipy.org*), Open3D (*https://www.open3d.org*), Matplotlib (*https://matplotlib.org*), and Seaborn (*https://oreil.ly/E5led*)):

```
pip install numpy pandas scipy open3d matplotlib seaborn
```

The last stage is to use an IDE; if you do not have one available, you can use Spyder:

```
pip install spyder
```

Now, you launch Spyder (type **spyder** in the console), and we can get on to setting up the script by first importing libraries:

```
#%% Importing libraries

# Base libraries
import numpy as np
import pandas as pd

# Visualization and plotting libraries
import matplotlib.pyplot as plt
import seaborn as sns

# Mathematical libraries
import scipy as sp

# 3D library
import open3d as o3d
```

First, we see the import statements for NumPy (np) and pandas (pd). These libraries are fundamental for working with numerical data (arrays, matrices) and tabular data (DataFrames), respectively. Then, the code imports SciPy (sp), a collection of algorithms and functions for scientific computing. SciPy complements NumPy by offering additional functionality for advanced mathematical operations, optimization, signal processing, and other scientific computing tasks.

As for 3D libraries, we use Open3D; there is no need to present the library. There are two choices for data visualization libraries: Matplotlib (plt) creates various plots and charts, while Seaborn (sns) builds on top of Matplotlib to provide a high-level interface for creating aesthetically pleasing and informative statistical graphics. These libraries allow us to explore and communicate insights from the data visually.

> I want to share some excellent functionalities to overhaul the visualizations that we create. We can use some styling prompts to ensure we have a dark background and coherent DPI outputs. Here are the prompts:
>
> ```
> plt.style.use('dark_background')
> plt.rcParams['figure.dpi'] = 600
> ```

We are ready to move on to the next stage: understanding our various variables.

Metadata Analysis and Data Profiling

We need to structure our approach moving forward. If you remember Chapter 1, there was logical thinking about how we proceed with any data processing pipeline. Let's make an exception and approach 3D DDA with a straightforward tactic.

First, we will go to the profiling stage to gain a broad perspective. To get a comprehensive overview of the data, let's first load it and store it in a pandas DataFrame named df_pcd:

```
df_pcd = pd.read_csv('../DATA/verviers_features.csv', delimiter= ' ')
```

This will allow us to utilize various pandas functionalities to summarize the data's characteristics quickly. The first helpful functionality provides a concise overview of the DataFrame, including information about data types, nonnull values (count) for each column, and memory usage:

```
df_pcd.info()
```

This helps identify potential data quality issues like missing values or unexpected data types. Then, we check the number of rows (observations) and columns (features) present in the data:

```
df_pcd.shape
```

This permits us to understand the overall vector shape of our matrix. To complement the first profile, I usually get a glimpse into the actual data points and their values with this line of code:

```
df_pcd.head()
```

This method displays the first few rows (default: 5) of the DataFrame. It allows us to ensure the format and content of the data. Then, we check the columns in the DataFrame:

```
df_pcd.columns
```

This clarifies the meaning associated with each feature in the data. Finally, we can use a powerful method that provides various summary statistics for numerical columns in the DataFrame:

```
df_pcd.describe()
```

It calculates statistics like mean, standard deviation, minimum, maximum, percentiles (quartiles), etc. This helps us understand the numerical features' central tendency, spread, and potential outliers. At this stage, we now have a clearer view of the following: our dataset consists of 500,000 points, each with 20 features on top of x, y, and z. Some features are normalized, and others are not. Their range is different, and the distribution is also uncommon.

The *.csv* file does not have external metadata associated with it. We can now move on to better delineate the 3D geometry and shape analysis.

Geometry and Shape Analysis

Let us start by creating an Open3D `PointCloud` object; we then fill in the points and color attributes of this new Python object and leverage implemented functions to speed up the analysis:

```
xyz_o3d = o3d.geometry.PointCloud()
xyz_o3d.points = o3d.utility.Vector3dVector(np.array(df_pcd[['X','Y','Z']]))
xyz_o3d.colors = o3d.utility.Vector3dVector(np.array(df_pcd[['R','G','B']]))/255
```

We are now ready to extract our overall dimensions.

Overall dimensions

The first stage is to compute the axis-aligned bounding box (`aabb`) to better delineate the overall extent of our scene/object:

```
aabb = xyz_o3d.get_axis_aligned_bounding_box()
aabb.color = [1,0,0]
```

We can now visualize our 3D point cloud with the axis-aligned bounding box:

```
o3d.visualization.draw([xyz_o3d, aabb], bg_color = (1, 1, 1, 1),
show_skybox = False, line_width = 5)
```

This also allows us to compute the center of the bounding box, as well as the point cloud centroid:

```
centroid = xyz_o3d.get_center()
aabb_center = aabb.get_center()
```

This returns:

```
centroid:           array([ 0.478817 , -2.8190636,  1.3072453])
aabb_center : array([-0.14883376, -3.33756063,  1.68849848])
```

As you can see, we have a difference, which is normal given the irregular distribution of our points.

Volume estimation

Now let's move onto volume estimation, as this is key toward our objective of understanding the space that can be navigated through:

```
# Volume Estimation (BB)
p_aabb_volume = aabb.volume()
print('Volume of AABB: ',p_aabb_volume)
```

This returns the following:

```
Volume of AABB: 605.82 m³
```

However, at this stage, it is interesting to note that the extracted volume is from the axis-aligned bounding box. If we are not in such a scenario (which is very likely), we would need to extract the object's shape based on the oriented bounding box with the smallest volume:

```
oobb = xyz_o3d.get_minimal_oriented_bounding_box()
```

The approach takes advantage of the fact that the minimum bounding box must have an edge that is collinear with at least one edge of the convex hull. For each triangle in the hull, find the minimal axis-aligned box in the triangle's frame. Finally, it returns the container with the least amount of space within. What does that mean? We can find out:

```
oobb.color = [0,1,0]
oobb.volume()
o3d.visualization.draw_geometries([xyz_o3d, oobb, aabb])
```

As you can see, this returns a light gray-oriented bounding box that differs from the dark gray axis-aligned bounding box (see Figure 8-7). As for the volume, the best-fitting one now holds: 475.83 m³.

Figure 8-7. Top view of a point cloud of a room, showing the arrangement of objects and spatial features

This is based on the convex hull that you can visualize:

```
# Convex hull
ch = xyz_o3d.compute_convex_hull()
# returns the triangle mesh and the list of idxs of the points

o3d.visualization.draw_geometries([ch[0], xyz_o3d.select_by_index(ch[1])])
ch[0].get_volume()
```

> 3D axis-aligned bounding boxes (AABBs) and 3D oriented bounding boxes (OBB) are leveraged in various applications. They are used for collision detection in video games, object tracking in airports, occlusion culling for rendering complex scenes, path planning, ray tracing, and the list goes on. The choice between AABB or OBB depends on the trade-off between accuracy and performance. AABBs are simpler to calculate and generally faster to test for overlap than OBBs but are less accurate.

After studying the extent and characteristics of our scene through a global analysis, let's move on to local neighborhood-based analysis.

Planarity

The first stage assesses the "planar" predominance in the scene. This is crucial as planar surfaces are ubiquitous in the real world, from the walls of buildings to organic shapes. Analyzing the planarity of 3D data allows us to understand the underlying structure and geometry of objects, which is essential for understanding their shape and function and identifying defects or anomalies. With Python, this translates into the following:

```
print(df_pcd['Planarity_(0.1)'].describe())
pla_as_colors = np.repeat(np.array(df_pcd[['Planarity_(0.1)']]), 3, axis=1)
xyz_o3d.colors = o3d.utility.Vector3dVector(pla_as_colors)

# Changing the colormap
viridis_cmap = plt.colormaps['viridis']
viridis_array = viridis_cmap(pla_as_colors)
xyz_o3d.colors = o3d.utility.Vector3dVector(viridis_array[:,0,:3])
o3d.visualization.draw_geometries([xyz_o3d])
```

This shows that most of our point clouds belong to planar regions, with some local variation (see Figure 8-8). Now, what about their curvature?

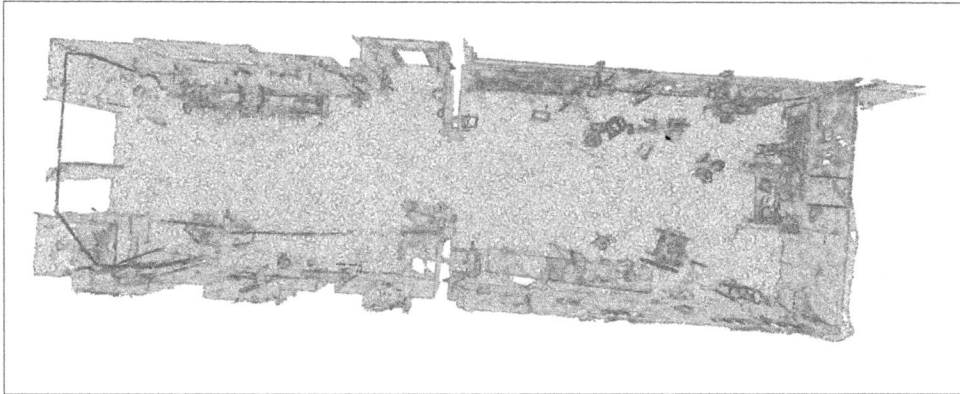

Figure 8-8. Alternative top view of a room point cloud: the shade variation suggests another attribute is being visualized

Curvature

Curvature is a fundamental property of 3D datasets that reveals essential information about their shape and structure. Analyzing curvature allows us to identify critical features and better understand local and global geometry. Let me take a step back.

In medical imaging, curvature analysis helps detect abnormalities in organs and tissues, aiding in diagnosis and treatment planning. In general applications, it is used for object recognition, pose estimation, and surface reconstruction and to enable accurate navigation and obstacle avoidance. Moreover, curvature can be used to segment 3D objects, extract features, and classify different types of surfaces. It can also be employed for surface smoothing, interpolation, and deformation.

This means that by understanding the curvature of our 3D dataset, we can gain valuable insights into its underlying properties, and we can decide on which downward processes should be used. First, let us profile the feature:

```
print(df_pcd['Mean_curvature_(0.1)'].describe())
```

This showcases something interesting: `max = 50.160545`. This hints that we need data normalization.

If you remember from Chapters 3 and 4, 3D data normalization is the process of transforming 3D data to a standard scale or range. This is often necessary before applying data analysis techniques to 3D data. The common 3D data normalization techniques include:

Min-max scaling
> Scales the data to a range between 0 and 1. It is simple to implement and works well for bounded data (Chapter 12).

Z-score normalization
> Transforms the data to have a mean of 0 and a standard deviation of 1. It is more robust to outliers than min-max scaling (Chapter 16).

Unit sphere normalization
> Scales data to fit within a unit sphere, preserving its shape.

Local normalization
> Normalizes each point in the data relative to its local neighborhood. It is helpful for data with nonuniform distributions (Chapter 11).

PCA normalization
> This method uses PCA to transform the data into a new coordinate system, which can help reduce dimensionality and remove noise (e.g., Chapters 6 and 7).

The best technique depends on the application and properties of the data. For example, min-max scaling is relevant for bounded data with no outliers, while z-score normalization is better for unbounded data with outliers. In our case, min-max scaling is perfectly adapted:

```
# Apply min-max scaling in pandas using the .min() and .max() methods
def min_max_scaling(df):
    # copy the dataframe
    df_norm = df.copy()
    # apply min-max scaling
    for column in df_norm.columns:
        df_norm[column] = (df_norm[column] - df_norm[column].min()) /
        (df_norm[column].max() - df_norm[column].min())

    return df_norm
```

We can now colorize using the returned normalized feature vector, as our function is now defined:

```
# Call the min_max_scaling function
df_pcd_normalized = min_max_scaling(df_pcd)
df_pcd_normalized['Mean_curvature_(0.1)'].describe()
```

```
# Colorize based on the new normalized curvature
curv_as_colors = np.repeat(np.array(df_pcd[['Mean_curvature_(0.1)']]), 3, axis=1)
viridis_array = viridis_cmap(curv_as_colors)
xyz_o3d.colors = o3d.utility.Vector3dVector(viridis_array[:,0,:3])
o3d.visualization.draw_geometries([xyz_o3d])
```

This computation showcases how the walls mix planar and curvature-full regions (see Figure 8-9). These new insights hint that any downward process (e.g., shape recognition, model fitting, 3D segmentation) has to account for unevenly distributed curvature zones. Let's push our set of revelations by trying to extract the primary orientation of any 3D dataset (a global view).

Figure 8-9. Top view of a point cloud representing a room's interior, gradient based on height or another attribute

Main orientation

The main orientation is very relevant, especially when we investigate single objects. Let's consider that the object is a 3D model. The main orientation helps to standardize the 3D model's position, making it easier to compare and analyze different models. It also optimizes the model for 3D printing or manufacturing processes and can be used to understand the object's shape and function.

Relying on PCA (see Chapter 6 or the 3D Resource Hub (*https://learngeodata.eu/3d-data-science-with-python*) for more info) is a great practice for getting such information. Let's define a PCA function that is slightly different from the one shown in Chapter 6:

```
xyz_np = np.array(xyz_o3d.points)
def pca(point_cloud):
    ...
    return eigenvalues, eigenvectors

eig_val, eig_vec = pca(xyz_np)
```

```
principal_dir = eig_vec[:, 0]
print(principal_dir)
```

This returns a vector (in our case, close to the x-axis): [0.999 -0.015 0.006].

With the main orientation, we obtain another "global signature" of our dataset. This can be further extended by analyzing the aspect ratio of the geometric model.

Aspect ratio

Determining the aspect ratio of a 3D model helps us understand the model's proportions and overall shape. For example, if a model's aspect ratio is incorrect, it may not fit into a desired space or look realistic. Additionally, if the aspect ratio is not optimized for 3D printing, producing the model may be difficult or impossible. So, let us work with the oriented bounding box and extract its length (L), width (l), and height (h):

```
#Aspect Ratio
L, l, h = oobb.extent
```

From here, we compute three ratios:

```
ratio_Lh, ratio_lh, ratio_Ll  = [L/h, l/h, L/l]
```

This can give us indications for downward processes, such as 3D shape classification. Indeed, these ratios can be leveraged to classify if a 3D model is a table or a chair.

At this stage, you have a great set of complementary information and global signatures to help you understand the geometric implications of our subject. To be as thorough as possible, let me quickly showcase how these aspect ratios can impact the analysis of 3D single object shapes.

The case of a single 3D mesh shape

In this section, let's take the example of a car and a plane. First, let's import two new datasets by defining the paths to the files:

```
meshes = ["../DATA/Car.ply", "../DATA/Soviet_Plane.ply"]
```

From there, we can use the following function to study the bounding-boxes differences as well as the aspect ratio implications:

```
for data in meshes:
    mesh = o3d.io.read_triangle_mesh(data)
    mesh.compute_vertex_normals()
    #Axis-Aligned BB
    aabb = mesh.get_axis_aligned_bounding_box()
    aabb.color = [1,0,0]

    # Oriented BB
    oobb = mesh.get_minimal_oriented_bounding_box()
    oobb.color = [0,1,0]
```

```
o3d.visualization.draw_geometries([mesh, oobb, aabb])

L, l, h = oobb.extent
ratio_Lh, ratio_lh , ratio_Ll  = [L/h, l/h, L/l]
print(ratio_Lh, ratio_lh, ratio_Ll)
```

As you can see, the car's OBB is elongated compared to the plane's (Figure 8-10A and B).

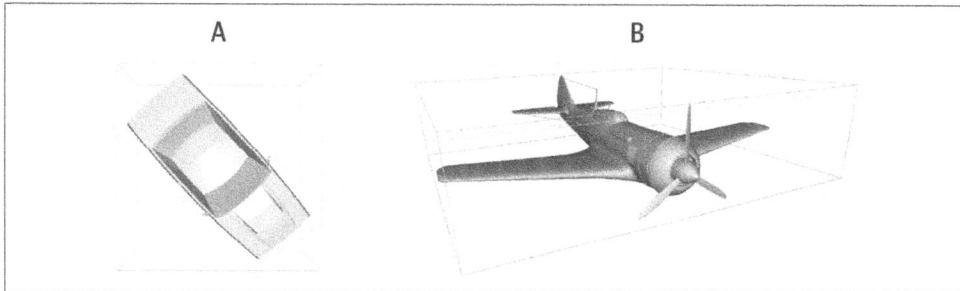

Figure 8-10. (A) A 3D car model with its bounding box; (B) a 3D mesh of an airplane enclosed within a bounding box, likely used for spatial analysis or alignment

How does this play out when we look at the ratios? Well, let's look at the results in Table 8-1.

Table 8-1. The ratio of the 3D model of the car versus the plane

Aspect ratio of the 3D car model			Aspect ratio of the 3D plane model		
L/h	l/h	L/l	L/h	l/h	L/l
3.25	1.46	2.22	4.02	3.54	1.13

As you can see, the difference between l/h and L/l efficiently relates to the varying ratios we observed visually. We can thus use these ratios to bypass human interaction through automated processes such as automatic 3D shape classification. Let's now illustrate this by giving you the tools for statistical analysis applied to 3D datasets.

Statistical Analysis

At this stage, we want to better understand the distribution of data points and features. In 3D data science, the trigger point is 3D geometry. Ensuring that we perfectly grasp the distribution of our spatial attributes is essential. Let's review a simple process to identify statistical insights quickly. To support this concept, let's work with a density-based pointer, the distance between points in our point cloud:

```
nn_distances = xyz_o3d.compute_nearest_neighbor_distance()
```

This permits us to record the mean distance to neighbors for each point in the point cloud.

> At this stage, you know how to leverage k-d trees, and you can easily create a function if you want to tweak the number of points used to compute the value or output the index of those points. Refer to Chapter 4 in that case.

First, let's study central tendencies. The idea is that one single value may best describe (to some extent) the feature under study. We can use central tendency measures to get a quick and simple idea of our feature's appearance. The center of our data can be a precious piece of information, telling us exactly how the feature is biased since whichever value the data revolves around is essentially a bias. We can think of the mean value as the average, or the point around which all of our data points cluster. To define the mean, we give equal weight to all of the values that go into the calculation:

```
mean_dist = round(np.mean(nn_distances),3)
print(f"Mean: {mean_dist} m")
```

In our case we have: `Mean: 0.017 m`. The median is the middle value of the dataset, i.e., if we sort the data from smallest to biggest (or biggest to smallest) and then take the value in the middle of the set, that's the median (50% is above, and 50% is below):

```
median_dist = round(np.median(nn_distances),3)
print(f"Median: {median_dist} m")
```

This returns: `Median: 0.016 m`.

> The median is more robust to outliers than the mean since the mean will be pulled one way or the other if there are some high-magnitude outlier values. Having a median close to the mean usually indicates that we are in a symmetrical distribution.

Now that we better understand central tendencies, quickly identifying the spread is usually good practice. The spread of our feature is the extent to which it is squeezed toward a single value or more spread out across a wider range. The standard deviation is the most common way of quantifying data spread:

```
std_dist = round(np.std(nn_distances),3)
print(f"Standard Deviation: {std_dist} m")
```

This returns a standard deviation of `0.008 m`. We can further describe the position of each data point throughout the range using percentiles. A percentile describes the exact position of the data point in terms of how high or low it is positioned in the range of values. More formally, the pth percentile is the dataset's value, which can be split into two main buckets. The first bucket, i.e., the lower part, contains p percent of the data, i.e., the pth percentile:

```
percentile25_dist = round(np.percentile(nn_distances, 25),3)
percentile50_dist = round(np.percentile(nn_distances, 50),3)
percentile75_dist = round(np.percentile(nn_distances, 75),3)
percentile95_dist = round(np.percentile(nn_distances, 95),3)
```

The returns are as follows: 25% percentile: 0.011; 50% percentile: 0.016; 75% percentile: 0.022; 95% percentile: 0.032. Finally, I would like to complete the dataset's statistical description by computing our data's skewness and mode. These statistical tools can be beneficial in understanding the shape of the distribution and identifying potential outliers. They will also make it easy to identify the dominant trend or pattern in the data, which can be crucial for understanding the underlying phenomena and making informed decisions. The skewness measures its asymmetry, while the mode is the value that appears most frequently in a data distribution:

```
skew_dist = sp.stats.skew(nn_distances)
mode_dist = sp.stats.mode(nn_distances)
print(f"Skew: {skew_dist}")
print(f"Mode: {mode_dist} m")
```

This returns a skew of `0.675` m and a mode: `ModeResult(mode=3.65e-05, count=2)`.

A positive value for the skewness means that values are concentrated on the left of the center of the data points; negative skewness means values focus on the right of the center. These are beautiful for building quantifiable analytical insights. However, it is usually good practice to accompany them with qualitative analysis supported by clear graphics. And this is where you can play on a wide range of statistical data visualization tools.

Histograms

Bar charts (or histograms) are an effective way to visualize data distribution. They can help compare categories, identify outliers, and track trends. They are handy for understanding the spatial distribution of data. For example, a bar chart could show the number of objects of different types in a 3D scene, the average distance, or the distribution of points in a point cloud. Let's illustrate how we can use histograms to visualize a single feature distribution with this little Python snippet:

```
plt.hist(nn_distances)
plt.xlabel("Values (m)")
plt.ylabel("Frequency")
plt.title("Distribution of Neighbor Distances")
plt.show()
```

The code results in a histogram accompanied by an image that showcases the distribution of neighbor distances that the nn_distances variable holds (see Figure 8-11A). But it may lack "resolution" in how fine we want to see the distribution. On top of that, it would be great to overlay the mean, the median, and the mode, as previously defined. For this purpose, you can leverage the following function:

```
def hist_styled(feature, f_name):
    # Create histogram
    n, bins, patches = plt.hist(feature, bins=100, facecolor='#2ab0ff',
    edgecolor='#e0e0e0', linewidth=0.5, alpha=0.7, zorder=4)
    n = n.astype('int') # it MUST be integer

    # Good old loop. Choose the colormap of your taste
    for i in range(len(patches)):
        patches[i].set_facecolor(plt.cm.cool(n[i]/max(n)))

    # Calculate statistics
    mean_val = np.mean(feature)
    median_val = np.median(feature)
    mode_val = bins[np.argmax(n)]

    # Add vertical lines for mean, median, and mode
    plt.axvline(mean_val, color='lightcoral', linestyle='dashed', linewidth=2,
    label=f'Mean: {mean_val:.3f}')
    plt.axvline(median_val, color='palegreen', linestyle='dashed', linewidth=2,
    label=f'Median: {median_val:.3f}')
    plt.axvline(mode_val, color='skyblue', linestyle='dashed', linewidth=2,
    label=f'Mode: {mode_val:.3f}')

    # Set title and labels
    plt.title(f'{f_name}-feature: Distribution', fontsize=10)
    plt.xlabel(f'{f_name} values', fontsize=10)
    plt.ylabel('Point Numbers', fontsize=10)

    # Add grid
    plt.grid(True, c = 'grey', ls = '--', lw = 0.2, zorder=0)

    # Add legend
    plt.legend(fontsize=7)

    plt.show()
    return
```

This allows us to use and create a qualitative analytical histogram combined with the mean, median, and mode under a specific distribution. Let's use the function hist_styled(nn_distances, 'NN_Distance'), which results in Figure 8-11B.

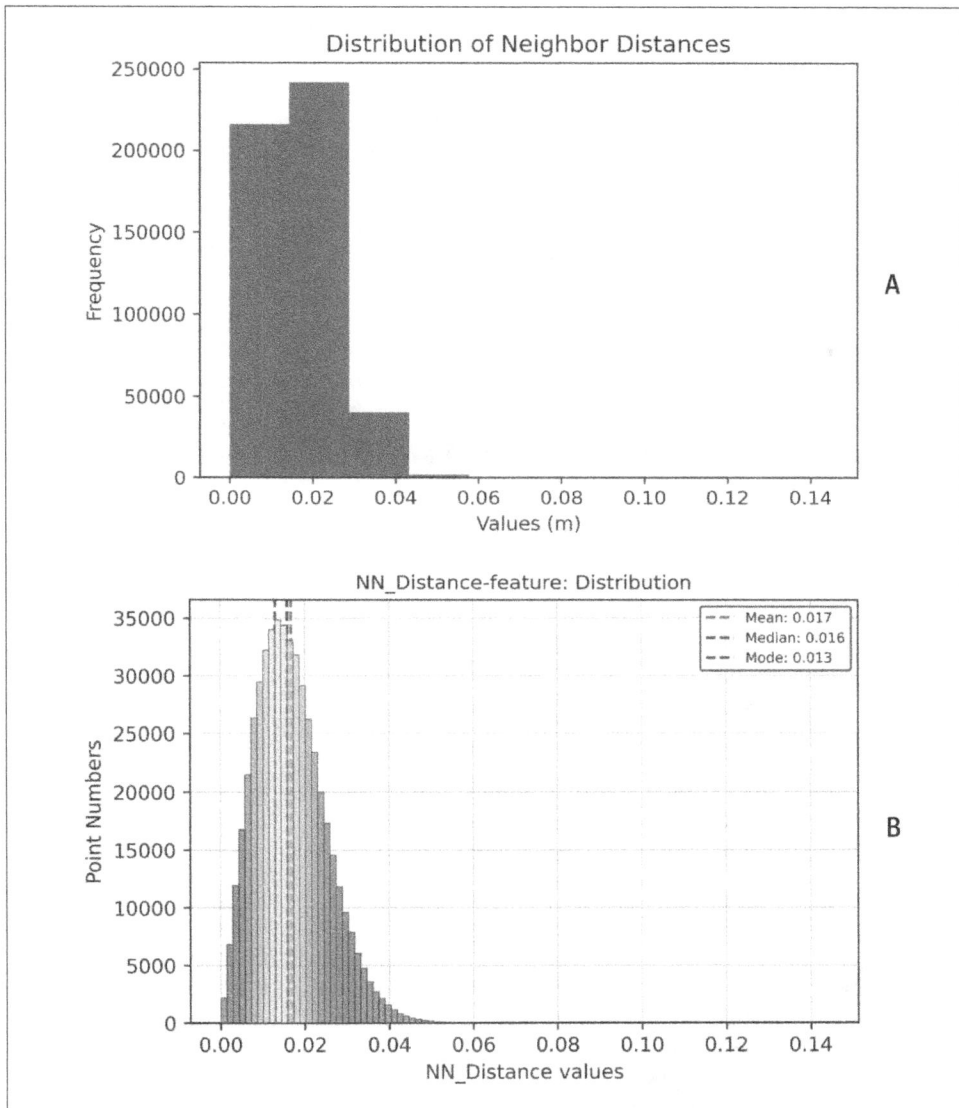

Figure 8-11. (A) Histogram illustrating the distribution of distances between neighboring points in a point cloud; (B) distribution of nearest neighbor distances within a point cloud, including mean, median, and mode

Another way to visualize spatial distribution is using the equivalent of histograms in 3D: 3D bar charts. By adding a third dimension to traditional bar charts, these visualizations enable us to explore complex relationships between variables and identify patterns that might be hidden in lower-dimensional representations. This is particularly useful for understanding data distribution across multiple dimensions. To create such a 3D bar chart with Python, we can use Matplotlib's bar3d and combine it with a script like the following:

```
import matplotlib.colors as colplt
import matplotlib.cm as cm

fig = plt.figure()
ax = fig.add_subplot(projection='3d')
hist, xedges, yedges = np.histogram2d(df_pcd['X'], df_pcd['Y'], bins=100)

xpos, ypos = np.meshgrid(xedges[:-1] + 0.1, yedges[:-1] + 0.1, indexing="ij")

#Shortening and making sure we comply to dimensions
xpos, ypos, zpos = [xpos.ravel(), ypos.ravel(), 0]

dx = dy = 0.1 * np.ones_like(zpos)
dz = hist.ravel()

offset = dz + np.abs(dz.min())
fracs = offset.astype(float)/offset.max()
norm = colplt.Normalize(fracs.min(), fracs.max())
colors = cm.cool(norm(fracs))

ax.bar3d(xpos,ypos,zpos,dx,dy,dz, color=colors)
plt.show()
```

This returns the bar plot in Figure 8-12A.

This Python code creates a 3D bar plot using Matplotlib. It starts by importing necessary modules, then creates a figure and 3D axis object. It then uses the np.histogram2d() function to generate a 2D histogram of the X and Y columns from a DataFrame called df_pcd. The code then sets up the necessary coordinates and dimensions for the 3D bars, including the bar heights dz, which are based on the 2D histogram values. It applies a color mapping to the bars using the cm.cool() colormap and finally displays the 3D plot using plt.show().

This is fantastic for refining our initial visualization by understanding where we have consistency in the data distribution. Using this feature as a filtering pointer would, for example, allow us to identify better the points that belong to the floor. Now, there is also the possibility of showcasing the density. But of course, monitoring if the distribution fits some classical distribution model is always essential. The most common is the normal distribution. We can compute the Gaussian to understand better how well we fit the data. You can add these lines to the previous function to have the plotting function compute the distribution model:

```
# Compute the Gaussian curve
mu, sigma = sp.stats.norm.fit(feature)
x = np.linspace(min(feature), max(feature), 100)
gaussian = sp.stats.norm.pdf(x, mu, sigma)

# Scale the Gaussian to match the histogram height
scaling_factor = np.max(n) / np.max(gaussian)
gaussian_scaled = gaussian * scaling_factor
```

This results in Figure 8-12B: a histogram that can provide valuable insights into the distribution of data. The histogram resembles a bell-shaped curve, suggesting that the data is normally distributed. As a reminder, key indicators include a symmetrical shape, where the left and right sides of the curve are approximately mirror images of each other. Additionally, the peak of the curve should be located near the center. While a perfect normal distribution is rare in real-world data, these characteristics provide strong evidence of a close approximation.

This hints at a homogeneous point density for the most part. If we can identify the outliers, we can have a close-to-perfect normal distribution, which is much easier to handle for downward processes.

As an example of what we achieved with these steps, we could use our previous pandas DataFrame and isolate another feature with a selector sel:

```
sel = feature_names[2]
print(f"{sel} statistics: \n {df_pcd[sel].describe()}")
```

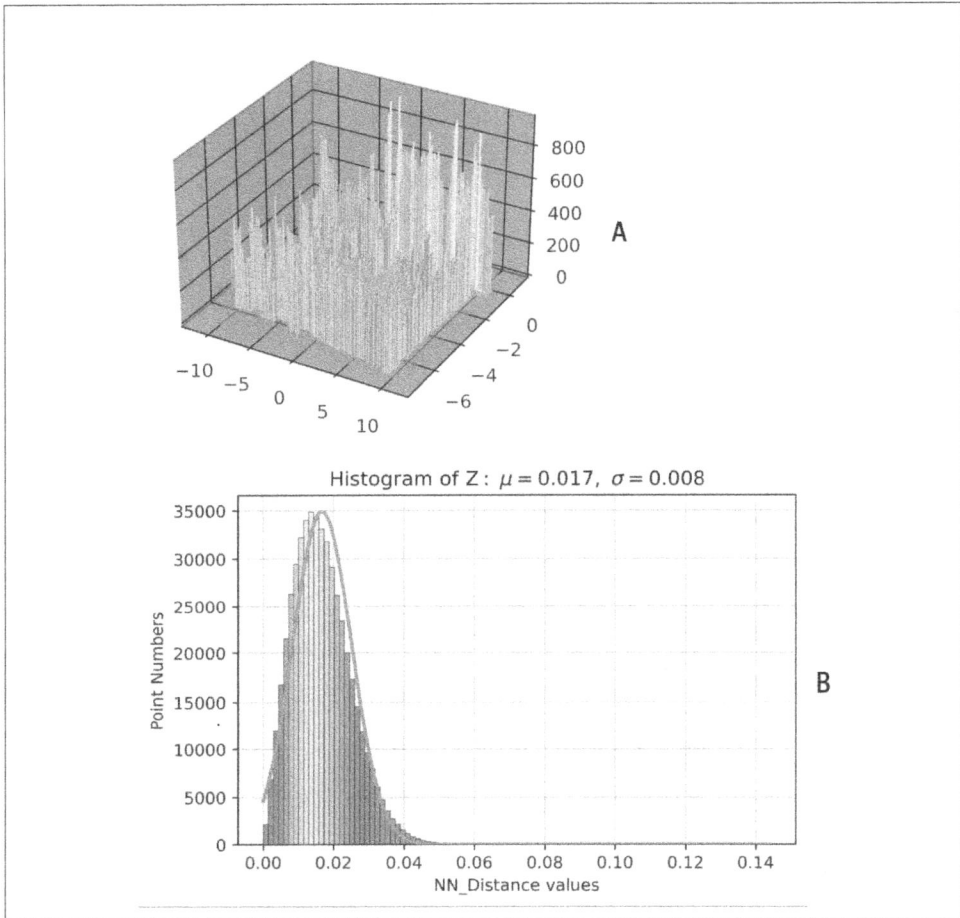

Figure 8-12. (A) 3D histogram showing the spatial distribution of points within a point cloud, colored by density; (B) histogram of nearest neighbor distances with a fitted Gaussian distribution

This includes the major statistics and we see that the `DataFrame` Z column contains 500,000 rows with a mean value of 1.3 and a standard deviation of 0.89, indicating moderate variability in the data. The minimum value is 0.09, while the 25th, 50th, and 75th percentiles are 0.44, 1.09, and 2.00, respectively. The maximum value in the Z column is 3.27, providing a comprehensive summary of the distribution of values in this dataset. Our plot function allows us to create the image in Figure 8-13A. This is highly informative, as we can immediately see that our feature Z does not correspond to a normal distribution. The mode is also very revealing as it could be used to identify potential peaks in our dataset that are likely linked to the floor. We can also study our distribution against other models (see Figure 8-13A and B).

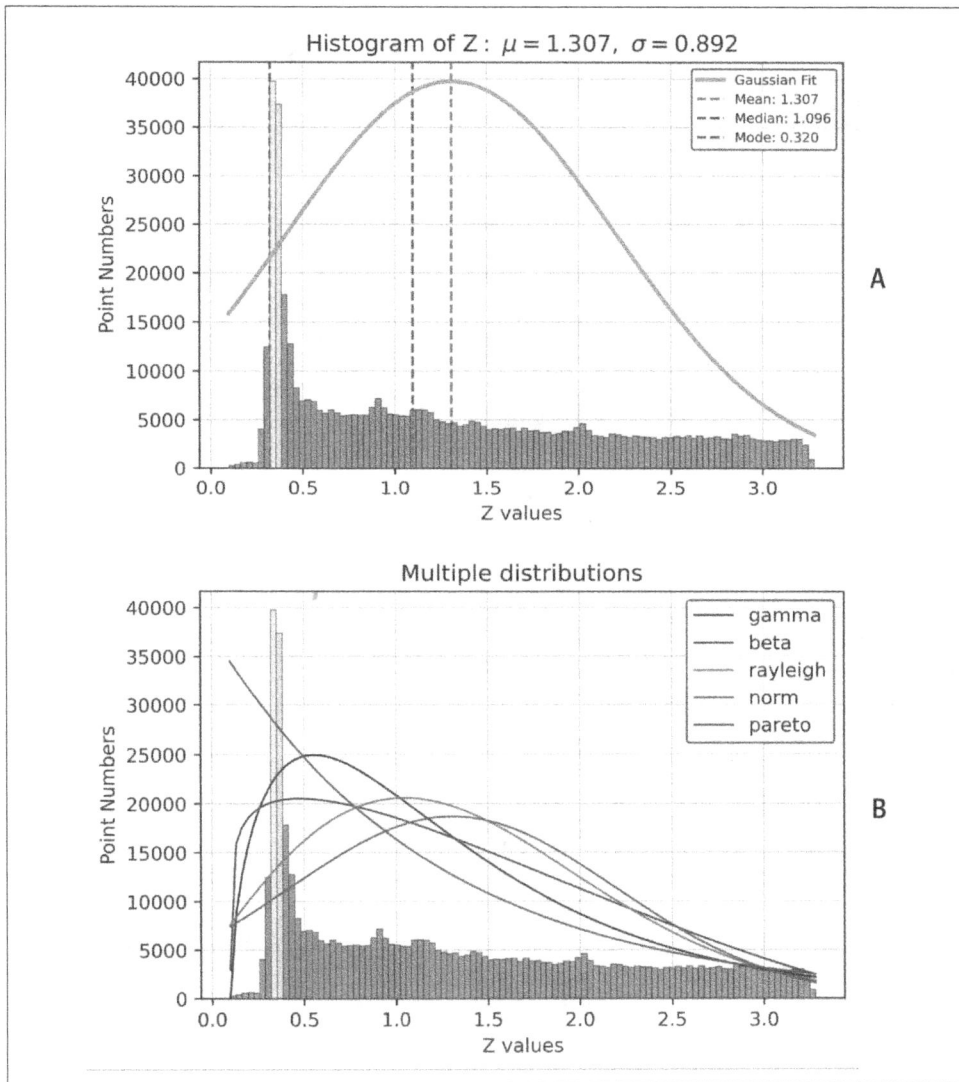

Figure 8-13. Histograms of (A) Z values fitted with a Gaussian distribution, displaying mean, median, and mode; (B) Z values overlaid with fitted gamma, beta, Rayleigh, normal, and Pareto distributions

Comparing a distribution to multiple models offers several advantages. First, it allows us to assess the fit of different theoretical distributions to the observed data, helping us understand the underlying data-generating process. Second, it enables us to identify potential outliers or anomalies that deviate significantly from the expected distribution. Third, it provides a quantitative measure of how well each model represents the data, allowing us to select the most appropriate model for further analysis. Finally, it helps us to make more accurate predictions and inferences based on the chosen model.

As you can see, leveraging histograms can be very handy for all 3D data analytical tasks. But we also have another great tool at our disposal: boxplots.

Boxplot for 3D analytics

Boxplots are a valuable tool for visualizing the distribution of 3D data. They concisely summarize the data's central tendency, spread, and potential outliers. By examining the boxplot, we can quickly identify patterns, anomalies, and significant variations within the data.

I have found this information to be crucial for understanding the underlying characteristics of any 3D dataset and for making informed decisions about further analysis or modeling tasks. It is only natural that we review a way to leverage boxplots for your analytical scenarios. Let's define a function that we can use to generate boxplots of single variables properly:

```python
def boxplot_styled(feature, f_name):
    plt.boxplot(feature, patch_artist=True,
    boxprops = dict(facecolor = '#08F7FE', alpha = 0.7),
    medianprops = dict(color = '#FE53BB', linewidth = 1), showfliers = False,
    labels=[f_name])

    plt.ylabel("Value")
    plt.title(f'{f_name}-feature: BoxPlot', fontsize=10)
    plt.grid(True, c = 'grey', ls = '--', lw = 0.1, zorder=0)
    plt.show()
    return
```

This Python function `boxplot_styled` creates a customized boxplot visualization for a given feature (dataset column) and feature name. It takes two parameters: `feature` (the data for the boxplot) and `f_name` (the name of the feature). The function uses Matplotlib to create the boxplot, with customized styling including a light blue box color, a pink median line, and no outliers displayed. It also adds a y-axis label, a title, and a gray grid background to the plot. We then display the box plot using `plt.show()` and return the plot. Let's use the function:

```
boxplot_styled(nn_distances, 'NN_Distance')
```

This returns Figure 8-14a. Boxplots can also be combined simultaneously for a first step toward correlation analysis. For this, multiple boxplots for several variables in a 3D dataset provide a powerful visual tool for comparing distributions and identifying relationships between attributes. By visualizing each variable's central tendency, spread, and outliers, we can quickly detect patterns, anomalies, and potential correlations. This can be particularly useful for understanding the impact of one variable on another, identifying influential factors, and tuning the downward data cleaning, preprocessing, and feature engineering steps. To create such a multiple boxplot figure, you can leverage the following Python code:

```
colors = ['#08F7FE', '#FE53BB']
plt.boxplot(df_pcd[['X', 'Y', 'Z']], patch_artist=True,
boxprops = dict(facecolor = colors[0], alpha = 0.7),
medianprops = dict(color = colors[1], linewidth = 1.5), labels=['X', 'Y', 'Z'])
plt.xlabel("Dimension")
plt.ylabel("Value")
plt.title("Boxplot of 3D coordinates")
plt.grid(True, c = 'grey', ls = '--', lw = 0.1, zorder=0)
plt.show()
```

This results in Figure 8-14A and B.

All these tools and strategies are currently applied to the entire point cloud. Naturally, their full power emerges when you combine them, resulting in multiple granularity levels of analysis, e.g., studying a local subset of your dataset. You can push the concept even more by creating multilevel analytical frameworks that leverage analysis at several granularities per variable.

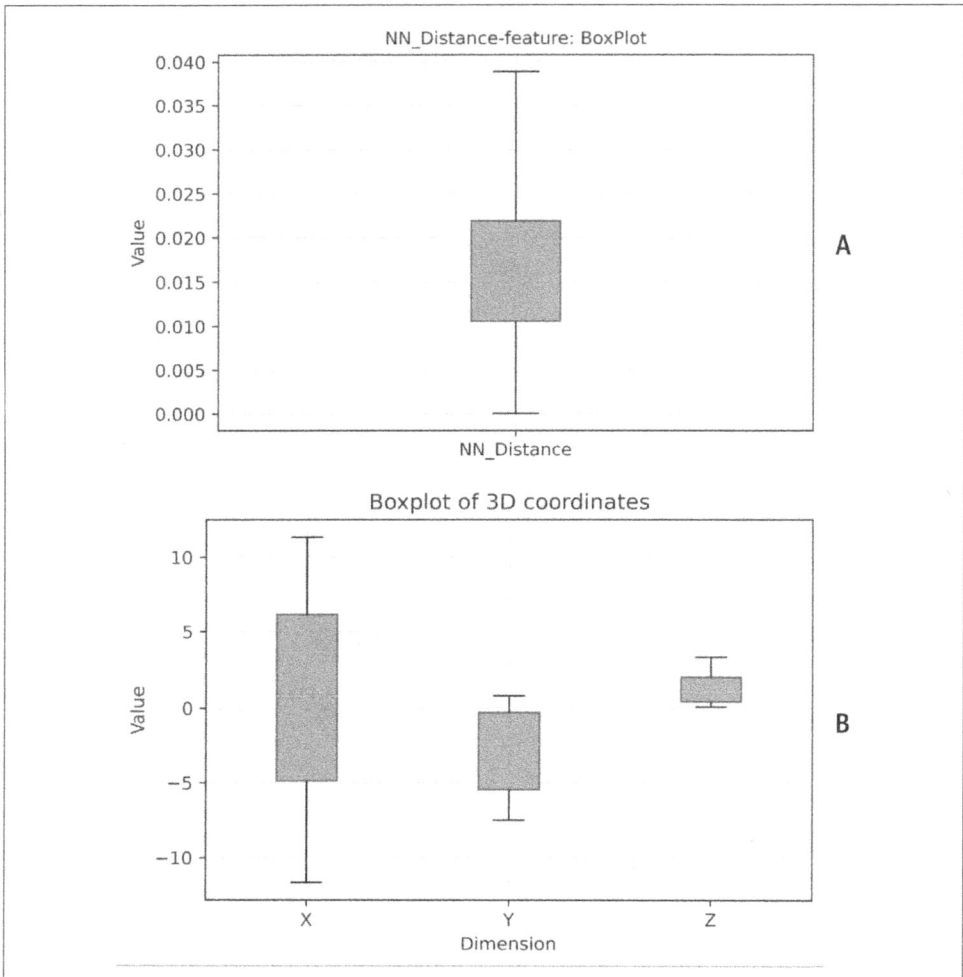

Figure 8-14. Boxplot visualizations displaying (A) the distribution of nearest neighbor distances within a point cloud and (B) the distributions of X, Y, and Z coordinates within a point cloud

Now, how can you acquire some hints about possible correlations within your dataset? First, let's plot the boxplot of all of our features within the base dataset (see Figure 8-15).

This specific statistical visualization would prove very useful, for example, when trying to understand which features of the underlying dataset play a large descriptive role. Features with a tight distribution usually have the potential to be great filters. But what about trying to understand feature correlations?

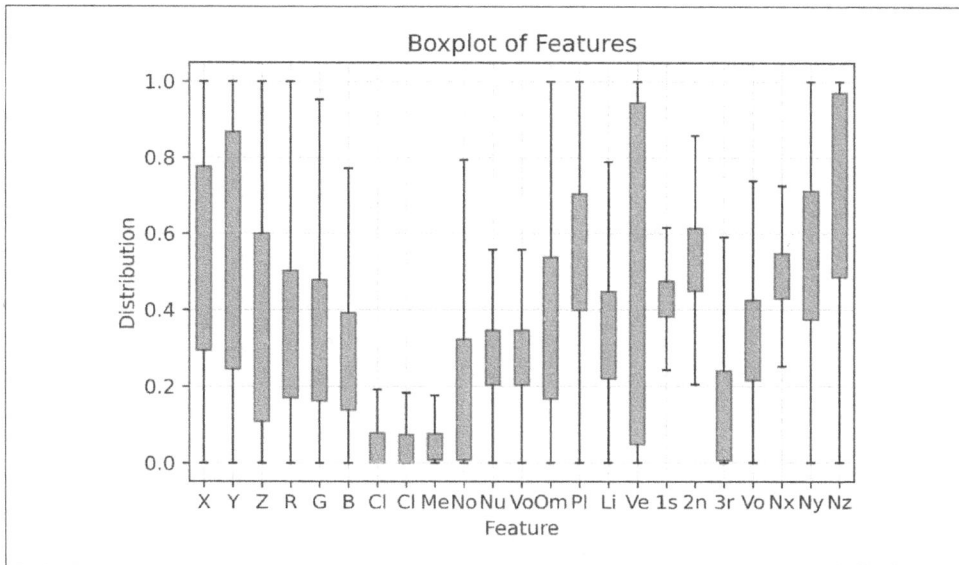

Figure 8-15. Boxplot visualization showing the distribution of all extracted features from the point cloud dataset

Correlations

We can leverage the Seaborn library to generate density plots of multiple features at once.

> A density plot is a visualization that shows the distribution of a dataset over a continuous interval. It is similar to a histogram but uses a smooth line to estimate the probability density function of the variable. The y-axis in a density plot represents the probability density, and the area under the curve always sums to 1.

To achieve this, we use a function that creates a 3D kernel density estimation (KDE) plot using the Seaborn and Matplotlib libraries:

```
sns.kdeplot(df_pcd[['X', 'Y', 'Z']], palette = 'cool', fill = True)
plt.grid(True, c = 'grey', ls = '--', lw = 0.2, zorder=0)
sns.kdeplot(df_pcd[['R', 'G', 'B']], palette = 'cool', fill = True)
plt.grid(True, c = 'grey', ls = '--', lw = 0.2, zorder=0)
sns.kdeplot(df_pcd[[ei1, ei2, ei3]], palette = 'cool', fill = True)
plt.grid(True, c = 'grey', ls = '--', lw = 0.2, zorder=0)
```

This code creates a 3D density plot of the point cloud data, showing where points are concentrated in 3D space, and adds a subtle gray dashed grid in the background for reference (see Figure 8-16A, B, and C).

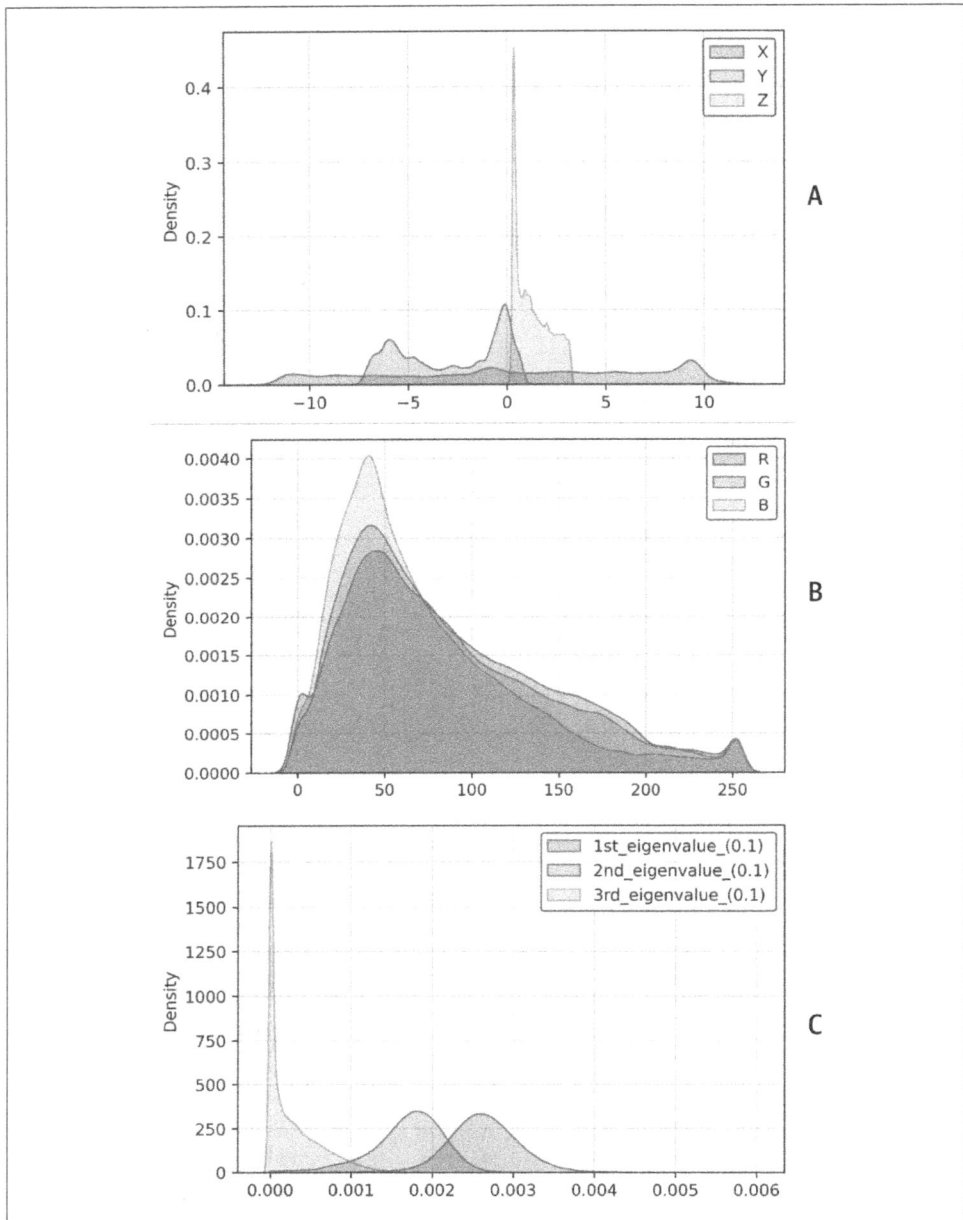

Figure 8-16. Density distributions within a 3D point cloud of (A) X, Y, and Z coordinates within a 3D point cloud; (B) red, green, and blue color channel values; (C) the first three eigenvalues

This visualization helps us uncover underlying structures or patterns that might not be apparent when examining individual distributions. It allows for identifying outliers or isolated regions within the point cloud. The plot's shape indicates a little symmetry in the data distribution across different dimensions. In the X, Y, and Z plot, multiple distinct peaks are present, suggesting separate clusters or modes within the 3D space. The R, G, and B showcase a stretch in the density plot that may indicate correlations. Lastly, the overall flatness or variability of the density plot provides information about the uniformity of point distribution throughout the 3D space.

These insights offer a comprehensive view of the data's structure and distribution in a single visualization and are a first step toward correlation analysis. Let's calculate the correlation matrix (pairwise correlations between all columns in the DataFrame) of the DataFrame 'df_pcd' using the corr() method, and visualize it using Seaborn's heatmap function:

```
corr = df_pcd.corr()# plot the heatmap
sns.heatmap(corr, xticklabels=corr.columns, yticklabels=corr.columns,
annot=False, cmap=sns.diverging_palette(220, 20, as_cmap=True))
```

A heatmap's color-coding system (black for highly correlated, white for weakly correlated) allows us to grasp the direction and intensity of any association between variables (see Figure 8-17).

This lets us quickly identify that the third eigenvalue highly correlates with several features and the Z component with verticality. This makes perfect sense as these features are derived from each other. Therefore, if we were to go on to select relevant features, we would prefer to use these and drop some of the derived features.

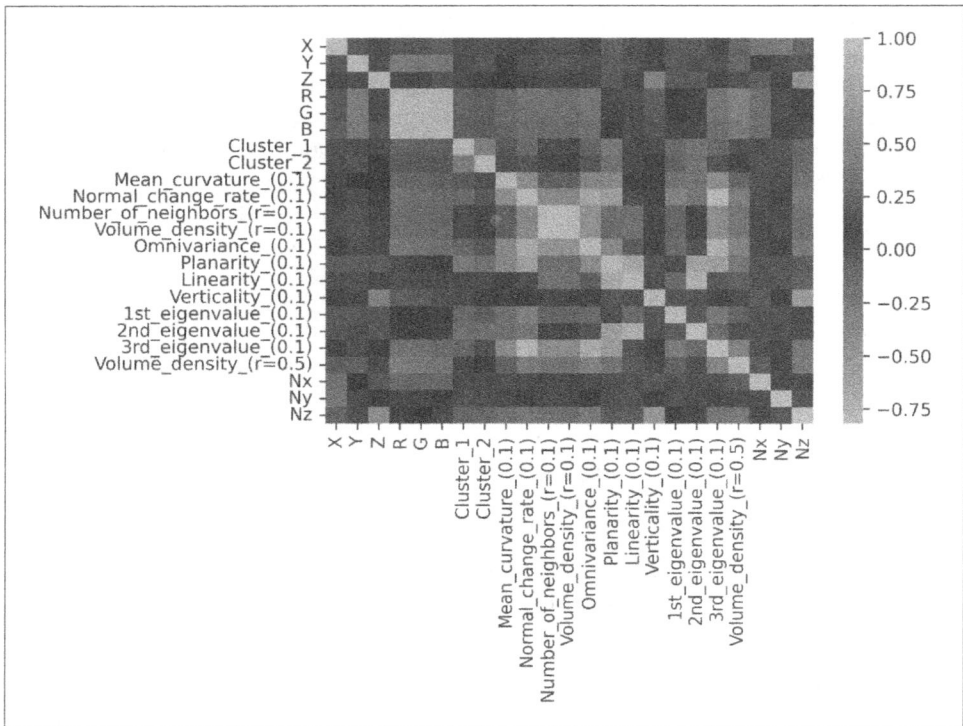

Figure 8-17. Correlation matrix displaying relationships between various geometric and colorimetric features extracted from a point cloud

Attribute Analysis

3D point clouds are rich datasets containing spatial information and various attributes associated with each point. These attributes provide valuable information about the scanned objects or environments' surface properties, geometry, and characteristics.

As we have already looked into the various features in previous steps, I want to explore three key attributes commonly found in datasets: RGB color, surface normals, and eigenvalues. Each attribute offers unique insights into the nature of the objects and plays a crucial role in advanced 3D data analysis. To best complement the learning, let's illustrate some interesting cases. First, we import libraries and a dataset from the chapter's folder:

```
import numpy as np
import open3d as o3d
import matplotlib.pyplot as plt
from matplotlib import cm
```

```
# Load a point cloud (ensure it has intensity information)
pcd = o3d.io.read_point_cloud("../DATA/chair_colored_notpure.ply")
o3d.visualization.draw_geometries([pcd])
```

RGB color

RGB (red, green, blue) color information is fundamental in many 3D datasets, especially those captured by RGB-D sensors or colorized using photogrammetry techniques. Color data significantly enhances the visual interpretation of 3D datasets, aiding in object recognition and segmentation tasks. Moreover, it provides essential texture information for creating realistic 3D models.

Several techniques have proven useful when analyzing RGB data in point clouds. Color histograms offer a way to understand the distribution of colors within the dataset, which can reveal patterns or anomalies. Color-based segmentation allows for grouping points with similar colors, facilitating the identification of distinct objects or regions. Additionally, analyzing color gradients can help detect edges or boundaries based on color-value changes such as:

```
def analyze_rgb(pcd):
    colors = np.asarray(pcd.colors)

    # Color histogram
    plt.figure(figsize=(15, 5))
    for i, color in enumerate(['red', 'green', 'blue']):
        plt.subplot(1, 3, i+1)
        plt.hist(colors[:, i], bins=50, color=color, alpha=0.7)
        plt.title(f'{color.capitalize()} Channel Distribution')
        plt.xlabel('Value')
        plt.ylabel('Frequency')
    plt.tight_layout()
    plt.show()

    # Color-based segmentation (simple example)
    red_mask = colors[:, 0] > 0.1  # Segment points with high red values
    red_cloud = pcd.select_by_index(np.where(red_mask)[0])
    o3d.visualization.draw_geometries([red_cloud])

analyze_rgb(pcd)
```

However, working with RGB data in point clouds is not without challenges. Ensuring color consistency across different lighting conditions can be problematic, especially when dealing with data captured at other times or under varying illumination. Handling shadows and reflections also presents difficulties, as these can introduce color distortions that may not reflect the actual properties of the scanned surfaces. This is where surface normals are helpful.

Surface normals

Surface normals are vectors perpendicular to the surface at each point, providing crucial information about the local geometry. These normals play an essential role in various applications, from surface reconstruction and mesh generation to feature extraction and object recognition. They are indispensable for accurate lighting and shading calculations. The analysis of surface normals typically begins with their estimation. Standard methods include PCA of local point neighborhoods or fitting local planes to nearby points. Once computed, normals can be used to assess surface smoothness and continuity, which is valuable for understanding the overall structure of scanned objects. Normal-based segmentation techniques group points with similar normal orientations, helping to identify distinct surfaces or shapes within the point cloud (we will study this in depth in Chapter 9):

```python
def analyze_normals(pcd, radius=0.1):
    pcd.estimate_normals(search_param=o3d.geometry.KDTreeSearchParamHybrid(radius=
radius, max_nn=30))
    normals = np.asarray(pcd.normals)

    # Visualize normal consistency
    consistencies = []
    pcd_tree = o3d.geometry.KDTreeFlann(pcd)
    for i in range(len(pcd.points)):
        [_, idx, _] = pcd_tree.search_knn_vector_3d(pcd.points[i], 10)
        neighborhood_normals = normals[idx[1:]]
        consistency = np.mean(np.abs(np.dot(neighborhood_normals, normals[i])))
        consistencies.append(consistency)

    plt.hist(consistencies, bins=50)
    plt.title('Normal Consistency Distribution')
    plt.xlabel('Consistency')
    plt.ylabel('Frequency')
    plt.show()

    # Visualize point cloud with normals
    pcd.paint_uniform_color([0.6, 0.6, 0.6])
    o3d.visualization.draw_geometries([pcd], point_show_normal=True)

analyze_normals(pcd)
```

Working with surface normals presents its own set of challenges. The estimation process can be sensitive to noise in the point cloud data and variations in point density. Additionally, there's often an inherent ambiguity in the normal direction—determining whether a normal faces inward or outward can be tricky, especially for complex or incomplete scans.

Eigenvalues

As we discussed in the previous chapter, we can calculate eigenvalues that reveal a point cloud's shape and local geometry by analyzing the covariance matrix of nearby points. These are useful for calculating geometric properties that characterize the local point distribution; they represent the variance of points along principal directions. Analyzing eigenvalues often involves computing ratios between them to determine the type of local geometry. For instance, specific ratios can indicate whether a local region is predominantly linear, planar, or volumetric in nature. Researchers have developed a range of dimensionality features based on eigenvalues, such as omnivariance and anisotropy, which provide nuanced descriptions of local point distributions. Knowing that adds another powerful technique to our 3D analytical toolbox: examining eigenvalues at different neighborhood sizes to detect multiscale features within the point cloud.

To help when you feel that you would benefit from eigenvalue-based analysis, I wrote some simple functions leveraging the libraries you are now used to. They are found in the 3D Data Science Hub (*https://learngeodata.eu/3d-data-science-with-python*) and you're welcome to incorporate them into your workflows:

```
def compute_eigenvalue_features(pcd, radius=0.1):
    …
    return linearity, planarity, sphericity
```

This allows us to extract and visualize useful structural information from a 3D point cloud dataset. The `radius` parameter determines the size of the neighborhood around each point that we use to compute the covariance matrix and eigenvalues. The choice of radius is kind of a balancing act—you want it to be large enough to capture meaningful local structure but not so large that you start losing important details. The primary challenge in eigenvalue analysis lies in determining the appropriate neighborhood size for computation. Too small a neighborhood may be overly sensitive to noise, while too large a neighborhood may obscure important local details. Additionally, areas with mixed geometry types can be challenging to characterize using eigenvalue-based methods alone, which may demand additional qualitative analysis.

While each attribute provides valuable information, the true power of 3D analysis often lies in combining multiple attributes. For example, integrating RGB color with intensity data can lead to more robust material classification. Combining surface normals with eigenvalue analysis can result in more accurate feature detection. The fusion of color and geometric information enables advanced semantic segmentation tasks, pushing the boundaries of what's possible in point cloud analysis. Understanding and effectively utilizing attributes like RGB color, surface normals, eigenvalues, and intensity in complex 3D data is not just a theoretical exercise. It is an essential process in obtaining significant data that can be directly utilized in practical situations, revealing vast information about the surveyed surroundings.

With these exploratory tools, let's develop some diagnostic strategies to assess our 3D datasets.

3D Diagnostic Tools

3D scene diagnosis is critical for analyzing and understanding complex 3D datasets (see Figure 8-18). By examining the geometric properties of 3D scenes, we can gain valuable insights into the underlying structure and relationships between objects and environments.

This information can help identify anomalies, optimize 3D geometries, and improve the performance of various 3D applications (e.g., autonomous vehicles, virtual reality, and robotics). By performing a comprehensive 3D scene diagnosis, we can ensure these systems' accuracy, reliability, and efficiency. Let me dive into the tools that can complete your toolbox.

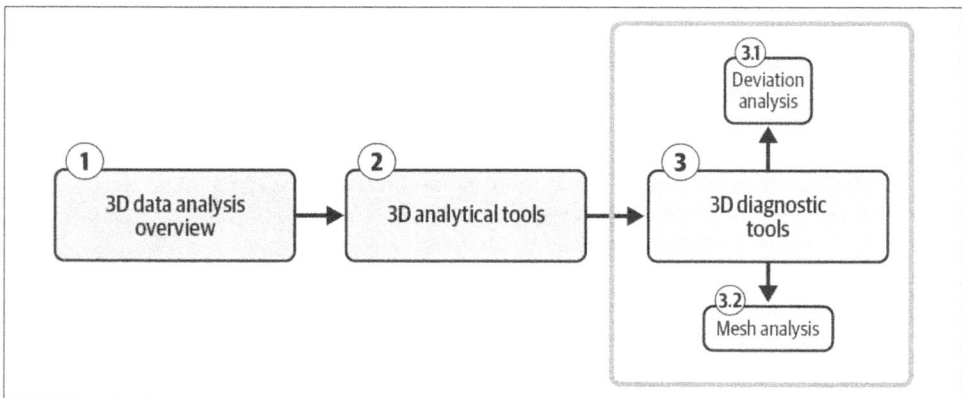

Figure 8-18. Overview of 3D data analysis tools, highlighting the diagnostic toolset

3D diagnostic tools allow us to use 3D data analysis to diagnose a current 3D state. This usually means that we try to understand our scene by studying its geometric decomposition: whenever we approach a 3D scene diagnosis, we often play with geometric shapes.

This is where my initial words in the intro should resonate with you. This analytical bucket actually combines several 3D data science methods including 3D shape detection, 3D modeling, 3D segmentation, recognition, and more. I will abstract as much as possible and focus on the fundamental understanding and skills to diagnose possible deviations from 3D shapes. Let's use the case of 3D planes as a diagnostic base.

3D Deviation Analysis: Planar Case

Let's look at a very quick way to monitor how close a 3D point cloud is to a specific plane. This is very useful in a number of scenarios, for example, to monitor how good a planar fit is, how leveled a 3D dataset is, or even if a modeled object nicely fits a planar predisposition.

> We study a 3D point cloud. This is usually the best way to move forward whenever you want to ensure an analytical workflow that combines both global and local understanding of geometries. Suppose your base dataset is a CAD model, a 3D mesh, or a voxel assembly. In that case, you usually have to sample your model with a sampling method, such as `mesh.sample_points_uniformly( number_of_points = 100)`, to derive a point cloud with 100 points from a mesh object.

Our approach is designed to be replicable. Therefore, let's define a simple workflow as shown in Figure 8-19.

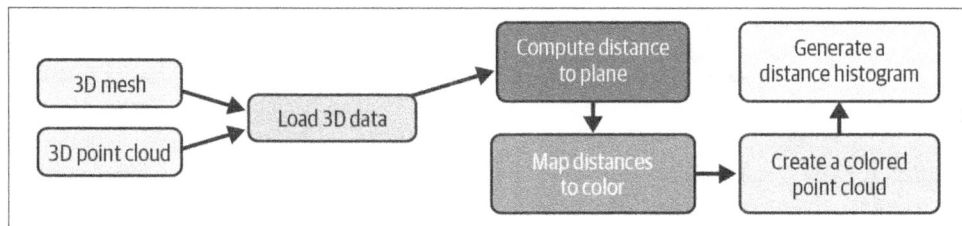

Figure 8-19. 3D deviation analysis workflow for planar use cases

The first step is to import the three libraries (NumPy, Open3D, and Matplotlib) and define our dataset paths:

```
import numpy as np
import open3d as o3d
import matplotlib.pyplot as plt

point_cloud_file = "../DATA/office.ply"
plane_file = "../DATA/ground_plane.ply"
```

Then, we define a `compute_distance_to_plane` function:

```
def compute_distance_to_plane(point_cloud_file, plane_file):
```

We then use the Open3D library to load the point cloud and plane mesh from the respective *.ply* files:

```
# in compute_distance_to_plane function
    point_cloud = o3d.io.read_point_cloud(point_cloud_file)
    plane_mesh = o3d.io.read_triangle_mesh(plane_file)
```

From there, we compute the distance to the plane. This is done by considering each point in the point cloud and computing the absolute value of the dot product between the vector from the plane's first vertex to the point and the plane's normal vector. This gives the distance from the point to the plane:

```
# Compute distances from point cloud to plane
    distances = []
    for point in point_cloud.points:
        min_distance = np.inf
        for i, triangle in enumerate(plane_mesh.triangles):
            v0 = plane_mesh.vertices[triangle[0]]
            v1 = plane_mesh.vertices[triangle[1]]
            v2 = plane_mesh.vertices[triangle[2]]
            normal = np.cross(v1 - v0, v2 - v0)
            normal /= np.linalg.norm(normal)
            distance = np.abs(np.dot(point - v0, normal))
            min_distance = min(min_distance, distance)
        distances.append(min_distance)

    distances = np.array(distances)
```

I opted for the dot product approach because it's a well-established and efficient way to calculate the perpendicular distance from a point to a plane. This method is commonly used in 3D computer graphics.

Let me walk you through the preceding example quickly. First, I initialize an empty list called distances. Then, I loop through each point in the point_cloud object. For each point, I find the minimum distance from that point to any of the triangles that make up the plane_mesh object. To do this, I iterate through each triangle in the mesh, compute the normal vector of the triangle, and then calculate the absolute value of the dot product between the vector from the point to the first vertex of the triangle and the normal vector. This gives me the perpendicular distance from the point to the plane defined by the triangle. I keep track of the minimum distance for each point and append it to the distances list. After processing all the points, I convert the distances list to a NumPy array. This allows me to perform further operations or visualizations on the distance values easily.

We can map distances to colors to communicate this analysis. For this, let's create a colored point cloud by mapping the distance values to the cool colormap using plt.cm.cool() and assigning the resulting colors to the point cloud. Note that you can use another colormap if needed:

```
# Create a colored point cloud based on the distance to plane
    point_cloud.colors = o3d.utility.Vector3dVector(plt.cm.cool(distances /
    np.max(distances))[:, :3])
```

Then, we export a histogram of the distance values to a file named *distance_histogram.png*:

```
# Export the distance histogram
    plt.figure()
    plt.hist(distances, bins=50, facecolor='#2ab0ff', edgecolor='#e0e0e0',
    linewidth=0.5, alpha=0.7, zorder=4)
    plt.xlabel("Distance to plane")
    plt.ylabel("Frequency")
    plt.savefig("../RESULTS/distance_histogram.png")
```

We close the function by returning the colored point cloud, the plane mesh, and the distance values:

```
    return point_cloud, plane_mesh, distances
```

To visualize the results, we then call the function and plot the results accompanied by stylized histograms:

```
    colored_point_cloud, plane_mesh,
    distances = compute_distance_to_plane(point_cloud_file, plane_file)

    # Visualize the colored point cloud
    o3d.visualization.draw_geometries([colored_point_cloud, plane_mesh])
```

This gives us the visualization in Figure 8-20A, and the accompanying histogram (hist_styled(distances, "Distance to plane")) in Figure 8-20B.

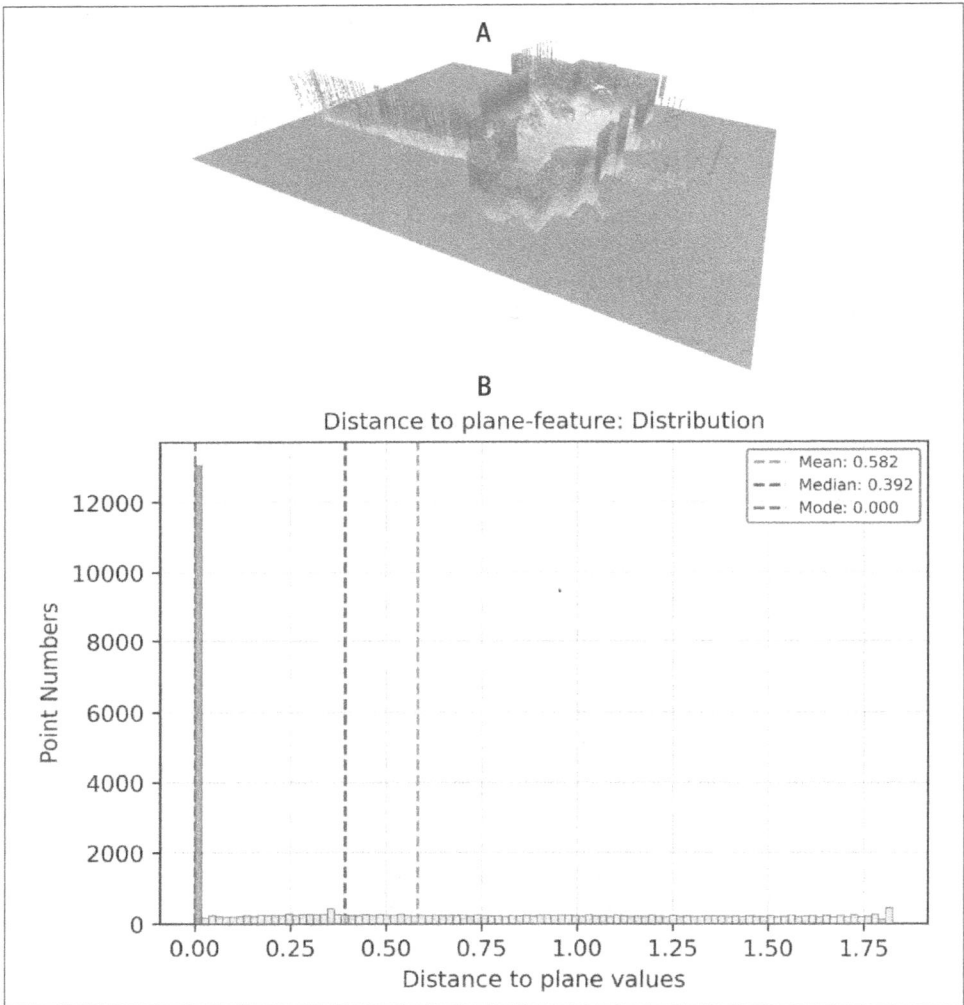

Figure 8-20. (A) Point cloud visualized with a reference plane, color-coded according to the distance between each point and the plane; (B) distribution of point distances to a plane feature, exhibiting a primary mode at 0

As you can see, the fit is interesting in delineating the ground point from the rest and could indicate that our dataset is perfectly leveled. As seen in previous chapters, this histogram is a strong element in a workflow, and we could use it to segment our dataset into two parts with a thresholding method.

3D Deviation Analysis: Mesh Case

Now that you know how to handle basic planar cases, what about the more complex 3D models? How could you tell if your motor aerodynamics perfectly fit your designed 3D CAD model? How can you ensure the building you constructed follows the as-designed building information model (BIM) (*https://oreil.ly/vgZ23*)?

Well, this pushes the deviation analysis to a more global view, where we want to study a 3D model in its entirety against a 3D point cloud. However, the underlying truth is that we sampled the 3D mesh to obtain a 3D point cloud that we could compare against the studied dataset. So, to simplify things, let me share the three main components of this approach. We will leverage a known data structure (the *k*-d tree) to accelerate the nearest point search (as a reminder from Chapter 4, recursively partitioning the 3D space into smaller regions enables us to accelerate searching operations, nearest neighbor queries, and range queries). Then, we will do iterative distance computation to find the minimal distance between the point clouds. Finally, we will color the point cloud and export the histogram, as seen previously. This means that we operate linearly in our function `compute_distance_to_mesh` as shown in Figure 8-21.

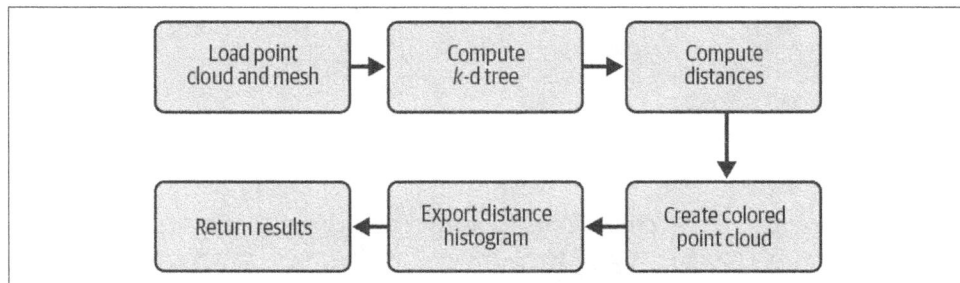

Figure 8-21. Process for calculating and visualizing point-to-plane distances, encompassing data loading, k-d tree construction, distance computation, point cloud coloring, and histogram generation

Let's write a Python function that computes the distances from a 3D point cloud to a 3D mesh surface and then creates a visualized point cloud where each point is colored based on its distance to the mesh:

```python
def compute_distance_to_mesh(point_cloud_file, mesh_file):
    # Load point cloud and mesh
    point_cloud = o3d.io.read_point_cloud(point_cloud_file)
    mesh = o3d.io.read_triangle_mesh(mesh_file)

    # Compute KDTree for the mesh
    # mesh.compute_triangle_normals()
    mesh.compute_vertex_normals()
    pcd_mesh = mesh.sample_points_uniformly(number_of_points=1000000)
    mesh_tree = o3d.geometry.KDTreeFlann(pcd_mesh)

    # Compute distances from point cloud to mesh
    distances = []
    for point in point_cloud.points:
        _, inds, dist = mesh_tree.search_knn_vector_3d(point, 1)
        distances.append(np.sqrt(dist[0]))

    distances = np.array(distances)

    # Create a colored point cloud based on the distance to mesh
    point_cloud.colors = o3d.utility.Vector3dVector(plt.cm.jet(distances /
    np.max(distances))[:, :3])

    # Export the distance histogram
    plt.figure()
    plt.hist(distances, bins=50, facecolor='#2ab0ff', edgecolor='#e0e0e0',
    linewidth=0.5, alpha=0.7, zorder=4)
    plt.xlabel("Distance to mesh")
    plt.ylabel("Frequency")
    plt.savefig("distance_histogram.png")

    return point_cloud, mesh, distances
```

Let's walk through the function. I first load the point cloud and mesh data using Open3D. Then, I compute the vertex normals for the mesh and sample 1 million points from the mesh to create a point cloud representation. I use this point cloud to build a k-d tree for efficient nearest-neighbor searches.

Sampling a 3D mesh as a 3D point cloud is a valuable technique for 3D diagnostic analysis. By converting the complex mesh structure into discrete points, we can simplify the data representation and reduce computational costs. Furthermore, we can leverage point-based algorithms and techniques, making them suitable for various 3D diagnostic applications that would not be possible otherwise.

Next, I loop through each point in the original point cloud and use the *k*-d tree to find the nearest point on the mesh. I calculate the Euclidean distance between the point and its nearest neighbor on the mesh and append this distance to a list. After processing all the points, I convert the distance list to a NumPy array. To visualize the results, I create a new point cloud where each point is colored based on its distance to the mesh, using the `jet` colormap. I also export a histogram of the distance values to a file called *distance_histogram.png*. The `jet` colormap is a common choice for visualizing scalar fields, but you could experiment with other colormaps depending on your preferences.

> I hardwired a special parameter in the function. Can you spot it? It is the number of points from the sampling strategy. This should be used with care, as it influences the quality of the result. A too-low number will suffer some imprecision, while a too-high count can create performance issues. Again, finding heuristics for unsupervised parameter setting is an excellent track to explore, which we do in Chapter 9 and onward.

Having defined our function, we can now call it and analyze the results:

```
point_cloud_file = "../DATA/pcd_car.ply"
mesh_file = "../DATA/mesh_car.ply"
colored_point_cloud, distances = compute_distance_to_mesh(point_cloud_file,
mesh_file)

# Visualize the colored point cloud
o3d.visualization.draw_geometries([colored_point_cloud])
```

Figure 8-22A showcases the results, with its accompanying histogram (Figure 8-22B). As you can see, we have a slight shift highlighted in our point cloud, as we have a slight mean, median, and mode shift. From earlier, you should be quite aware of how analyzing this histogram will help you identify the design problems of the car as compared to its base model. This is a useful "tool" in your 3D data scientist journey. Also, this can be extended by other algorithms, like M3C2 analysis (*https://oreil.ly/j3xsj*), using Hausdorff distance metrics (*https://oreil.ly/7OqCm*) or leveraging more advanced techniques that you can find on the 3D Data Science Hub (*https://learngeo data.eu/3d-data-science-with-python*).

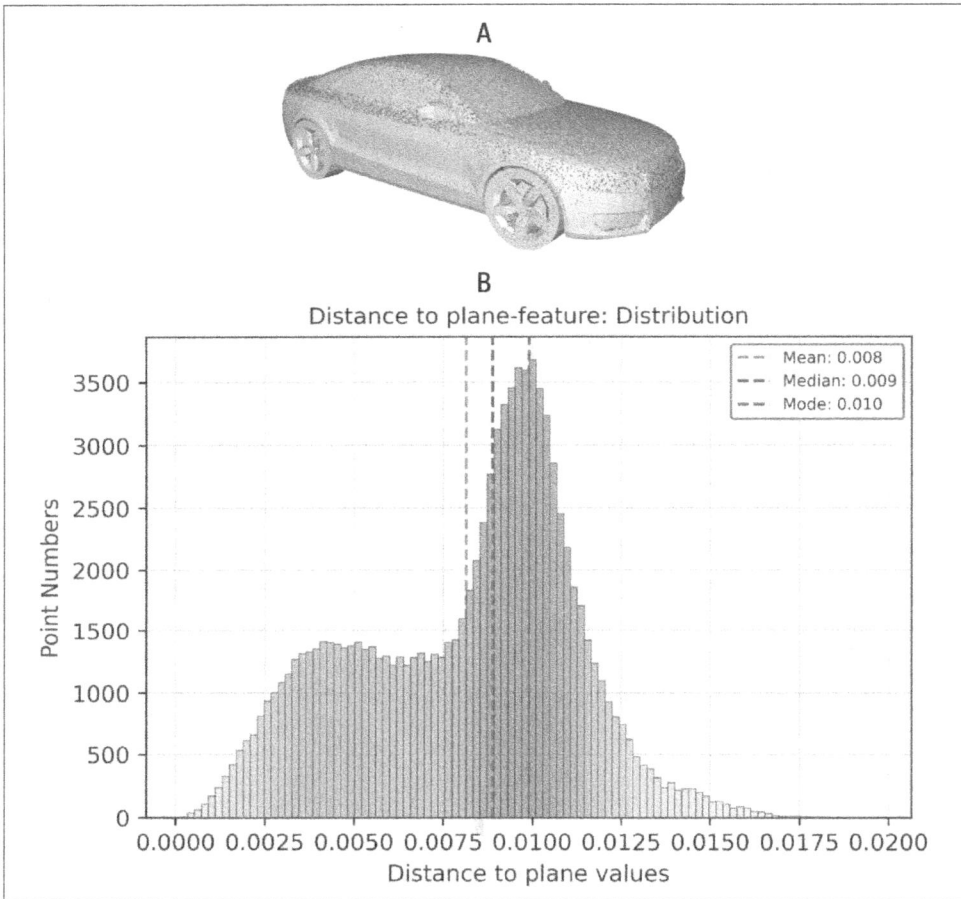

Figure 8-22. (A) 3D mesh of a car, color-coded based on proximity to a reference plane and (B) accompanying histogram

> This specific case relates very closely to what we did in the previous chapter when we wanted to address 3D data registration (Chapter 7). Our approach actually looks at the distance fit between two point clouds, but to a data processing end.

With these newfound 3D analytical tools at your disposal, we are ready to dive into hands-on projects, especially to communicate findings, simulations, and solutions through immersive experiences in external solutions.

Summary

3D data analysis is an essential aspect of 3D data science, providing robust tools and methods to derive significant insights from intricate three-dimensional information. This field spans a broad spectrum of methodologies, ranging from fundamental statistical analysis to sophisticated geometric calculations and visualization methods. This also showcased the need for geometric analysis, where the role of shapes and distance computation between datasets heavily influence the quality of our analysis.

As 3D scanning technologies become more accessible and datasets more complex, the need for sophisticated analysis tools increases. You can expect to see advancements such as:

- Incorporating machine learning and artificial intelligence methods with 3D data analysis to enhance automation and generate more intelligent observations
- Development of more efficient algorithms for handling larger and more complex 3D datasets
- Improved visualization techniques for better communication of 3D analytical results
- Enhanced 3D computing capabilities for analytical tasks in robotics, autonomous vehicles, and augmented reality

To approach these perspectives, make sure you are comfortable with:

- Understanding the importance of central tendencies, spread, and distribution in 3D datasets
- Recognizing the value of different visualization techniques such as histograms, boxplots, and 3D scatter plots in understanding 3D data
- Learning to leverage Python's libraries like Open3D, NumPy, Matplotlib, and Seaborn for 3D data analytical tasks
- Grasping PCA's role in characterizing local geometry for 3D analytics
- Understanding how to analyze deviation between point clouds and reference geometries (planes or meshes).

So, naturally, what's next is to venture into the realm of 3D shape detection to leverage our analytical skills.

Hands-on Project

Before moving on to the next stage, I have a practical example for you. You can try the following:

1. Load a 3D point cloud dataset of your choice (you can use one provided in the chapter or find an open source dataset online).

2. Perform basic descriptive analysis of the dataset: calculate and visualize the distribution of points along the x, y, and z axes using histograms; compute and visualize the nearest neighbor distances using a boxplot; and, if color information is available, create a scatter plot of the points colored by their RGB values.

3. Estimate surface normals for the point cloud and visualize them.

4. Compute eigenvalue-based features (linearity, planarity, sphericity) for the point cloud and visualize these features on the point cloud.

5. If possible, perform a deviation analysis of your point cloud against a simple geometric shape (like a plane or sphere) and visualize the results.

Remember to document your process, interpret the results, and reflect on what insights you've gained about the dataset through this analysis. This exercise will help reinforce the concepts and techniques covered in the chapter while giving you hands-on experience with 3D data analysis.

3D Shape Recognition

You may remember the introductory words from the start of the book: manmade environments are predominantly geometric assemblies of known shapes (planes, spheres, cylinders, etc.). Naturally, we ask ourselves if we can leverage this characteristic as it has the potential to benefit 3D scene understanding tasks.

What if we design a system capable of identifying such 3D shapes with precision? We could help automate manufacturing quality control processes by rapidly inspecting products for defects. We could enable autonomous vehicles to perceive their surroundings accurately, enhancing safety and efficiency. We could help reconstruct ancient artifacts for archeological simulations.

Giving machines 3D shape recognition capabilities enables them to perceive and interact with the physical world. As a first stage, it helps to understand the environment. By recognizing objects and their spatial relationships, systems can navigate complex environments and make informed decisions.

Furthermore, 3D shape recognition allows machines to perform tasks autonomously. Automated systems can execute functions like object manipulation, assembly, and inspection with remarkable precision and efficiency. This technology also aids in detecting abnormalities and analyzing geometrical structures in a given 3D dataset. And suppose you reason with enough of a low-level view. In that case, you can extract a scene decomposition without going into explicit functional definition (i.e., the scene seems planar, but you do not need to find out what the ground is).

The question that arises from these statements is then straightforward. Do you know how to detect 3D shapes in your voxel, mesh, depth image, or point cloud dataset?

This chapter provides comprehensive knowledge of 3D shape recognition techniques, focusing on two widely used methods: random sample consensus (RANSAC) (*https://oreil.ly/jH7vG*) and region growing (*https://oreil.ly/_5yn8*). We aim to equip you with the knowledge and tools necessary to apply these algorithms to real-world 3D datasets, primarily point clouds, recognizing your integral role in the learning process.

> If your dataset was not originally a point cloud, you can check out Chapter 4 to unlock the canonical mapping methodologies to transform your 3D dataset into a point cloud before applying processing methods. Reverting to the original 3D data representation is then possible using bidirectional mapping with the newfound insights.

In the first part of this chapter, we explore the RANSAC algorithm, explaining its principles, steps, and applications in 3D shape recognition. Then, we discuss the region-growing algorithm, its underlying concepts, and its use in segmenting 3D point clouds. Finally, I will provide a hybrid scene modeling implementation that plays on both shape recognition strategies.

By the end of this chapter, you will have a solid understanding of the RANSAC and region growing algorithms, their strengths and limitations, and how to apply them effectively to 3D shape recognition problems.

> You can find all the materials for this chapter on the 3D Data Science Hub (*https://learngeodata.eu/3d-data-science-with-python*). To access the files (code, datasets, articles) and resources, you may need to share an email address, a personal password, or proof that you are the book's owner.

RANSAC from Scratch: 3D Planar Shape Recognition

As we dive deeper into 3D shape recognition, it is important to remember that we work with real-world data. This means the data is far from perfect. Noise and outliers can easily corrupt real-world data, making it difficult for algorithms to identify shapes accurately (Figure 9-1).

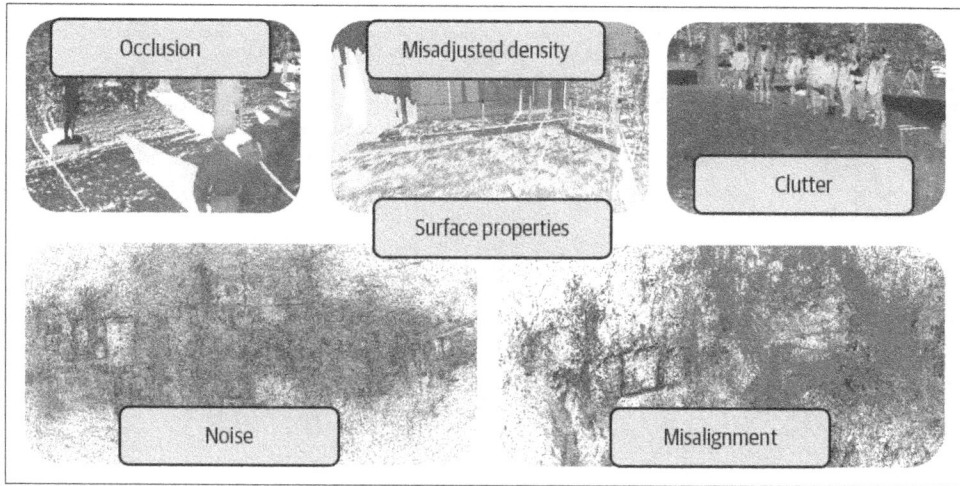

Figure 9-1. Example of various real-world phenomena that impact the quality of the 3D data

Occlusions and clutter can further complicate matters, as objects may be partially hidden or surrounded by other objects. And let's not forget the variability of objects in the real world. The same object can look completely different depending on the angle, lighting, or position.

To top it off, 3D shape recognition algorithms can be incredibly demanding, especially when dealing with large datasets or complex shapes. This is how we arrive at the first approach: RANSAC.

RANSAC

RANSAC is not rocket science but rather a super-practical approach for noisy, real-world datasets. It is a trial-and-error approach that will group your data points into two segments: an inlier set and an outlier set. Then, you can forget about the outliers and work with your inliers (or vice versa). So let me use a tiny but simple example to illustrate how RANSAC works. Let's say that we want to detect a 3D plane from the point cloud in Figure 9-2. How can we do that?

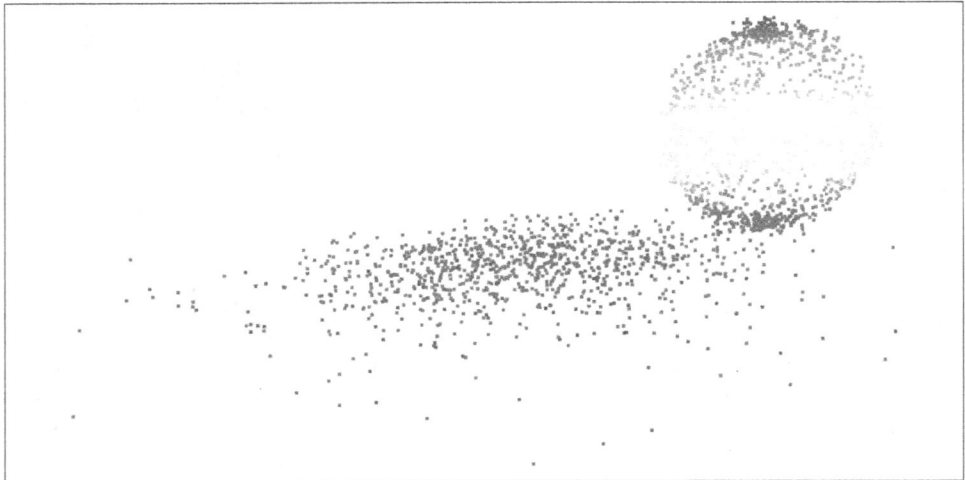

Figure 9-2. A collection of points in three-dimensional space forms a point cloud

First, we create a plane from the data, and for this, we randomly select three points from the point cloud, which is necessary to establish a plane. Then, we simply check how many of the remaining points fall on the plane (to a certain threshold), which will give the proposal a score. Then, we repeat the process with three new random points and see how we are doing. Is it better? Is it worse?

We repeat this process over and over again, let's say, 10 times, 100 times, 1,000 times, and then we select the plane model with the highest score (i.e., which has the best "support" of the remaining data points). And that will be our solution: the supporting points plus the three points we have sampled constitute our inlier point set, and the rest is our outlier point set.

For the skeptics, does a question arise? How do we determine how many times we should repeat the process? How often should we try it? Well, that's something we can compute, but let's put it aside for now to focus on the matter at hand: how can we implement RANSAC for 3D point cloud shape detection? Well, to achieve this goal with RANSAC, we proceed in four steps (see Figure 9-3).

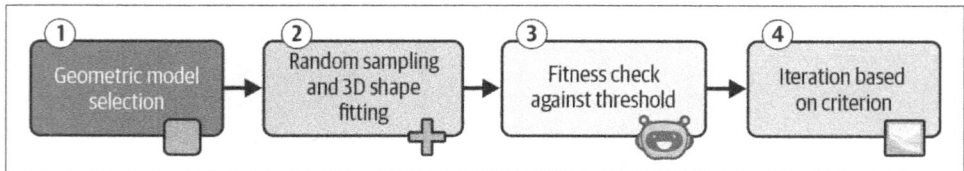

Figure 9-3. RANSAC, a robust method for shape detection in 3D data, involves four key steps

First, we select one geometrical model (a plane, a sphere, etc.) that we want to detect in the data. Second, we fit this model to the tiniest random sample from our dataset (three points taken randomly if we're going to define a plane). Third, we want to estimate how good the fit is by checking how many points are close to the surface of interest to get an inlier count. This means working in two phases: defining a 3D shape detection parameter, the threshold, and computing a distance to the 3D shape that is compared against this threshold. Finally, we repeat this process over a certain amount of iterations and keep the geometry that maximizes the inlier count as our best 3D shape candidate.

Now that we are set up, let's gather some datasets for our tests.

Data and Environment Setup

In this section, there are not one but three datasets that you can choose from, download, and use for your experiments. These were chosen to illustrate three different scenarios and provide the base data with which to play. The different point clouds available can be found in the chapter's directory (the researcher desk, the car, the playground). The real-life scenarios that we want to showcase are the following:

Robotics
We are designing a robot that needs to clean both the ground and the table and avoid obstacles when cleaning. We will also want to detect the position of elements of interest and use that as a basis for future cleaning tasks to determine if we need to reposition them initially.

ADAS (Advanced Driver-Assistance System)
We are interested in allowing a vehicle to drive itself: an autonomous vehicle. For this purpose, we use one epoch of a Velodyne VLP-16 scan, on which we usually do real-time analysis for object detection.

Construction
A playground constructed some years ago presents problems due to unstable groundwork. Therefore, we want to assess the planarity of the element and determine if a leveling operation is necessary.

Geometric Model Selection

Let's design a method that is easily extendable to different use cases. For a more straightforward experience, we chose the plane geometry. It is the best way to quickly make sense of the extensive array of datasets out there. But bear with me; I will now give you some crucial mathematical understanding we use to describe a plane in Euclidean space. The general form of the equation of a plane in $\mathbb{R}^3$ is $ax + by + cz + d = 0$.

In a normal vector $\tilde{n} = (a, b, c)$ which is either perpendicular to the plane or any vector parallel to it, the a, b, and c constants make up the vector. When you grasp this, playing with transformations (translations, rotations, scaling) and fitting it is super easy. The d constant will shift the plane from the origin.

This means that a point $p = (x, y, z)$ belongs to the plane guided by the normal vector $\tilde{n}$, if it satisfies the equation. If you understand this, you understand the first-hand principle that guides geometric fitting.

> Geometric fitting involves finding the best-fitting geometric model, such as lines, planes, or curves, to a set of 3D data points. This process is crucial for object recognition, motion tracking, and 3D reconstruction. By accurately fitting geometric models to data, we can extract meaningful information, reduce noise, and improve the overall quality of analysis and visualization.

3D Shape Fitting

3D shape fitting allows us to model and analyze complex 3D objects accurately. We can extract meaningful information by fitting mathematical models to 3D data, such as the object's shape, size, and orientation. We can then leverage shape fitting to identify and quantify deviations from ideal shapes, which is essential, as seen in Chapter 8.

Let's develop a RANSAC plane detection algorithm for 3D point clouds from scratch to better understand what's under the hood. We achieve this with three libraries: Open3D (*https://www.open3d.org*), NumPy (*https://numpy.org*), and scikit-learn (*https://scikit-learn.org*) (for the k-d tree part). It's hard to be more minimalistic. The first stage is to import libraries:

```
import open3d as o3d
import numpy as np

from sklearn.neighbors import KDTree
```

Then, we load a point cloud to address one of the scenarios hinted at in the data and environment setup:

```
pcd = o3d.io.read_point_cloud("../DATA/the_researcher_desk.ply")
points = np.asarray(pcd.points)
```

Now, let us dive into the parameters.

Manual parameter setup

We must define a threshold parameter to determine whether a point belongs to the fitted planar shape (inlier) or is an outlier. We will base our discrimination on a point-to-plane distance; we thus need to grasp the unit in our point cloud quickly. Let's display the point cloud with open3D:

```
o3d.visualization.draw_geometries([pcd])
```

The result is illustrated in Figure 9-4.

Figure 9-4. Setting up threshold parameters for point cloud analysis

By leveraging interactive tools from Chapters 3 and 5 (and analytical tools from Chapter 8), we can extract that, on average, we have neighboring points every 5 mm. Thus, we could set the threshold parameter 10 times higher (absolutely empirical): threshold = 0.05.

In this same vein, we could set the number of iterations to a considerable number, let's say, 1,000 iterations. But would it not be helpful to compute these parameters by leveraging data-driven heuristics?

Automatic parameter setup

We may be limited by the need for some domain knowledge to set up the threshold. Therefore, it would be exciting to try bypassing this to open the approach to nonexperts. Here's a useful thought: what if we computed the mean distance between points in our datasets and used this as a base to set up our threshold?

To expedite the process of asking the nearest neighbors for each point, we might utilize a k-d tree to try and determine such a value. At this stage of the process, I recommend using a scikit-learn implementation and separating the k-d tree into two hyperplanes at each node:

```
tree = KDTree(np.array(points), leaf_size=2)
```

Next, we can use the straightforward query method to find out what each point in the point cloud's k nearest neighbors are:

```
nearest_dist, nearest_ind = tree.query(points, k=8)
```

This outputs the point distance and the point indexes, respectively. If we investigate the nearest neighbor distance:

```
nearest_dist_mean = np.mean(nearest_dist[:,1:],axis=0)
```

We see that the eight neighbors are, on average, between 4 mm and 9 mm from their neighbors:

```
array([0.00462562, 0.0052954 , 0.00599643, 0.00672068, 0.00744045,
       0.0081288 , 0.00875479])
```

> The first distance value of the nearest neighbor is always equal to 0. Weird, you ask? This is because we query the whole point cloud against itself; thus, each point has a distance to itself. Therefore, we need to filter the first element per row: `nearest_dist, nearest_ind = tree.query(points, k=8)`.

Great! So now, if we were in a scenario where we wanted to get a local representation of the mean distance of each point to its nth closest neighbors, we use:

```
nearest_dist_m = np.mean(nearest_dist[:,1:])
```

We get 6.7 mm. Careful parameter selection is crucial to ensuring the success of RANSAC 3D shape detection experiments. We can effectively handle noisy data and outliers by setting a query range of 8 to 15 points and averaging their distances while maintaining sensitivity to subtle shape variations. This offers a good local representation of the noise ratio in the point cloud. This approach helps improve the RANSAC algorithm's accuracy and robustness, leading to more reliable 3D shape detection results.

> There is an automatic method for getting the correct iteration number every time. If we wish to succeed with a probability p (e.g., 99%), our data has an outlier ratio e (e.g., 60%), and we need s points to create our model (here 3), the following formula provides the number of trials (iterations) we need:

$$T = \frac{\log{(1-p)}}{\log{\left(1-(1-e)^8\right)}}$$

Now, we are ready to move on to model fitting with RANSAC.

Model fitting with RANSAC

The moment has come to investigate RANSAC as an iterative procedure for estimating mathematical model parameters from a dataset that contains outliers. The beginning point: this dataset likely includes observations of the model. As we saw earlier, we begin the process with a random selection of data points, followed by their fitting to a model. The number of residual data points that fall within a defined tolerance is then determined. The process is repeated multiple times, and the model that fits the most data points (inliers) is chosen as the best estimate, effectively identifying and excluding outliers from the final model fit.

Let's simulate an iteration before automating over the specified number of iterations. First, we will want to grasp three random points from the point cloud:

```
sample = points[np.random.choice(points.shape[0], 3, replace=False)]
```

Then, we want to determine the equation of the plane, e.g., find the parameters a, b, c, and d of the equation $ax + by + cz + d = 0$. For this, we can play with a fantastic linear algebra property that says that the cross product of two vectors generates an orthogonal one.

Next, we need to define two vectors from the same point on the planes v1 and v2, and then compute the normal to these, which will then be the normal of the plane:

```
v1 = sample[1] - sample[0]
v2 = sample[2] - sample[0]
normal = np.cross(v1, v2)
```

The cross product of two vectors always yields a perpendicular vector (orthogonal) to both input vectors, and this isn't just a mathematical coincidence—it's deeply rooted in geometry and algebra. Geometrically, when you take two vectors, they define a plane, and their cross-product points are perpendicular to this plane. The magnitude of this new vector equals the area of the parallelogram formed by the original vectors ($|a||b|\sin(\theta)$), while its direction follows the right-hand rule. This property makes the cross-product especially useful when dealing with rotational motion, where the angular momentum must be perpendicular to the plane of rotation.

From there, we can normalize our normal vector. Since we typically only care about the direction of the vectors (not their magnitude) for calculations, we normalize it to a length of 1. This normalization ensures consistency across calculations regardless of the dataset size. Then get the a, b, and c that define the vector, and find d using one of the three points that fall on the plane: $d = -(ax + by + cz)$:

```
a,b,c = normal / np.linalg.norm(normal)
d = -np.dot(normal, sample[1])
```

And now, we are ready to attack the computation of any remaining point to the plane we just defined.

Point-to-plane distance

If you are up to taking my word for it, here is what we need to implement. The point-to-plane distance is expressed using the following notation:

$$D = \frac{|ax_1 + by_1 + cz_1 + d|}{\sqrt{a^2 + b^2 + c^2}}$$

This is the orthogonal distance between a point and the plane (i.e., the shortest Euclidean distance between these entities). The projection of vector w onto the unit normal vector n provides the magnitude of the distance between point p and the specified plane. It is a known fact that the magnitude of vector n is equal to one, and the distance between point p and the plane is equal to the absolute value of the dot product of vectors w and n. This distance is the shortest, as it is the orthogonal distance between the point and the plane (see Figure 9-5).

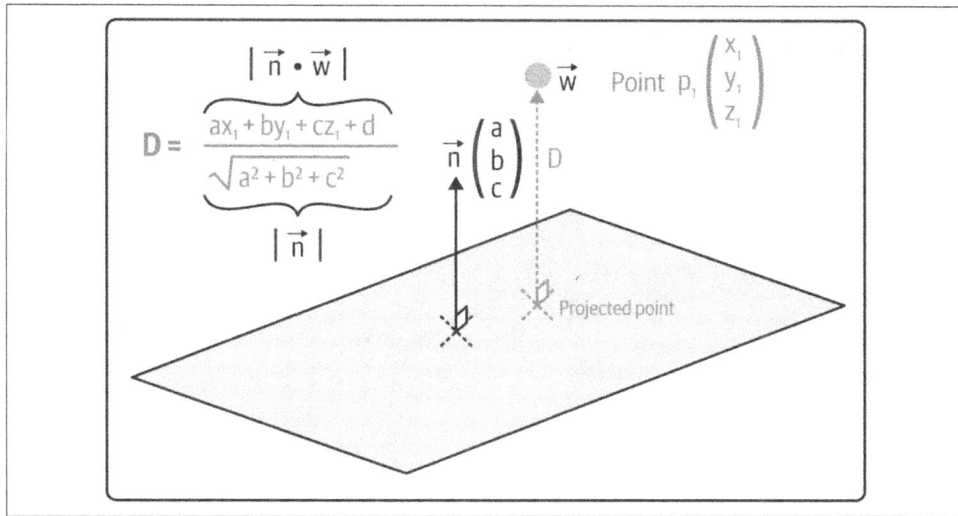

Figure 9-5. The point-to-plane distance measures the shortest distance between a point and a plane

A normal n defines the plane, and we can see what the distance D point to plane we want to compute looks like. We can compute this distance by taking each point in the point cloud that is not part of the three we used to establish a plane in one RANSAC iteration, like this:

```
distances = (np.dot(points, [a, b, c]) + d) / np.sqrt(a**2 + b**2 + c**2)
```

For our random choice and plane fit, this outputs:

```
array([-1.39510085, -1.41347083, -1.410467 , …, -0.80881761, -0.85785174,
-0.81925854])
```

Do you see the negative values? When dealing with distances and normals, we need to consider that a normal vector to a plane can point in two opposite directions—either up or down relative to the surface, both being mathematically valid. This ambiguity matters because signed distances (distances with direction) can be positive or negative depending on which way the normal points. For example, if we're calculating the distance from a point to a plane, the sign tells us which side of the plane the point is on.

However, since we can flip the normal 180° and still have a valid normal vector, this sign could arbitrarily change. In many applications (like collision detection or point-in-mesh tests), we often need consistent, positive distances regardless of the normal's orientation. That's why we might take the absolute value or use unsigned distances—it removes the ambiguity introduced by the normal's arbitrary orientation while still giving us the useful measurement of how far something is from the plane. Think of it like measuring the height of a table—whether you measure from the floor up or from the ceiling down, you want the same positive number representing the actual distance.

From there, we can check against the threshold and filter all points that answer the criterion to only keep the points with a point-to-plane distance under the threshold as inliers:

```
inliers = np.where(np.abs(distances) < threshold)[0]
```

> The [0] allows us to work only with indexes at this step and not overflow our system with unnecessary point coordinates.

RANSAC plane detection, in a single iteration, uses three randomly selected points to fit a plane. The points represent inliers detected by the algorithm (see Figure 9-6). Do you already know what the next substep is about?

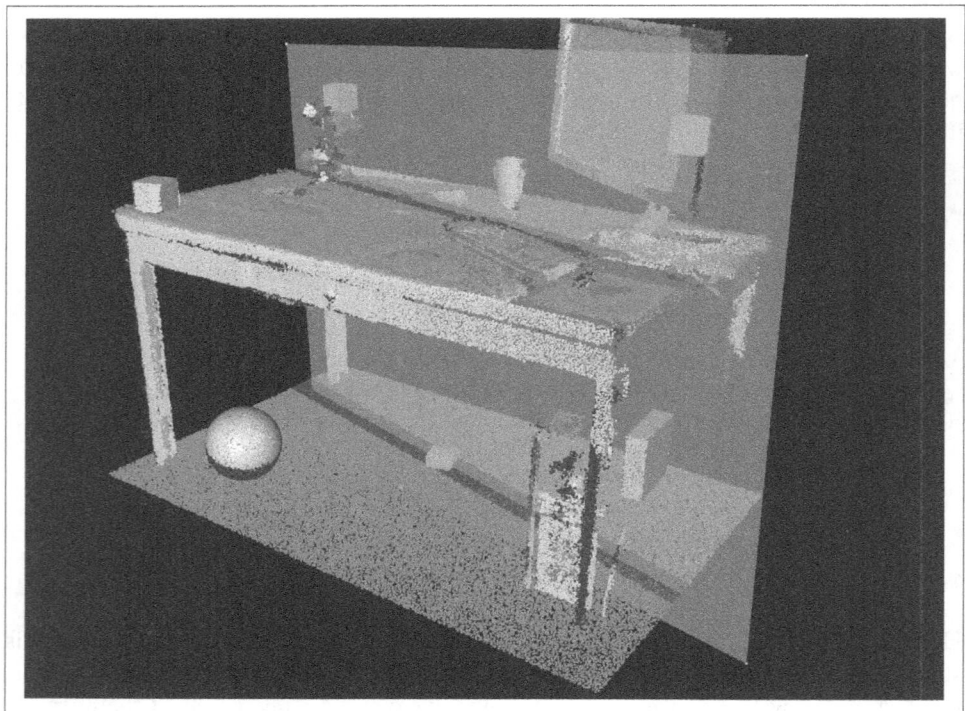

Figure 9-6. RANSAC plane detection in a single iteration

Iteration and Function Definition

We will need to repeat a certain number of times to discover the best plane! In order to do this, we will build a function that takes as input point coordinates, the threshold, and the number of iterations and returns the plane equation and the point inliers indexes:

```
def ransac_plane(points, num_iterations=1000, threshold=0.1):
    ...
    return best_plane, best_inliers
```

We create the RANSAC loop over the iteration parameter. For each loop, we will compute the best-fitting RANSAC plane and retain both the equation and the inliers indexes. With the if statement, we then check if the score of the current iteration is the largest, in which case we switch the point indexes.

Now, let's fill out our RANSAC function:

```
def ransac_plane(points, num_iterations=1000, threshold=0.1):
    best_inliers = []
    best_plane = None

    for _ in range(num_iterations):
        # Randomly sample 3 points
        sample = points[np.random.choice(points.shape[0], 3, replace=False)]

        # Calculate plane equation ax + by + cz + d = 0
        v1 = sample[1] - sample[0]
        v2 = sample[2] - sample[0]
        normal = np.cross(v1, v2)
        a, b, c = normal / np.linalg.norm(normal)
        # d = -np.dot(normal, sample[1])
        d = -np.sum(normal*sample[1])

        # Calculate distances of all points to the plane
        distances = np.dot(points, [a, b, c]) + d / np.sqrt(a**2 + b**2 + c**2)

        # Count inliers
        inliers = np.where(np.abs(distances) < threshold)[0]

        if len(inliers) > len(best_inliers):
            best_inliers = inliers
            best_plane = (a, b, c, d)

    return best_plane, best_inliers
```

And here we are: we created a RANSAC function for plane detection that eats 3D point clouds. It is sufficiently optimized to allow you to attack large point clouds without the risk of your computer burning! Let's extend this to a practical case where we want to segment our point cloud.

Application 1: RANSAC for Segmentation Tasks

We start with the initial principle: RANSAC is a robust algorithm for estimating model parameters from noisy data. Now, what is interesting is when we use RANSAC as part of an application in a workflow. When used as a preprocessing step for 3D segmentation, RANSAC helps to identify and remove outliers or noise from the input data. By fitting a model to a subset of the data points and iteratively refining the model parameters, RANSAC can help us detect the main shape(s) that we want to hold out from a series of subsequent steps. In our application, we want to hold out the ground from the other points.

Thus, let's retain the results of the defined function in two variables:

```
plane_params, plane_inliers = ransac_plane(points, 1000, 0.14 )
```

Now, we want to split the original point cloud and grasp the outliers efficiently (i.e., the points that are not the ground). Here is an excellent turnaround. We create a mask that will act as a quick filter to get the rest of the points that do not belong to inliers:

```
mask = np.ones(len(points), dtype=bool)
mask[plane_inliers] = False
outliers = points[mask]
```

We now have an inlier set and an outlier set! Let's check out the results with Open3D:

```
plane_cloud = o3d.geometry.PointCloud()
plane_cloud.points = o3d.utility.Vector3dVector(points[plane_inliers])
plane_cloud.paint_uniform_color([0.8, 0.2, 0.6])
#this gives a uniform color to the points

plane_outliers = o3d.geometry.PointCloud()
plane_outliers.points = o3d.utility.Vector3dVector(outliers)
plane_outliers.paint_uniform_color([0.3, 0.1, 0.9])

o3d.visualization.draw_geometries([plane_cloud, plane_outliers])
```

RANSAC accurately detects and segments the ground plane from the rest of the point cloud (see Figure 9-7).

Figure 9-7. Ground plane segmentation using RANSAC

We found the ground in this scene. This means that we can segment our dataset into two groups of points: the ground and the remaining elements. You can continue iteratively to refine the various components in your scene.

You have implemented a complete RANSAC model fitting algorithm for plane detection and 3D point cloud segmentation from scratch. On top of that, you can now automatically set RANSAC parameters so that you have not a 99% automatic solution but 100% automation. And this is a significant step! Now that you have grasped how to use RANSAC, it's time to explore how leveled our scene is.

Application 2: RANSAC for Analytical Tasks

Let's explore another application that leverages RANSAC. We can effectively identify and fit planes to large planar surfaces, such as floors and walls. By detecting these dominant planes, we can assess the levelness of a 3D scene. This information is crucial for various 3D analytical tasks, including object detection, segmentation, and reconstruction.

A well-leveled scene simplifies these tasks by providing a stable reference frame and reducing the complexity of the analysis. Additionally, detecting the levelness can help identify potential issues or challenges in the 3D data, such as tilt or distortion.

For this case study, we first load a point cloud, such as the playground:

```
pcd = o3d.io.read_point_cloud("../DATA/the_playground.ply")
points = np.asarray(pcd.points)
```

We can then execute RANSAC, as well as automatic parameters (see "Automatic parameter setup" on page 276), by being very accommodating to the noise level:

```
# Compute threshold
tree = KDTree(points, leaf_size=2)
nearest_dist, nearest_ind = tree.query(points, k=8)
nearest_dist_mean = np.mean(nearest_dist[:,1:],axis=0)
threshold = np.max(nearest_dist_mean) + np.std(nearest_dist_mean)

# RANSAC shape detection
t0 = time.time()
plane_params, plane_inliers = ransac_plane(points, 1000, threshold)
t1 = time.time()
```

This is where it is interesting to study how we can extract the planarity of the detected shape. For this, let's define a function that computes the angle between two vectors:

```
def angle_between_vectors(v1, v2):
    # Convert lists to numpy arrays
    v1 = np.array(v1)
    v2 = np.array(v2)

    # Compute the dot product
    dot_product = np.dot(v1, v2)

    # Compute the magnitudes
    v1_mag = np.linalg.norm(v1)
    v2_mag = np.linalg.norm(v2)

    # Compute the cosine of the angle
    cos_angle = dot_product / (v1_mag * v2_mag)

    # Use arccos to get the angle in radians
    angle_rad = np.arccos(np.clip(cos_angle, -1.0, 1.0))
```

```
# Convert to degrees
angle_deg = np.degrees(angle_rad)

return min(angle_deg, 180 - angle_deg)
```

Let me briefly explain what happens in the function. First, I convert two vectors (v1 and v2) to NumPy arrays (these vectors could represent the normal direction of a detected plane or floor).

Then I calculate the dot product between these vectors. This tells me how "similar" the directions are. The dot product is large if the vectors point in the same direction. If they point in opposite directions, it's negative.

I also need to know the length of each vector (their magnitude). I use NumPy's norm function for this. It's like measuring the length of those imaginary sticks. Next, I divide the dot product by the product of the magnitudes. This gives me the cosine of the angle between the vectors I prepare (in degrees, which is more intuitive). If two vectors are almost parallel but pointing in opposite directions, I'll get a small angle instead of something close to 180 degrees.

We can now use this to check if detected planes are level by comparing their normal vectors to the up direction [0,0,1]. The surface is probably horizontal, like a floor, if the angle is close to 0 or 180 degrees. If it's close to 90 degrees, it might be a wall:

```
angle = angle_between_vectors(plane_params[0:3], [0,0,1])
print(f"The angle of the plane is {angle:.2f} degrees")
```

This returns an angle of 9.41°, verified on the dataset. This gives us how tilted or rotated our plane is compared to the perfectly leveled one. Think of it like measuring the angle between two sticks pointing in different directions. You can also use this to check if a surface is level with the ground or if two planes are perpendicular. The results of RANSAC for analytical tasks in 3D scenes are illustrated in Figure 9-8.

Finally, let's explore a third scenario that leverages RANSAC: what if we want to use the inliers as a base for 3D modeling tasks?

Figure 9-8. Application of RANSAC for analytical tasks in a 3D scene

Application 3: RANSAC for Modeling Tasks

Now, what if we wanted to model a geometric scene with RANSAC? Well, this is also feasible! By repeatedly sampling and fitting, RANSAC identifies several model parameters that best explain the underlying shape while disregarding outliers and noise.

Once a robust model is obtained, the inlier points can be used to reconstruct the 3D shape, resulting in a refined and accurate 3D model. We can model an entire scene if we repeat this process for all found 3D shapes. To execute this idea, we'll move in several steps, as illustrated in Figure 9-9. We start by importing the necessary libraries (see Figure 9-10).

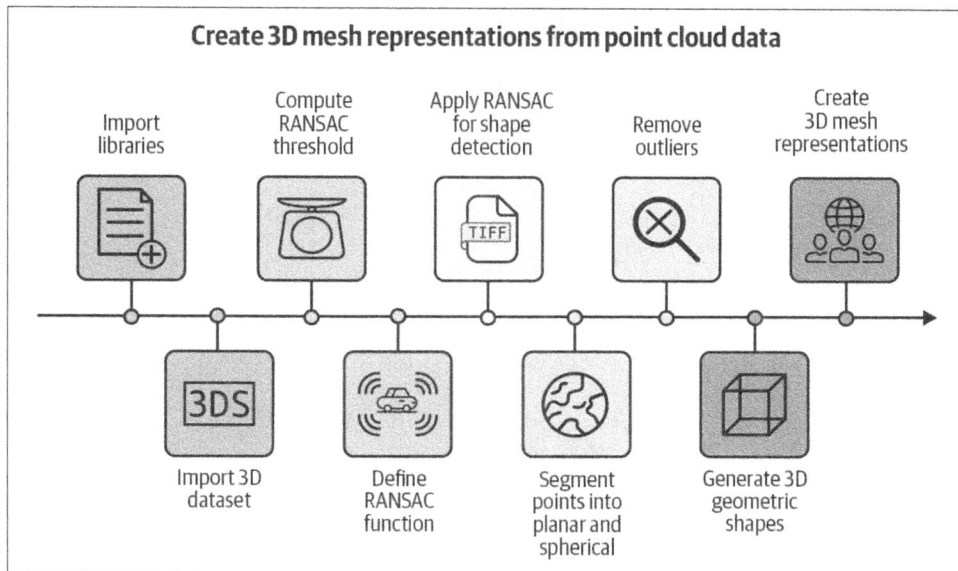

Figure 9-9. Application 3: RANSAC for modeling tasks

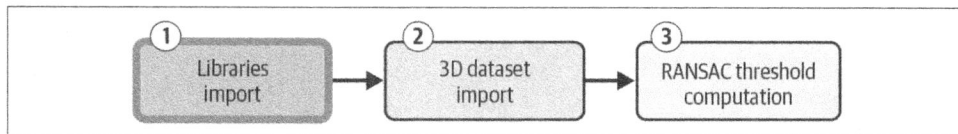

Figure 9-10. Step 1: Importing libraries

Open3D is used for 3D data processing and visualization, NumPy for numerical operations, and scikit-learn for additional machine learning functionalities like k-d tree estimation:

```
import numpy as np
import open3d as o3d

from sklearn.neighbors import KDTree
```

Then, we load a point cloud dataset from a PLY file (see Figure 9-11).

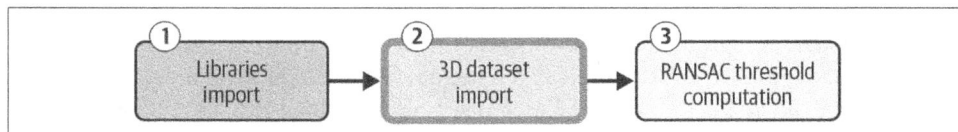

Figure 9-11. Step 2: 3D dataset import

The point cloud is read using Open3D's io.read_point_cloud() function. The points are then extracted as a NumPy array for easier manipulation. To give us a visual understanding of the data we're working with, the point cloud is visualized using Open3D's visualization function:

```
pcd = o3d.io.read_point_cloud("../DATA/pcd_synthetic.ply")

points = np.asarray(pcd.points)
o3d.visualization.draw_geometries([pcd])
```

We proceed to calculate an adaptive threshold that will be used later in the shape detection process (see Figure 9-12).

> Let me explain why an adaptive threshold is really useful here for working with point cloud data. We are working with 3D scans of different environments and objects. Sometimes these scans are very detailed, with lots of points close together. Other times, they're sparse, with points spread far apart. This often happens when the scanner is closer to some parts of the object than others, or when different scanners are used. For these scenarios, it becomes critical to have an approach that can account for nonuniform sampling.

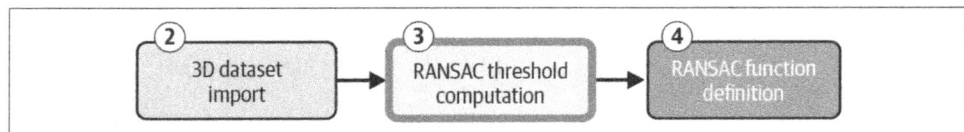

Figure 9-12. Step 3: RANSAC adaptive threshold

The little Python snippet provides an adaptive approach that allows the script to work effectively with point clouds of varying densities:

```
tree = KDTree(points, leaf_size=2)
nearest_dist, nearest_ind = tree.query(points, k=8)
nearest_dist_mean = np.mean(nearest_dist[:,1:],axis=0)
threshold = np.min(nearest_dist_mean) + np.std(nearest_dist_mean)
```

The code first creates a KDTree, which helps to quickly find the closest points to any given point. I calculate the mean distance to these neighbors for every point. Some areas will have small mean distances (dense regions) and others will have large mean distances (sparse regions). This is where the adaptive part becomes important. Instead of using a fixed threshold like "0.1 meters," I let the data tell me what's appropriate. I take the smallest mean distance discovered (representing the densest area) and add one standard deviation to it.

The standard deviation explains how much variation there is in these distances across the whole scan. This adaptive approach saves us from a common headache. If I used a fixed threshold that works well for a dense scan, it might fail completely on a sparse scan. Or if I set it for sparse scans, it would be too loose for dense scans. The algorithm can handle both situations automatically by adapting to each specific point cloud.

For example, if I'm scanning a building, the walls close to the scanner might have points every few millimeters, while distant corners might have points several centimeters apart. An adaptive threshold adjusts to both areas naturally, making my shape detection more reliable across the entire scan.

Now, let's define two RANSAC functions (see Figure 9-13). The first, `ransac_plane()`, is designed to detect planes in the point cloud. The second, `ransac_sphere()`, detects spheres.

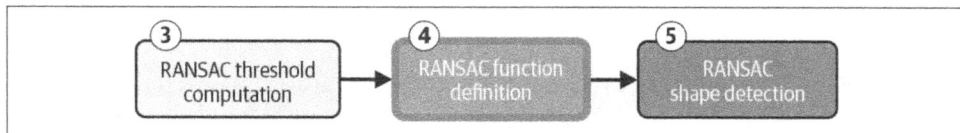

Figure 9-13. Step 4: Defining RANSAC functions

Both functions work on the same principle: they randomly sample points from the cloud, fit a shape to these points, and then find all the points (inliers) that match this shape within the previously calculated threshold. This process is repeated many times, and the shape with the most inliers is considered the best fit:

```
def ransac_plane(points, num_iterations=1000, threshold=0.1):
    best_inliers = []
    best_plane = None

    for _ in range(num_iterations):
        # Randomly sample 3 points
        sample = points[np.random.choice(points.shape[0], 3, replace=False)]

        # Calculate plane equation ax + by + cz + d = 0
        v1 = sample[1] - sample[0]
        v2 = sample[2] - sample[0]
        normal = np.cross(v1, v2)
        a, b, c = normal / np.linalg.norm(normal)
        # d = -np.dot(normal, sample[1])
        d = -np.sum(normal*sample[1])

        # Calculate distances of all points to the plane
        distances = (np.dot(points, [a, b, c]) + d) / np.sqrt(a**2 + b**2 + c**2)

        # Count inliers
        inliers = np.where(np.abs(distances) < threshold)[0]

        if len(inliers) > len(best_inliers):
            best_inliers = inliers
```

```
        best_plane = (a, b, c, d)

    return best_plane, best_inliers

def ransac_sphere(points, num_iterations=1000, threshold=0.1):
    best_inliers = []
    best_sphere = None

    for _ in range(num_iterations):
        # Randomly sample 4 points
        sample = points[np.random.choice(points.shape[0], 4, replace=False)]

        # Calculate sphere equation (x-a)^2 + (y-b)^2 + (z-c)^2 = r^2
        A = np.array([
            [2*(sample[1][0]-sample[0][0]), 2*(sample[1][1]-sample[0][1]),
            2*(sample[1][2]-sample[0][2])],
            [2*(sample[2][0]-sample[0][0]), 2*(sample[2][1]-sample[0][1]),
            2*(sample[2][2]-sample[0][2])],
            [2*(sample[3][0]-sample[0][0]), 2*(sample[3][1]-sample[0][1]),
            2*(sample[3][2]-sample[0][2])]
        ])

        B = np.array([
            [sample[1][0]**2 + sample[1][1]**2 + sample[1][2]**2 -
            sample[0][0]**2 - sample[0][1]**2 - sample[0][2]**2],
            [sample[2][0]**2 + sample[2][1]**2 + sample[2][2]**2 -
            sample[0][0]**2 - sample[0][1]**2 - sample[0][2]**2],
            [sample[3][0]**2 + sample[3][1]**2 + sample[3][2]**2 -
            sample[0][0]**2 - sample[0][1]**2 - sample[0][2]**2]
        ])

        center = np.linalg.solve(A, B).flatten()
        radius = np.sqrt(np.sum((sample[0] - center)**2))

        # Calculate distances of all points to the sphere surface
        distances = np.abs(np.sqrt(np.sum((points - center)**2,
        axis=1)) - radius)

        # Count inliers
        inliers = np.where(distances < threshold)[0]

        if len(inliers) > len(best_inliers):
            best_inliers = inliers
            best_sphere = (*center, radius)

    return best_sphere, best_inliers
```

Let's break down how RANSAC finds both planes and spheres through these functions. For planes, I randomly pick three points and use vector math to find a plane equation that fits through them—imagine stretching a piece of paper between three pins. For spheres, I need four points and need to solve a system of equations to find the center and radius—like finding a bubble that touches four specific points in space.

In both cases, I run this process many times (1,000 by default) and track which shape fits the most points within a small distance threshold. This threshold is like a tolerance. If a point is closer than this distance to my shape, I count it as part of that shape (an inlier). After trying all iterations, I return the shape parameters and the list of points that fit best. The beauty of RANSAC is that it can find shapes even when there's noise or when points belong to multiple shapes, as we observed before, finding a sphere sitting on the floor in a 3D scan. Now that we have our RANSAC functions defined, we apply them to detect a plane and a sphere in the point cloud (see Figure 9-14).

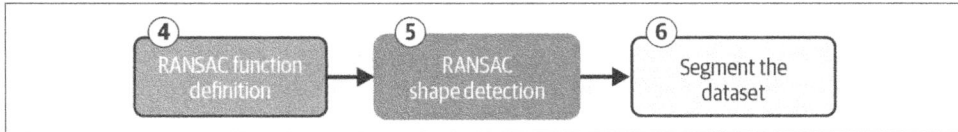

Figure 9-14. Step 5: Detecting geometric shapes

The resulting parameters describe the detected shapes mathematically. For the plane, we get the coefficients of the plane equation ($ax + by + cz + d = 0$). For the sphere, we get its center coordinates and radius:

```
# Find plane and sphere using RANSAC
plane_params, plane_inliers = ransac_plane(points, 1000, threshold)
sphere_params, sphere_inliers = ransac_sphere(points, 1000, threshold)

print(f"Plane equation: {plane_params[0]}x + {plane_params[1]}y +
{plane_params[2]}z + {plane_params[3]} = 0")
print(f"Sphere equation: (x - {sphere_params[0]})^2 +
(y - {sphere_params[1]})^2 + (z - {sphere_params[2]})^2 = {sphere_params[3]**2}")
```

With the shapes detected, we can now segment the point cloud. We separate point clouds for the planar and spherical points (see Figure 9-15).

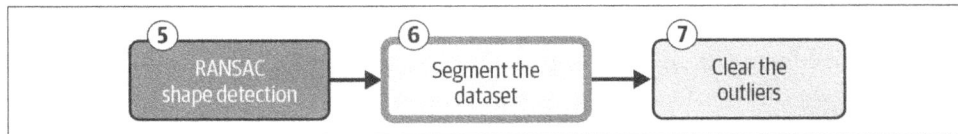

Figure 9-15. Step 6: Point cloud segmentation

To visually distinguish them, the planar points are dark gray and the spherical points are light gray. Each segmented point cloud is then visualized separately, allowing us to see how well the shape detection is performed (see Figure 9-16):

```
# Select and segment the planar points
plane_cloud = o3d.geometry.PointCloud()
plane_cloud.points = o3d.utility.Vector3dVector(points[plane_inliers])
plane_cloud.paint_uniform_color([1, 0, 0])  # Red
o3d.visualization.draw_geometries([plane_cloud])
```

```
# Select and segment the spherical points
sphere_cloud = o3d.geometry.PointCloud()
sphere_cloud.points = o3d.utility.Vector3dVector(points[sphere_inliers])
sphere_cloud.paint_uniform_color([0, 1, 0])  # Green
o3d.visualization.draw_geometries([sphere_cloud])
```

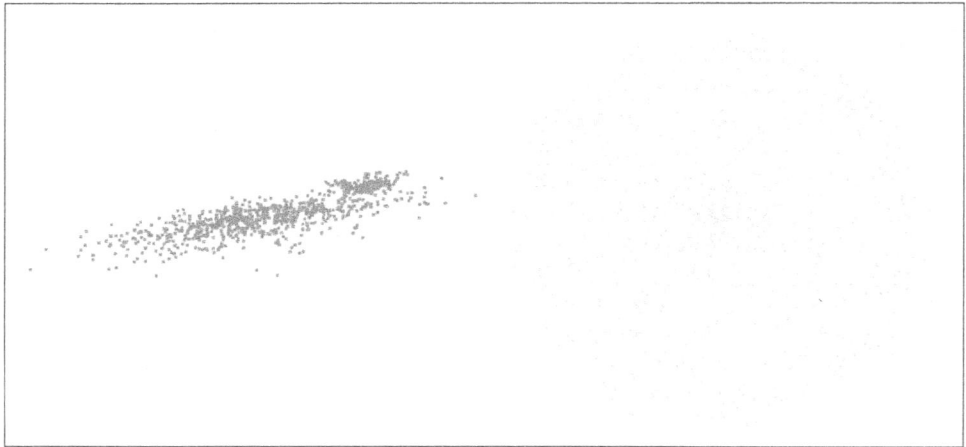

Figure 9-16. The planar points are dark gray, and the spherical points are light gray

Not all points in our original cloud will belong to either the detected plane or sphere (see Figure 9-17).

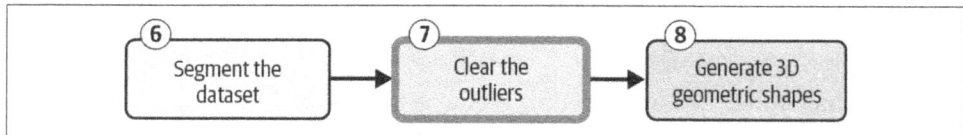

Figure 9-17. Step 7: Clear the outliers

We need to identify these remaining points, which don't belong to our detected shapes, and we color them dark gray:

```
from functools import reduce

# Select and segment the remaining points
other_points = np.delete(points, reduce(np.union1d, (plane_inliers,
sphere_inliers)), axis=0)

other_cloud = o3d.geometry.PointCloud()
other_cloud.points = o3d.utility.Vector3dVector(other_points)
other_cloud.paint_uniform_color([0.7, 0.7, 0.7])
```

If working with a large point cloud, `np.delete` may not be my recommendation. In this case, prefer the Boolean masking approach. Here's why: `np.delete` creates a new copy of the entire array and then removes elements, which uses a lot of memory, especially with large point clouds. Boolean masking is more efficient because it only creates one Boolean array the size of points and then performs a single indexing operation. This means if I have a million points, I only need an array of a million Booleans rather than copying the entire points array first.

Think of it like this: instead of copying a huge book and then tearing out pages (`np.delete`), I'm just putting sticky notes on the pages I want to keep (Boolean masking) and then reading only those pages.

Let's now create a visualization that includes all three point sets: the plane points, the sphere points, and the other points. This gives us a complete view of how the original point cloud has been segmented (see Figure 9-22A):

```
o3d.visualization.draw_geometries([plane_cloud, sphere_cloud, other_cloud])
```

At this stage, we have a nice delineation between the gray points (outliers), the dark gray points (plane), and the light gray points (sphere). To approach our 3D modeling goal, let's take the same approach but use the parameters that we extracted to generate 3D models as meshes. This is our next step (see Figure 9-18).

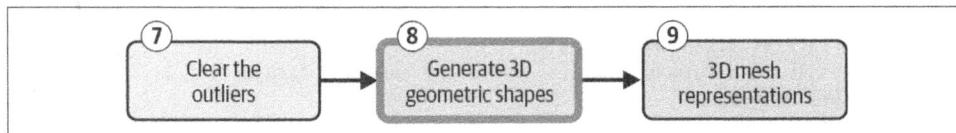

Figure 9-18. Step 8: Generate 3D geometric shapes

To this end, let's define two functions. The `create_plane_mesh()` function creates a mesh representation of the detected plane, while `create_sphere_mesh()` does the same for the sphere. These functions take the parameters of the detected shapes and generate 3D mesh objects that can be visualized alongside our point clouds:

```
def create_plane_mesh(plane_params, size=2, resolution=20):
    a, b, c, d = plane_params

    # Create a grid of points on the plane
    x = np.linspace(-size/2, size/2, resolution)
    y = np.linspace(-size/2, size/2, resolution)
    X, Y = np.meshgrid(x, y)

    # Calculate Z coordinates
    Z = (-d - a*X - b*Y) / c
```

```
# Create vertices and triangles
vertices = np.column_stack((X.ravel(), Y.ravel(), Z.ravel()))
triangles = []
for i in range(resolution-1):
    for j in range(resolution-1):
        v0 = i * resolution + j
        v1 = v0 + 1
        v2 = (i + 1) * resolution + j
        v3 = v2 + 1
        triangles.extend([[v0, v2, v1], [v1, v2, v3]])

# Create Open3D mesh
mesh = o3d.geometry.TriangleMesh()
mesh.vertices = o3d.utility.Vector3dVector(vertices)
mesh.triangles = o3d.utility.Vector3iVector(triangles)
mesh.compute_vertex_normals()

return mesh

def create_sphere_mesh(sphere_params, resolution=50):
    center_x, center_y, center_z, radius = sphere_params

    # Create a UV sphere
    u = np.linspace(0, 2 * np.pi, resolution)
    v = np.linspace(0, np.pi, resolution)
    x = center_x + radius * np.outer(np.cos(u), np.sin(v))
    y = center_y + radius * np.outer(np.sin(u), np.sin(v))
    z = center_z + radius * np.outer(np.ones_like(u), np.cos(v))

    # Create vertices and triangles
    vertices = np.column_stack((x.ravel(), y.ravel(), z.ravel()))
    triangles = []
    for i in range(resolution-1):
        for j in range(resolution-1):
            v0 = i * resolution + j
            v1 = v0 + 1
            v2 = (i + 1) * resolution + j
            v3 = v2 + 1
            triangles.extend([[v0, v2, v1], [v1, v2, v3]])

    # Create Open3D mesh
    mesh = o3d.geometry.TriangleMesh()
    mesh.vertices = o3d.utility.Vector3dVector(vertices)
    mesh.triangles = o3d.utility.Vector3iVector(triangles)
    mesh.compute_vertex_normals()

    return mesh
```

A short note: the choice of size and resolution values will depend on the specific needs of your application, balancing the desired level of detail (LoD) against the available computational resources. Generally, you'll want to use larger sizes and higher resolutions for more accurate and visually appealing meshes while considering the trade-offs regarding performance and memory usage. The default parameters should work wonders for most applications.

When creating a plane mesh, the key parameters are the `plane_params` tuple, which defines the equation of the plane, and the size and resolution parameters. The `size` parameter determines the physical extent of the plane, setting the range of the x and y coordinates. The `resolution` parameter controls the level of detail, with a higher resolution resulting in more vertices and triangles in the mesh. The function uses these inputs to generate a grid of points on the plane and then constructs the vertices and triangles to form the triangle mesh. Similarly, when creating a sphere mesh, the important parameters are the `sphere_params` tuple, which specifies the center coordinates and radius of the sphere, and the `resolution` parameter. The resolution sets the number of points along the longitude and latitude axes of the sphere, with a higher value producing a more detailed mesh but also increasing the computational and memory requirements.

Using the functions defined, we now create mesh representations of the detected plane and sphere (see Figure 9-19). To differentiate them from the point clouds, the plane mesh is gray, and the sphere mesh is light gray (see Figure 9-22b).

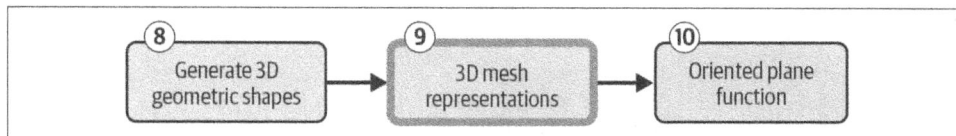

Figure 9-19. Step 9: 3D mesh representations

Computing an oriented plane representation can be more useful than just the plane equation in many scenarios. The plane equation gives us the normal vector and offset, but an oriented plane also includes two in-plane vectors that define the full 3D coordinate frame of the plane.

This more complete information makes it easier to reason about the plane's position and orientation, simplifies various geometric computations, and enables more accurate visualization and integration with other 3D data structures. While determining the in-plane vectors requires some additional computation, the benefits of having the full oriented plane representation are often worth it, especially for tasks that involve aligning, transforming, or working with the plane in a 3D context. Therefore, to approach these goals, let's define a new function called `create_oriented_plane_mesh()` (see Figure 9-20).

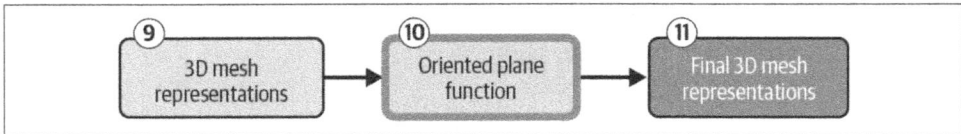

Figure 9-20. Step 10: Oriented plane function

We can use PCA to orient the plane based on the inlier points. By doing so, we can create a plane mesh that better fits the actual data, especially in cases where the detected plane might not be perfectly aligned with the coordinate axes:

```
def create_oriented_plane_mesh(plane_params, inlier_points):
    a, b, c, d = plane_params
    normal = np.array([a, b, c])

    # Project inlier points onto the plane
    projected_points = inlier_points - (np.dot(inlier_points, normal) +
    d).reshape(-1, 1) * normal / np.dot(normal, normal)

    # Perform PCA to find the principal directions on the plane
    mean = np.mean(projected_points, axis=0)
    centered_points = projected_points - mean
    _, _, vh = np.linalg.svd(centered_points)

    # The first two right singular vectors are the principal directions
    u = vh[0]
    v = vh[1]

    # Calculate the extent of the points along the principal directions
    u_coords = np.dot(centered_points, u)
    v_coords = np.dot(centered_points, v)
    u_range = np.max(u_coords) - np.min(u_coords)
    v_range = np.max(v_coords) - np.min(v_coords)

    # Create a grid of points on the plane
    resolution = 20
    u_grid = np.linspace(-u_range/2, u_range/2, resolution)
    v_grid = np.linspace(-v_range/2, v_range/2, resolution)
    U, V = np.meshgrid(u_grid, v_grid)

    # Calculate the 3D coordinates of the grid points
    grid_points = mean + U.reshape(-1, 1) * u + V.reshape(-1, 1) * v

    # Create vertices and triangles
    vertices = grid_points
    triangles = []
    for i in range(resolution-1):
        for j in range(resolution-1):
            v0 = i * resolution + j
            v1 = v0 + 1
            v2 = (i + 1) * resolution + j
```

```
        v3 = v2 + 1
        triangles.extend([[v0, v2, v1], [v1, v2, v3]])

    # Create Open3D mesh
    mesh = o3d.geometry.TriangleMesh()
    mesh.vertices = o3d.utility.Vector3dVector(vertices)
    mesh.triangles = o3d.utility.Vector3iVector(triangles)
    mesh.compute_vertex_normals()

    return mesh
```

As a final step, let's create the ultimate visualization of our processed data (see Figure 9-21).

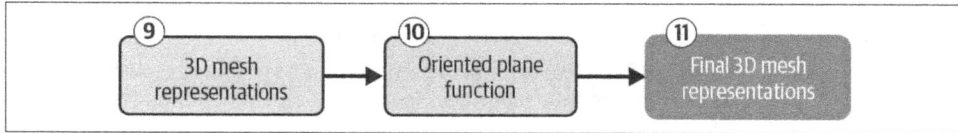

Figure 9-21. Step 11: Final 3D mesh representations

Let's generate the oriented plane mesh using our new function. These optimized meshes, along with all the point clouds (plane points, sphere points, and other points), are combined into a final visualization. This gives us a comprehensive view of the original point cloud data, the shapes we've detected within it, and high-quality mesh representations of those shapes (see Figure 9-22C).

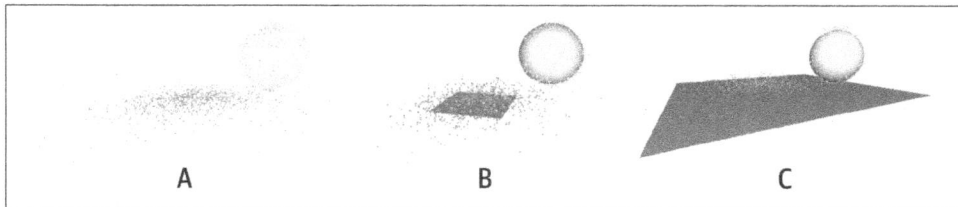

Figure 9-22. (A) Full point cloud segmentation visualization, (B) differentiated mesh visualization, (C) final visualization with oriented meshes

This now demonstrates a complete workflow for processing 3D point cloud data, from loading and analyzing the data to detecting shapes and creating visual representations of those shapes. It showcases the power of combining the mathematical techniques of RANSAC and PCA with modern 3D processing libraries to extract meaningful information from complex 3D data. Now, before moving on to the next stage, let me issue some clear boundaries linked with using RANSAC for 3D shape detection in Table 9-1.

Table 9-1. Limitations of RANSAC for 3D shape detection

Limitations	Description
Sensitivity to noise	RANSAC is sensitive to noise and outliers in the point cloud data. High noise levels or random points can cause the algorithm to fail to detect the accurate shapes.
Difficulty with complex shapes	RANSAC works best for simple geometric shapes like planes, spheres, and cylinders. It can struggle to detect more complex or irregular 3D shapes accurately.
Parameter tuning	RANSAC has several parameters that need to be carefully tuned, such as the distance threshold and number of iterations. Finding the optimal parameters can be time-consuming and dataset dependent.
Computational complexity	The RANSAC algorithm involves random sampling and iterative model fitting, which can be computationally expensive, especially for large point clouds or high-resolution shapes.
Bias toward larger shapes	RANSAC tends to favor detecting larger shapes in the point cloud, as they will have more inlier points. Smaller or occluded shapes may be overlooked.
Handles only one shape type at a time	The standard RANSAC implementation can only detect one type of shape (e.g., planes or spheres) at a time. Detecting multiple shape types simultaneously requires more advanced techniques.
Lack of uncertainty quantification	RANSAC does not provide explicit information about the uncertainty or confidence in the detected shapes, which can be important for downstream applications.
Sensitivity to initial Conditions	The RANSAC solution can be sensitive to the initial randomly sampled points, potentially leading to different results across runs on the same data.

Overall, while RANSAC is a powerful and widely used algorithm for 3D shape detection, these are the limitations that users should be aware of when applying it to real-world 3D datasets. Careful parameter tuning, data preprocessing, and the use of more advanced techniques may be necessary to overcome these challenges. Therefore, let us put a pin on that to provide some solutions in "A Hybrid Approach: RANSAC and Region Growing" on page 308. But first, let's move on to 3D region growing solutions.

Region Growing for 3D Shape Detection

Let me share with you one of my favorite scene understanding techniques: region growing. It is so fundamental that it can act as a strong shape detection approach for complex scenes where one wants to delineate its geometrical constituents. Region growing is a segmentation technique that starts with a seed point and iteratively adds neighboring points to the region if they meet certain criteria.

In the context of 3D datasets, these criteria often involve spatial proximity, similarity in feature values (e.g., intensity, color, or normal), or other relevant properties. As a first-hand approach, we saw that measuring spatial proximity can be accomplished by using defined metrics such as the Euclidean distance (*https://oreil.ly/EEMbN*), the Manhattan distance (*https://oreil.ly/OgImb*), or the Mahalanobis distance (*https://oreil.ly/Jm4-4*).

Euclidean distance is the most straightforward metric. It measures the direct, straight-line distance between two points. Imagine a bird flying directly from point A to point B. Manhattan distance, also known as city block distance or L1 distance, is calculated by summing the absolute differences between the corresponding coordinates of the two points. Imagine a taxi driver navigating a city grid. The distance it travels is the Manhattan distance, as it must follow the streets and can't cut through buildings. Mahalanobis distance is a more sophisticated metric that takes into account the covariance of the data. This means it considers the correlation between different dimensions of the data. It's particularly useful when dealing with data that is not uniformly distributed or has correlated features.

The second aspect is using a measure of similarity within the chosen feature(s). Common features for 3D datasets include intensity, color (if applicable), surface normals, curvature, or point descriptors. We also can grow regions based on connectivities, such as 4-connectivity (adjacent points in the same plane) or 6-connectivity (adjacent points in the same plane or directly above/below). But how does region growth work?

Region Growing Principles

We basically follow a six-step process, as illustrated in Figure 9-23. First, we select a seed point. This is the starting point of the region's growing process. You can either manually choose a seed point or use an algorithm to select one based on certain criteria automatically. The choice of seed point can significantly impact the quality of the segmentation. Then, once you have your seed point, you create an empty region to store the points that will be added. This region acts as a container for the segmented object.

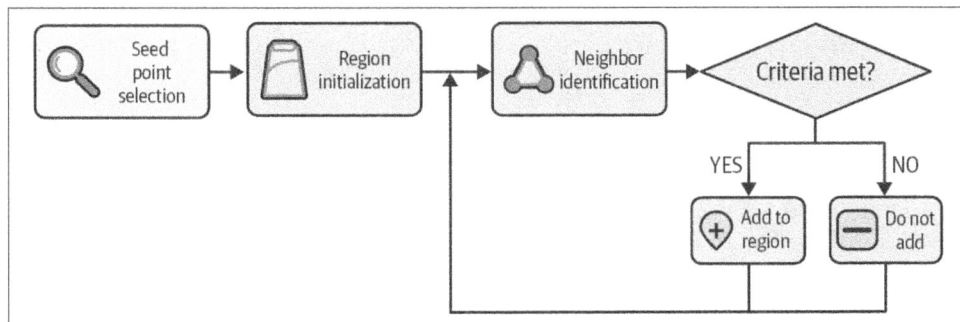

Figure 9-23. Principles of region growing

We then identify the neighbors of the current region points. In 3D, this involves looking at points that are adjacent to a certain search structure. The definition of neighbors can vary depending on the specific application and the similarity measure used (for spatial datasets, it is usually a Euclidean distance). Then, for each neighbor, we check if it meets the specified criteria. This could be solely based on distance or similarity in normals, intensity, color, or other relevant factors.

The criteria should be chosen carefully to ensure that points belonging to the same shape are grouped while points from different objects are separated. Then, we move on to the region expansion stage. If a neighbor meets the criteria, we add it to the region. This process is repeated iteratively until no new points can be added. Finally, we repeat steps 3 to 5 until no new points can be added to the region. This ensures that the region has grown to its full extent.

While region growing is a powerful technique, it's not without its challenges. The first observation is that the quality of a segmentation can heavily depend on the seed point you choose. A bad seed point can lead to a bad segmentation. For instance, if the seed point is located within a small, isolated object, the region growing process may not capture the entire object.

Also, we usually heavily consider two phenomena when tuning the parameters of our region-growing approach. If our threshold criteria are too lenient, we might end up with too many small regions, which qualify for oversegmentation. This can happen when points belonging to the same object are not correctly merged.

Then, we have undersegmentation. This is when we use a threshold criterion that is too strict, where we might miss important parts of the object we are trying to segment. This can occur when points from different objects are mistakenly grouped.

Finally, noise in our data can make it difficult for region growing to work effectively. Noise can introduce false positives or negatives, leading to incorrect segmentation results.

Despite these limitations, region growing remains a valuable tool for 3D segmentation and shape recognition. Ongoing research is exploring ways to address these challenges, such as:

Adaptive criteria
  Developing criteria that can adapt to different regions of the data

Hybrid approaches
  Combining region growing with other segmentation techniques (e.g., thresholding, clustering) to improve performance

Parallel processing
  Utilizing parallel computing techniques to accelerate region growing for large datasets

Now that you grasp the potential let's put it to the test and create our region growing for 3D datasets from scratch.

Region Growing: Real-World Setup

In this session, we delve into the practical application of region growing for 3D point cloud segmentation and shape recognition. This means first targeting our real-world setup (see Figure 9-24).

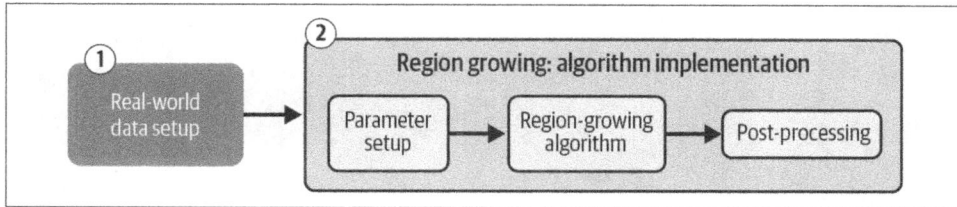

Figure 9-24. Region growing: real-world setup

The process involves several critical stages, starting with the import of necessary libraries such as NumPy (*https://numpy.org*) and SciPy (*https://scipy.org*) for computation (utilizing the KDTree module for accelerated point queries) and Open3D (*https://www.open3d.org*) for visualization:

```
import numpy as np
from scipy.spatial import KDTree
import open3d as o3d
```

The choice of these libraries stems from their robust functionalities combined with minimal overhead, allowing for streamlined and efficient processing. This foundation enables us to handle point cloud data with precision and flexibility, ensuring that we can focus on the core shape detection task without being bogged down by peripheral concerns. The datasets used include AI-generated point clouds and those acquired via a mobile laser scanning system in industrial environments. This is a good way to test its robustness and potential for integration into production workflows. Let's load our point cloud datasets, which can vary from AI-generated models to real-world scans from mobile laser systems (see Figure 9-25A):

```
point_clouds = ["../DATA/darth_vader.ply", "../DATA/depth_anything.ply",
"../DATA/industrial_room_part.ply", "../../DATA/NAAVIS_EXTERIOR.ply"]
pcd = o3d.io.read_point_cloud(point_clouds[-1])
o3d.visualization.draw_geometries([pcd])
```

Visualizing the point clouds at this juncture (see Figure 9-25B and C) provides a preliminary insight into their structure and complexity, guiding our parameter choices for the segmentation algorithm. This stage is crucial as it sets the stage for subsequent processing by converting the raw point clouds into NumPy arrays, ready for manipulation:

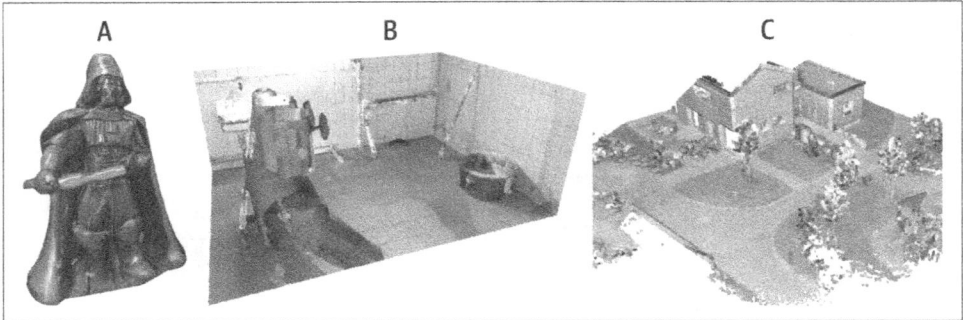

Figure 9-25. (A) A generated model of Darth Vader, created with generative AI; (B) An industrial room from a terrestrial laser scanner; (C) An outdoor scene from a mobile mapping LiDAR system

```
# Convert Open3D object to NumPy for processing stages
input_points = np.asarray(pcd.points)
normals = np.asarray(pcd.normals)
```

With our data ready, we proceed to the next stage: the implementation of region growing.

Region Growing: Implementation

This image visualizes the critical parameters in the region growing algorithm (see Figure 9-26).

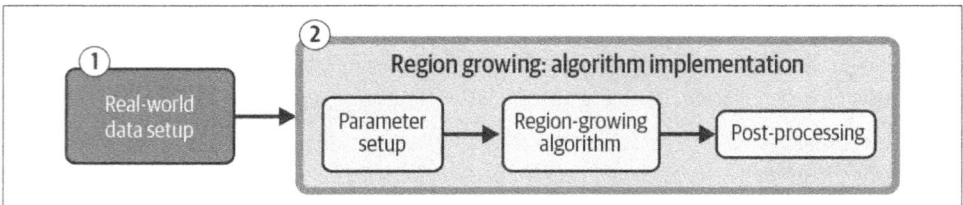

Figure 9-26. Region growing: Implementation

As we saw previously, to implement our region growing algorithm, we start by defining the parameters. Two pivotal parameters dictate the segmentation efficacy: the angle threshold, which determines if two points belong to the same region based on the angle between their surface normals, and the distance threshold, which confines the vicinity of points considered for region growing:

```
angle_threshold = np.pi/6
distance_threshold = 0.1
```

These parameters are knowledge driven, reflecting the resolution and characteristics of the point cloud data, and are instrumental in ensuring that the segmentation captures meaningful structures. They define an angle threshold of pi/6 radians (which is equivalent to 30 degrees) and a distance threshold of 0.1 meters. These thresholds will be used to determine whether points should be included in the same segment during the region-growing process.

> As with RANSAC, we could automatically determine our distance threshold to skip this definition and move closer to a fully unsupervised approach. The current parameter, 0.1 m, is good for the outdoor laser-scanned scene but may prove limiting for the others.

Next, we move on to the initialization phase. We determine the total number of points in the input point cloud and store this value in the variable N:

```
N = len(input_points) # Total number of points
```

Using the input points, a k-d tree data structure is then created. This k-d tree is instrumental in performing efficient nearest-neighbor searches throughout the algorithm:

```
kd_tree = KDTree(input_points)
```

Two key data structures are also initialized: segments, an empty list that will store the resulting segmented regions, and unsegmented, a set containing the indices of all points in the point cloud. Initially, unsegmented includes all point indices, as no segmentation has occurred yet:

```
segments = [] # Output of the algorithm
unsegmented = set(range(N)) # Algorithm worker
```

The main segmentation loop begins after this initialization and continues as long as unsegmented points remain in the point cloud. Each iteration represents the growth of a new segment:

```
while unsegmented:
```

At the start of each new segment, the algorithm randomly selects a seed point from the unsegmented set. This seed point becomes the first point in the new segment. The algorithm then uses the k-d tree to efficiently find all neighbors of this seed point that are within the specified distance threshold. These neighboring points are added to a stack, which will be used to process points in a depth-first manner:

```
# Start a new segment
    seed_index = unsegmented.pop()
    segment = [seed_index]
    # stack of points to be processed (neighbor indices)
    stack = list(kd_tree.query_ball_point(input_points[seed_index],
    distance_threshold))
```

The algorithm then enters a nested loop that processes points from the stack until it's empty. For each point taken from the stack, the algorithm first checks if it's still unsegmented. If the point has already been segmented (i.e., it's not in the unsegmen ted set), it's skipped, and the algorithm moves on to the next point in the stack:

```
while stack:
        point_index = stack.pop()
        if point_index not in unsegmented:
            continue # Makes sure that we consider "free" points
```

For points that are still unsegmented, the algorithm performs a key check: it calculates the angle between the current point's normal vector and the seed point's normal vector. This is done using the dot product of the two normal vectors, followed by the arccos function to get the angle. If this angle is less than the specified angle threshold, the point is part of the same segment as the seed point:

```
normal_angle = np.arccos(np.dot(normals[seed_index], normals[point_index]))
```

When a point meets the angle criterion, several actions occur. First, the point is removed from the unsegmented set, marking it as processed. It's then added to the current segment. Finally, the algorithm uses the k-d tree again to find all neighbors of this newly added point that are within the distance threshold. These neighbors are added to the stack for future processing. This process allows the segment to grow organically and include all connected points that meet the distance and angle criteria:

```
if normal_angle < angle_threshold:
        unsegmented.remove(point_index)
        segment.append(point_index)
        stack.extend(kd_tree.query_ball_point(input_points[point_index],
        distance_threshold))
```

Once the stack is empty, it means that all connected points meeting the criteria have been added to the current segment. At this point, the complete segment is added to the segments list, and the algorithm returns to the main loop to start a new segment with a new seed point:

```
segments.append(segment)
```

This region growing implementation effectively segments the point cloud into regions where points are close to each other (within the distance threshold) and have similar normal orientations (within the angle threshold). The resulting segments represent areas of the point cloud with consistent geometric properties.

The efficiency of this loop is bolstered by the use of *k*-d tree structures, which expedites nearest-neighbor searches and significantly reduces computational time: `Region Growing Successful in: 2.682 seconds`.

Yes, I know, we do not yet want to visualize the results. But, if you were to explore the `segments` variable, you would get a sense as to how the grouping is handled. For example:

```
segments[8]
```

This returns:

```
[1580, 1614, 1669, 1626, 1607, 1653, 1672, 1611, 1624, 1602, 1589, 1659, 1681, 1608, 1662, 1627]
```

These are the indexes of the points in the point cloud that belong to the new segment.

Post segmentation, we assign labels and colors to the identified segments for visual verification with a new function:

```python
def coloring_segments(input_points, segments):
    colors_l = np.zeros_like(input_points)
    labels = np.zeros(len(input_points), dtype=int)

    count = 0
    for segment in segments:
        # color = np.random.rand(3)  # Generate a random color
          # Assign the color to all points in the segment
        # Optional
        if len(segment) <= 1:
            labels[segment] = -1
            colors_l[segment] = [0,0,0]
            count-=1
        elif len(segment) > 1 and len(segment) < 10:
            labels[segment] = 0
            colors_l[segment] = [1,0,0]
            count-=1
        else:
            labels[segment] = count
            colors_l[segment] = np.random.rand(3)
        count+=1
    return labels, colors_l

labels, colors_l = coloring_segments(input_points, segments)
```

This step not only aids in qualitative assessment but also paves the way for further postprocessing tasks. For instance, small segments can be filtered out, and segments can be analyzed for specific characteristics such as planarity or spatial extent. These attributes can be used for higher-level tasks such as object recognition or scene understanding, enhancing the utility of the segmented data.

To visualize our results and also ensure we can use them, we assign the colors to our point cloud with Open3D and then create the equivalent NumPy point cloud for export purposes:

```
pcd.colors = o3d.utility.Vector3dVector(colors_l)
pcd_segmented = np.hstack((input_points, np.atleast_2d(labels).T))
```

What's left? Well, let's visualize our results:

```
o3d.visualization.draw_geometries([pcd])
```

This returns what we see in Figure 9-27.

Figure 9-27. Visualization of region growing results

This allows us to very efficiently retrieve the ground and most planar elements as single segments, which could then be used as part of shape fitting purposes. Also, this hints that varying the parameters (search distance and threshold) will actually largely impact the results. For example, in Figure 9-28 I vary the distance threshold.

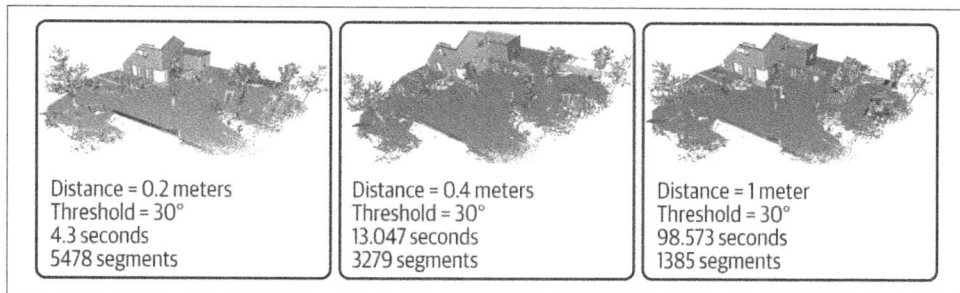

Figure 9-28. Impact of varying parameters on segmentation

As you can see, the higher the distance, the less segments we have, and the more time it takes. But what happens if I vary the angle threshold (see Figure 9-29)?

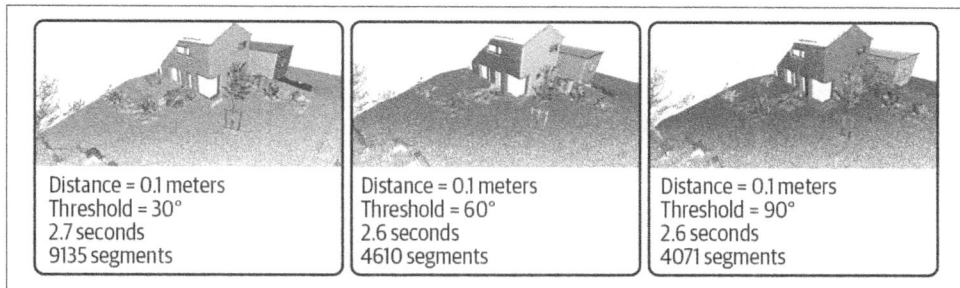

Figure 9-29. Effect of varying the angle threshold

Notice that we get bigger segments, with a limit where 90° is no longer very relevant. Finally, we export the segmented point clouds for external use, ensuring interoperability with other software tools like CloudCompare:

```
o3d.io.write_point_cloud("../RESULTS/NAAVIS_EXTERIOR_segmented.ply", pcd)
np.savetxt("../RESULTS/NAAVIS_EXTERIOR_segmented.xyz", pcd_segmented,
fmt='%1.6f', delimiter=';', header='X;Y;Z;Segment')
```

This step underscores the practical application of our method, enabling seamless integration into broader workflows. By transforming our segmented data into open formats such as PLY or ASCII, we facilitate further analysis beyond the confines of our Python environment.

We used distance and normals to measure how distant and similar points are so we could estimate and group them. But what if we want to leverage color information? Well, this is possible. Only, there are some tricks, as we cannot directly use R, G, B information the same way we use normal components (Nx, Ny, Nz). I designed an RGB region-growing solution that you can directly find in the 3D Resource Hub (*https://learngeodata.eu/3d-data-science-with-python*).

Now, let's design a hybrid solution that combines both RANSAC and region growing.

A Hybrid Approach: RANSAC and Region Growing

First, let's import two libraries and the point cloud and preprocess, lightly, our point cloud:

```
import numpy as np
import open3d as o3d

pcd = o3d.io.read_point_cloud('../DATA/office.ply')
center = pcd.get_center()
pcd.translate(-center)
```

We read a point cloud from a PLY file named *office.ply* using Open3D's I/O functionality. After loading the point cloud, we calculate its center using the `get_center()` method. This gives us the average position of all points in the cloud. We then translate the entire point cloud so that its center is at the origin (0, 0, 0). Finally, we use Open3D's visualization module to display the centered point cloud, allowing us to inspect our data visually (see Figure 9-30a):

```
o3d.visualization.draw_geometries([pcd])
```

We can then compute the nearest-neighbor distance for each point in the point cloud. This operation calculates the distance to each point's closest neighboring point:

```
nn_distance = pcd.compute_nearest_neighbor_distance()
```

The resulting information is crucial for setting appropriate parameters in subsequent steps, particularly for normal estimation and planar patch detection. Understanding the density and distribution of points in the cloud helps us make informed decisions about search radii and other distance-based parameters.

Now, we can define several parameters to use in the subsequent steps of our pipeline:

```
p_radius = np.mean(nn_distance)*3

p_normal_variance_threshold_deg = 45
```

```
p_coplanarity_deg = 75
p_outlier_ratio=0.75
p_min_plane_edge_length=0
p_min_num_points=0
```

p_radius is set to three times the mean nearest neighbor distance, providing a reasonable search radius for local operations. We also set thresholds for normal variance and coplanarity in degrees, which will be used to determine how points are grouped into planar patches. The outlier ratio determines the maximum proportion of points that can be considered outliers in a patch. Lastly, we set minimum values for plane edge length and number of points, both of which are set to 0 in this case. This means we don't impose any minimum size or point count for detected planes. Finally, as I took a point cloud without normals, let me estimate the normal vectors for each point in the point cloud:

```
pcd.estimate_normals(search_param=o3d.geometry.KDTreeSearchParamHybrid(radius =
p_radius, max_nn=30))
```

Normal vectors are perpendicular to the surface at each point and provide important information about the local geometry. We use a hybrid search method that combines radius search with a maximum number of neighbors. The search radius is set to our previously calculated p_radius, and we limit the search to a maximum of 30 nearest neighbors. This hybrid approach helps ensure consistent normal estimation across varying point densities (see Figure 9-30b).

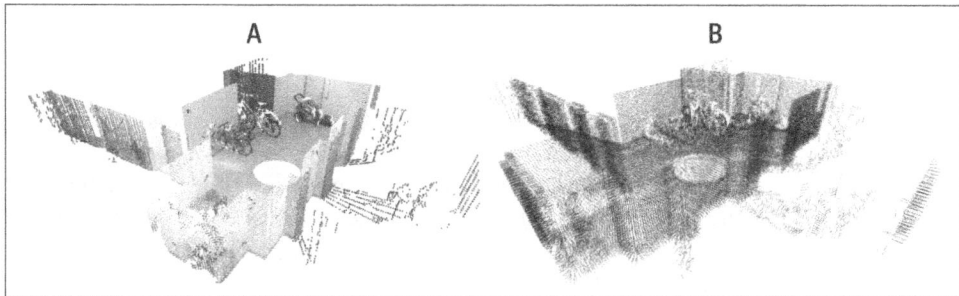

Figure 9-30. (a) Point cloud centering and visualization; (b) Normal estimation using hybrid search

We are now ready to detect planar patches in the point cloud using a sophisticated algorithm that combines RANSAC with region growing:

```
oboxes = pcd.detect_planar_patches(
    p_normal_variance_threshold_deg,
    p_coplanarity_deg,
    p_outlier_ratio,
    p_min_plane_edge_length,
    p_min_num_points,
```

```
search_param=o3d.geometry.KDTreeSearchParamKNN(knn=30)
)
```

This method is effective at identifying flat surfaces in the point cloud, such as walls, floors, or tabletops. The function uses the parameters we defined earlier to determine how points should be grouped into planar patches. It returns a list of oriented bounding boxes (oboxes) that encapsulate each detected planar region. Finally, we print the number of detected patches, giving us an idea of how many distinct planar surfaces were found in the point cloud:

```
print("Detected {} patches".format(len(oboxes)))
```

This returns 31 patches.

The final stage is to create 3D meshes for each detected planar patch. We iterate through each oriented bounding box and create a triangle mesh from it:

```
p_obox = 0.5

geometries = []
obox_volumes = []

for obox in oboxes:
    mesh = o3d.geometry.TriangleMesh.create_from_oriented_bounding_box(obox,
    scale=[1, 1, p_obox])
    mesh.paint_uniform_color(obox.color)
    mesh.compute_vertex_normals()
    geometries.append(mesh)
    obox_volumes.append(obox.volume())
    geometries.append(obox)
```

The mesh is scaled in the z direction by p_obox (set to 0.5), which helps visually distinguish the planar patches. We then color the mesh using the color assigned to the bounding box and compute its vertex normals for proper lighting in visualization. Both the mesh and the original bounding box are added to a list of geometries. Additionally, we calculate and store the volume of each bounding box, which will be used to estimate the total occupied volume of the detected planar regions.

All that is left is to visualize the results of our algorithm:

```
o3d.visualization.draw_geometries(geometries+[pcd])

struct_vol = np.sum(obox_volumes)
print(f"Estimated Occupied Volume: {struct_vol} m3")
```

This allows us to see how well our planar patch detection has performed and how the detected planes relate to the original data:

```
o3d.visualization.draw_geometries(geometries+[pcd])
```

This results in Figure 9-31.

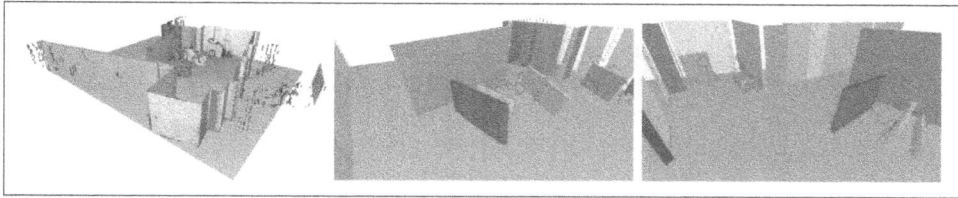

Figure 9-31. Visualizing the planar patch detection results

As you can see, each planar element is then modeled with an oriented bounding box, which is a very interesting stage in the creation of 3D models or analytical tasks.

Indeed, after visualization, we can calculate the total volume of all detected planar patches by summing the volumes of the individual bounding boxes. This gives us an estimate of the occupied volume in the scanned structure, which we print out in cubic meters:

```
struct_vol = np.sum(obox_volumes)
print(f"Estimated Occupied Volume: {struct_vol} m3")
```

This final step provides both a visual and quantitative summary of our point cloud analysis, with an estimated occupied volume of 10.76 m³.

Accurately determining the volume of objects or regions of interest within a 3D scanned dataset provides invaluable insights. In construction, it helps estimate material quantities, optimize resource allocation, and monitor progress. It assists in level design, pathfinding, collision detection, and physics simulations for video game development. In medical imaging, it aids in quantifying tissue volumes, tracking disease progression, and planning surgical interventions. This is just a fraction of the possible use case of such a solution.

Let's now take a step back and review what we have learned.

Summary

You can be proud of your two new 3D data spells: RANSAC and region growing. While RANSAC excels at fitting predefined models to data, even with significant noise and outliers, region growing focuses on grouping points based on shared local features, making it ideal for segmentation tasks where explicit shape models aren't necessary.

RANSAC's strength lies in its iterative model-fitting process. It randomly samples subsets of data points and attempts to fit the desired model, such as a plane or a sphere. Points aligning well with the fitted model are categorized as inliers, while those deviating significantly are labeled as outliers. Its ability to handle datasets with high noise and outliers is remarkable. This makes RANSAC suitable for applications like detecting dominant planes (walls, floors) in a cluttered point cloud.

In contrast, region growing adopts a bottom-up approach, systematically exploring local neighborhoods and groupings based on specific similarity criteria. These criteria can be based on various factors like proximity to the seed point, color similarity, or the alignment of surface normals, making it highly adaptable to different segmentation tasks. Region growing has the ability to effectively segment objects or areas within a point cloud that share common local attributes, even when those objects don't conform to simple geometric shapes (see Table 9-2).

Table 9-2. Comparison of RANSAC and region growing for point cloud segmentation

	RANSAC	Region growing
Goal	Model fitting	Segmentation
Noise robustness	High	Moderate
Shape predefinition	Required	Not required
Starting point	Random samples	Seed points
Growth	Model-driven	Feature similarity and proximity-driven

RANSAC can be employed initially to robustly identify major planar surfaces in a point cloud, even within noisy data. Once these planes are detected, region growing can be applied to refine the segmentation within each plane. This refinement process could leverage additional features like color information, leading to more accurate and detailed segmentation results. This hybrid approach effectively leverages the strengths of both techniques, with RANSAC providing robustness to outliers and initial shape detection and region growing contributing its segmentation prowess based on local feature similarity.

At this stage, you clearly understand the principles behind RANSAC and region growing for 3D shape recognition tasks. You have also investigated proper parameter selection, which significantly impacts the effectiveness of these algorithms. Real-world data often necessitates robust and adaptive approaches for accurate shape recognition.

You can be proud of your newfound skills, giving you the ability to:

- Implement RANSAC for plane and sphere detection from scratch using Python.
- Understand inliers, outliers, and the iterative model fitting process.
- Determine the optimal number of RANSAC iterations and how to automatically set parameters like the distance threshold.
- Implement region growing for 3D point cloud segmentation using geometric features (normals, distance) and color information.
- Implement a hybrid approach for planar-based modeling (RANSAC with region growing)

Hands-on Project

Now, if you feel up to the task, here is a project to help you refine your skills:

1. Experiment with the hybrid approach combining RANSAC and region growing on the provided datasets or other 3D point clouds.

2. Adjust the parameters for both algorithms, such as the distance and angle thresholds, and observe their impact on the segmentation results.

3. Explore the effects of different seed point selection strategies on the performance of region growing.

4. Investigate the use of different color spaces and color difference metrics for color-based region growing.

5. Apply these techniques to extract shape features for knowledge-based object recognition of your house's objects.

We primarily utilize geometric features like normals and distances. The integration of richer feature sets involves developing new similarity metrics that incorporate these additional features or using machine learning techniques to learn feature representations that are more discriminative for specific shape recognition tasks. This will be studied in Chapter 14. Finally, it may be interesting to explore hybrid and advanced techniques. For example, developing strategies for automatically determining the best sequence of operations (RANSAC then region growing, or vice versa) and parameters for different datasets and shape recognition tasks. Integrating other segmentation techniques, such as graph-based segmentation or clustering, could lead to more powerful and versatile 3D shape recognition pipelines. This is the goal of Chapter 12.

While these approaches are super useful for shape detection mechanisms, know that they are highly sought after in modeling, segmentation, and classification workflows. The good news? We'll dive into these and begin with 3D modeling in Chapter 10.

3D Modeling: Advanced Techniques

3D modeling is essential for creating a digital version of real-world objects and spaces. It allows for accurate analysis, simulation, and visualization, which are important for architectural, scientific, or entertainment applications. As engineers and researchers, we always aim for precision and clarity in our 3D models. Traditional 3D modeling techniques in Chapter 4 may not work well with complex shapes or large point cloud datasets. Poor-quality 3D data representations can cause visual issues, incorrect simulations, and slow data processing, making it harder to get the best results.

In this chapter, we want to advance 3D modeling techniques. Let's investigate solid solutions using Python through four distinct yet powerful approaches: high-fidelity meshing, voxelization techniques, parametric modeling, and image-based 3D modeling (see Figure 10-1).

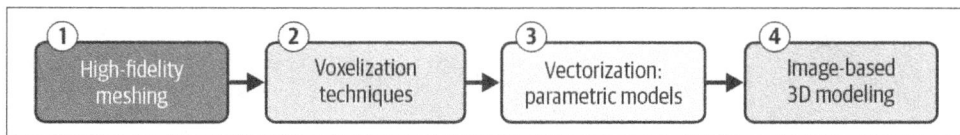

Figure 10-1. Four essential 3D modeling approaches

This section explores the main techniques for making accurate, efficient, and beautiful 3D models. We learn how to improve mesh quality, create voxel-based models from point clouds, and use parametric modeling to better control geometry.

In the last section, we also look into image-based 3D reconstruction, which involves using AI models to estimate depth from single images and create 3D objects. This chapter is designed for hands-on exploration. Practical Python code underpins each technique, allowing you to experiment. Let's dive right in.

You can find all the materials for this chapter on the 3D Data Science Resource Hub (*https://learngeodata.eu/3d-data-science-with-python*). To access the files (code, datasets, articles) and resources, you may need to share an email address, a personal password, or proof that you are the book's owner.

High-Fidelity Meshing

Allow me to demonstrate a 3D surface reconstruction procedure using Python, which efficiently generates a mesh from point clouds. You have the ability to export, visualize, and incorporate the results into your preferred 3D software (see Figure 10-2).

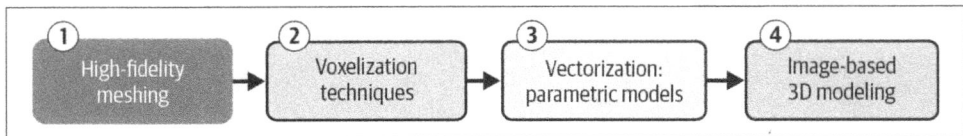

Figure 10-2. High-fidelity 3D surface meshing

In addition, I offer a straightforward method for generating numerous levels of detail (LoDs), which is useful for creating real-time applications such as virtual reality applications.

General Overview of High-Fidelity 3D Meshes

To complement the previous chapters, I want to emphasize how meshes are constituted. 3D meshes are geometric data structures often composed of connected geometric entities (usually triangles) that explicitly describe a surface. A mesh is the vertices, edges, and faces that define a 3D model's surface.

The quality of a mesh significantly impacts its usability and accuracy for 3D data science. Indeed, it affects:

Visual reliability
  High-fidelity mesh ensures smooth shading and avoids visual artifacts like jagged edges during rendering or animation.

Simulation accuracy
  Complex simulations, such as fluid flow analysis or structural mechanics, rely heavily on mesh quality.

Data processing efficiency
  A well-structured mesh benefits many data processing algorithms used in data science, like mesh segmentation or registration.

Therefore, aligning the goal with the underlying meshing strategy is crucial. The first consideration is whether we want to create a polygon model or another way to represent a 3D model. I invite you to refer to Chapter 4 for an extended view, and Table 10-1 to complement the view.

Table 10-1. The main characteristics of various 3D model representation (Polygonal, NURBS, SDF, Boolean assemblies)

Feature	Polygonal	NURBS (B-rep)	SDF (B-rep)	Boolean assemblies
Representation	Mesh of vertices, edges, and faces	Mathematical function defining a curve or surface	Function defining the distance to the surface	Combination of primitive shapes
Flexibility	Versatile, can represent a wide range of shapes	Highly flexible, can represent complex curves and surfaces	Can represent complex shapes, especially organic ones	Limited to the shapes of the primitives used
Editing	Direct manipulation of vertices and edges	Manipulation of control points and weights	Requires specialized techniques and tools	Boolean operations on existing shapes
Rendering efficiency	Generally efficient	Can be computationally expensive, especially for complex models	Can be computationally expensive, especially for complex shapes	Depends on the complexity of the assembled shapes
Level of detail	Can be adjusted by adding or removing polygons	Can be adjusted by modifying control points and weights	Can be adjusted by modifying the SDF function	Depends on the resolution of the primitive shapes
Common use cases	Game development, 3D printing, CAD	Industrial design, automotive design, architecture	Procedural generation, fluid simulation, level design	Creating complex shapes from simpler ones

Let's move backward, from right to left. Boolean operations on 3D models (*https://oreil.ly/MH1Vz*) are best suited for creating complex shapes by combining simpler primitives. This technique is instrumental in industrial design, architecture, and manufacturing, where precise control over the shape and topology of objects is essential (I invite you to read Chapter 4 for creating 3D models with Boolean operations).

Signed distance functions (SDFs) (*https://oreil.ly/5s97G*) excel at representing organic shapes and implicit surfaces, making them ideal for fluid simulation and level design in video games. SDF usually works well with Boolean assemblies for internal processes. They are then transformed into a polygonal 3D mesh model for the same needs (visualization and interoperability). Following is an example using this `sdf` library (*https://oreil.ly/tL5Eq*):

```
from sdf import *

f = sphere(1) & box(1.5)

c = cylinder(0.5)
f -= c.orient(X) | c.orient(Y) | c.orient(Z)

f.save('out.stl')
```

This results in Figure 10-3.

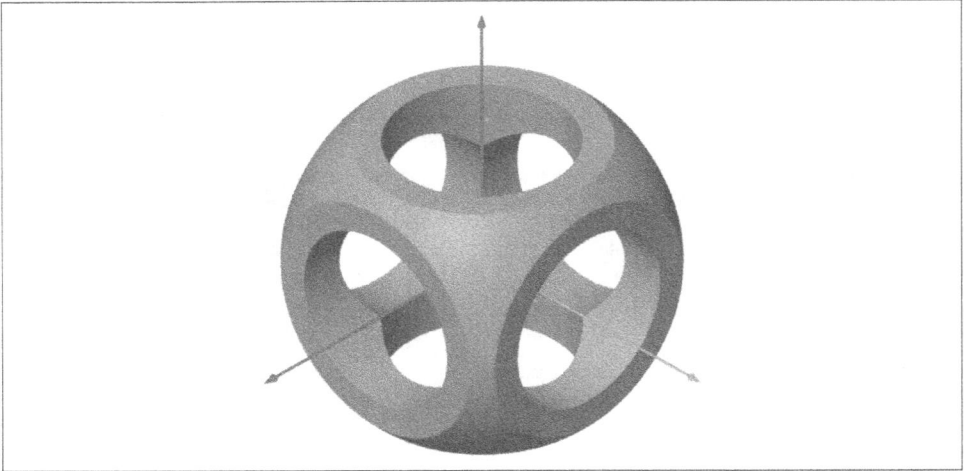

Figure 10-3. The 3D geometry exported from the preceding code block.

Nonuniform rational B-splines (NURBS) (*https://oreil.ly/wXLp2*) are highly flexible and can represent a wide range of curves and surfaces, making them suitable for automotive design, shipbuilding, and aerospace engineering, where precise control over the shape and smoothness of surfaces is critical. PyNURBS (*https://oreil.ly/O4nT0*) is the Python library designed for creating, manipulating, and analyzing NURBS curves and surfaces. It provides tools for working with B-spline functions, curve and surface fitting, and geometric modeling like what you might find in professional CAD software. This Python snippet is an example of how a curve object is defined with control points:

```
from pynurbs import Curve

# Define control points for a NURBS curve
control_points = [[0, 0, 0], [1, 2, 0], [2, 1, 0], [3, 0, 0]]

# Create a NURBS curve object
curve = Curve(control_points=control_points, degree=3)
```

These are then usually transformed into a polygon mesh model for visual purposes and interoperability:

```
# Generate a surface mesh from the curve
surface_mesh = curve.extrude(height=1.0)
```

3D polygonal mesh models are the most widely used type of 3D model due to their versatility, simplicity, and efficiency. They are composed of vertices, edges, and faces, making them easy to represent, manipulate, and render. This simplicity allows for efficient algorithms for operations like smoothing, subdivision, and collision detection. Additionally, polygonal meshes can be easily exported and imported between different software applications, ensuring compatibility and interoperability. Their ability to approximate complex shapes with varying LoDs makes them suitable for a wide range of applications, from video game development and 3D printing to medical imaging and scientific visualization.

> Polygon models are the most commonly used mesh models in 3D modeling. Therefore, we focus on them. Also, while Boolean assemblies are not a specific 3D model representation, they are a powerful technique for combining different shapes to create more complex ones. Finally, if you want to extend on NURBS (*https://oreil.ly/YRg6d*) and SDFs (*https://oreil.ly/dxTVp*), I invite you to explore the 3D Data Science Resource Hub (*https://learngeo data.eu/3d-data-science-with-python*) for tutorials with these strategies.

While other representations like NURBS and SDFs offer higher precision and flexibility, polygonal meshes often provide a good balance of accuracy and performance for most practical use cases. Therefore, let me showcase how we can generate 3D polygonal mesh models by creating and editing the mesh model's vertices, edges, and faces. This process can be done from scratch, or by sculpting mechanisms (the mesh model is shaped like clay or mud by directly manipulating the object's surface. Sculpting is often used in combination with other modeling techniques to create realistic and complex models), or through 3D reconstruction methods (converts a physical object into a 3D digital model).

Procedural modeling is an exciting way to generate 3D meshes at scale. It generates 3D mesh models using algorithms and mathematical functions. This process is beneficial for creating complex and detailed models that are difficult to model manually, such as natural landscapes, clouds, or fluids:

```
from procedural import mountain_terrain, generate_forest

# Define parameters for a mountainous terrain
mountain_params = {"peak_height": 2.0, "valley_depth": 1.0, "size": (20.0, 20.0)}

# Generate a mountain terrain mesh
mountain_mesh = mountain_terrain.generate(mountain_params)

# Define parameters for a forest
forest_params = {"tree_type": "pine", "density": 50, "area": mountain_mesh.bbox}

# Generate a forest on the mountain terrain
forest_mesh = generate_forest.generate(forest_params, base_mesh=mountain_mesh)

# Combine and visualize the generated meshes (visualization libraries needed)
combined_mesh = mountain_mesh + forest_mesh
combined_mesh.visualize()
```

Now that you have a better view of what mesh is, let's leverage Python to define a high-fidelity meshing workflow.

The Mission

You visited a museum in Denmark and found a statue that genuinely resonates with you. Without waiting, you started shooting several photos following the 3D reconstruction methodology approach shown in Chapter 3. You want to turn that into a physical replica using 3D printing. After some research, you know that you need to transform your set of images into a 3D model suitable for 3D printing. Thus, you need to generate a consistent mesh from your data. The mission is to design an efficient workflow to turn 3D point clouds from 3D reconstruction workflows into printable objects.

Data Preparation

Open your Python scripting tool (such as Spyder GUI, Jupyter, or Google Colab) and import two libraries:

```
import numpy as np
import open3d as o3d
```

Next, we establish variables to store the data pathways:

```
input_path="your_path_to_file/"
output_path="your_path_to_output_folder/"
dataname="sample.xyz"
point_cloud= np.loadtxt(input_path+dataname,skiprows=1)
```

Lastly, we convert the `point_cloud` variable's data type from NumPy to the Open3D `o3d.geometry.PointCloud` type for subsequent processing:

```
pcd = o3d.geometry.PointCloud()
pcd.points = o3d.utility.Vector3dVector(point_cloud[:,:3])
pcd.colors = o3d.utility.Vector3dVector(point_cloud[:,3:6]/255)
pcd.normals = o3d.utility.Vector3dVector(point_cloud[:,6:9])
```

The following command instantiates the Open3D point cloud object and adds points, colors, and normals from the original NumPy array. To obtain a rapid visual representation of the content you have loaded, you might use the subsequent command:

```
o3d.visualization.draw_geometries([pcd])
```

Choose a Meshing Strategy

We are prepared to commence the surface reconstruction process by generating a mesh from the point cloud. I will provide my preferred method for achieving efficient results. However, before we get there, we must understand the fundamental processes. I adhere to two specific meshing strategies.

3D meshing: The Ball-Pivoting Algorithm

The Ball-Pivoting Algorithm (BPA) (*https://oreil.ly/yHS6k*) utilizes a virtual ball to create a mesh from a point cloud through simulation. Initially, we assume that the provided point cloud is composed of points sampled from the surface of an object. The points must accurately depict a smooth and clear surface explicitly defined in the reconstructed mesh (see Figure 10-4). Assuming this premise, envision rolling a little sphere across the surface of the point cloud.

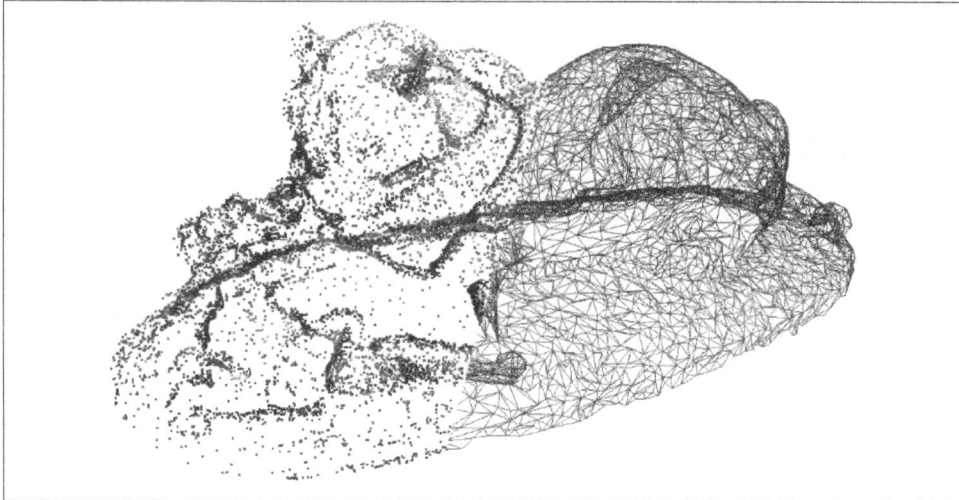

Figure 10-4. 3D Meshing with the Ball-Pivoting Algorithm

The size of this small sphere is determined by the scale of the mesh and should be somewhat bigger than the typical gap between points. When a ball is released onto the surface, it becomes trapped and comes to rest on three specific places, creating a triangular shape known as the seed triangle. The ball moves along the triangular edge created by two points starting from that spot. The ball comes to rest in a different position, resulting in the formation of a new triangle using two existing vertices. Additionally, a new triangle is added to the mesh. As we progressively rotate and shift the ball, additional triangles are generated and incorporated into the mesh. The ball persists in rolling until the mesh is completely constructed (see Figure 10-5).

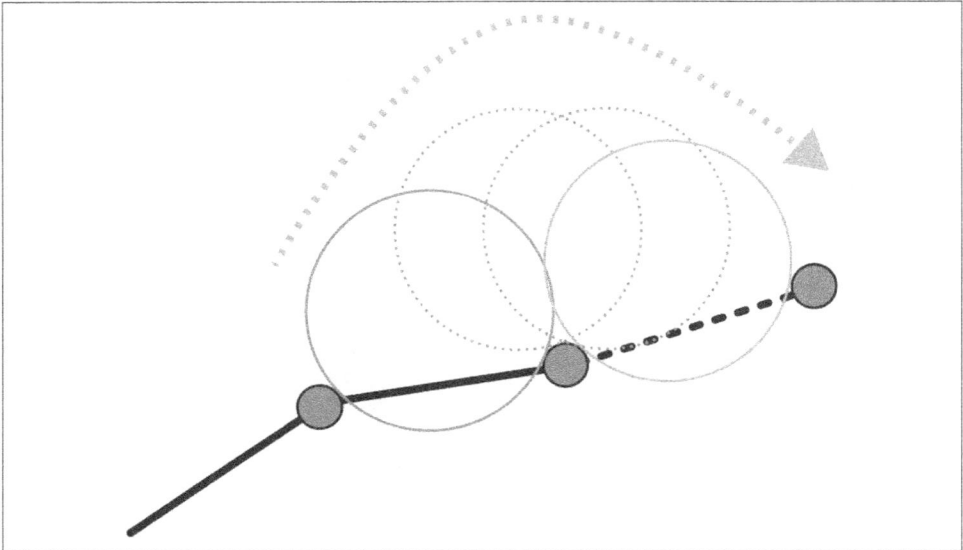

Figure 10-5. The Ball-Pivoting Algorithm simulates a virtual ball rolling over a set of 3D points, creating a triangular mesh

The concept underlying the BPA is straightforward, yet there are other considerations to take into account during the process:

What are the criteria for selecting the ball radius?
> The radius is determined through empirical analysis, taking into account the size and scale of the input point cloud. Ideally, the ball's diameter should be somewhat more than the average distance between points.

What happens if there is a significant distance between the points in some areas, causing the ball to pass through?
> When the ball rotates around an edge, it may fail to strike the correct point on the surface and instead collide with another point on the item or even the exact same three locations it had previously hit. In this scenario, we verify that the normal vector of the newly formed triangle facet is consistently aligned with the normal vectors of the associated vertices. If the condition is not met, we discard the triangle and produce a void instead.

What happens if the surface has a fold or depression, where the gap between the surface and itself is smaller than the size of the ball?
> In this scenario, the ball would traverse the crease and disregard any points located inside the crease. Nevertheless, this behavior is suboptimal since the rebuilt mesh deviates from the actual shape.

What if the surface is divided into several regions of points in such a way that the ball is unable to roll smoothly between these regions?

The virtual ball is repeatedly released onto the surface at different positions. This procedure guarantees that the ball encompasses the whole mesh, even in cases where the points are unevenly distributed.

The graphics in Figure 10-6 demonstrate the extent of the radius's influence in five different ways. The ideal mesh achieves a harmonious compromise between the most accurate geometry alignment and the minimal amount of triangles.

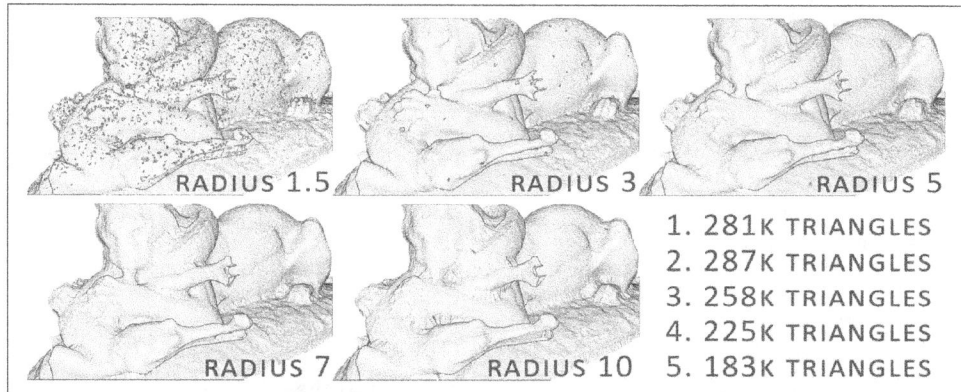

Figure 10-6. Impact of radius on mesh quality

Ultimately, finding the perfect radius comes down to understanding the unique characteristics of the point cloud you work with, like its density, noise levels, and underlying geometry. A bit of experimentation, combined with interactive adjustments based on visual feedback, usually hits the sweet spot and produces high-quality meshes.

When meshing 3D point clouds using the BPA, determining the appropriate radius can be tricky but crucial. I like to start with an adaptive approach, where I adjust the radius based on the local point density across the cloud. By analyzing the distribution of nearest-neighbor distances, I can set the radius as a multiple of the mean or median distance—large enough to connect points in dense regions but not so large that it creates unwanted connections in sparser areas. Another technique I've found helpful is to leverage an octree data structure to partition the 3D space and compute the average distance between points at each octree node. I then use the minimum or average of these per-node distance estimates to set the ball radius, which helps account for local variations in point density.

However, at this stage, you can also use another approach: the Poisson algorithm (*https://oreil.ly/4kuE8*) for 3D Reconstruction.

The Poisson algorithm for 3D reconstruction

The Poisson reconstruction (*https://oreil.ly/4kuE8*) is a more intricate and mathematical process. The strategy utilized is an implicit meshing method, which attempts to enclose the data within a smooth fabric-like structure. In summary, our approach involves building a new set of points that represents an isosurface (*https://oreil.ly/nbhXG*) connected to the normals, to fit a watertight surface from the original point set. Multiple parameters exist that influence the outcome of the meshing process:

Which depth?
> The tree depth is utilized during the reconstruction process. The greater the elevation, the more intricate the mesh will be (default: 8). When dealing with noisy input, the algorithm fails to identify outliers. Hence, they are included as vertices in the resulting mesh. A lower value, often ranging from 5 to 7, produces a smoothing effect while sacrificing fine details. The higher the depth value, the higher the resulting vertices of the generated mesh (see Figure 10-7).

Which width?
> This defines the desired width of the most detailed level of the tree structure, which is an octree (see Chapter 4 for a refresher). If the depth is determined, this procedure parameter is disregarded.

Which measurement scale are you referring to?
> This pertains to the ratio of the diameter of the reconstruction cube to the diameter of the bounding cube of the samples. The concept is extremely theoretical, and the default parameter typically functions optimally with a value of 1.1.

Which fit?
> If the `linear_fit` parameter is enabled, the reconstructor will utilize linear interpolation to estimate the positions of iso-vertices.

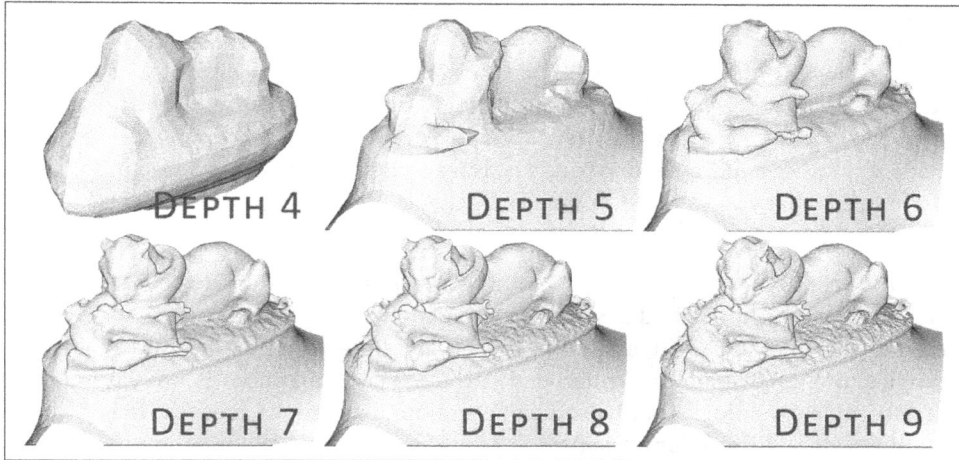

Figure 10-7. Poisson reconstruction algorithm

We can study the impact of the scale parameters on the outcomes in Figure 10-8. We can especially observe how it surrounds the original point cloud.

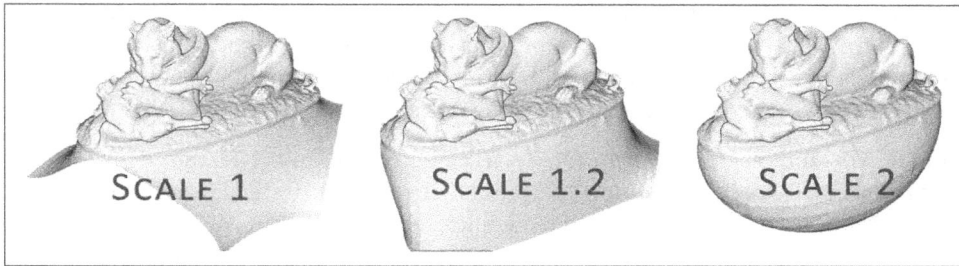

Figure 10-8. Impact of Poisson reconstruction parameters

Let's recap: the BPA works by starting with a seed triangle and "pivoting" a ball of a specific radius around the edges, connecting points within the ball's radius. This creates a triangular surface mesh that closely follows the contours of the input point cloud. This makes it well suited for applications where you have clean, well-sampled point data and need an efficient way to generate a surface mesh.

The Poisson surface reconstruction takes a different approach. It first builds a spatial data structure, like an octree, and then uses a global optimization process to find the surface that best fits the input points. This results in a watertight, high-quality triangular mesh that can handle complex topologies. It is popular in 3D scanning and medical imaging applications where the input point clouds may be imperfect or have missing data.

Other 3D Meshing Strategies

Several other 3D meshing strategies exist besides Poisson surface reconstruction and the BPA. But with the book being already quite large, we decided not to include these other approaches. Nevertheless, let me give you a quick comparison to help you decide if another can better suit your needs, as well as point you toward complimentary tutorials that you can find in the 3D Data Science Resource Hub (*https://learngeo data.eu/3d-data-science-with-python*).

First, the marching cubes algorithm (*https://oreil.ly/KdmsT*) operates on a 3D scalar field, such as a volumetric dataset or an implicit surface representation. It works by iterating through a 3D grid, analyzing the scalar values at each cube, and generating triangles to approximate the underlying surface. Marching cubes is known for handling complex topologies and producing smooth, watertight meshes. It is a go-to choice for visualizing volumetric data in scientific visualization and computer graphics fields. You can find a detailed Python tutorial on the 3D Data Science Resource Hub (*https://learngeodata.eu/3d-data-science-with-python*).

Then, the Delaunay triangulation (*https://oreil.ly/Cfm--*) is a method for creating a mesh by connecting a set of 3D points to maximize the minimum angle of all the triangles. This results in high-quality, well-shaped triangular elements, which is essential for finite element analysis and other computational methods that rely on the mesh. Delaunay triangulation–based meshing is widely used in computational geometry and engineering applications where the input data is a 3D point cloud.

Finally, Voronoi diagrams (*https://oreil.ly/FWK35*) partition the 3D space into a set of polyhedral cells, each containing the points closest to a particular generator point. This volumetric approach can handle complex topologies and noise well and is useful for spatial partitioning, centroidal Voronoi tessellation, and other applications that benefit from a polyhedral mesh representation rather than a triangular one. I summarize these strategies in Table 10-2.

Table 10-2. Different 3D meshing strategies

Algorithm	Handling of noise	Computational complexity	Typical applications
Ball Pivoting Algorithm (BPA)	Moderate	O(n log n)	CAD, computer graphics
Poisson surface reconstruction	Good	O(n log n)	3D scanning, medical imaging
Marching cubes	Moderate	O(n)	Volumetric data visualization, isosurface extraction
Delaunay triangulation	Moderate	O(n log n)	Finite element analysis, computational geometry
Voronoi diagrams	Moderate	O(n log n)	Spatial partitioning, centroidal Voronoi tessellation

Armed with these approaches, we can target a very large array of solutions. Let's explore the bolts of making it all work with Python.

3D Meshing with Python

Let's dive into the two selected meshing approaches: the BPA and the Poisson reconstruction.

The Ball-Pivoting Algorithm

Initially, we calculate the required radius parameter by utilizing the average lengths derived from all the distances between locations:

```
distances = pcd.compute_nearest_neighbor_distance()
avg_dist = np.mean(distances)
radius = 3 * avg_dist
```

Using a single command line, we can generate a mesh and save it in the variable named bpa_mesh:

```
bpa_mesh = o3d.geometry.TriangleMesh.create_from_point_cloud_ball_pivoting(pcd,
o3d.utility.DoubleVector([radius, radius * 2]))
```

Prior to exporting the mesh, we have the option to reduce the resolution of the output to a suitable quantity of triangles, such as 100,000 triangles:

```
dec_mesh = mesh.simplify_quadric_decimation(100000)
```

In addition, if you suspect that the mesh may include unusual artifacts, you can use the following instructions to verify its consistency:

```
dec_mesh.remove_degenerate_triangles()
dec_mesh.remove_duplicated_triangles()
dec_mesh.remove_duplicated_vertices()
dec_mesh.remove_non_manifold_edges()
```

I find these postprocessing operations to be beneficial for cleaning up and optimizing the output. remove_degenerate_triangles() gets rid of any flat or zero-area triangles that can sometimes slip through during the meshing process. Then, I run remove_duplicated_triangles() to identify and remove any redundant faces that are exactly the same. Going a step further, remove_duplicated_vertices() helps consolidate any vertices that occupy the exact physical location, simplifying the mesh geometry. Finally, remove_non_manifold_edges() is great for dealing with edge cases where the mesh connectivity doesn't quite make sense, removing any nonmanifold elements that could cause problems down the line. By executing this suite of cleanup operations, you can refine 3D meshes, improve their quality, reduce file size, and ensure they're ready for downstream modeling, visualization, or high-level tasks.

3D reconstruction with Poisson

Obtaining findings using the Poisson distribution is a simple and direct process. You must modify the arguments that you provide to the function as previously explained:

```
poisson_mesh = o3d.geometry.TriangleMesh.create_from_point_cloud_poisson(pcd,
depth=8, width=0, scale=1.1, linear_fit=False)[0]
```

The function returns a sequence consisting of an `o3d.geometry` object and a NumPy array. To only choose the `o3d.geometry`, you need to justify the [0] at the end.

In order to achieve a pristine outcome, it is frequently imperative to incorporate a cropping procedure to eliminate undesired artifacts prominently marked in gray in the image on the left (see Figure 10-9).

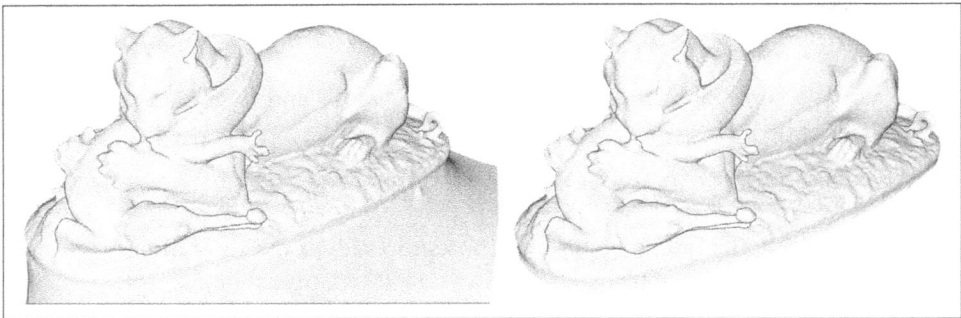

Figure 10-9. The initial Poisson reconstruction (before, left) and the trimmed mesh (after, right)

To accomplish this, we calculate the initial bounding box that encompasses the unprocessed point cloud. We then utilize this bounding box to exclude any surfaces from the mesh that fall outside of its boundaries:

```
bbox = pcd.get_axis_aligned_bounding_box()
p_mesh_crop = poisson_mesh.crop(bbox)
```

Congratulations on successfully obtaining one or more variables that store the mesh geometry. To complete the process of including it in your application, we can export it. This is where it becomes interesting to study.

Levels of Detail Creation

Exporting the data is simple and easy using the `write_triangle_mesh` function. When creating a file, we need to include the desired extension, which can be either *.ply*, *.obj*, *.stl*, or *.gltf*, along with the mesh we wish to export.

Here, we store both the BPA and Poisson reconstruction as *.ply* files:

```
o3d.io.write_triangle_mesh(output_path+"bpa_mesh.ply", dec_mesh)
o3d.io.write_triangle_mesh(output_path+"p_mesh_c.ply", p_mesh_crop)
```

But we can go one step further by generating 3D meshes with multiple LoDs. This is very useful for optimizing performance and visual quality in various applications. By creating simplified versions of a complex 3D model, you can tailor the LoD to the viewer's distance and the available computational resources.

This allows you to render more detailed models close-up while using simpler, lower-resolution models at a distance. This technique significantly improves frame rates and reduces rendering time, especially in real-time applications like video games and virtual reality. Additionally, LoDs can help manage memory usage, as you can selectively load and unload different LoD based on the current view and the available system resources.

To efficiently produce several LoDs, we proceed by creating your initial function. The function requires a mesh, a list of LoDs (specified as a target number of triangles), the file format for the resultant files, and the file directory where the files will be written. The function, which is to be written in the script, has the following structure:

```
def lod_mesh_export(mesh, lods, extension, path):
  mesh_lods={}
  for i in lods:
    mesh_lod = mesh.simplify_quadric_decimation(i)
    o3d.io.write_triangle_mesh(path+"lod_"+str(i)+extension, mesh_lod)
    mesh_lods[i]=mesh_lod
  print("generation of "+str(i)+" LoD successful")
  return mesh_lods
```

The function has two main functionalities. First, it allows you to export the data to a specific location of your choosing in the desired file format. Second, it provides the option to keep the results in a variable, which might be helpful for further processing within Python. It performs complex operations, although when executed, it appears no visible changes occur. Rest assured, your application has now been updated with the definition of lod_mesh_export. As a result, you can easily use it in the console to modify parameters with the values you want:

```
my_lods = lod_mesh_export(bpa_mesh, [100000,50000,10000,1000,100], ".ply",
output_path)
```

What is particularly intriguing is that now there is no longer a need to repeatedly rewrite a significant amount of code for varying LoDs. The function requires the input of various parameters:

```
my_lods2 = lod_mesh_export(bpa_mesh, [8000,800,300], ".ply", output_path)
```

To visualize a specific LoD containing 100 triangles in Python, you can access and visualize it using the following command:

```
o3d.visualization.draw_geometries([my_lods[100]])
```

Congratulations on having put in place a system that can generate 3D meshes with various LoDs. Let's now move onto the visualization part.

Visualization and Software

To view data outside of Python, you can utilize your preferred program, such as open source tools like Blender (*https://blender.org*), MeshLab (*https://meshlab.net*), or CloudCompare (see Chapter 2 for a full list of software). Simply load the output files into the GUI of the software. Through WebGL, you may directly access the mesh using the Three.js editor on the web. If you want to develop a virtual world experience, I recommend starting with Blender by following the tutorial series that I offer in the 3D Data Science Resource Hub (*https://learngeodata.eu/3d-data-science-with-python*). Ultimately, you have the ability to integrate it into any 3D printing software and obtain estimates regarding the cost by utilizing online printing providers (see Figure 10-10A).

> To ensure successful 3D printing, proper mesh preparation is essential. First, check for watertightness to avoid printing errors. You can use MeshLab (*https://meshlab.net*) to identify and repair any holes or gaps in the mesh. Second, optimize the mesh for 3D printing by ensuring sufficient thickness and removing unnecessary details. Tools like Blender (*https://blender.org*) or MeshLab can help with these tasks. Third, orient the model correctly on the print bed to minimize support structures and improve print quality. Finally, choose the appropriate file format, such as STL or OBJ, and export your model for 3D printing. Remember to consult your 3D printer's specific requirements and limitations.

We discussed the process of configuring an automated Python tool for generating a three-dimensional mesh from a collection of data points. This procedure is an exquisite tool that is quite useful in numerous 3D automation applications. Nevertheless, we assumed that the point cloud is already devoid of noise and that the normals are properly aligned (see Figure 10-10B).

If this is not the case, then further action is required, and you should refer back to 4 for preprocessing, and Chapters 6 and 7 for feature extraction. You have just acquired the skills to import, mesh, export, and view a complex 3D point cloud dataset with various LoDs. This allows you to deepen the possibilities of transferring reality into virtual assets and back to modified or high-fidelity physical prints. Now, it is time to extend the process with voxel datasets.

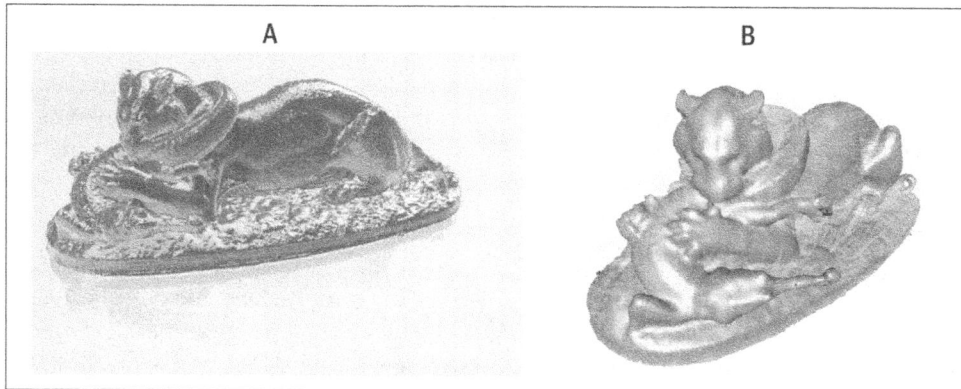

Figure 10-10. (A) This is a possible "gold print" of our 3D mesh, which is 20 cm long; (B) The visualization of the normals on the 3D mesh

3D Voxels and Voxelization

We have explored various methods for working with 3D data, from point clouds to intricate meshes. Now, we delve into the captivating world of voxels, the 3D equivalent of pixels, and how they empower us to represent and analyze complex 3D structures (see Figure 10-11).

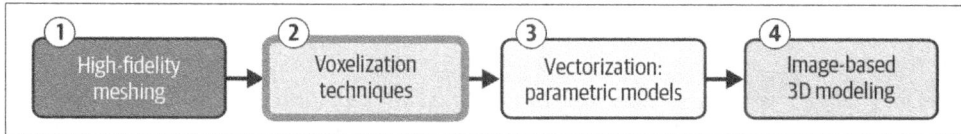

Figure 10-11. Introduction to 3D voxels and voxelization

Voxelizing 3D datasets transforms any 3D dataset (continuous or not) into discrete 3D grids. This process simplifies complex geometries, making them more amenable to analysis, processing, and visualization. By converting 3D point clouds, meshes, or other representations into voxels, you can easily apply volumetric operations like erosion, dilation, and smoothing.

Voxelization also enables the integration of 3D data with other data modalities, such as medical images or geophysical data. Additionally, it facilitates machine learning techniques, particularly deep learning, by providing a regular, grid-based representation suitable for training and inference. But if you have played *Minecraft* before, you know how handy it is to build an interactive digital environment.

This section provides you with the knowledge and Python code to build your 3D voxel representations from point cloud data. We start with setting up the Python environment and preparing our point cloud dataset. We then create a voxel grid, the foundation upon which our 3D voxel world resides. This will allow us to explore the construction of individual voxel entities and establish techniques to automate the generation of a complete 3D voxel dataset from your point cloud.

Python Environment Initialization

When faced with an identified problem, there are often multiple viable paths forward, each with its own strengths, weaknesses, and nuances. As a programmer, it's important to maintain an open and adaptive mindset, exploring different approaches and leveraging the full functionality of the libraries and frameworks available to you.

To emphasize, no unique direction exists when coding to solve an identified problem. The solution I provide relies on a combination of clever techniques and strategic use of the tools available in the Open3D library. It incorporates some lesser-known features and optimizations within Open3D, demonstrating how a deep understanding of the available function stack can lead to more innovative, efficient, and effective problem solving. Again, there is often more than one way to "skin a cat." The key is being resourceful, creative, and willing to think outside the box when coding to address complex challenges.

We rely on only three functional libraries: laspy (*https://oreil.ly/lFSC8*), Open3D (*https://open3d.org*), and NumPy (*https://numpy.org*). If you are unfamiliar with them, which is unlikely at this stage of the book, you can refer to Chapter 3. In our environment, let's use these three simple lines:

```
import laspy as lp
import numpy as np
import open3d as o3d
```

All that remains now is identifying a point cloud we want to be voxelized. Luckily, you can see the provided point cloud in Figure 10-12. Once set up, we can start loading the data in our environment.

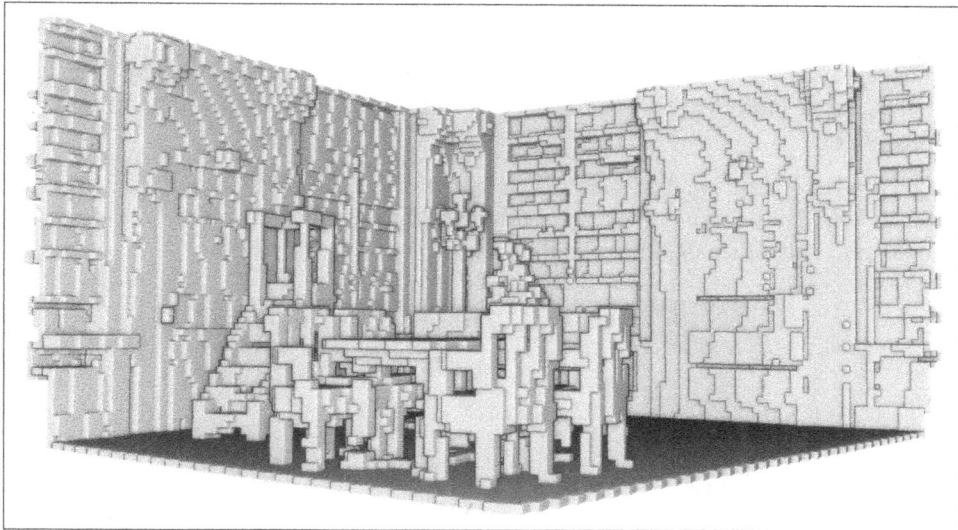

Figure 10-12. Point cloud selection for voxelization.

Loading the Data

First, we create two variables to handle the input path (that you should adapt to your case) and the `dataname`, as follows:

```
input_path="C:/DATA/"
dataname="heerlen_table.las"
```

Now, it's time to load the data in our program. We first store the point cloud as a laspy object in a `point_cloud` variable:

```
point_cloud=lp.read(input_path+dataname)
```

Then, to use the data stored in the `point_cloud` variable, we transform it into the Open3D point cloud format. If you remember from previous tutorials, we separate what color is from the spatial coordinates (x, y, and z):

```
pcd = o3d.geometry.PointCloud()
pcd.points = o3d.utility.Vector3dVector(np.vstack((point_cloud.x, point_cloud.y,
point_cloud.z)).transpose())
pcd.colors = o3d.utility.Vector3dVector(np.vstack((point_cloud.red,
point_cloud.green, point_cloud.blue)).transpose()/65535)
```

> If you look closely, you can see a weird 65535. This is because, in the laspy format, the colors are integers coded in a 16-bit unsigned fashion, which means that numbers range from 0 to +65535; we want to scale to an [0,1] interval.

Creating the Voxel Grid

We have a point cloud and want to fit an assembly of voxel cubes to approximate it. For this, we generate voxels only in parts with points on an established 3D grid (see Figure 10-13).

To get the voxel unit, we first need to compute the point cloud's bounding box, which delimits our dataset's spatial extent. Only then can we discretize the bounding box into an assembly of small 3D cubes: the voxels.

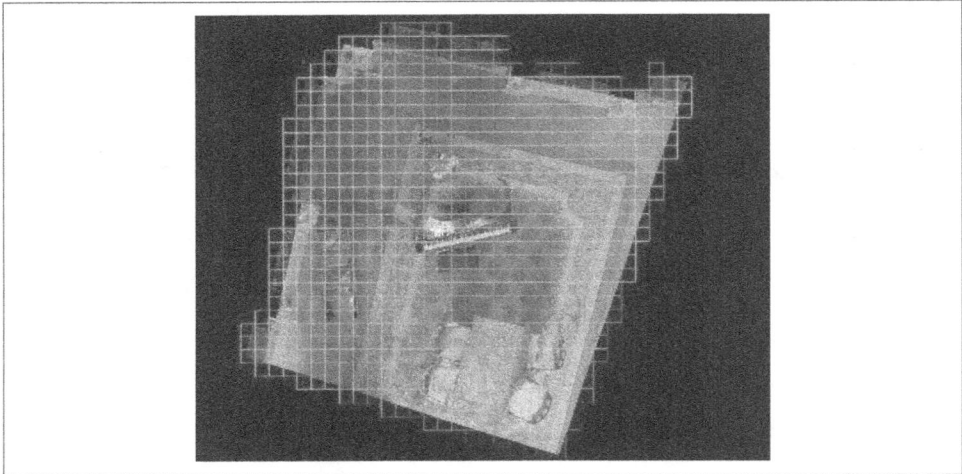

Figure 10-13. Voxel grid creation

We compute the voxel size by providing a relative value linked to the initial bounding box to "generalize" the approach if you switch the input point cloud. For this, you can see in the following that we extract the bounding box of the point cloud, take the maximum edge, and decide to set the voxel size to 0.5% of its value (this is absolutely arbitrary).

Finally, we round up the obtained value to four digits without suffering from imprecise calculation:

```
v_size=round(max(pcd.get_max_bound()-pcd.get_min_bound())*0.005,4)
```

> This is perhaps one of the most borderline code lines I wrote, but it provides an excellent example of what can be achieved with quick empirical statements. A lot can be improved there, especially concerning rounding errors and arbitrary thresholds.

Now that we have defined our voxel unit, we switch to a much more efficient "representation" linked to the spatial information, but more efficient: a binary table (false or true, 0 or 1). For this, the first trick, using Open3D, is to generate the voxel grid using this command line:

```
voxel_grid=o3d.geometry.VoxelGrid.create_from_point_cloud(pcd,voxel_size=v_size)
```

Awesome, you now are the owner of a voxel representation of your point cloud, which you can visualize (if outside Jupyter environments) with:

```
o3d.visualization.draw_geometries([voxel_grid])
```

You obtain the dataset shown in Figure 10-14. This is a remarkable feat. But if we stopped there, we would be stuck with a minimal scope of utilization. Indeed, you would be limited to using the Open3D library to play with the voxel structure, with a limited number of functions that, depending on your need, may not fit your application. Therefore, let's explore the process of creating a 3D mesh from this data structure. We can export it in an open format (*.ply* or *.obj*) and load it into other software, such as MeshLab, Blender, CloudCompare, MagicaVoxel, Unity, Unreal Engine, and all the other software listed in Chapter 2.

Figure 10-14. Voxelization level of detail

Generating the Voxel Cubes (3D Meshes)

Now that we have the voxel_grid, we extract the filled voxels so that we can use them as separate entities later:

```
voxels=voxel_grid.get_voxels()
```

If we check what our new `voxels` variable looks like, we get a list containing `open3d.cpu.pybind.geometry` geometries. `Voxel` types hold the voxel information with `grid_index` (19, 81, 57) and `color` (0.330083, 0.277348, 0.22266).

OK, now we translate this into an assembly of 3D cubic meshes (8 vertices and 12 triangles describing the six faces). See the trick that we are trying to achieve? First, let's initialize our triangle mesh entity that holds this cube assembly:

```
vox_mesh=o3d.geometry.TriangleMesh()
```

Now, we iterate over all voxels with `for v in voxels`, and for each voxel, we generate a cube of size one, stored in the variable cube. We then paint it with the voxel color at hand and position the voxel on the grid using the index provided by the `grid_index` method.

Finally, we add our newly colored and positioned cube to the mesh entity with:

```
for v in voxels:
    cube=o3d.geometry.TriangleMesh.create_box(width=v_size, height=v_size,
    depth=v_size)
    cube.paint_uniform_color(v.color)
    cube.translate(v.grid_index, relative=False)
    vox_mesh+=cube
```

The translation method takes as an argument whether or not the translation should be done relative to the first argument given (the position). Because we set it to `False`, the first argument, `v.grid_index`, acts as the absolute coordinates in our system. Now, we can export the voxel-based polygonal mesh 3D model.

Export the Mesh Object (.ply or .obj)

We now have a 3D mesh object ready for the I/O operation…almost. Indeed, we are yet in an arbitrary system with arbitrary integer units. To situate ourselves in the initial frame of reference imposed by the input point cloud, we must apply a rigid transformation (translation, rotation, and scale) to return to the original position. For clarity, I decompose this ambition into three lines of code.

First, we relatively translate the 3D mesh (voxel assembly) by half the voxel unit. This process is explained by the fact that when we created our initial voxel grid, the reference was the lowest left point of the voxel instead of the barycenter (which is positioned at [0.5,0.5,0.5] relatively in the unit cube):

```
vox_mesh.translate([0.5,0.5,0.5], relative=True)
```

Then, we scale our model by the voxel size, to transform each cube unit into its real size. This process makes use of the `scale` method, which takes two arguments. The first is the scaling factor, and the second is the center used when scaling:

```
vox_mesh.scale(voxel_size, [0,0,0])
```

Finally, we need to translate our voxel assembly to its true original position by translating using the voxel grid origin relatively:

```
vox_mesh.translate(voxel_grid.origin, relative=True)
```

Great, now we have our final cube assembly positioned correctly. An optional command is to merge close vertices. Whenever we generate a cube, we can be in a configuration where one of the corner vertices is overlapping another cube-corner vertex. Thus, it is better to clean that up while preserving the topology:

```
vox_mesh.merge_close_vertices(0.0000001)
```

And finally, the last element in our pipeline is to export our *.ply* (or *.obj* depending on the extension you prefer) file to the folder of choice in your OS browser:

```
o3d.io.write_triangle_mesh(input_path+"voxel_mesh_h.ply", vox_mesh)
```

From there, you are free to use the output file in your chosen software. If, when importing, you get a rotated file, you can also add the following lines to your Python code and change the exported variable to vox_mesh.transform(T):

```
T=np.array([[1, 0, 0, 0], [0, 0, 1, 0], [0, -1, 0, 0], [0, 0, 0, 1]])
o3d.io.write_triangle_mesh(input_path+"4_vox_mesh_r.ply", vox_mesh.transform(T))
```

This simply creates a transform matrix T that defines a rotation counterclockwise around the y-axis. Some software (e.g., Blender) usually shows an inverted z- and y-axis. This process means you have another trick up your sleeve if that happens to you. Of course, this is just a hint of what you can do. You just learned how to import point clouds, turn them into a voxel grid, trick the system into making them 3D meshes, and then export them fully automatedly! Interestingly, using voxels in Python also permits better understanding of the relationships and topology of any point cloud scene. But let's leave that to Chapter 12 and continue with parametric modeling.

Parametric Modeling

In the world of 3D data science, the concepts of vectorization and parametric modeling play a crucial role in efficiently processing and analyzing complex 3D data. Vectorization sounds like a dangerous science-fiction, laser-related disintegration weapon. But it is more subtle than that, luckily (see Figure 10-15).

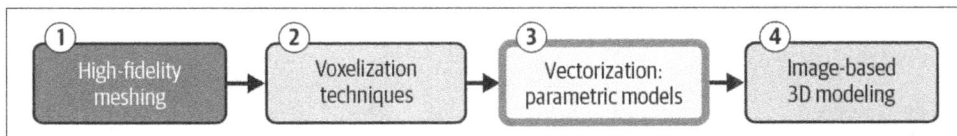

Figure 10-15. Vectorization and parametric modeling in 3D data science

Vectorization is the process of going from some data modality, such as raster imagery, to a vector or set of entities. It is the default go-to process for creating geospatial information systems in many applications. Indeed, vector data has properties that are very useful for handling analytical and decision-making tasks. However, one specific branch of vectorization is hardcore: 3D point cloud to vector datasets. And this is precisely the kind of challenge that we like and that we are going to address. This chapter seeks to offer a thorough overview of these core techniques, which are crucial for any aspiring 3D data scientist. Let's start with some brief words on vectorization. Traditionally, 3D graphics and visualization have relied on the rasterization process, where 3D scenes are converted into a grid of pixels (a raster image) for display. However, as the complexity and scale of 3D data have grown, the limitations of rasterization have become increasingly apparent.

Vectorization, on the other hand, offers a more efficient and flexible approach to 3D data processing. Instead of representing 3D data as a collection of pixels, vectorization involves converting data into a format that can be efficiently processed by modern computer hardware, particularly by taking advantage of the parallel processing capabilities of GPUs and vectorized CPU instructions.

At the heart of 3D vectorization is the representation of 3D data as multidimensional arrays or vectors. This allows for the application of powerful linear algebra operations, which can be executed in a highly optimized and parallel manner by modern hardware. By leveraging vectorization, we can dramatically improve the performance and scalability of our workflows to work with larger and more complex datasets.

This creates opportunities for modeling techniques that rely on vector notions, particularly parametric modeling. Parametric modeling is a highly effective technique in the armory of 3D data research. It entails the depiction of three-dimensional objects by utilizing a collection of factors that establish the object's form, dimensions, and additional characteristics.

Parametric models offer several advantages over traditional 3D representations, such as improved compression, efficient data storage, and the ability to perform advanced geometric operations. By parameterizing the 3D data, 3D data scientists can easily manipulate and modify the shape of objects, enabling tasks like design optimization, shape exploration, and statistical analysis (see Figure 10-16).

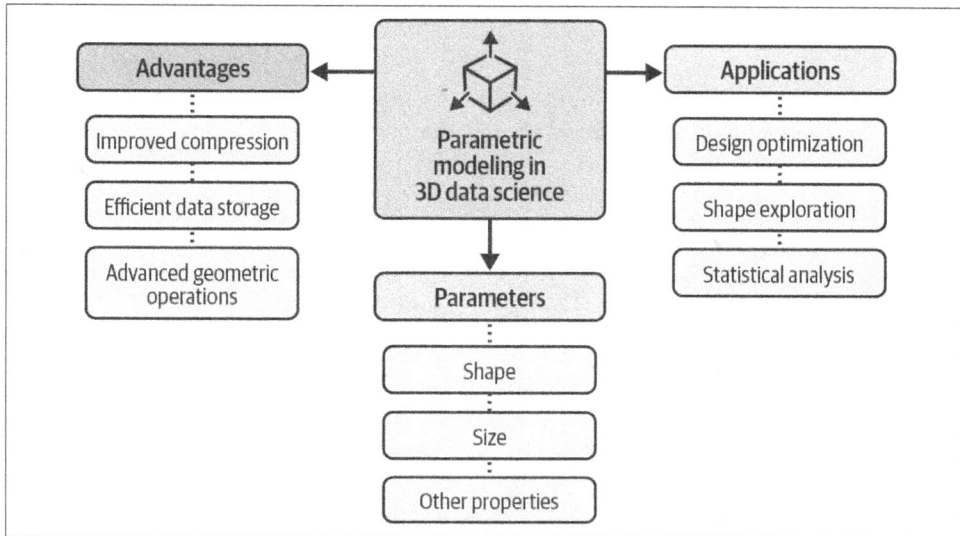

Figure 10-16. Advantages of parametric modeling in 3D data science

In this section, we study the fundamental concepts of 3D parametric modeling, including common parametric representations, techniques for constructing and manipulating parametric models, and practical examples of how to work with parametric data in Python using the CadQuery library (*https://oreil.ly/37bS5*). By the end, you will have a strong understanding of how to use CadQuery to create, manipulate, and analyze 3D data in your Python-based 3D data science workflows.

CadQuery and Environment Setup

This section extensively uses the CadQuery library, a powerful open source Python-based tool for parametric 3D modeling and design. CadQuery provides a user-friendly and intuitive interface for creating, manipulating, and analyzing complex 3D geometries, making it a nice addition to your arsenal.

> CadQuery is built on top of the Open Cascade geometry kernel, which is a robust and feature-rich library for 3D modeling and engineering. By using CadQuery, we can leverage the power of Open Cascade while benefiting from the simplicity and flexibility of a Python-based API.

One of CadQuery's key advantages is its support for parametric modeling. This means we can define 3D objects using a set of parameters, which can then be easily modified and updated to explore different design options or adapt to changing requirements. This aligns perfectly with the parametric modeling concept we discussed earlier.

Consequently, we will employ boundary representations (B-reps) to define objects, namely by their enclosing surfaces. In a B-rep system, fundamental constructions define a shape, with each construct building upon the previous one:

Vertex
 An individual point located in three-dimensional space

Edge
 A link between two or more vertices that follows a specific path, often referred to as a curve

Wire
 A grouping of interconnected edges

Face
 A collection of edges or wires that form a boundary around a surface

Shell
 A composite of faces that are joined together along certain edges

Solid
 An object that has a completely enclosed interior

Compound
 An aggregation of solid objects

When utilizing CadQuery, these objects are generated, ideally with minimal effort. Within the current CAD kernel, an additional collection of geometrical constructions exists. For instance, an arc-shaped edge retains a connection to a fundamental curve that forms a complete circle, while each linear edge encompasses the equation of a line. CadQuery protects you from these constructs. Finally, with CadQuery, we are going to leverage workplanes, constructions (2D and 3D), selectors, and construction geometry.

Workplanes

The majority of CAD applications employ the notion of workplanes. If you have prior knowledge of other CAD applications, you are likely to find CadQuery's workplanes familiar and easy to work with. However, if you lack experience, it is crucial to grasp the fundamental notion of workplanes.

Workplanes define a two-dimensional surface in three-dimensional space, which serves as a reference point for locating other features. The object possesses a central point and a coordinate system specific to its location. Most object creation procedures are performed with respect to the current workplane. The initial workplane often established is the "XY" plane, alternatively referred to as the "front" plane. After defining a solid, the typical method for creating a workplane is to choose a face on the solid that you wish to modify and generate a new workplane in relation to it. It is possible to create additional workplanes in any location inside the global coordinate system or in relation to other planes by utilizing offsets or rotations.

One of the most influential aspects of workplanes is their ability to enable work in a 2D space using the workplane's coordinate system. CadQuery then converts these points from the workplane's coordinate system to the world coordinate system, ensuring that your 3D features are accurately positioned as intended. This greatly simplifies the process of creating and managing scripts.

Constructions

After establishing a workplane, you can operate in a two-dimensional space. Subsequently, you can utilize the elements you generate to construct three-dimensional entities. The software includes all the standard 2D elements such as circles, lines, arcs, mirroring, and points.

It is possible to directly create 3D basic shapes like boxes, wedges, cylinders, and spheres. In addition, you have the option to perform sweeping, extrusion, and lofting operations on 2D geometry in order to create 3D features. Indeed, the fundamental foundational processes are also accessible.

Sweeping, extrusion, and lofting are fundamental operations in 3D modeling that allow you to create complex 3D shapes from 2D geometry. Sweeping involves taking a 2D profile and moving it along a path to generate a 3D object. Extrusion extends a 2D shape along a perpendicular axis, creating a 3D solid. Lofting creates a 3D surface by interpolating between two or more 2D profiles, resulting in a smooth transition between the shapes. These techniques are widely used in industrial design, architecture, and video game development to create a wide range of 3D objects.

Selectors

Selectors enable the user to choose and define new features by selecting one or more existing features. For instance, you can extend a box by selecting the top face as the position for a new element. Alternatively, you can create a box via extrusion and choose all the vertical edges to apply a fillet to them (a fillet is a rounded edge or corner in a 3D geometry used to smooth sharp edges). Selectors can be considered your hand and mouse counterparts while constructing an object using a traditional CAD system.

Construction geometry

Construction geometry refers to features that are not inherent to the thing itself but are specifically defined to assist in its construction. An illustrative instance could involve establishing a rectangle and utilizing its corners to determine the precise positions of a series of apertures.

Environment setup

Before we dive into the hands-on examples, let's ensure that we have the necessary Python environment setup. For this chapter, we use our book environment setup, and leverage three libraries:

- NumPy (`pip install numpy`)
- Open3D (`pip install open3d`)
- CadQuery (*https://oreil.ly/nxVES*) (`pip install cadquery`)

Once you are within your environment with your IDE open, let's get started:

```
import cadquery as cq
import open3d as o3d
import numpy as np
```

We are ready to define parametric models.

I/O for Parametric Models: 2D (DXF) and 3D (STL)

We are now ready to dive into the code! Let's briefly touch on two file formats that are very useful for working outside of Python with your software: DXF (Drawing Exchange Format) for 2D drawings and STL (stereolithography) for 3D models.

For this, let's do some sketching operations. I define a list of points that form the vertices of a special polygon:

```
pts = [(0,0),(0,20),(12,20),(12,18),(2,18),(2,16),(8,16),(8,14),(2,14),(2,0)]
```

Then, I create a new workplane named "front" and draw a polyline using the points from `pts`:

```
flo = cq.Workplane("front").polyline(pts).close()
```

The `close()` method ensures that the polyline forms a closed shape. Finally, I export the resulting shape as a DXF file named *letter.dxf* in the *../RESULTS* directory. All you have to do is open your CAD software, or CloudCompare, and visualize the nice first letter of my first name:

```
cq.exporters.export(flo, "../RESULTS/letter.dxf")
```

This image demonstrates the use of Python's `close()` method to ensure a polyline forms a closed shape, then exporting the shape as a DXF file (see Figure 10-17).

Figure 10-17. Exporting 3D shapes as DXF files

DXF is a vector-based format primarily used for exchanging 2D CAD drawings between different software applications. It stores geometric entities such as lines, arcs, circles, and text as mathematical equations. It is composed of a series of groups, each containing a code and a value. These groups define the various elements of the drawing, including entities, attributes, and header information.

Advantages:

- Widely supported by various CAD software applications
- Can store a wide range of 2D drawing elements
- Can be edited and modified using text editors or specialized DXF editing tools

Disadvantages:

- Can become complex and difficult to understand for large or intricate drawings
- May not be ideal for storing 3D models, as it primarily focuses on 2D geometry

Now, let's move onto STL files. They are a simple 3D model format that represents a surface as a series of triangular facets. Each facet is defined by its three vertices and a normal vector. STL files consist of a header and a series of facets. The header contains information about the model, such as its name and units. The facets section lists the coordinates of each vertex and the normal vector for each facet.

With CadQuery (*https://oreil.ly/kX6_E*), let's define a 3D shape known as an "IPN" profile, which is commonly used in structural engineering. The code starts by defining four variables: L, H, W, and t. These represent the length, height, width, and thickness of the IPN profile, respectively:

```
(L, H, W, t) = (100.0, 20.0, 20.0, 1.0)
```

Then, a list of points pts is defined. These points represent the vertices of a 2D polygon that will form the cross-section of the IPN. The coordinates of each point are calculated based on the dimensions defined in the variables. The points are arranged in a specific order to create a shape with rounded corners:

```
pts = [
    (0, H / 2.0),
    (W / 2.0, H / 2.0),
    (W / 2.0, (H / 2.0 - t)),
    (t / 2.0, (H / 2.0 - t)),
    (t / 2.0, (t - H / 2.0)),
    (W / 2.0, (t - H / 2.0)),
    (W / 2.0, H / -2.0),
    (0, H / -2.0),
]
```

Finally, we create the shape:

```
result = cq.Workplane("front").polyline(pts).mirrorY().extrude(L)
```

And export as an STL file:

```
cq.exporters.export(result, "../RESULTS/IPN.stl")
```

This gives us Figure 10-18 in CloudCompare.

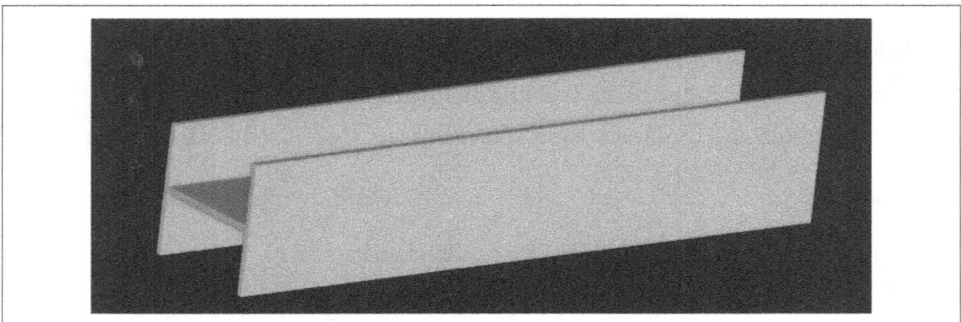

Figure 10-18. Creating and exporting 2D/3D shapes as STL files

Beautiful, we have a way to create 2D and 3D shapes and export them outside of Python very simply. If I had to summarize STL models, this is what I would like you to retain:

Advantages:

- Simple and easy to understand
- Widely supported by 3D printing and manufacturing software
- Can be utilized to depict a diverse array of three-dimensional forms

Disadvantages:

- May not be suitable for complex or detailed models, as the triangular facets can introduce artifacts or reduce accuracy
- Limited to representing surfaces, without information about internal structure or material properties

Great. Now that we firmly know we are not stuck in Python (i.e., any data output can be exported with standard file formats), we can explore the various parametric modeling techniques that we can areuse.

Parametric Modeling Techniques

Parametric modeling techniques empower you to create complex 3D models efficiently and precisely. By starting with simple geometric primitives and applying operations like sketching, extruding, revolving, filleting, mirroring, cutting, and trimming, you can build intricate designs with ease (see Figure 10-19).

Let's leverage a parametric approach to modify our model's geometry by changing the underlying parameters, making it highly flexible and adaptable. Then, we can automate repetitive tasks, reduce design time, and minimize errors. This is particularly valuable in automotive, aerospace, and product design, where precision and efficiency are paramount. Let's start with sketching.

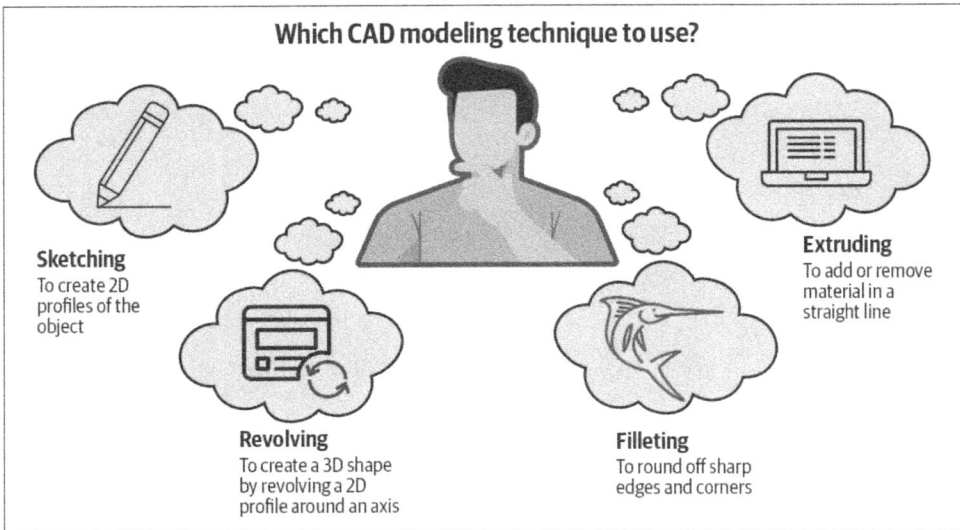

Figure 10-19. Overview of parametric modeling techniques

Sketching

Sketching is the foundation of solid modeling: it is the process of creating 2D outlines that serve as the basis for 3D features. These outlines can be simple or complex, depending on the desired shape. Sketches often involve applying constraints to ensure geometric accuracy and consistency. Constraints define relationships between geometric elements, such as parallelism, perpendicularity, tangency, and dimensions. You can sketch the first letter of your name as we did earlier.

Now, what about extruding?

Extruding

Extrusion is a technique for creating 3D solids by extending a sketch along a specified direction. Depending on the sketch geometry, the resulting solid can be a prism, cylinder, or more complex shape. Extrusion often involves parameters that control the height or depth of the extruded feature, allowing for dynamic adjustments.

Let me bring back my first letter to showcase this with CadQuery:

```
flo_e = cq.Workplane("front").polyline(pts).close().extrude(1)
cq.exporters.export(flo_e, "../RESULTS/letter.stl")
```

Breaking down the code, this is what I do:

- `cq.Workplane("front")` creates a new workplane named "front" where the shape will be defined.
- The `polyline(pts)` method creates a polyline using the points from the `pts` list (which is not shown in the provided code). This polyline represents the cross-section of the shape.
- The `close()` method ensures that the polyline forms a closed shape.
- The `extrude(1)` method extrudes the shape along the z-axis to create a 3D object with a thickness of 1 unit.

All that is left is to export the shape and visualize it:

```
cq.exporters.export(flo_e, "../RESULTS/letter.stl")
```

This image demonstrates the final step of exporting a parametric model, using Python to save the shape as an STL file for visualization in 3D modeling software (see Figure 10-20).

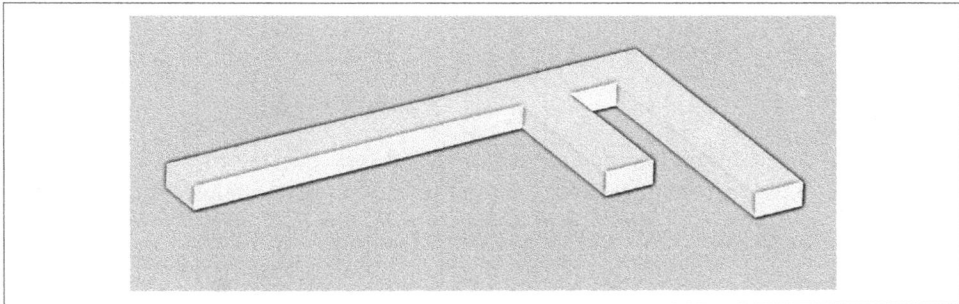

Figure 10-20. Exporting the shape to STL format

You now have a very quick way to generate 3D "logo" letters that you can leverage for your next startup venture. Now we can move on to filleting and chamfering.

Filleting and chamfering

Filleting and chamfering are techniques used to smooth sharp edges and corners in a 3D model. Filleting involves creating a rounded edge, while chamfering creates a beveled edge. The radius of the fillet or the angle of the chamfer can be specified as parameters.

But first, let's define a really nice Python helper to visualize our results within Python with Open3D:

```
def visualize_object(cq_object):
    temp_path = "../RESULTS/temp.stl"
```

```
cq.exporters.export(cq_object, temp_path)
o3d_mesh = o3d.io.read_triangle_mesh(temp_path)
o3d_mesh.compute_vertex_normals()
o3d.visualization.draw_geometries([o3d_mesh])
```

And now, let me showcase the fillet with the following:

```
result = cq.Workplane("XY").box(3, 3, 0.5).edges("|Z").fillet(0.125)
visualize_object(result)

result = cq.Workplane("XY").box(3, 3, 0.5).faces("+Z").chamfer(0.1)
visualize_object(result)
```

This permits us to go from a classical cube to a filleted cube and a chamfered cube (see Figure 10-21).

Figure 10-21. Fillet and chamfer example

And what about mirroring?

Mirroring

Mirroring is a technique for creating symmetrical features by reflecting a part of a model across a plane. It can be useful for creating objects with bilateral symmetry.

Here is an example from CadQuery that creates a cupboard handle:

```
r = cq.Workplane("front").hLine(1.0)  # 1.0 is the distance, not coordinate
r = (r.vLine(0.5).hLine(-0.25).vLine(-0.25).hLineTo(0.0))
result = r.mirrorY().extrude(0.25)  # mirror the geometry and extrude
visualize_object(result)
```

Here, I'm starting on the "front" plane and using the hLine() function to draw a horizontal line with a length of 1.0 units. From there, I chain together a series of other 2D sketching operations—vLine() for vertical lines, hLine() for horizontal lines, and hLineTo() to draw a horizontal line to a specific x-coordinate.

The real magic happens when I call mirrorY() on the resulting 2D sketch. This reflects the geometry across the y-axis, essentially giving us a symmetrical shape. Then, by calling extrude(0.25), I can take this 2D shape and turn it into a 3D object with a depth of 0.25 units.

Finally, I pass the resulting 3D object to a visualize_object() function, which renders a 3D preview of the geometry I've created. This is a really handy way to quickly validate the shape and dimensions of what I've built before potentially using it in a larger CAD or engineering project.

I find CadQuery to be a wonderfully expressive and intuitive library for 3D modeling. The ability to chain together 2D sketching operations, leverage symmetry, and extrude into the third dimension makes constructing complex geometries with just a few lines of code easy. It's a great tool to have in your arsenal when you need to rapidly prototype or iterate on 3D designs.

You could also explore patterns to create multiple instances of a feature or a group of features and cut/trim to remove portions of a solid or modify its shape. This can be done using cutting planes, trimming curves, or Boolean operations. This may be the best way to explore Boolean operations.

The Boolean Operations

If you remember from early chapters, Boolean operations, usually wielded through constructive solid geometry (CSG), are powerful techniques that allow you to combine or subtract geometric objects to create complex shapes. By performing union, intersection, and difference operations, you can create intricate models from simpler components.

Let me showcase two models: a box and a sphere. To generate them, you can use the following code:

```
Box = cq.Workplane("XY").box(1, 1, 1, centered=(False, False, False))
Sphere = cq.Workplane("XY").sphere(1)
```

Now, let's explore the operations with CadQuery.

Union: Combining objects

The union operation combines two or more objects into a single object. This is useful for creating assemblies or adding features to an existing model.

To execute a union (|) with CadQuery, type the following:

```
result = Box | Sphere
visualize_object(result)
```

Intersection: Finding common areas

The intersection operation (&) finds the common area between two or more objects. This can be used to create holes, cutouts, or intersections between different shapes:

```
result = Box & Sphere
visualize_object(result)
```

Difference: Subtracting objects

The difference operation (-) subtracts one object from another, leaving behind the remaining portion. This is useful for creating voids or removing material from a model:

```
result = Box - Sphere
visualize_object(result)
```

Figure 10-22 illustrates the result of subtracting a sphere from a box using the difference operation in Python, showcasing how Boolean operations can create voids or modify shapes in 3D modeling.

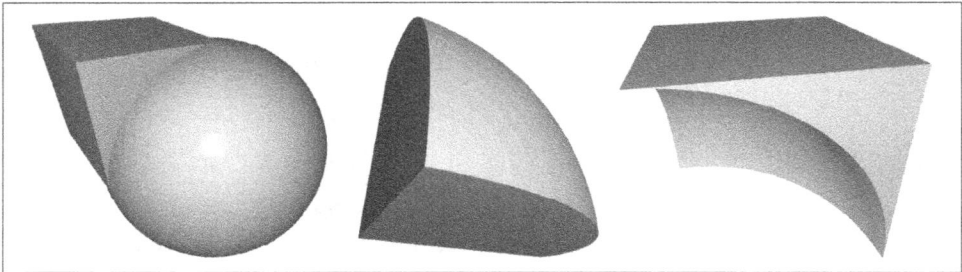

Figure 10-22. An example of CSG using the subtraction operation to create a complex shape from simpler primitives

Boolean operations are essential for creating complex models and assemblies. By combining different shapes and performing various operations, you can achieve a wide range of geometric outcomes. With all of these nice tools in your hands, you can easily model multiple pieces.

Modeling Various Pieces

Now that we have a solid understanding of basic CAD modeling techniques and Boolean operations, let's explore how to apply these concepts to model various pieces of a room. First, let's define walls:

```
# Define parameters (modify these to customize the house)
length = 5  # Length of the house in m
width = 3  # Width of the house in m
height = 2.5  # Height of the house in m

wall_thickness = 0.2

door_width = 1  # Width of the door in m
door_height = 4  # Height of the door in m
window_width = 1.5  # Width of the window in m
window_height = 1  # Height of the window in m
roof_pitch = 45  # Roof pitch angle in degrees
```

```
result = cq.Workplane("front").rect(length+wall_thickness,
width+wall_thickness).rect(length, width).extrude(height)
visualize_object(result)
```

Then, we can create doors and place them on opposite sides:

```
s = cq.Sketch().rect(door_width, door_height)

result2 = (
    cq.Workplane("front")
    .rect(length+wall_thickness, width+wall_thickness)
    .rect(length, width)
    .extrude(height)
    .faces(">X")
    .workplane()
    #.transformed((0, 0, -90))
    .placeSketch(s)
    .cutThruAll()
)

visualize_object(result2)
```

In this code, I'm starting off on the "front" plane and using the rect() function to create a rectangular base with dimensions of length + wall_thickness and width + wall_thickness. I then subtract a smaller rectangle with just the length and width dimensions, creating a recessed area within the larger base.

Next, I call extrude(height) to give this 2D shape some depth, turning it into a 3D object. From there, I select the faces on the positive X side using the faces(">X") method, and create a new workplane on those faces using workplane().

Initially, I had rotated this workplane 90 degrees around the z-axis by uncommenting the transformed((0, 0, -90)) line, but I've since removed that to keep things simpler. On this new workplane, I then placeSketch(s), which I assume is placing some kind of 2D sketch that I've defined separately.

Finally, I call cutThruAll() to remove the portion of the 3D object that overlaps with the sketch, leaving me with the desired result. I store this modified 3D object in the result2 variable and then pass it to the visualize_object() function to preview the final geometry.

Finally, we can add the ground with Boolean operations:

```
ground = cq.Workplane("XY").box(length+0.2, width+0.2, 0.1)
result_total = result2 | ground
visualize_object(result_total)
```

Figure 10-23 demonstrates the process of adding a ground plane to the 3D model using Boolean operations, specifically the union operation, to combine the existing shape with a new base.

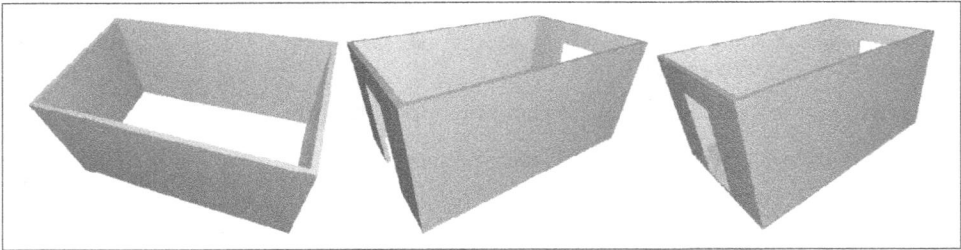

Figure 10-23. Adding the ground: using Boolean operations

How nice it is to model a simple room with some lines of code. But know that you can model more complex examples like Figure 10-24.

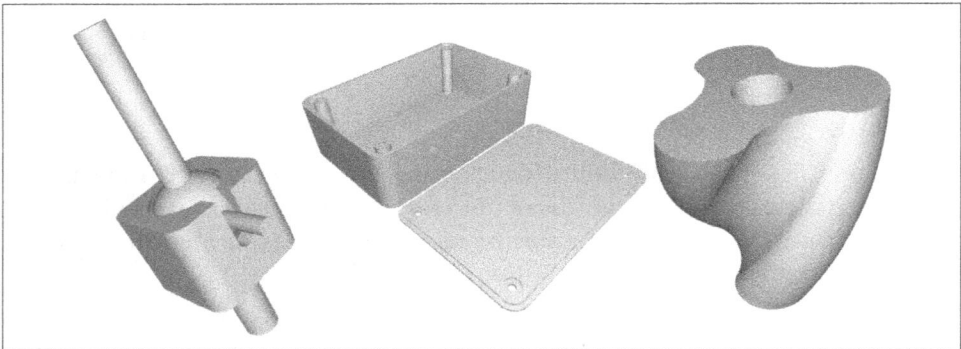

Figure 10-24. Modeling complexity with code

Conclusion

3D parametric modeling is a powerful technique that allows you to create complex 3D models by defining their shapes using mathematical equations and parameters. This approach offers several advantages over traditional 3D meshing and voxelization. By using parametric models, you can easily modify the shape of an object by adjusting its parameters, enabling rapid design iterations and exploration. Additionally, parametric models can generate a wide range of variations of a base model, providing flexibility and creativity in design. Furthermore, parametric models can be used to create highly accurate and detailed 3D models essential for engineering, architecture, and product design applications.

To solidify your understanding, try modeling a simple object, such as a chair or a table, using the techniques discussed in this chapter. Experiment with different parameters and Boolean operations to create variations of the object. You can also import existing 3D models and modify them using parametric modeling techniques. Completing this exercise will give you valuable hands-on experience and further develop your vectorization and parametric modeling skills. Then, you will be ready for the project related to this chapter.

The next frontier in 3D modeling involves generating 3D models from non-3D data modalities, such as monocular images. This emerging field, known as 3D reconstruction from single images, leverages techniques like deep learning to infer 3D structures from 2D information. By combining the power of parametric modeling (or other 3D modeling approaches unlocked in this chapter) with 3D reconstruction, we can create highly detailed and realistic 3D models from a single image or a sequence of images, opening up new possibilities for creating digital environments.

Monocular Image-based 3D Modeling: Depth Estimation and Reconstruction

Let's now focus on a groundbreaking field that allows us to infer 3D structure from a single 2D image: monocular image-based 3D reconstruction (see Figure 10-25). This is a whole other paradigm, where our input does not directly measure accurate spatial phenomenon. We want to accurately reconstruct 3D scenes from images so that we can create applications with an elementary entry point: a smartphone (or a basic camera). This can change how we develop immersive virtual and augmented reality experiences, enable robots to navigate complex environments, or help self-driving cars perceive their surroundings. Additionally, this technology can be used to analyze historical artifacts, medical images, and satellite imagery, providing valuable insights into the past, present, and future.

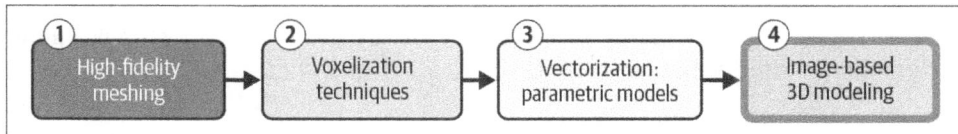

Figure 10-25. Monocular image-based 3D modeling: depth estimation and reconstruction

This capability significantly expands the possibilities beyond traditional multiview reconstruction techniques like photogrammetry (See Chapters 1 and 18). If you remember, with photogrammetry, we need multiple images of an object taken from different viewpoints to create a 3D model. This can be time-consuming, requires specialized equipment, and may not be feasible in all situations.

In contrast, monocular reconstruction enables us to create 3D models from a single image, making it much more accessible and efficient. This opens up new opportunities, and I am excited to share a method that leverages some Python libraries and straightforward steps. We can use either a photo we take or an AI-generated image. From a single image we aim to generate a 3D model, following the workflow illustrated in Figure 10-26. Without further ado, let's dive right in.

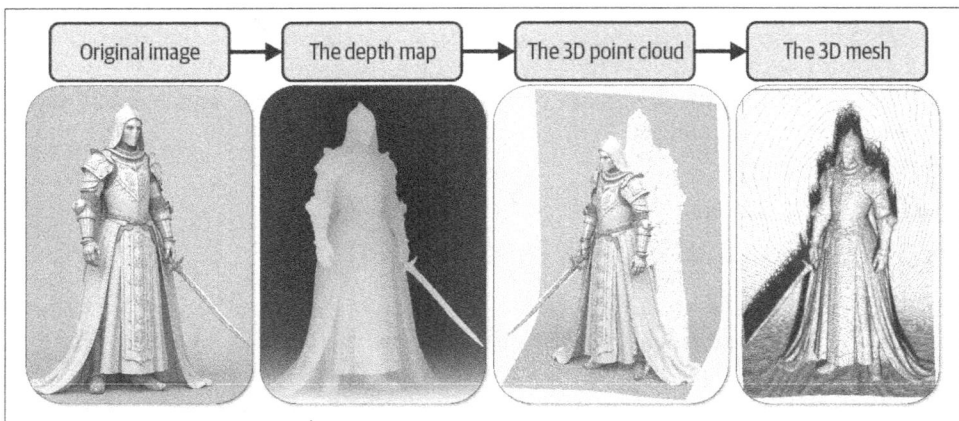

Figure 10-26. The full workflow and results, from the original image to the 3D mesh

Setting Up the Environment and Installing the Libraries

The first step to setting up the environment is creating a virtual environment using Anaconda. This process ensures that the libraries you install for this project won't conflict with other projects on your system. After that, you can install the necessary libraries and turn on the virtual environment. We use five libraries: PyTorch, Pillow, Matplotlib, Transformers, and Open3D.

You can install these libraries using pip:

```
pip install torch pillow matplotlib transformers open3d
```

The Python Transformers library (*https://oreil.ly/e57f1*) is a powerful toolkit for working with state-of-the-art deep learning models, especially those based on the Transformer architecture (*https://oreil.ly/dA5bm*). It allows us to utilize pretrained models, uses multiple frameworks, and is very easy to use. While primarily known for its applications in natural language processing, it's increasingly finding use in the 3D domain. By leveraging the attention mechanism (*https://oreil.ly/xvklP*), we can effectively process data (sequences), which can be very useful for point cloud analysis, 3D object detection, and 3D shape generation.

Once you have installed all the libraries, you can move on to the next stage: gathering data.

Gathering a Dataset

The easiest way to gather an image to which you have rights is to take a picture of your (private) environment that does not include anything other than your objects. So, this is what I did with my phone; you can find it in the resource folder (see the 3D Data Science Resource Hub (*https://learngeodata.eu/3d-data-science-with-python*)).

Another possibility is to generate an image using generative AI tools. Without going into too much detail, generative models have emerged as a groundbreaking approach for generating realistic and diverse data. Two prominent models are currently used: generative adversarial networks (GANs) (*https://oreil.ly/4NXJG*) and diffusion models (*https://oreil.ly/kqEYH*). I encourage you to check out the 3D Data Science Resource Hub (*https://learngeodata.eu/3d-data-science-with-python*) for hands-on solutions using these approaches for image generation.

Again, within the resource folder, you can find an image of a knight generated using Stable Diffusion (*https://oreil.ly/lf4Aj*) with the following prompt: "A proud knight, in a simple, minimalist setup, with a large cape and a sword." Now that you have at least two images, we can proceed to the next stage: image preprocessing and model setup.

Image Preprocessing and Model Setup

Inside our IDE, we first import libraries:

```
import matplotlib
from matplotlib import pyplot as plt
matplotlib.use('TkAgg')
from PIL import Image
import torch
from transformers import GLPNImageProcessor, GLPNForDepthEstimation
```

Note that we also select the model we want to use with transformers. Then, we go on to the second step: setting up our model parameters. For this, we leverage the very high abstraction capabilities of transformers, and we put our hands on the GLPN image processor (*https://oreil.ly/7tMj2*).

> GLPNImageProcessor in the Transformers library is a class specifically designed to preprocess images with models trained on the Global Local Patch Network (GLPN) architecture (*https://oreil.ly/QX8ez*). GLPN is a deep learning architecture focused on depth estimation tasks. These transformations include resizing, rescaling, and managing channels and act as a translator between your raw image data and the format that a GLPN model expects. It ensures your images are presented so that the model can understand and process them effectively for depth estimation tasks.

So let's create an instance of the GLPNImageProcessor class by calling the from_pretrained() method and passing in the model name "vinvino02/glpn-nyu". This allows us to preprocess images in a way that's compatible with the pretrained model:

```
feature_extractor = GLPNImageProcessor.from_pretrained("vinvino02/glpn-nyu")
```

Next, I create an instance of the GLPNForDepthEstimation class, also by calling the from_pretrained() method with the same model name. This gives me access to the pretrained depth estimation model, which I can use to predict the depth of objects in an image. The specific model I'm using, "vinvino02/glpn-nyu", is a depth estimation model that was trained on the NYU-Depth dataset:

```
model = GLPNForDepthEstimation.from_pretrained("vinvino02/glpn-nyu")
```

With these two objects set up, we are now ready to use them to perform depth estimation on images. The heavy lifting is massive here because it would take much longer if you were to do this without this transformer library. Now, concerning the image, we use a "Pillow" to load the image:

```
image = Image.open("../DATA/gen_AI_knight.jpg")
```

We then manipulate the dimensions of the image to ensure both the width and height are multiples of 32 while maintaining the original aspect ratio (ratio of width to height), as this is what our model expects:

```
new_height = 480 if image.height > 480 else image.height
new_height -= (new_height % 32)
new_width = int(new_height * image.width / image.height)
diff = new_width % 32
```

Then, we finalize the image resizing process based on the calculations:

```
new_width = new_width - diff if diff < 16 else new_width + 32 - diff
new_size = (new_width, new_height)
image = image.resize(new_size)
```

These lines take the calculated dimensions, potentially adjust the width for stricter divisibility by 32 (optional), and then use those dimensions to resize the original image. This process ensures the final image has a height no larger than 480 pixels and that both width and height are multiples of 32.

The final stage is to prepare the image for our specific deep learning model (GLPN) by using the pretrained image processor to handle the image formatting and preprocessing steps. This process ensures the model receives the image data in the format it expects for optimal performance:

```
inputs = feature_extractor(images=image, return_tensors="pt")
```

This line utilizes the loaded `feature_extractor` to process the provided image. The `images` argument specifies the image data to be processed. The `return_tensors="pt"` argument instructs the function to return the processed data as PyTorch tensors (assuming PyTorch is being used for the model). Essentially, this line performs any necessary preprocessing steps on the image based on the knowledge of the pretrained processor, likely including resizing and format adjustments, and then converts the result into a format suitable for the model.

Everything is now ready to go for depth estimation.

Depth Estimation Predictions from the Model

Monocular depth estimation is a computer vision technique that aims to infer the depth information of a scene from a single 2D image. This process involves estimating the distance of each pixel in the image from the camera. This information can then be used for our 3D reconstruction goal (among other things). Therefore, let's move on to the crucial step in the process: obtaining the model's prediction based on the prepared image data.

For this, we use the following lines:

```
with torch.no_grad():
    outputs = model(**inputs)
    predicted_depth = outputs.predicted_depth
```

This code snippet allows us to leverage the pretrained model to make a prediction. It first ensures efficiency by temporarily turning off training-related calculations (we do not use gradients). Then, it sends the image data through the model's layers, capturing its output. Finally, it isolates the specific prediction we're looking for.

From there we refine the raw prediction we got from the model:

```
pad = 16
output = predicted_depth.squeeze().cpu().numpy() * 1000.0
output = output[pad:-pad, pad:-pad]
image = image.crop((pad, pad, image.width - pad, image.height - pad))
```

> The first line, `output = output.squeeze().cpu().numpy()`, focuses on the output variable, which holds the model's unprocessed prediction for depth. It removes unnecessary dimensions, then transfers data to the CPU, and finally converts to a NumPy format. The next line deals with scaling the depth values and slicing with padding. Finally, we crop the image to match.

In essence, this postprocessing step takes the raw prediction from the model, converts it into a more manageable format, scales the values if necessary, removes potential border issues, and ensures the prediction aligns with the final version of the original image. This process prepares the processed depth data for downward tasks.

If you want to visualize the predictions, use the following code snippet (results in Figure 10-27):

```
fig, ax = plt.subplots(1, 2)
ax[0].imshow(image)
ax[0].tick_params(left=False, bottom=False, labelleft=False, labelbottom=False)
ax[1].imshow(output, cmap='plasma')
ax[1].tick_params(left=False, bottom=False, labelleft=False, labelbottom=False)
plt.tight_layout()
plt.pause(5)
```

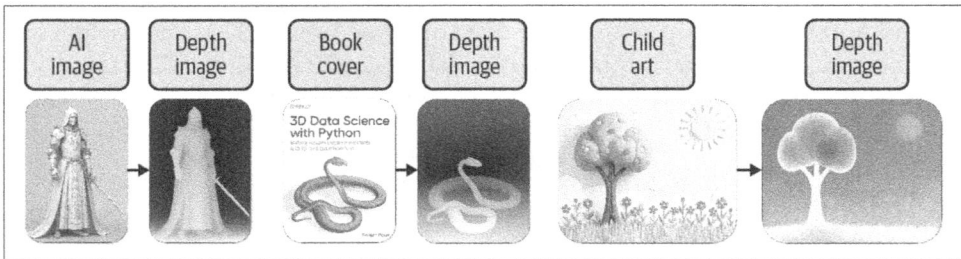

Figure 10-27. The results of our depth predictor on various images

Great. Now that we have our depth image let's turn it into a 3D point cloud.

Point Cloud Generation

Generating 3D point clouds from depth images is a fundamental step that we covered in Chapter 4. We essentially leverage the depth image information about the distance of objects from the camera sensor to reconstruct the 3D geometry of a scene.

To achieve this, the first thing is to import the necessary libraries:

```
import open3d as o3d
import numpy as np
```

After that, we get the width and height from the image size and store the depth in a variable as well as the image from the previous output:

```
width, height = image.size
depth_image = (output * 255 / np.max(output)).astype('uint8')
image = np.array(image)
```

We are now ready to leverage Open3D, following the guidelines in Chapter 4:

```
depth_o3d = o3d.geometry.Image(depth_image)
image_o3d = o3d.geometry.Image(image)
rgbd_image = o3d.geometry.RGBDImage.create_from_color_and_depth(image_o3d,
depth_o3d, convert_rgb_to_intensity=False)
```

This code snippet leverages Open3D to create a particular data structure that combines the color information from a regular image with the processed depth information. The two Open3D objects are combined in an RGB-D image that can be used for more 3D processing and visualization tasks within the Open3D library.

Defining the Camera Intrinsics

Hold on now. Defining camera intrinsics is crucial for accurately generating 3D point clouds from depth images. If you remember, camera intrinsics, such as focal length and principal point, describe the camera's internal parameters and how it projects 3D points onto the 2D image plane. By knowing these parameters, you can correctly map the depth values from the image to 3D coordinates in real-world space. Without accurate camera intrinsics, the generated point cloud will suffer from distortions and inaccuracies, making it unreliable for further processing and analysis. Therefore, careful calibration and estimation of camera intrinsics are essential for obtaining high-quality 3D point clouds.

A Note on Intrinsic/Extrinsic Parameters

Defining both camera intrinsics and extrinsics is crucial for accurately generating 3D point clouds. As you know, camera intrinsics describe the camera's internal parameters and how it projects 3D points onto the 2D image plane. Camera extrinsics, on the other hand, define the camera's pose in the world coordinate system, including its rotation and translation. This can be inferred from the depth values. By accurately estimating these parameters, we can correctly map the depth values from the image to 3D coordinates in the real world.

To obtain accurate intrinsic parameters for real-world camera images, I would use a well-defined calibration pattern, like a checkerboard, and capture multiple images from different angles. Then, I'd employ a calibration method such as the method in OpenCV (*https://oreil.ly/WhupV*) to detect the pattern's corners and estimate the camera's intrinsic parameters. For AI-generated images we are estimating intrinsic parameters without EXIF data, which is more challenging, but still possible. One approach is to use a self-calibration technique (*https://oreil.ly/1AAij*). This involves analyzing multiple images of the same scene from different viewpoints. By identifying corresponding points in these images and applying geometric constraints, you can simultaneously estimate the camera's intrinsic and extrinsic parameters. Another method is to use known object dimensions. If you have an object with known dimensions in the image, you can use its projected size and the known distance to estimate the camera's focal length. However, this method is less accurate and relies on precise measurements.

It's important to note that these methods are often less accurate than a calibration pattern. Additionally, factors like lens distortion and image noise can influence the accuracy of the estimated parameters. It's generally recommended to use a calibration pattern whenever possible for more reliable results.

If you want to deepen these aspects beyond the scope of the book, you can find guided tutorials on the 3D Data Science Resource Hub (*https://learngeodata.eu/3d-data-science-with-python*).

So, as you can imagine, we have a depth image. We don't know the camera's parameters, so we are going to use a basic model for now (we could use a calibration file for more precise modeling). For this task, let's leverage Open3D:

```
camera_intrinsic = o3d.camera.PinholeCameraIntrinsic()
```

This code block sets up the intrinsic parameters (*https://oreil.ly/8xrvU*) of a virtual camera. Understanding camera intrinsics is crucial for tasks like 3D reconstruction, where the relationship between 3D points in the real world and their corresponding pixels in an image needs to be established. The code first creates a `PinholeCamera Intrinsic` object using `o3d.camera`. With the assumption that light beams travel in straight lines and converge at a single point (the pinhole), this process depicts a simplified model of a camera. Now that we have a mode, we need to define some parameters:

```
camera_intrinsic.set_intrinsics(width, height, 500, 500, width/2, height/2)
```

The `set_intrinsics` method takes several arguments that define the intrinsic properties of the virtual camera:

width
  This argument corresponds to the image's width in pixels.

height
  This argument corresponds to the height of the image in pixels.

`500, 500`

These values represent the virtual camera's focal length fx and fy, typically specified in pixel units. The focal length defines the distance between the camera's center and the image plane where light rays converge. It's set to a fixed value of 500 in the horizontal and vertical directions, assuming a square sensor.

`width/2, height/2`

These values define the principal point of the virtual camera, also known as the optical center. This process point represents the intersection of the optical axis with the image plane. Here, it's set to the center of the image by specifying half the width and half the height.

> Intrinsic parameters are a set of parameters that describe the internal characteristics of a camera. The focal length is the distance between the lens's optical center and the image sensor. It determines the camera's field of view, with longer focal lengths resulting in narrower fields of view and shorter focal lengths resulting in wider fields of view. On the other hand, the sensor pixel size refers to the physical size of individual pixels on the image sensor. Smaller pixel sizes generally allow for higher-resolution images but can also introduce more noise.
>
> The principal point is the point where the optical axis intersects the image plane. As a reminder:
>
> - Focal length [pixels] = focal length [mm] / sensor pixel size [μm/pixels]
> - Sensor pixel size [μm/pixels] = sensor size along edge [mm] / pixels along edge [pixels]

By setting these intrinsic parameters, we establish a basic model of how the virtual camera captures and projects 3D points onto a 2D image plane. And now it is time to generate our 3D point cloud:

```
pcd_raw = o3d.geometry.PointCloud.create_from_rgbd_image(rgbd_image,
camera_intrinsic)
o3d.visualization.draw_geometries([pcd_raw])
```

This is great! But we want to push an extra kilometer/mile and provide a 3D mesh. For this reason, we need to do some clean-up and postprocessing operations—a great way to reapply what we learn in previous chapters.

3D Modeling: 3D Point Cloud to Mesh

It is fantastic to have the ability to take a picture and generate 3D from it, just with Python and open source libraries. We now move on to the super-exciting step, which is 3D mesh generation. Interestingly, we use and leverage Open3D here as well. So, we create this little piece of code, which postprocesses the 3D point cloud:

```
# Outliers removal
cl, ind = pcd_raw.remove_statistical_outlier(nb_neighbors=20, std_ratio=6.0)
pcd = pcd_raw.select_by_index(ind)

# Estimate normals
pcd.estimate_normals()
pcd.orient_normals_to_align_with_direction()

o3d.visualization.draw_geometries([pcd])
```

After removing outliers based on a very high standard ratio multiplier, we estimate the normals to use Poisson reconstruction with a depth of 10 on our 3D point cloud. Note the [0] to keep the biggest component. Before visualizing, let's generate a rotation matrix to rotate the mesh, and then visualize:

```
mesh = o3d.geometry.TriangleMesh.create_from_point_cloud_poisson( pcd, depth=10,
n_threads=1)[0]
# Rotate the mesh
rotation = mesh.get_rotation_matrix_from_xyz((np.pi, 0, 0))
mesh.rotate(rotation, center=(0, 0, 0))

# Visualize the mesh
o3d.visualization.draw_geometries([mesh], mesh_show_back_face=True)
```

As we discussed earlier, we have various depth possibilities; the deeper you go, the more refined you are. The result is very interesting because we also have the color associated with our mesh. I could pass that in inpainting methods (*https://oreil.ly/ zwj94*) or generative AI, change everything, and color our mesh.

The last step is actually exporting the mesh and visualizing it:

```
o3d.io.write_triangle_mesh('../RESULTS/gen_AI_knight.jpg, mesh)
```

This concludes our solution for transforming from a 2D monocular image to a 3D reconstruction. We showed how to use data from one image, or possibly several photographs, to build a digital model of a three-dimensional scene. We utilized libraries like Transformers and Open3D to process pictures, extract depth information, and even combine color and depth data for further analysis. Specifically, you have a clear understanding and the know-how to leverage pretrained depth estimation models to analyze images, potentially refining the raw predictions and utilizing specialized libraries to work with the extracted 3D data.

To extend the capabilities showcased here, you can utilize the state-of-the-art monocular depth estimation models listed here:

- MiDaS (*https://oreil.ly/pbl6w*): Known for its accuracy and speed.
- Depth Anything (*https://oreil.ly/om7rT*) (and Depth Anything V2 (*https://oreil.ly/4aYy8*)): A powerful foundation model for monocular depth estimation, it offers impressive performance and flexibility.
- MoGe: A powerful model for recovering 3D geometry from monocular open-domain images.
- Depth Pro (*https://oreil.ly/n1DIw*): Foundation model for zero-shot metric monocular depth estimation.

Summary

Having explored various techniques for crafting 3D data models, you are now equipped with a powerful toolkit for bringing objects and environments to digital life. We studied high-fidelity 3D meshing to create smooth and detailed object representations. Then, we shifted gears to understand 3D voxelization, using voxels to define an object's volume. Then, we built 3D parametric models constituting very complex datasets that can be changed with parameter adjustments. Finally, we explored how to turn 2D images into 3D models with AI models.

You can consider yourself an experienced modeler if you can check off these concepts:

- Understanding the trade-offs between different 3D representation methods (meshes, voxels, parametric models) and choosing the right approach for a given problem
- Proficiency in working with various file formats (e.g., STL, DXF, PLY) and understanding interoperability between different 3D modeling tools and workflows
- Combining multiple techniques, such as meshing and voxelization, for robust and versatile 3D modeling pipelines
- Developing a 3D parametric modeling workflow for 3D object modeling
- Leveraging pretrained AI models for depth estimation to accelerate 3D reconstruction

As 3D data science continues to evolve, these techniques will play an increasingly crucial role in bridging the gap between physical and digital worlds. The ability to create accurate, flexible, and efficient 3D models opens up new possibilities for data analysis, simulation, and decision making across industries.

Hands-on Project

Here is a scenario to put your skills to the test. Can you create a 3D modeling pipeline that combines the multiple techniques learned in this chapter? Here's a suggested workflow:

1. Start with a 2D image of an object (you can take a photo or use an AI-generated image).

2. Use the image-based 3D reconstruction technique to generate a point cloud from the image.

3. Apply voxelization to the resulting point cloud, experimenting with different voxel sizes.

4. Convert the voxelized representation back into a mesh using a surface reconstruction algorithm.

5. Use parametric modeling techniques to add or modify features of the reconstructed object (e.g., add a base or handle to a reconstructed cup).

6. Export your final model in both STL and PLY formats.

7. Visualize the results at each step and compare the final model with the original image.

This will help you integrate multiple 3D modeling techniques and gain hands-on experience with the entire pipeline from 2D image to parameterized 3D model.

Now, it is time to extend your 3D modeling knowledge with a real-world project to reconstruct a neighborhood from aerial LiDAR data.

3D Building Reconstruction from LiDAR Data

As experts in cutting-edge applications, you understand the importance of accurate 3D city models for various tasks, such as urban planning, disaster management, and infrastructure development. You need reliable methods to extract valuable information from complex 3D datasets, especially aerial LiDAR point clouds. Current methodologies for generating these models often rely on manual processing or semiautomated techniques that are time-consuming and prone to errors. We need a robust, automated solution to handle large datasets and provide accurate 3D models.

In this chapter, we'll follow a fictional example where you are a real estate investor evaluating a neighborhood's potential. You want to determine the feasibility of extending existing houses horizontally and vertically. To make informed decisions, you need precise information about each building's footprint and height. What if we could curate LiDAR data from existing open data portals and extract a neighborhood from the area we believe has potential? From there, we could abstract the messy LiDAR point cloud to detect the houses, extract their footprints, and then create a detailed model for each house to give us a sense of its footprint and height.

In this chapter, I will describe the four-phase plan (see Figure 11-1) for reconstructing 3D models of buildings from aerial LiDAR data and address the limitations of existing methods. We establish a workflow by setting up the Python coding environment and installing essential libraries. Next, we prepare the LiDAR data by classifying building points using masks and translating coordinates for optimal visualization. We then automatically identify individual building segments within the point cloud. This allows the extraction of a 2D building footprint from a selected segment and enriches it with semantic information such as building height, area, and perimeter. We then introduce an automated process to streamline this workflow, enabling the efficient processing of an entire neighborhood from a single dataset.

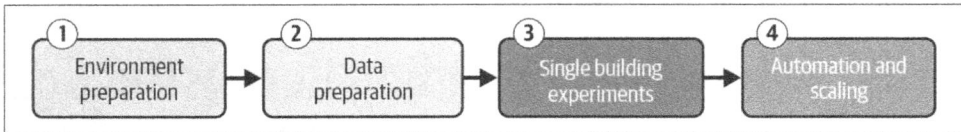

Figure 11-1. A workflow that outlines the four main phases for reconstructing 3D buildings from LiDAR data, starting with environment preparation and ending with automation and scaling

> As always, you can find all the materials for this chapter on the 3D Data Science Resource Hub (*https://learngeodata.eu/3d-data-science-with-python*). To access the files (code, datasets, articles) and resources, you may need to share an email address, a personal password, or proof that you are the book's owner.

Let's start with the Python setup.

Phase 1: 3D Python Setup

By now, setting up your local Python environment should not involve too many secrets (see Figure 11-2). The first thing we want to do is set up a virtual environment with Python.

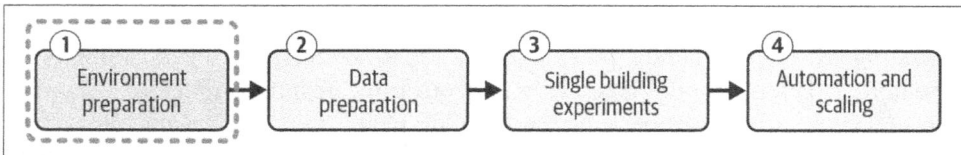

Figure 11-2. Phase 1: 3D Python setup

Project Environment Setup

You can review Chapter 3 for our book's overall setup. As a reminder, here are the initialization steps:

1. Download Anaconda or Miniconda.

2. Install and then launch a virtual environment with Anaconda:

   ```
   conda env list
   conda create -n YOUR_ENV python=3.10
   conda activate YOUR_ENV
   ```

And this is it: you are in the environment. From there, you can install various libraries we need for the experiments. So first, we install three base libraries with the package manager pip (the libraries we want are NumPy (*https://numpy.org*), pandas (*https://oreil.ly/946mS*), and Matplotlib (*https://matplotlib.org*) to plot):

```
pip install numpy matplotlib pandas
```

Now, we can move on to installing 3D libraries. We are going to install two 3D libraries. First, let's install Open3D (*https://open3d.org*) to manipulate 3D data:

```
pip install open3d
```

Finally, to be close to production setups, we install laspy (*https://oreil.ly/qQOPO*) with its extensions lazrs and LASzip to manipulate 3D point clouds in the LAS and LAZ file formats:

```
pip install laspy[lazrs, laszip]
```

laspy (*https://oreil.ly/4NdI1*) provides an excellent interface for reading, writing, and manipulating LiDAR data stored in the LAS and LAZ formats. With laspy, you can easily access and extract essential information from point clouds, such as point coordinates, intensity values, and classification labels. This allows us to perform a wide range of tasks, including filtering, cleaning, and visualizing point cloud data. laspy's efficient data handling and processing capabilities make it a valuable tool for large-scale 3D point cloud analysis.

We have the base libraries (NumPy, pandas, and Matplotlib) and two 3D libraries (Open3D and laspy). Now, let's load several geospatial libraries. First, we install Rasterio (*https://oreil.ly/UMxtk*) to manipulate raster data:

```
pip install rasterio
```

Then, we install GeoPandas (*https://oreil.ly/lgwG3*), a pandas extension for geospatial data processing:

```
pip install geopandas
```

Finally, we install Shapely (*https://oreil.ly/OA_pF*) to handle the 2D vector data and the alphashape library, a small library for computing alpha shapes (*https://oreil.ly/-kzaY*):

```
pip install shapely alphashape
```

Now we are almost ready. Our environment is isolated we have loaded a bunch of libraries. The last stage is to install an IDE onto our environment so that we can code correctly. To change our usual setup, let us use JupyterLab to work within a notebook. Its interactive nature allows experimentation with different algorithms and visualization techniques quickly and efficiently. This allows you to work with code cells, markdown cells, and rich output formats like 3D visualizations, making it easy to document your work and share insights with others. This is usually a great way to communicate intermediate results with tech-oriented stakeholders. To install JupyterLab (*https://jupyter.org*), we run the following command:

```
pip install jupyterlab
```

We can launch JupyterLab by typing the following in the Anaconda prompt terminal:

```
jupyter lab
```

> Before launching JupyterLab (if you did, just close the Explorer window and press Ctrl + C in the Anaconda prompt to kill the server), it is better to launch the server from your code folder by first changing the directory: cd C:/path_to_code_folder. This way, you can ensure proper file management.

Project Notebook Setup

Now, let's initialize our script in our notebook. This means loading the libraries:

```
# Base libraries
import numpy as np
import pandas as pd

# Plotting library
import matplotlib.pyplot as plt

#3D libraries
import open3d as o3d
import laspy
print(laspy.__version__)

# Geospatial libraries
import rasterio
import alphashape as ash
import geopandas as gpd
import shapely as sh

from rasterio.transform import from_origin
from rasterio.enums import Resampling
from rasterio.features import shapes
from shapely.geometry import Polygon
```

To best use Jupyter notebooks, you can explore Table 11-1 to find dedicated resources to help you.

Table 11-1. The Jupyter Notebook "cheat sheet" to guide your development productivity

Feature	Description	How to use
Running cells	Executes the code or renders the text within a cell	Ctrl + Enter
Running and move to next cell	Runs the current cell and selects the cell below	Shift + Enter

Feature	Description	How to use
Run and insert below	Runs the current cell and inserts a new cell below	Alt + Enter
Changing cell types		
Code cell	Contains code to be executed	Y (in command mode)
Markdown cell	Contains text formatted using Markdown	M (in command mode)
Inserting and deleting cells		
Insert cell above	Adds a new cell above the currently selected cell	A (in command mode)
Insert cell below	Adds a new cell below the currently selected cell	B (in command mode)
Delete cell	Removes the currently selected cell	D + D (twice, in command mode)
Cell navigation and operations		
Select cell above	Moves the selection to the cell above	Up arrow key (in command mode)
Select cell below	Moves the selection to the cell below	Down arrow key (in command mode)
Split cell	Divides the active cell at the cursor position	Ctrl + Shift + -
Multiple cursors	Allows simultaneous editing at multiple points within a cell	Ctrl + Click (in edit mode)
Kernel operations		
Restart	Restarts the kernel, clearing all variables and definitions	Menu: Kernel > Restart
Restart and clear output	Restarts the kernel and clears the output displayed below cells	Menu: Kernel > Restart and Clear Output
Restart and run all	Restarts the kernel and runs all cells sequentially	Menu: Kernel > Restart and Run All
Interrupt	Stops the kernel's current computation	Menu: Kernel > Interrupt
Markdown formatting		
Headings	Create headings	# for H1, ## for H2, etc.
Bold	Turn bold	**bold** or __bold__
Italic	Turn italic	*italic* or _italic_
Lists	Create lists	Unordered: * or -, Ordered: 1., 2., etc.
Hyperlinks	Add a hyperlink	[Link Text](URL)
Code	Write code	Inline: `code`, Block: `code` or indent by 4 spaces
Images	Insert images	![Alt Text](Image URL)
Magic commands	Special commands that provide additional functionality	Prefix with % (line magic) or %% (cell magic)
%run	Executes an external script file within the cell	%run myscript.py
%timeit	Measures and reports the execution time of a code snippet	%timeit my_function()
%writefile	Saves the cell content to a file	%writefile myfile.txt
%store	Saves a variable for use in a different notebook	%store my_variable
%pwd	Prints the current working directory	%pwd
%%javascript	Executes the cell contents as JavaScript code	%%javascript console.log("Hello");

We are all set up and ready for phase 2: data preparation.

Phase 2: Data Preparation

In this phase, we curate the data, prepare it, profile it, and then preprocess it so that we can directly use it to run follow-up actions (see Figure 11-3).

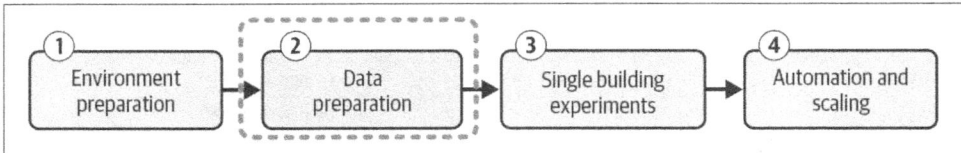

Figure 11-3. Phase 2: Data preparation

Aerial LiDAR Data Curation

The data I chose is from Vancouver, a region in Canada that I love. You can go to the Vancouver portal (Open Data Portal (*https://oreil.ly/0jnG9*)), and select one tile from there (see Figure 11-4).

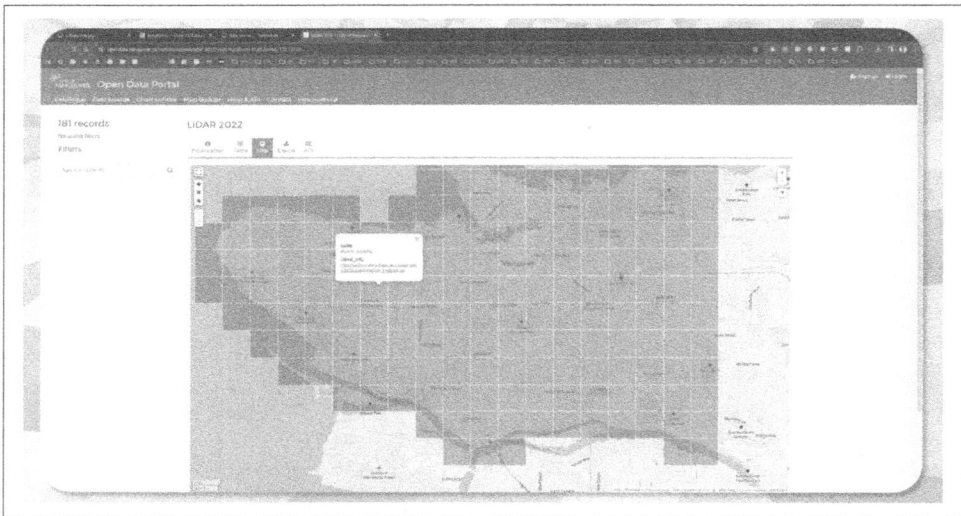

Figure 11-4. The Vancouver Open Data Portal with a specific LiDAR dataset tile highlighted for download

You can choose a tile and download it; you'll get a LAZ file. You can find the neighborhood file in the chapter's resource directory. Some metadata can also be helpful. If you go to the information tab, you can access some exciting information. We see that the data contains classification information. Now, let's load the data. To do that, we use laspy and the `read` function and pass the path to our dataset. I express everything relatively:

```
# Neighborhood point cloud
las = laspy.read('../DATA/neighborhood.laz')
```

From here, we can explore the classification field by getting the last classification and printing its unique values:

```
# Explore the classification field
print(np.unique(las.classification))
```

Then, finally, we print whatever we have in our header:

```
print([dimension.name for dimension in las.point_format.dimensions])
```

Finally, I like to print out the coordinate system and projection system. For this, we will use a VLR, or Variable Length Record, in the context of LAS files. This is a flexible way to store additional metadata, ranging from simple text descriptions to complex data structures. VLRs are particularly useful for storing information that doesn't fit neatly into the standard LAS format, such as sensor-specific data, project-specific metadata, or custom classifications. In this particular case, we find out that the projection information (coordinate reference system, or CRS) is stored in the third VLR:

```
# Explore CRS info
crs = las.vlrs[2].string
print(las.vlrs[2].string)
```

We can output a table showing classes ranging from 1 to 7, XYZ, and many features. The projection is in EPSG 26910 (*https://epsg.io/26910*), which is NAD_1983_UTM_Zone_10N (*https://oreil.ly/Ae5dR*). The projection information is essential if you want to fuse data sets together.

> A projection is a system that defines how we represent the curved surface of the Earth on a flat surface like a map. It's like flattening a globe into a 2D image. This process inevitably introduces some distortion, but different projections minimize different types of distortion. For example, a projection like EPSG:26910, or NAD_1983_UTM_Zone_10N, is a Universal Transverse Mercator (UTM) (*https://oreil.ly/GhZVy*) projection that minimizes distortion in areas with a north-south orientation. By understanding and selecting the appropriate projection, we can represent geographic data for spatial analysis tasks.

That's it for curating the data. Now, we are going to do some preprocessing.

Aerial LiDAR Data Preprocessing

First, we want to create a mask on our dataset to retain only the points that belong to a building. Because we benefit from the classification information, it is trivial to use the laspy data structure to retain only the building's points in pts_mask:

```
# Create a mask to filter points
pts_mask = las.classification == 6
```

Then we transform that into a NumPy array and use `np.vstack()` to stack the points vertically:

```
xyz_t = np.vstack((las.x[pts_mask], las.y[pts_mask], las.z[pts_mask]))
```

In the preceding cell, I use a little trick to pass the mask and filter out only the points belonging to the building class. Once we have created this NumPy array (which is transposed, thus the name of the variable `xyz_t`), we create an Open3D geometry of type `PointCloud` and pass the data points to this new geometry:

```
# Transform to Open3D.o3d.geometry.PointCloud and visualize
pcd_o3d = o3d.geometry.PointCloud()
pcd_o3d.points = o3d.utility.Vector3dVector(xyz_t.transpose())
```

From there, we can ensure we work with local coordinates to avoid truncated data due to large georeferenced coordinates, which implies laggy and imprecise visualizations. So, to do just that, we translate our point cloud using its center with these little lines of code:

```
# Transform to Open3D.o3d.geometry.PointCloud and visualize
pcd_o3d = o3d.geometry.PointCloud()
pcd_o3d.points = o3d.utility.Vector3dVector(xyz_t.transpose())
```

The final operation is to visualize the results with this function:

```
# Visualize the results
o3d.visualization.draw_geometries([pcd_o3d])
```

In the Open3D window (see Figure 11-5), we see the little buildings extracted using the classification field.

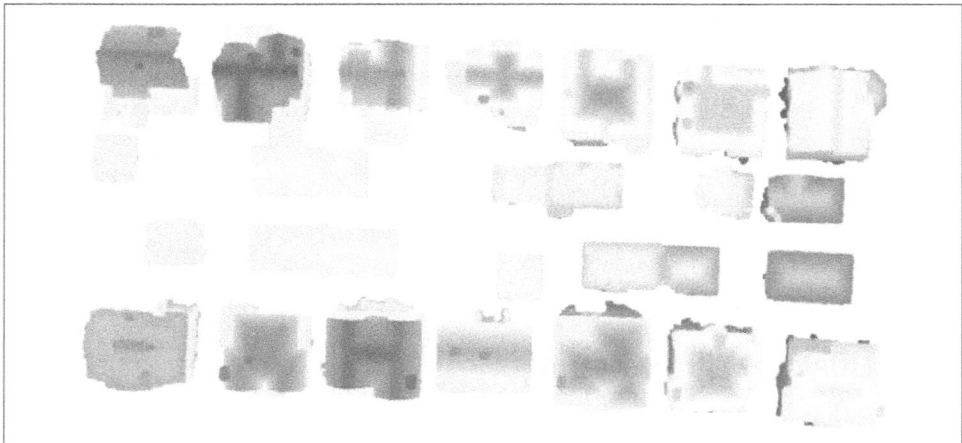

Figure 11-5. The point cloud of the selected buildings

As you can see, we have some "label noise" at the border of the houses that makes the data imperfect (i.e., real world) in the semantic space. In the various experiments, we have to account for that additional imperfection.

> Label noise in 3D LiDAR data refers to inaccuracies or errors in the labels assigned to individual points in the point cloud. These errors can arise from various sources, including sensor noise, environmental factors, or human annotation mistakes. Label noise can significantly impact the performance of 3D processing tasks. Incorrectly labeled points can mislead machine learning models, reducing accuracy and robustness.

By using the same process, let's isolate only the ground points. You may have a hint as to why we are doing that at this stage. If you do not, you must trust me unquestioningly (which is dangerous); in phase 3, we are going to unveil why we are keeping ground points in our experiments. The process is very similar, except that the translation is not activated again:

```
# Isolating ground points
pts_mask = las.classification == 2
xyz_t = np.vstack((las.x[pts_mask], las.y[pts_mask], las.z[pts_mask]))

ground_pts = o3d.geometry.PointCloud()
ground_pts.points = o3d.utility.Vector3dVector(xyz_t.transpose())
ground_pts.translate(-pcd_center)

# Visualize the results
o3d.visualization.draw_geometries([ground_pts])
```

The last step is a bit of profiling to better understand the distribution of selected points. For this, we use the same approach as in Chapter 8. Let's compute the nearest-neighbor distance between points and get the average distance for each point. We can do that with the following code snippet:

```
# Identifying the average distance between building points
nn_distance = np.mean(pcd_o3d.compute_nearest_neighbor_distance())
print("average point distance (m): ", nn_distance)
```

This shows that we have an average of 11 cm between points. This data-driven parameter can then be saved for later use in our automation experiments. It's time for phase 3, which is the single-unit experiment.

Phase 3: Experiments

We move to the experimental side, i.e., trying to extract the 3D model for a single building (see Figure 11-6). What does this phase hold? Well, as you can see from Figure 11-7, we go through a series of steps. It starts with an unsupervised segmentation to identify clusters that would compose the building units.

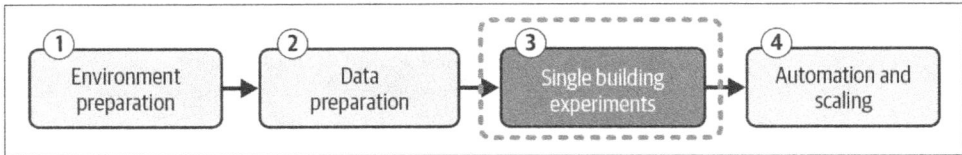

Figure 11-6. Phase 3: Experiments

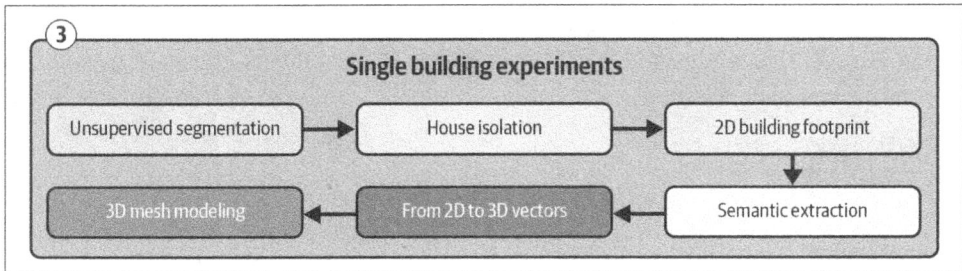

Figure 11-7. Unit-based experiment: 3D model creation

Then, we select just one house from that and explore the feasibility of our solution as a unit-based experiment. This means we want to extract a 2D building footprint, which can contain beneficial semantic and attribute information. Then, we convert the 2D footprint to a 3D vector, the base for creating the model with both the meshes and the faces. If things are clear, we can dive into the coding side of making this work.

> The 3D processing modules in Chapter 2 are the building blocks of complex 3D data science pipelines. Tying these processes, like noise removal, segmentation, 2D vectorization, semantic extraction, and 3D modeling, into a single 3D data science workflow is incredibly powerful. By integrating these steps, you can automate complex tasks, reduce manual effort, and improve the overall efficiency and accuracy of your 3D data analysis solution. This integrated approach enables you to tackle very challenging problems; this chapter is one example of the results of such a 3D workflow.

Unsupervised Point Cloud Segmentation

You can approach unsupervised point cloud segmentation in many exciting ways. However, it largely goes beyond the scope of this chapter. Luckily, you can unlock a deep understanding and know-how through Chapters 12 and 13.

In this section, we will explore a solution leveraging Density-Based Spatial Clustering of Applications with Noise (DBSCAN) (*https://oreil.ly/D5vRr*). It is very beneficial in a Euclidean space with clear spatial noncontiguity. On top of that, you can filter out noise and only return relevant clusters.

> One other way could be to use *k*-means (*https://oreil.ly/dSu4T*). The problem is that you must filter out outliers and know how many clusters you have. So, it's not "fully" automatic because you need to supervise the parameters tightly. While there are ways to define this parameter automatically, it may not be the most accurate solution. This is shown in Chapter 12. Another solution is to use the Segment Anything Model (SAM) (*https://oreil.ly/ZusKF*). This is an exciting option. However, you must go through a projection phase that may be heavy on the computation side. This is shown in Chapter 13.

DBSCAN is a clustering algorithm that groups together closely packed points. It defines a neighborhood around each data point, with a radius of ε (epsilon). If a point has at least a minimum number of points (min_cluster_points) within its ε-neighborhood, it's considered a core point. Points that are reachable from core points, either directly or indirectly, are assigned to the same cluster. Points that are not reachable from any core point are classified as noise.

Here we define two parameters to make DBSCAN work:

```
epsilon = 2
min_cluster_points = 100
```

A larger ε value leads to larger clusters, while a smaller value results in smaller, more tightly packed clusters. Similarly, a higher min_cluster_points value requires denser clusters, while a lower value allows for looser clusters. By tuning these parameters, you can effectively identify clusters of varying densities and shapes within your data.

> The DBSCAN algorithm is beyond the scope of the current chapter. But be reassured that it is covered in depth in Chapter 12 and beyond. DBSCAN is a central algorithm that powers many successful workflows because of its nice trade-off between simplicity, accuracy, and computational efficiency.

Having defined the parameters, let's compute the DBSCAN cluster labels. Each point will receive a label corresponding to a specific cluster or to what is tagged as "noise" (–1):

```
labels = np.array(pcd_o3d.cluster_dbscan(eps=epsilon,
min_points=min_cluster_points))
```

These two lines aim to get the number of clusters created:

```
max_label = labels.max()
print(f"point cloud has {max_label + 1} clusters")
```

Finally, let's color each of our clusters differently. For that, we use Matplotlib. We take the `tab20` discrete color range and each label that is above zero because the noise is tagged "-1" (hence, we color it black):

```
# We use a discrete color palette to randomize the visualization
colors = plt.get_cmap("tab20")(labels / (max_label if max_label > 0 else 1))
colors[labels < 0] = 0
pcd_o3d.colors = o3d.utility.Vector3dVector(colors[:, :3])
```

Now, let's plot the results:

```
# Point cloud visualization
o3d.visualization.draw_geometries([pcd_o3d])
```

We obtain nicely detected houses with various shades, making them a strong base for individual modeling (see Figure 11-8).

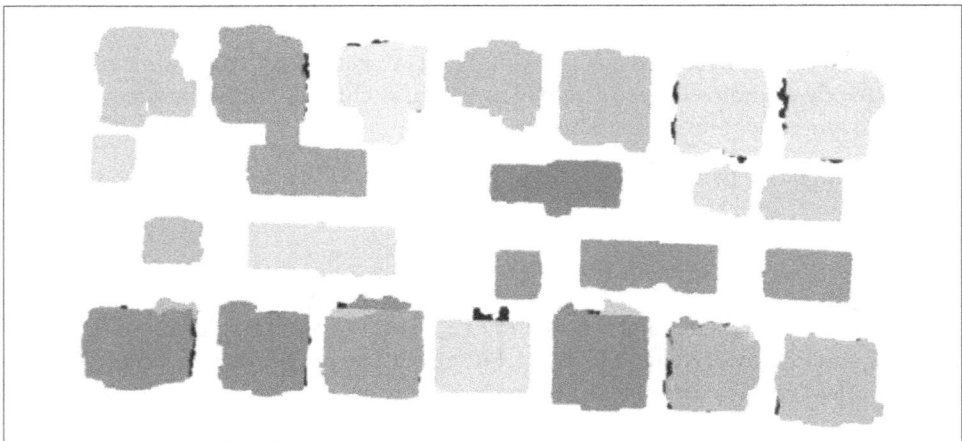

Figure 11-8. The results of the DBSCAN algorithm on points with the building label (each color is an independent cluster)

In the current experiment, we have 31 clusters. That is the base stage. We can now move on to single 3D house isolation.

3D House Segment Isolation

Why do we want to isolate a single house? We want to model and inject semantics for each building. Therefore, we need to reason at the object level, the object being a house (see Figure 11-9).

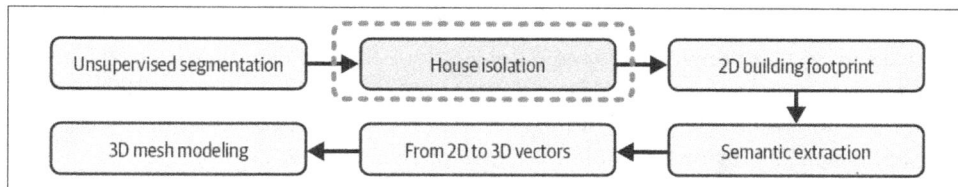

Figure 11-9. The second step of phase 3: isolating one house from the clustering output

This means all our processes should be tested on a single building unit before thinking neighborhood level. To get the instance (segment), we use the function `select_by_index` from Open3D:

```
# Selecting a segment to be considered
sel = 1
segment = pcd_o3d.select_by_index(np.where(labels==sel)[0])
```

The function returns an array with more than one element, but we are only interested in the point cloud object; thus, I take the first element with [0].

From this stage, you can draw the selected building unit to verify your base unit of thinking:

```
o3d.visualization.draw_geometries([segment])
```

This results in a 3D point cloud such as the ones shown in Figure 11-10.

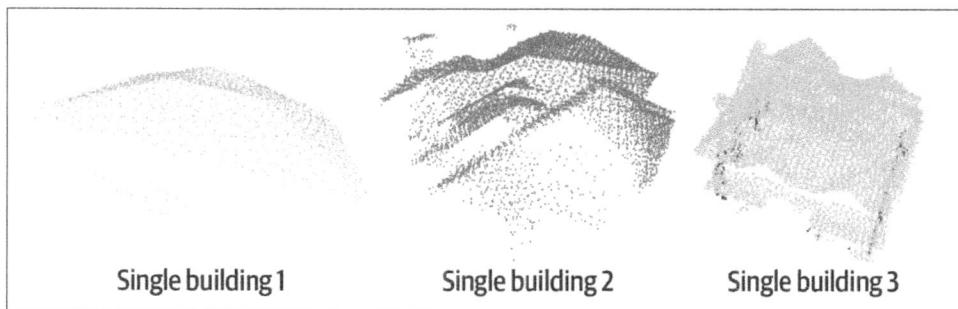

Single building 1   **Single building 2**   **Single building 3**

Figure 11-10. Three example buildings that can be selected using the methodology provided

Our base unit, a building, is the subject for the next stage: 2D building footprint extraction.

2D Building Footprint Extraction

Let's extract the 2D outline of the selection's building footprint. First, we extract the x and y coordinates of our point cloud:

```
# We extract only the x and y coordinates of our point cloud (Note: it is local)
points_2D = np.asarray(segment.points)[:,0:2]
```

We need to go from the initial Open3D object to a NumPy array. That's why I added the transformation expression. Now, we can move on to constituting an alpha shape. To use the alpha shape of the building vector, we use the library we imported previously, passing the previous variable with an `alpha` parameter:

```
# We compute the shape (alpha shape) and return the result with shapely
building_vector = ash.alphashape(points_2D, alpha=0.5)
```

> In computational geometry, an alpha shape, or α-shape, is a family of curves that captures the essential shape of a set of points in Euclidean space. It stems from a Delaunay triangulation and provides a way to represent the overall outline of the points while allowing for varying levels of detail. The α parameter controls the level of detail. A smaller alpha value creates a more detailed and intricate shape, while a more significant one results in a more straightforward and smoother shape (a convex hull at its extreme).

We use Shapely to print our results. In this cell, type the name of your variable at the end. Then, press Enter to check out the magic that happens:

```
building_vector
```

Now, let's create some way to store this information using GeoPandas. We use a Geo-DataFrame. You pass the geometry, the building vector, and the CRS:

```
#%% Store the 2D polygon in a GeoDataFrame
building_gdf = gpd.GeoDataFrame(geometry=[building_vector], crs='EPSG:26910')
building_gdf.head(1)
```

> If you print the first line, you will see that we have geometry, a polygon, stored in a data frame. It's a GeoDataFrame entity, a data structure combining the best of both worlds: a pandas DataFrame's tabular format and the spatial capabilities of a geographic information system (GIS). By integrating geographic data, such as points, lines, and polygons, with tabular data, GeoDataFrames enable us to perform spatial analysis, mapping, and visualization.

We are ready to move on to the next stage: semantic and attribute extraction.

Semantic and Attribute Extraction

We have already accomplished a lot, but it still needs to be finished, and it may be interesting to push a bit more before going into the next phase. For this, we want to address semantic and attribute extraction (see Figure 11-11).

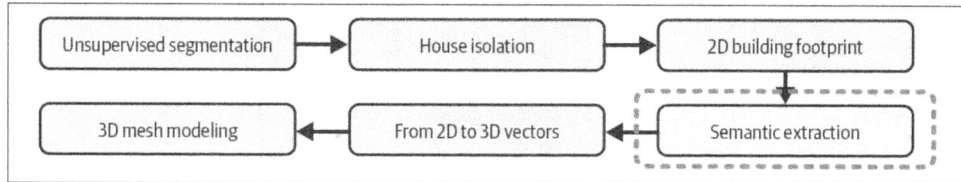

Figure 11-11. Semantic extraction step

Extending geometric understanding and extracting semantics from a single building unit within a 3D LiDAR point cloud is incredibly useful. By understanding the building's geometry and main constituents, we can gain valuable insights into the building's structure, function, and condition.

This information can then be leveraged to create detailed 3D models in our case, generate accurate building footprints, and identify potential maintenance issues. Therefore, we first want to use the house's height as a relative measure. Let me first show you something by computing the altitude:

```
# What is wrong with this line?
altitude = np.asarray(segment.points)[:,2]+pcd_center[2]
height_test = np.max(altitude)- np.min(altitude)
print('Is this correct: ',height_test)
```

What do you think is wrong if I do that? I compute the altitude by taking only the z-value of all the points. I add a translation of z to get to the real coordinates of z. After that, I take the maximum of the segmented point cloud and the minimum, which is the height.

We are dealing with imperfect data: a real-world dataset. You may be in a scenario where there are no measured points close to the ground within the segment describing the building or close to where your house is standing. This means that you may have a shift where if you were to use only the segment points as a base to measure the height, you may have a result that is 2 meters instead of 10 (i.e., the result would be wrong). That is something we want to avoid. How can we solve this issue?

Let's connect the dots. Do you remember that we extracted only the ground points at the beginning of the workflow? Well, it is time to put this in the basket and leverage the information we can gather from it.

We have to define the ground level in the local area. To achieve this, we define a query point based on the center of the segment. We can then take the center of this segment while dropping the z because we do not want the center of the segment there but the z-minimal segment value (i.e., the center of the potential footprint).

Let's define the ground level in our local area by first getting a query point:

```
query_point = segment.get_center()
query_point[2] = segment.get_min_bound()[2]
```

We can leverage this query point to select all the neighborhood points from the ground, to then compute the mean elevation value in the local area. As I say neighborhood, do you remember our data structure from Chapter 4?

Well, let's once again leverage a k-d tree that we construct from the ground points:

```
pcd_tree = o3d.geometry.KDTreeFlann(ground_pts)
```

From there, we do a k-NN search, where we retain the indexes of our points (pass the query point and take 200 neighbors):

```
[k, idx, _] = pcd_tree.search_knn_vector_3d(query_point, 200)
```

Now, for qualitative control we create a sample of points that we select using this search, using only ground points, and coloring the selection (Figure 11-12):

```
# From the nn search, we extract the points that belong to the ground and paint
# them grey
sample = ground_pts.select_by_index(idx, invert=False)
sample.paint_uniform_color([0.5, 0.5, 0.5])
o3d.visualization.draw_geometries([sample, ground_pts])
```

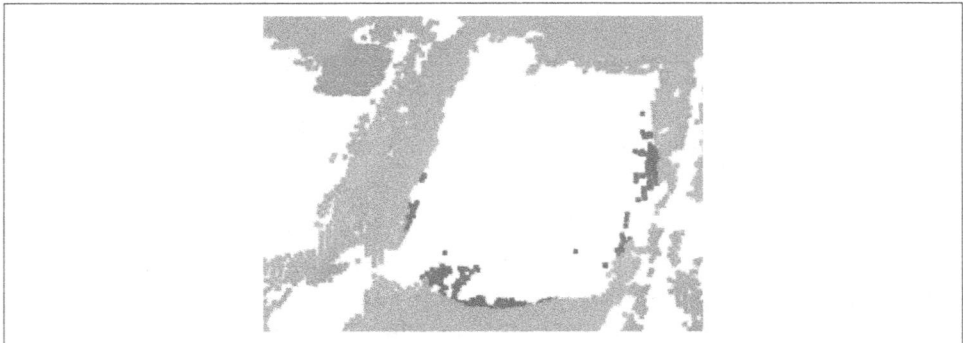

Figure 11-12. The gray box shows all the darker gray points selected to compute the building's minimum value

The points close to the building are the only data points that could hint at where the building stands with enough relevancy. So we use that to define the lowest bound to finally extract the height of the building:

```
# Extract the mean value of the ground in this specific place
ground_zero = sample.get_center()[2]

# Compute the true height of the building, roof included:
height = segment.get_max_bound()[2] - ground_zero
print('True Height: ', height)

# Check the difference:
print('Height Difference: ', height - height_test)
```

If we were to check the difference, we would obtain a shift of –12 centimeters for the sample. So, in this case, we were pretty close, but for some other cases where you have large occlusions, you may have much more significant differences.

Now, let's move on to computing some quantifiable attributes. The idea is to gain insights into building size, shape, and orientation. This information helps us understand the density and layout of the neighborhood, identify potential development opportunities, and assess the desirability of specific areas.

First, let's make sure to structure and curate the clustering label:

```
building_gdf[['id']] = sel
```

Now, let's compute some object-based properties, namely the height, area, and building perimeter (derived from the alpha shape):

```
building_gdf[['height']] = segment.get_max_bound()[2] - sample.get_center()[2]
building_gdf[['area']] = building_vector.area
building_gdf[['perimeter']] = building_vector.length
```

Finally, let's retrieve the local centroid coordinates as well as the segment center and number of points, followed by printing the results:

```
building_gdf[['local_cx','local_cy','local_cz']] =
np.asarray([building_vector.centroid.x, building_vector.centroid.y,
sample.get_center()[2]])
building_gdf[['transl_x','transl_y','transl_z']] = pcd_center
building_gdf[['pts_number']] = len(segment.points)

#print
building_gdf.head(1)
```

Now, I want to show you some ways to obtain some extra attributes to better describe our scene. For example, analyzing local maxima and minima can reveal architectural features like roof peaks and ground-level entrances, providing valuable context for property assessments and urban planning.

Let's first extract `points_1D` along the x-axis and print the local minima and the local maxima:

```
# Extra attributes
points_1D = np.asarray(segment.points)[:,2]
print('The local minima (along the Z axis)',np.min(points_1D))
print('The local maxima (along the Z axis)',np.max(points_1D))
```

Then, we plot the histogram. We could leverage that to extract some exciting ways to model the roof layers, for example. Let's see that right away:

```
# Extra attributes
points_1D = np.asarray(segment.points)[:,2]
print('The local minima (along the Z axis)',np.min(points_1D))
print('The local maxima (along the Z axis)',np.max(points_1D))

plt.hist(points_1D, bins='auto') # arguments are passed to np.histogram
plt.title("Histogram with 'auto' bins")
plt.show()
```

This returns Figure 11-13.

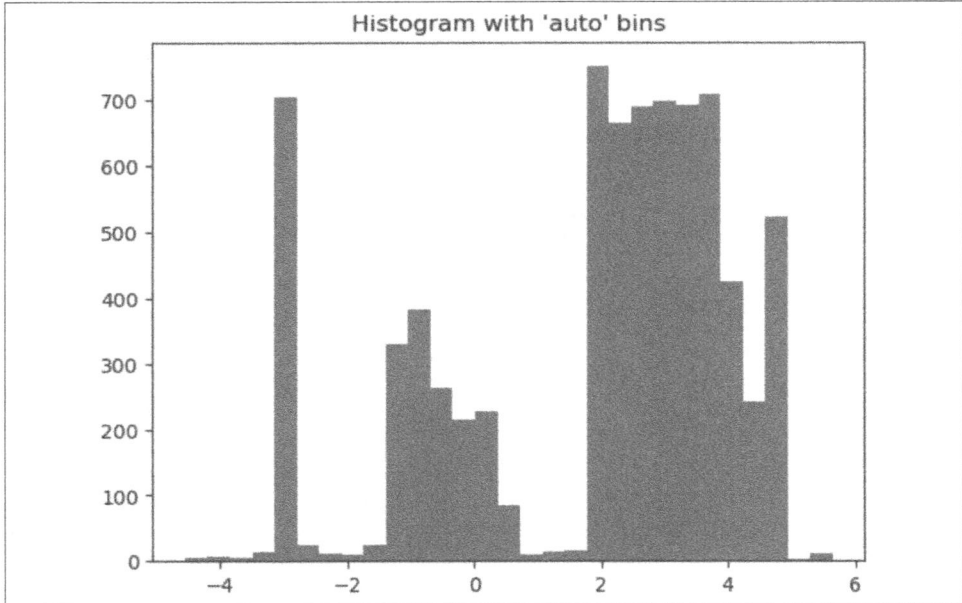

Figure 11-13. The histogram along the z-axis of the point distribution for our selected house

As you can see, with this histogram along the z-axis, the three roofs pop out through this chart. This process means we could use some breakpoints and identify the breaks to further segment our point cloud.

This information helps us understand the building's potential use, estimate renovation costs, and identify potential challenges during construction or redevelopment. Additionally, by comparing the attributes of different buildings, we can identify trends and patterns in the real estate market, informing investment decisions and urban planning strategies. However, for this, we need to extract and reconstruct each single entity further.

2D to 3D Vectors

We can now move on to the art of transforming our 2D shape into a 3D model (Figure 11-14). Our strategy is simple. We are going to get the vertices from our 2D shape and leverage our additional attributes, mainly the building height, to duplicate the vertices at the appropriate height.

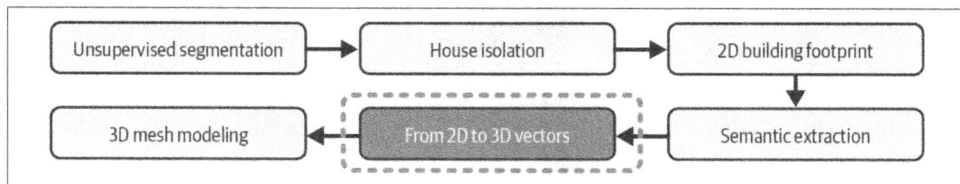

Figure 11-14. From 2D to 3D vectors

First, let's compute the 2D polygon:

```
# The base layer
# Generate the vertex list
vertices = list(building_vector.exterior.coords)

# Construct the Open3D object
polygon_2d = o3d.geometry.LineSet()
polygon_2d.points = o3d.utility.Vector3dVector([point + (0,) for point in
vertices])
polygon_2d.lines = o3d.utility.Vector2iVector([(i, (i + 1) % len(vertices)) for
i in range(len(vertices))])
```

This generates our 2D polygon as an Open3D object, taking GeoPandas as the seed data frame. We still have to compute a 3D vector.

Let me break down this Python code for you. First, I extract the exterior coordinates of the building vector into a list of vertices. Then, I create an Open3D LineSet object, which is essentially a way to represent a polygon as a set of connected lines. To do this, I convert each 2D point into a 3D point by adding a zero z-coordinate (that's what the + (0,) does). I then define the lines of the polygon by connecting each point to the next point in sequence, with a clever trick to connect the last point back to the first point using the modulo operator: % len(vertices). This ensures the polygon is closed, creating a complete outline of the building's footprint. The resulting polygon_2d object is a light-weight representation of the building's exterior shape that can be used to further our 3D geometric processing.

Now, we have to replicate the process for the top layer. We could simply translate the object (making a copy first), or use the same approach, with a shift using the height:

```
# Generate the same element for the extruded
extrusion = o3d.geometry.LineSet()
extrusion.points = o3d.utility.Vector3dVector([point + (height,) for point in
vertices])
extrusion.lines = o3d.utility.Vector2iVector([(i, (i + 1) % len(vertices)) for
i in range(len(vertices))])
```

In the end, we use Open3D to visualize (see Figure 11-15):

```
o3d.visualization.draw_geometries([polygon_2d, extrusion])
```

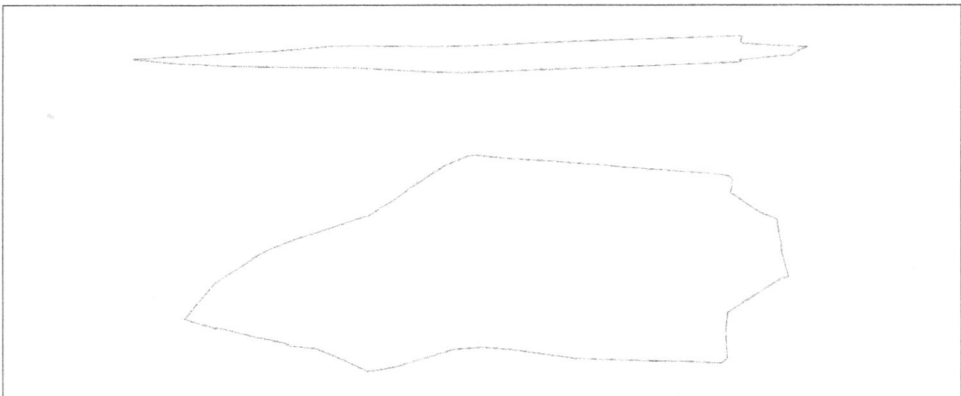

Figure 11-15. The 2D polygons from the set of vertices

Now, let me digress and show how to work with the vertices. Here is how you get to them:

```
# Plot the vertices
temp = polygon_2d + extrusion
temp.points
temp_o3d = o3d.geometry.PointCloud()
temp_o3d.points = temp.points
o3d.visualization.draw_geometries([temp_o3d])
```

We could be more efficient in vectorizing our dataset, which involves using a NumPy approach. I first define three variables: a, b, and c. a holds the coordinates of my alpha shape, b takes the alpha shape and drops the height, and c creates a single array that is the same size as a but filled with the center of the sample with height. This process allows us to define the footprint height and vertices, which are then combined and visualized:

```
# Generating the base vertices for the 3D mesh with NumPy
a = np.array(building_vector.exterior.coords)
b = np.ones([a.shape[0],1])*sample.get_center()[2]
c = np.ones([a.shape[0],1])*(sample.get_center()[2] + height)

# Define the ground footprint and the height arrays of points
ground_pc = np.hstack((a, b))
up_pc = np.hstack((a, c))

# Generate an Open3D "point cloud" made of the major points
temp_o3d = o3d.geometry.PointCloud()
temp_o3d.points = o3d.utility.Vector3dVector(np.concatenate((ground_pc, up_pc),
axis=0))
o3d.visualization.draw_geometries([temp_o3d])
```

Let's approach the 3D mesh model creation part with these lines and vertices.

3D Model Creation: Mesh

We are now ready to compute the alpha shape of the base building, and we can use Open3D to do that. Here, I pass the following code snippet:

```
# Computing the alpha shape of the 3D base points
alpha = 20
print(f"alpha={alpha:.3f}")
mesh = o3d.geometry.TriangleMesh.create_from_point_cloud_alpha_shape(temp_o3d,
alpha)
```

The alpha parameter is crucial here—it controls the "tightness" of the mesh reconstruction. By setting alpha to 20, I'm essentially telling the algorithm to create a surface that connects points within 20 units of each other. A smaller alpha would create a more conservative, tighter mesh with fewer connections between points, while a larger alpha would create a more expansive surface that includes more distant points.

Think of alpha like a "connection threshold"—points closer than this distance will be linked together to form triangles.

The alpha shape method is particularly interesting because it offers a flexible and adaptive approach to surface reconstruction. Unlike methods we've reviewed like ball pivoting or Poisson's reconstruction, the alpha shape allows us to dynamically control surface reconstruction by adjusting a single parameter. We can essentially "dial in" the level of detail we want, creating a mesh that captures the underlying geometry without being overly sensitive to outliers and sparse regions or making assumptions about point distributions that might not hold. Do not hesitate to leverage the other mesh strategies from Chapter 10 to see the differences visually.

To show your result, you can use the following:

```
mesh.compute_vertex_normals()
mesh.paint_uniform_color([0.5,0.4,0])
o3d.visualization.draw_geometries([temp_o3d, mesh, segment],
mesh_show_back_face=True)
```

Our model emerges from this. You see the vertices. If we zoom in, you can see that our point cloud is enclosed in our mesh. If you had warnings in the console, you can ignore them now; they are linked to an invalid tetra mesh within our creation. Before exporting, we want to reposition the mesh to its world coordinates. You can use Open3D's `translate` function and pass the center we just saved:

```
# Postprocessing operations & export
# Repositioning the mesh from local to the world coordinates
mesh.translate(pcd_center)
```

Then, you have the ability to export the new model, alongside the shapefile from GeoPandas:

```
# Export the mesh
o3d.io.write_triangle_mesh('../RESULTS/house_sample.ply', mesh,
write_ascii=False, compressed=True, write_vertex_normals=False,
write_vertex_colors=False, write_triangle_uvs=False)

# Export the shapefile
building_gdf.to_file("../RESULTS/single_building.shp")
```

If you open external software such as CloudCompare (see Chapter 2), you can load the input data that we had on top of the sample house and the shapefile to see the result of our little experiment (Figure 11-16).

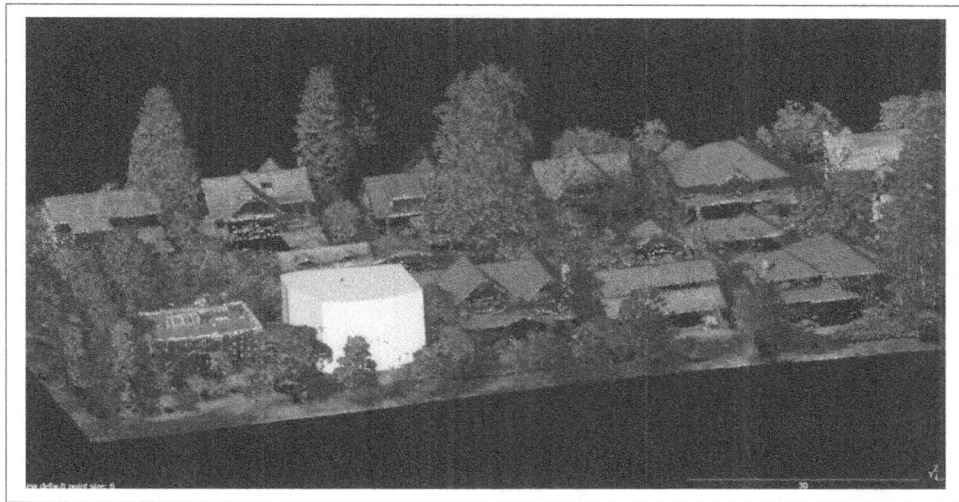

Figure 11-16. Postprocessing and export of 3D model

As you can see, our shape is nicely positioned where our building was standing. This makes sense and sticks nicely to the ground. While the geometry approximates the point cloud, our needs as real estate developers are fulfilled. This is a great time to insist that framing the goal is crucial to developing a fitted workflow. Indeed, in this case, it is better to use an alpha-shape modeling approach, which provides less accurate geometries but permits a reduction of the overhead, compared to 2.5D Delaunay or ball pivoting. We have one single building; what about automating the process for the entire neighborhood?

Phase 4: Automation and Scaling

Let's move on to phase 4: 3D automation and scaling (see Figure 11-17). Let's look at how we should approach this automation challenge. We're processing point cloud data to extract building information segment by segment. To construct our loop, we initialize our iterator as follows:

```
for sel in range(max_label+1):
```

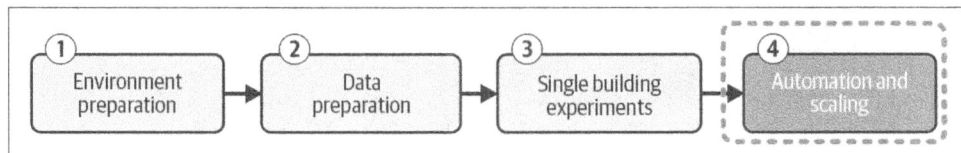

Figure 11-17. Phase 4: automating 3D model scaling and segmentation

In each iteration, we select a specific segment of point cloud data, compute its 2D building footprint using an alpha shape (with an alpha of 0.5 for shape reconstruction), and then determine the segment's height by querying nearby ground points:

```
segment = pcd_o3d.select_by_index(np.where(labels==sel)[0])
# o3d.visualization.draw_geometries([segment])

# Compute the building footprint
# altitude = np.asarray(segment.points)[:,2]+pcd_center[2]
points_2D = np.asarray(segment.points)[:,0:2]
building_vector = ash.alphashape(points_2D, alpha=0.5)

# Compute the height of the segment (house candidate)
query_point = segment.get_center()
query_point[2] = segment.get_min_bound()[2]
pcd_tree = o3d.geometry.KDTreeFlann(ground_pts)
[k, idx, _] = pcd_tree.search_knn_vector_3d(query_point, 50)
sample = ground_pts.select_by_index(idx, invert=False)
ground_zero = sample.get_center()[2]
```

We create a GeoDataFrame for each building segment, capturing key attributes like ID, height, area, perimeter, local and translated centroids, and point count. We then concatenate these building entries into a comprehensive data frame:

```
# Create the geopandas with attributes entry
building_gdf = gpd.GeoDataFrame(geometry=[building_vector], crs='EPSG:26910')
building_gdf[['id']] = sel
building_gdf[['height']] = segment.get_max_bound()[2] - sample.get_center()[2]
building_gdf[['area']] = building_vector.area
building_gdf[['perimeter']] = building_vector.length
building_gdf[['local_cx','local_cy','local_cz']] =
np.asarray([building_vector.centroid.x, building_vector.centroid.y,
sample.get_center()[2]])
building_gdf[['transl_x','transl_y','transl_z']] = pcd_center
building_gdf[['pts_number']] = len(segment.points)

# Add it to geometries entries
buildings_gdf = pd.concat([buildings_gdf, building_gdf])
```

For 3D geometry reconstruction, we generate ground and upper point clouds by extruding the 2D footprint vertically, create an Open3D point cloud, and then use alpha shape triangulation (with an alpha of 20) to generate a mesh representation of the building. We apply a random color to each mesh, translate it to its original point cloud center, and finally write the 3D mesh to a PLY file:

```
# Compute the 3D vertex geometries
a = np.array(building_vector.exterior.coords)
b = np.ones([a.shape[0],1])*sample.get_center()[2]
c = np.ones([a.shape[0],1])*(sample.get_center()[2] + height)
ground_pc = np.hstack((a, b))
up_pc = np.hstack((a, c))
temp_o3d = o3d.geometry.PointCloud()
```

```
temp_o3d.points = o3d.utility.Vector3dVector(np.concatenate((ground_pc, up_pc),
axis=0))

# Compute the 3D geometry of a house
alpha = 20
mesh = o3d.geometry.TriangleMesh.create_from_point_cloud_alpha_shape(temp_o3d,
alpha)
mesh.translate(pcd_center)
mesh.paint_uniform_color(random_color_generator())

o3d.io.write_triangle_mesh('../RESULTS/house_'+str(sel)+'.ply', mesh,
write_ascii=False, compressed=True, write_vertex_normals=False)
```

Note that I defined an optional utility function to randomly paint each building mesh with a color, which you can find here:

```
import random
def random_color_generator():
  r = random.randint(0, 255)
  g = random.randint(0, 255)
  b = random.randint(0, 255)
  return [r/255, g/255, b/255]
```

Therefore, to recap, this is what our automation loop does:

1. For each segment in our point cloud, we first select it using its label.

2. We then compute the building footprint using an alpha-shape approach, creating a simplified 2D boundary of the points.

3. We determine the ground level by finding the nearest 50 ground points to our segment's lowest point and calculating their average height.

4. We create a GeoPandas DataFrame to store building attributes like ID, height, area, perimeter, and local and translation coordinates. The height is calculated by finding the difference between the segment's maximum point and the ground level.

5. We also generate 3D vertices for each building by creating point clouds at the ground level and a specified height. These points then create a triangle mesh representing the building's geometry.

6. We translate the mesh to its original point cloud location and give it a random color for visualization.

7. Finally, we write each building's mesh to a PLY file in the ../RESULTS/ directory, allowing for later visualization.

This process allows us to systematically extract, characterize, and reconstruct individual building structures from our LiDAR point cloud dataset (see Figure 11-18).

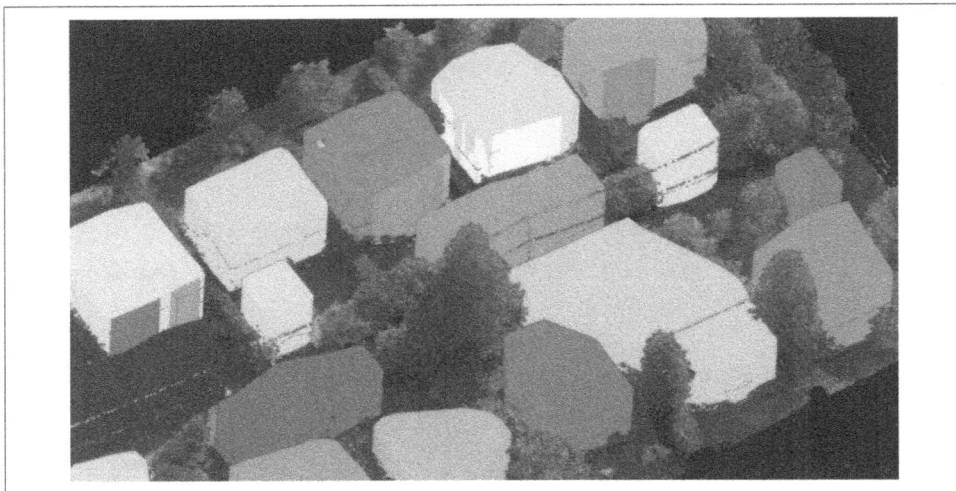

Figure 11-18. Instantiated 3D reconstructed building from the aerial LiDAR dataset

Before wrapping up, you can export the shapefile with all the footprints, with the following code snippet:

```
# Export the shapefile
building_gdf.to_file("../RESULTS/all_building.shp")
```

Now, if we look at our folder with all the houses, you can see that all of our buildings appear and instantiate so that they can be selected independently and loaded in external software.

> We would need to do some postprocessing, for example, to merge buildings, correct uncertainties, and address clustering artifacts. This is a great direct extension. How would you modify the approach to achieve sharper results?

Summary

The workflow discussed in this chapter offers a practical solution to reconstruct 3D building footprints from LiDAR point cloud data. It covers the essential steps from setting up the environment to data preparation, segmentation, vectorization, and basic postprocessing.

As a fictional real estate buyer looking to understand the patterns of residential construction in a neighborhood you are considering investing in, how did this workflow help you? Regardless of the industry you apply it to, you can now leverage Python to extract data from LiDAR point clouds. Your step-by-step solution isolates buildings from a LiDAR point cloud and extracts 2D and 3D information, including footprints, height, area, and perimeter. If I had to summarize the key takeaways, you should now feel pretty comfortable with:

- Setting up a virtual environment with Python using Anaconda
- Loading and exploring LiDAR point cloud data using libraries like laspy and NumPy
- Preprocessing LiDAR data by creating masks, transforming it into a NumPy array, and using unsupervised segmentation algorithms
- Extracting the building footprints from the point cloud data using the alpha shape algorithm
- Automating the process of extracting building footprints from multiple point cloud files
- Computing building parameters like height, area, and perimeter and export the results as 3D models and shapefiles

The resulting 3D models and shapefiles provide a rich foundation for subsequent analysis, visualization, and decision making in various applications. This automated approach empowers you to handle large datasets and easily extract accurate building information, saving time and resources.

The next stage in our 3D data science journey is to explore unsupervised segmentation techniques, such as DBSCAN. Let's move on to Chapter 12.

While the workflow provides a solid foundation, it is essential to note that more advanced segmentation techniques, like those covered in Chapter 12 and onward, might be needed for complex scenarios. Depending on the specific dataset and desired output, the postprocessing steps might also require adjustments or additional tools. Finally, I encourage you to extend the current workflow to your application, which should reveal exciting new tricks.

3D Machine Learning: Clustering

Extracting meaning from 3D data is not a simple objective. We start with massive datasets rich in geometric details from which we want to extract a semantic understanding (Chapter 1) needed for high-level tasks. The first challenge is the disparity between raw data and usable knowledge that constrains the potential of 3D machine learning across multiple disciplines, including robotics, autonomous navigation, and scene comprehension.

Indeed, while we benefit from an immense accessible corpus for training supervised models for image or text modalities, 3D openly accessible labeled datasets are rare. The lack of inherent labels makes supervised learning methods (*https://oreil.ly/ wumFO*) challenging or too expensive to implement on a large scale.

We require a method to bridge this gap without depending extensively on costly and labor-intensive human labeling. This feeds the idea to enable machines to autonomously discern the intrinsic structure within 3D data. This would enhance efficiency and allow us to address numerous difficulties. I want to teach you how to achieve this using the power of unsupervised learning through clustering algorithms (see Figure 12-1).

In this chapter, we focus on clustering through unsupervised learning mechanisms. Chapters 6 and 7 provide dimensionality reduction solutions; Chapters 8 and 9 showcase the use of regression techniques; and Chapters 14, 15, and16 are a deep dive into classification approaches.

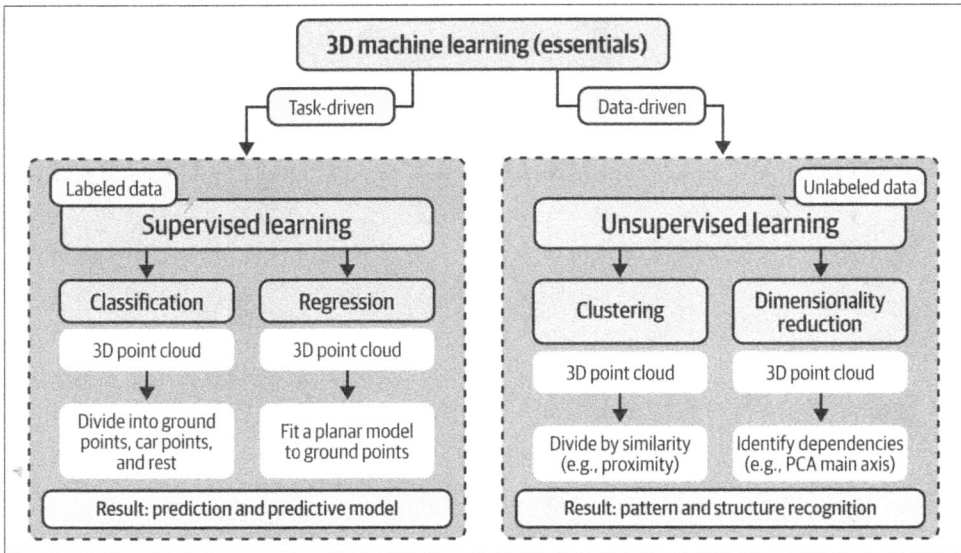

Figure 12-1. The main differences between supervised learning (https://oreil.ly/wumFO) and unsupervised learning (https://oreil.ly/bof5U) when it comes to 3D machine learning (https://oreil.ly/omXLF)

These methods facilitate categorizing analogous data points according to their intrinsic characteristics (geometry and attributes), enabling automatic object detection, scene comprehension, and subsequent semantic enhancement.

> Understanding the nature of things is beautiful. This is a trait that we share as lifelong learners. Expanding our knowledge in specific areas, like 3D data science, is a journey that we usually start with a goal in mind, but that triggers new journeys, like branches in a tree. This triggers creativity and innovation and pushes our technological evolution (among other things). This is why I am very invested and especially excited by what I share with you in this chapter. Let's walk one step closer to 3D scene understanding, but without going so close that we specialize to a point of no return.

Let's understand our 3D world beyond its physical structure and extract some kind of base meaning of the various constituting parts. Let's marry the realm of 3D geometry with a first approach to semantics, with one objective: trying to achieve a "conceptualization" of objects in a scene that can be generalized between applications.

Unsupervised learning with clustering models plays a crucial role in achieving this. 3D clustering approaches aim to partition a three-dimensional model (e.g., a point cloud) into multiple clusters or regions, each representing a distinct object or part.

They are unsupervised as they build on methods that don't require prelabeled training data. The goal is to group similar units together based on their characteristics, such as position, color, intensity, or geometric properties.

Take, for example, the challenge of analyzing a 3D point cloud acquired from a terrestrial laser scanner documenting a historical building façade. You are tasked with identifying individual architectural elements like windows, doors, columns, and decorative sculptures for documentation and restoration purposes. The raw point cloud contains millions of points, each represented by spatial coordinates (x, y, z) and attributes like color and intensity. The lack of semantic labels makes manual identification a tedious and time-consuming endeavor. We can automatically extract distinct clusters representing different architectural components by grouping similar points based on their spatial proximity, color, and geometric features.

The methods provide a high degree of autonomy and do not need any labeled datasets to function. As such, they are often used as a seed process to speed up the generation of labeled datasets for semantic segmentation tasks such as 3D object recognition. They are usually used before supervised learning and provide the ability to extract a semantic-aware structure on top of the base unit, which is promising for 3D scene understanding of digital environments.

In this chapter, we'll explore three main steps illustrated in Figure 12-2: we use (1) clustering principles with two main approaches that leverage (2) centroid-based and (3) density-based clustering.

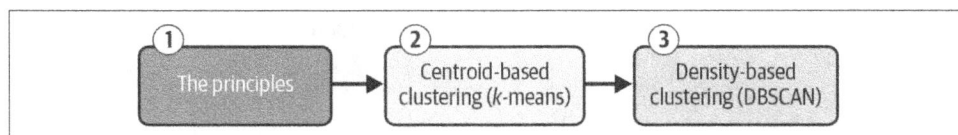

Figure 12-2. The general organization of the chapter, in three stages

We explore the mechanisms of clustering models like k-means and DBSCAN and their hybridization with RANSAC (see Chapter 9) for 3D shape recognition methods. This then lays the groundwork for further analysis and understanding, potentially leading to more advanced unsupervised segmentation methods (Chapter 13), and the injection of semantics (Chapter 14).

This chapter focuses on two clustering approaches among the ones detailed in "Types of Clustering Algorithms" on page 408. They are the most fundamental and, due to their flexibility, allow for the creation of solutions for various applications. Nevertheless, for the sake of completeness, I provide a look at connectivity-based and distribution-based (through image projection and deep learning) solutions in Chapter 13. This allows us to break down chapters into more manageable chunks. We leverage NumPy (*https://numpy.org*), Matplotlib (*https://matplotlib.org*), scikit-learn (*https://scikit-learn.org*), and Open3D (*http://open3d.org*). As always, on the 3D Data Science Resource Hub (*https://learngeodata.eu/3d-data-science-with-python*) you can find all the materials for this chapter. To access the files (code, datasets, articles) and resources, you may need to share an email address, a personal password, or proof that you are the book's owner.

Clustering for Unsupervised Segmentation

What is the reason that unsupervised segmentation and clustering are considered the primary basis of artificial intelligence? What factors should be considered when utilizing them? How can one assess or appraise performance? Let's review the principles (Figure 12-3).

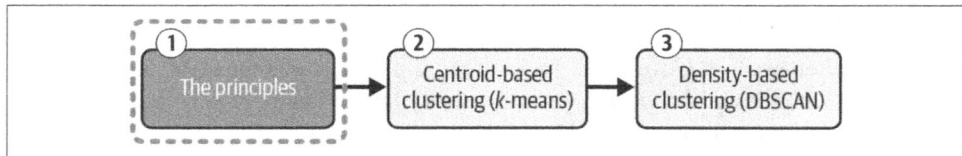

Figure 12-3. The principles of unsupervised segmentation and clustering

Clustering techniques enable the division of data into subgroups, known as clusters, without the need for supervision. These parts naturally group comparable observations together. Clustering methods rely heavily on the definition of similarity, which is often tailored to the specific field of application. For example, if we study a noisy point cloud representation of a room (see Figure 12-4), it is evident that the proximity in space is a determining factor in establishing the similarity required to form segments.

Figure 12-4. Clustering results on a point cloud representing a room, with points clustered based on spatial proximity

Clustering techniques are commonly employed for EDA. They also make up most of the procedures in AI classification pipelines, which generate well-labeled datasets in an unsupervised/self-learning manner. Let's take some time to ensure we perfectly grasp their potential.

Clustering Fundamentals

The original LeCun cake analogy presentation (*https://oreil.ly/m8Sno*), given at NIPS 2016, has been updated by highlighting specific sections (see Figure 12-5). Clustering techniques (*https://oreil.ly/BEkhw*) for unsupervised segmentation allow for creation of a segmented collection of data points that serve as the foundation for other processes, including feature extraction, classification, and 3D modeling. In addition to their use in 3D applications, they detect other entities, such as customers exhibiting similar behaviors for market segmentation purposes, users who utilize a tool similarly, communities within social networks, and recurring patterns in financial transactions.

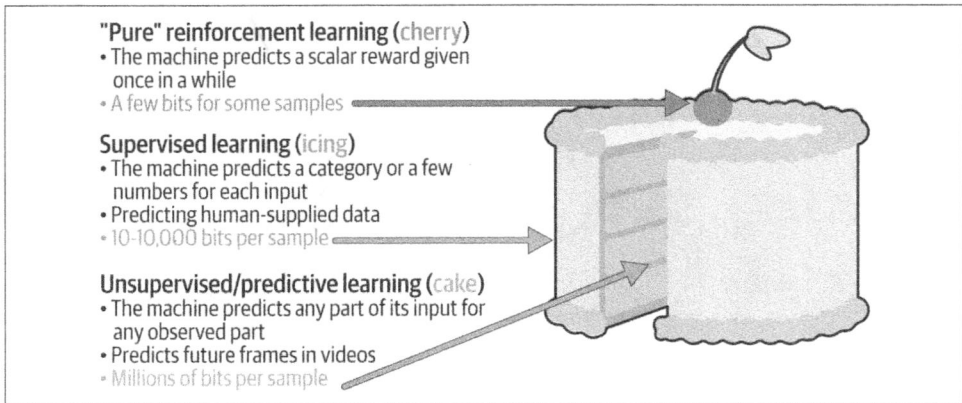

Figure 12-5. A modified "LeCun cake analogy," emphasizing the role of clustering as a foundational step in AI classification pipelines

They often function alongside dimensionality reduction technology that enables the visualization of various qualities (referred to as dimensions) in two or three dimensions. If a "view" exhibits significant decorrelation, a clustering method can be employed to create subgroups of these data points—known as clusters—as depicted in Figure 12-6.

An uncomplicated demonstration of identifying two clusters involves generating a line to divide the dataset into two subgroups. Thus, we can graphically depict the connections between the spots. Alternatively, simply one representative point per cluster might be shown instead of displaying the whole dataset. After identifying clusters, data can be viewed by selecting a single representative from each cluster and eliminating the remaining ones.

Essentially, clustering is a technique for categorizing related data points from a diverse dataset utilizing metrics such as Euclidean distance, cosine similarity, and Manhattan distance (we study this imminently). It can create arbitrary clusters with various shapes, and we can tailor algorithm choices based on their proficiency at identifying these clusters.

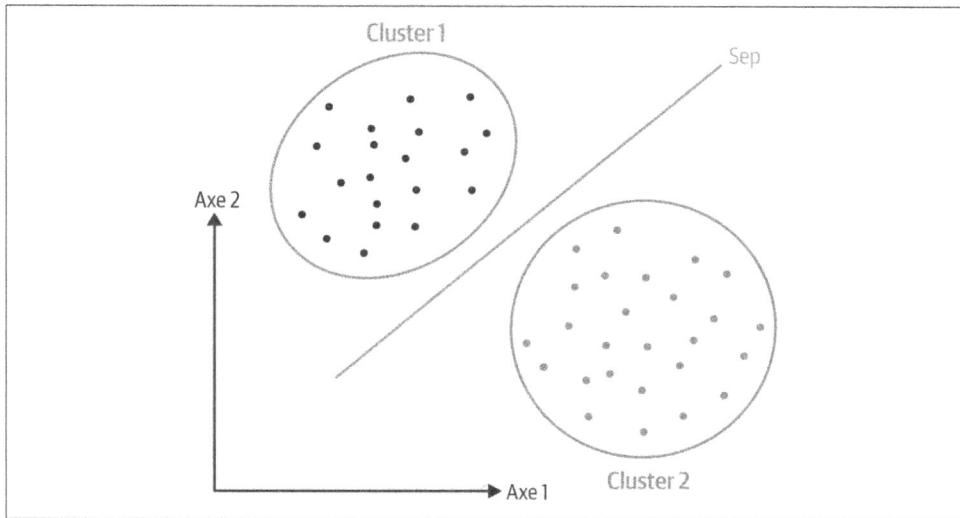

Figure 12-6. Clustering and dimensionality reduction in 3D segmentation

Clustering can be categorized into two types: hard and soft. Hard clustering allocates each data point to a specific cluster, ascertaining its membership in one cluster or another. Soft clustering assesses the likelihood of a data point's membership in multiple clusters, establishing its affiliation with either cluster. This probability is calculated for all data points.

Clustering techniques are very beneficial in situations where labeling data is costly. Consider the case of annotating a voluminous point cloud. Providing annotations for each point based on its representation can be a time-consuming and laborious task, to the extent that the individuals performing it may inadvertently introduce inaccuracies due to lack of focus or tiredness.

Utilizing a clustering algorithm to group similar points and involving a human operator solely for assigning labels to the clusters is a more cost-effective and perhaps more effective approach. As illustrated in Figure 12-7, you can see the benefit of using clusters as an entry point to labeling rather than taking and labeling each point. In this specific case, you can estimate that the labeling effort is cut by six (labeling 3 clusters instead of 18 points). These methods can be utilized to generalize a property of one point within a cluster to all other points within the same cluster (as seen with the chair object).

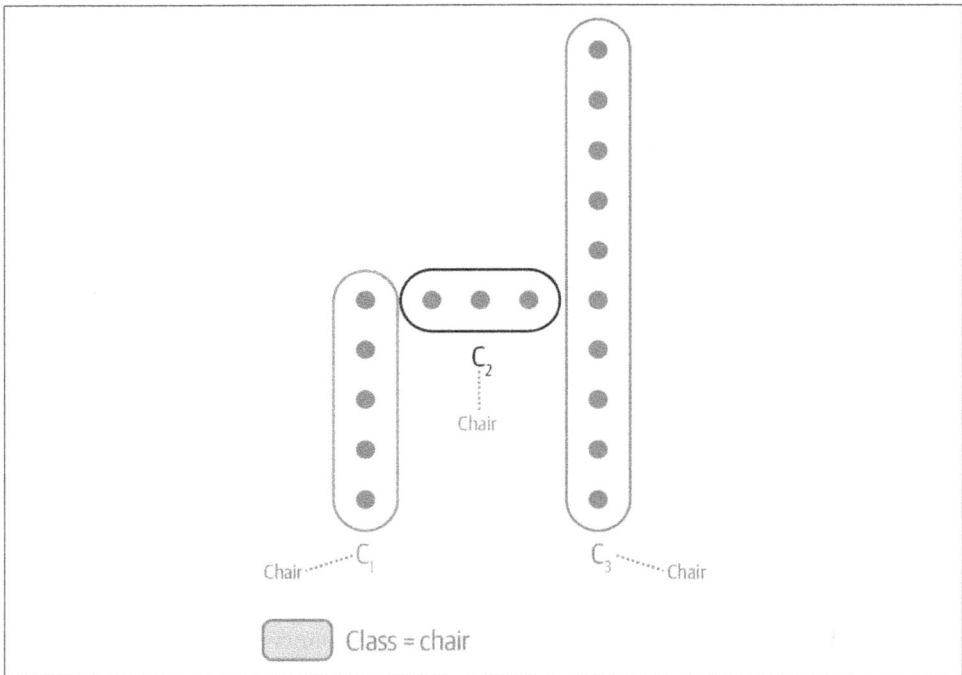

Figure 12-7. Cost-effectiveness and efficiency in semantic segmentation: a chair example

Outside the scope of geodata, inferring data properties is beneficial for several purposes. These include identifying similar images that likely depict the same object, animal, or person; extracting similar texts that likely discuss the same subject; and searching for pixels within an image that belong to the same object, known as segmentation. The preceding examples illustrate the representation of topics (pictures, text, pixels) as points in 2D, 3D, or nD space that are subsequently organized into clusters. Hence, if a single image within a cluster depicts a duck, it is probable that all the other images belonging to the same cluster also depict ducks.

We can establish multiple criteria that must be optimized to determine a compelling data division. These concepts are then employed to develop some of the most efficient clustering algorithms, which we will detail later.

Clustering Representativity

Unsupervised algorithms lack a distinct objective compared to supervised algorithms, where the task at hand, such as classification or regression, is well defined. Thus, the model's success is more subjective. Despite the challenge of defining the objective, there are three main performance measures:

- Distances and similarities
- The cluster's shape
- The stability of clusters

Let's detail them independently.

Distances and similarities

Clustering helps group together points that are near or most comparable to each other. The concept of clustering largely depends on the notions of distance and similarity. The formalization of these principles is thus highly beneficial (see Figure 12-8). We can leverage three principles:

- Proximity refers to the degree of closeness between two observations.
- Proximity also refers to the closeness between an observation and a cluster.
- Additionally, proximity can also describe the degree of closeness between two clusters.

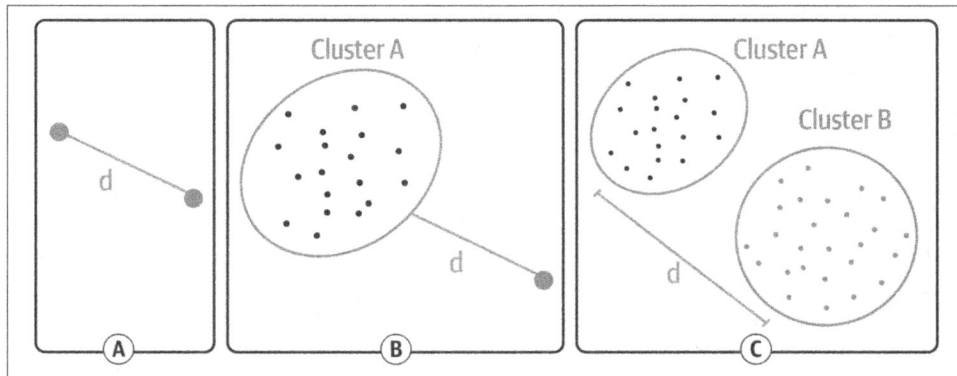

Figure 12-8. A simplified representation of the distances between (A) two individual observations; (B) an individual observation and a group of observations; and (C) two groups of observations

The Euclidean distance (*https://oreil.ly/rsl6h*) and the Manhattan distance (*https://oreil.ly/9lAep*) are commonly used as distance metrics. The Euclidean distance refers to the shortest and most direct distance between two places in Euclidean space. The Manhattan distance is named after the distance covered by a taxi on the streets of Manhattan, where all roads are either parallel or perpendicular to each other in a two-dimensional area (see Figure 12-9).

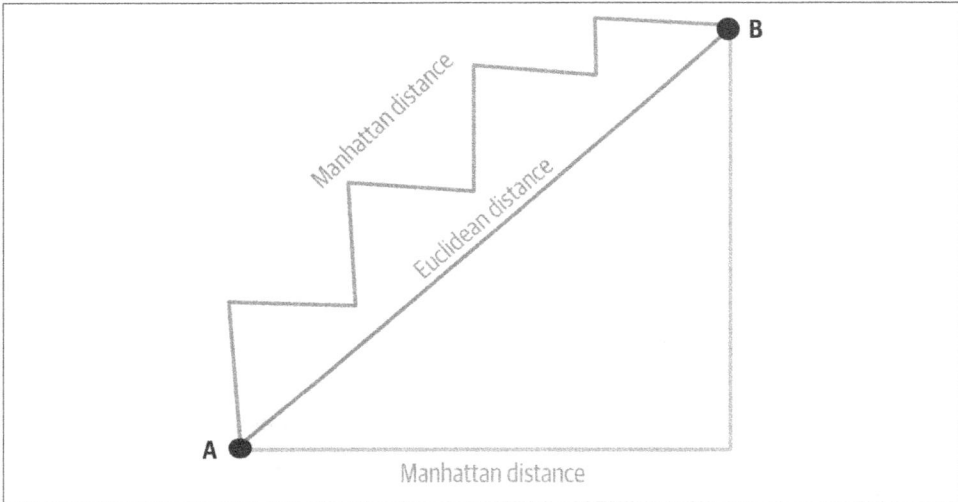

Figure 12-9. Basic depiction of the Euclidean and Manhattan distances

Hence, a measure of distance can be employed to establish a measure of similarity: the greater the distance between two places, the lower their similarity, and vice versa. To incorporate a minuscule amount of mathematics, we can effortlessly convert a distance d between x and y into a similarity measure s using the formula $s(x, y) = 1 / 1 + d(x, y)$.

One commonly utilized approach to measure similarity is the Pearson correlation (*https://oreil.ly/eUL27*). This computes the cosine of the angle between vectors x and y after the data has been centered. However, it is worth mentioning that the Pearson correlation considers the shape of the distribution rather than its amplitude, whereas the Euclidean distance primarily considers amplitude. Hence, the selection of the distance metric is crucial.

What other "tools" do we have at our disposal? One is cluster shape analysis.

Cluster shape

The cluster's shape is a crucial aspect we initially characterize (see Figure 12-10). The degree of tightness can provide a clue about the establishment of cohesive clusters. Frequently, we seek clusters that are tightly bound together. Let's illustrate these characteristics with an example, employing the Euclidean distance.

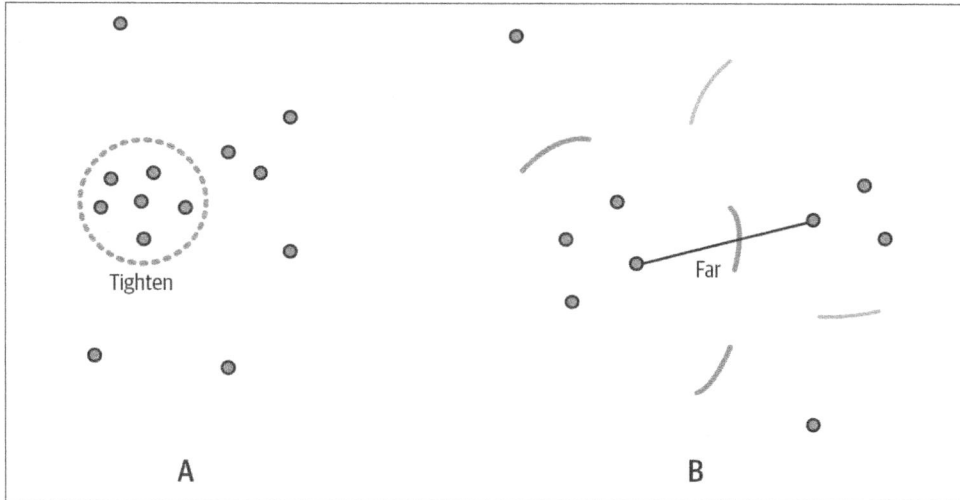

Figure 12-10. (A) When points are tightly grouped together, they must be assigned to the same cluster; (B) if two points are far apart from each other, they must be assigned to distinct clusters

Initially, it is relatively simple to calculate the centroid of a cluster, which refers to the barycenter of the points within that cluster. The homogeneity of a cluster can be defined as the mean distance from each location in the cluster to the centroid. A cluster that is tightly grouped will have reduced heterogeneity compared to a cluster consisting of points that are spread out. To characterize all clusters in our dataset, we can calculate the average homogeneity of each cluster rather than just one cluster (see Figure 12-11).

Furthermore, we desire the clusters to be situated at a considerable distance from one another. To measure this, we typically define the distance between the centroids of two clusters to quantify their separation. We can compute the mean of these quantities for all the pairs of clusters collected (see Figure 12-12).

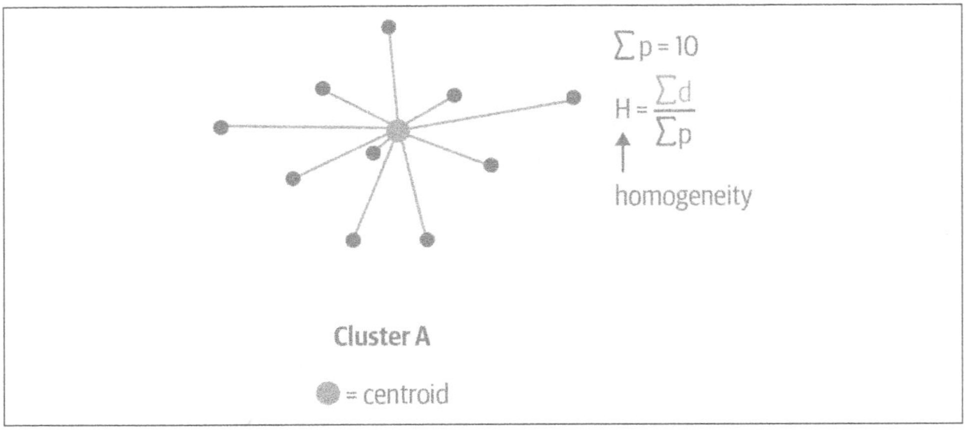

Figure 12-11. A straightforward example demonstrating how homogeneity provides an intuitive understanding for effectively characterizing clusters

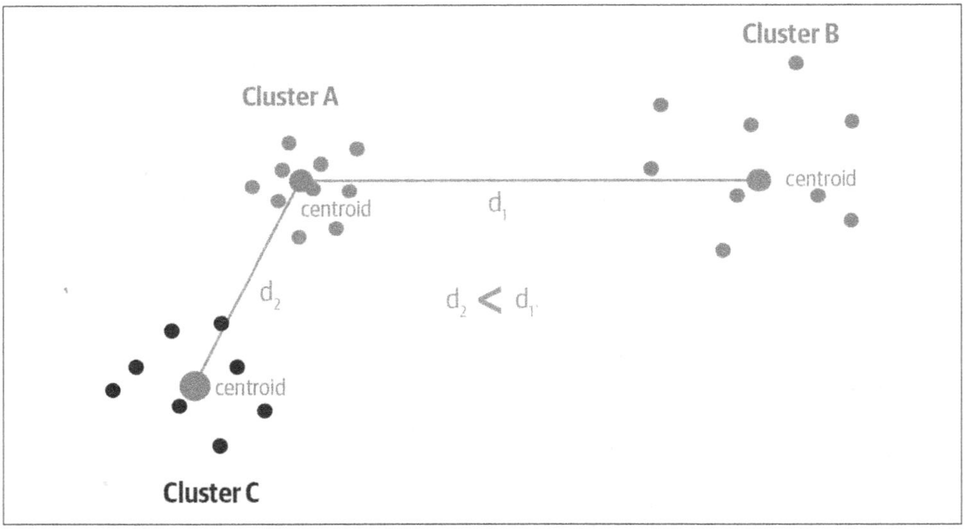

Figure 12-12. An uncomplicated depiction demonstrates how separation can achieve cohesive clustering

Currently, we have two factors to enhance: uniformity and differentiation. To simplify the process, we can categorize them under a single criterion known as the Davies–Bouldin index (*https://oreil.ly/UUbC4*).

This index measures the difference in distances within clusters (homogeneity) and between clusters (separation). We aim for low intracluster distances and large intercluster distances. This index is less effective for a given cluster when all the clusters are homogeneous and well separated.

We can use the silhouette coefficient (*https://oreil.ly/-gDPp*) as an alternate approach to assess the degree of homogeneity and separation in clustering. It is utilized to evaluate if a given point p belongs to the appropriate cluster. In this endeavor, we want to address two inquiries:

Is the proximity of p to the points in its cluster high?
The average distance, represented as $a(p)$, can be calculated by measuring the mean distance between point p and all other points inside its cluster.

Is the distance between this place and the other points significant?
We determine the minimum value of $b(p)$ that $a(p)$ could have if p were reassigned to a different cluster. Assuming p has been accurately allocated, it follows that $a(x)$ is less than $b(x)$. The silhouette coefficient, denoted as $s(x)$, is calculated using the formula $s(x) = b(x) - a(x) / max(a(x), b(x))$. It is a numerical value that falls within the range of -1 to 1. The assignment of p to its cluster is more satisfactory as the value approaches 1.

> There are other ways to assess the quality of the clustering results, with other metrics that measure the coherence within clusters and the separation between clusters, but this is beyond the scope of the book. Nevertheless, for the sake of completeness, allow me to give you two approaches: the Rand index (*https://oreil.ly/HtAhJ*) and the adjusted Rand index (*https://oreil.ly/nGeW1*). The Rand index compares how pairs of data points are grouped in the predicted cluster versus the "true" cluster. It provides a single score that indicates the proportion of agreements between the two clusters. The adjusted Rand index is a variation that adjusts for probabilities when evaluating the similarity between two clusters of data.

To assess clustering quality, one can compute its mean silhouette coefficient. This can be done using the scikit-learn (*https://scikit-learn.org*) library and the `sklearn.metrics.silhouette_score` (*https://oreil.ly/fx2T3*) function. But let's keep that in mind for later and move on to the third aspect to monitor: the stability of clusters.

Cluster stability

Another crucial aspect to consider is the uniformity of the grouping. Will the procedure yield consistent results when applied iteratively to the same data with different initializations, or when used with distinct subsets of the data, or when applied to data that contains some level of noise?

This criterion is especially significant for determining the number of clusters. If the chosen number of clusters aligns with the data's inherent structure, the clustering will exhibit more stability than when it does not.

To this end you can use the cluster reassignment rate (*https://oreil.ly/AMv-w*) to quantify stability. We calculate this by tracking the percentage of points that change cluster membership across multiple computational iterations. A low reassignment rate (typically below 5–10%) indicates robust clustering, suggesting that the spatial configuration remains consistent under slight algorithmic perturbations. This methodology involves generating multiple clustering solutions using controlled stochastic variations and then computing the proportion of points that migrate between clusters.

If you want to tie that to the centroid, then studying centroid movement (*https://oreil.ly/1tdn_*) between iterations provides a precise quantitative lens into cluster structural integrity. We measure this by calculating the Euclidean distance traversed by cluster centroids across different computational runs. Small centroid movements (less than 1–2% of the dataset's spatial extent) signify stable cluster geometries. This approach involves tracking centroid trajectories in high-dimensional spaces, employing distance metrics that can capture both translational and rotational variations. This technique allows us to distinguish between genuine cluster shifts and computational noise.

Finally, what I have found helpful is the use of cluster boundary variance as the most nuanced stability assessment technique for 3D datasets. We analyze this by examining the morphological changes in cluster boundaries across iterations, computing metrics like boundary point density fluctuations and local geometric deformation (you can use the Chapter 10 learnings on modeling to extract the shape for geometric boundaries). A key indicator includes the standard deviation of boundary point distributions of cluster interfaces. Stable clusters maintain consistent boundary characteristics.

These techniques transform abstract clustering metrics into interpretable spatial representations, allowing us to assess our clustering quality when needed. Let me quickly illustrate that on a dataset for which we have a well-defined data division, such as Figure 12-13. As an illustration, we can manipulate a collection of points organized into distinct planar shapes. The subsequent phase involves assessing if the clusters generated by the clustering algorithm align with the preestablished groupings.

Figure 12-13. Extracting a subset of a point cloud and generating a dataset annotated with planar information for the purpose of comparing it with clustering results

As a last stage to perfectly understand the principles of clustering for 3D unsupervised segmentation, let's define compatibility with domain-specific knowledge.

The Case of Oversegmentation and Undersegmentation

In the complex landscape of 3D data science, I need to highlight a key concept for relevant segmentation tasks. Oversegmentation (*https://oreil.ly/6ChzO*) and undersegmentation (*https://oreil.ly/c0Dqb*) represent critical challenges that fundamentally impact our ability to extract meaningful spatial information.

Oversegmentation occurs when clustering algorithms fragment a coherent object into excessive, granular segments, creating an artificially fragmented representation that obscures the underlying geometric structure. This phenomenon typically emerges when clustering algorithms prioritize local point variations over global object coherence.

Undersegmentation represents the inverse challenge, where distinct geometric entities are incorrectly merged into a single cluster (we observe a "bleeding" over natural frontiers). This occurs when clustering algorithms fail to capture nuanced spatial variations, resulting in oversimplified representations that collapse multiple distinct objects or structural components into a single, inadequate segment.

Current research demonstrates that undersegmentation can lead to catastrophic information loss, particularly in complex 3D environments like urban mapping, geological surveys, or medical imaging. To address these challenges, you can use a multi-scale clustering approach that dynamically adapts segmentation granularity based on local geometric complexity. This involves implementing adaptive clustering algorithms incorporating local point neighborhood characteristics and global spatial context. We can effectively mitigate oversegmentation and undersegmentation risks by developing sophisticated metrics that balance point-wise variance with geometric continuity.

While these concepts are beyond the book's scope, some exciting techniques include developing robust boundary detection mechanisms that preserve spatial relevance.

Types of Clustering Algorithms

Let's take a deeper dive. Clustering assists in the processing of unstructured data. Graphing, the shortest distance, and the density of the data points are a few of the elements that drive cluster formation.

We want to establish a process that assesses how connected objects are based on our similarity measure. While similarity measurements are easier to discover in smaller collections of features, it gets tougher to generate similarity metrics as the number of features increases.

Depending on the type of clustering algorithm being utilized, numerous strategies are employed to group the data from the datasets. Let's review the various types of clustering algorithms that you can leverage:

Centroid-based clustering (partitioning methods)
> The algorithms that fall into this category cluster data points by proximity. They employ three main similarity measures (Manhattan, Minkowski, Euclidean distances) and separate the input into a predetermined number of clusters. This algorithm's fundamental problem is that we must guess or use the elbow method to determine k, the number of groups. Only then can the clustering algorithm allocate data points. Centroid-based clustering thrives in scenarios with well-defined, spherical, or ellipsoidal cluster geometries. Algorithms like k-means (*https://oreil.ly/cMsnN*) excel when spatial distributions exhibit compact, roughly equidistant point configurations for the implementation for 3D data. Critical limitations emerge when dealing with clusters of irregular shapes or varying densities, complicating preprocessing and initial parameter selection.

Density-based clustering (model-based methods)
> This method finds clusters based on the density of data points: it represents my preferred approach for complex, nonlinear spatial distributions. Density-based clustering automatically calculates the number of clusters and is less sensitive to

their initial seed than centroid-based clustering. They perform well with clusters of all shapes and sizes, making them ideal for datasets with overlapping clusters. These approaches can handle dense and sparse data and distinguish groups with varied shapes by focusing on local density. The best-known density-based clustering approach is DBSCAN: it provides robust segmentation by prioritizing point density and local neighborhood characteristics. DBSCAN methods dynamically adapt to cluster shapes, effectively handling noise and capturing subtle geometric transitions, solving centroid-based difficulties. "DBSCAN for Unsupervised Segmentation" on page 421 delves more into this approach later in the chapter.

Connectivity-based clustering

Connectivity-based clustering emerges as a powerful approach for capturing hierarchical spatial relationships. The algorithms of this family focus on understanding how points interconnect, revealing intricate geometric structures through their neighborhood connections. Connected components represent a fundamental construct in this approach, where we identify spatially coherent point clusters based on proximity and neighborhood connectivity. We implement this through graph-theoretical algorithms (*https://oreil.ly/nnOlO*) that traverse point cloud neighborhoods, establishing cluster boundaries by examining local point relationships. Critical to this method is defining a connectivity threshold that balances local point interactions with global spatial context. In the last section of this chapter, we provide an implementation that leverages k-NN graphs and epsilon-neighborhood connections to decompose complex 3D datasets systematically.

Distribution-based clustering

Distribution-based clustering groups data points by their probability of belonging to the same probability distribution (Gaussian, binomial). The methodology leverages statistical distributions to capture complex geometric patterns that traditional clustering techniques cannot adequately represent. Gaussian mixture models (GMMs) are the primary computational mechanism for distribution-based clustering. They model 3D data as a combination of multiple probabilistic Gaussian distributions, enabling nuanced representation of spatial complexity. This approach excels in scenarios with overlapping or intricate geometric configurations, such as those in medical imaging. By assigning probabilistic cluster memberships, we can capture subtle spatial variations that deterministic clustering methods typically overlook. The main issue is in overfitting distribution models.

Table 12-1 shows the key points of each family of techniques.

Table 12-1. The various clustering approaches for 3D datasets

Clustering approach	Best use cases	Most used algorithm	Key characteristics
Centroid-based	Uniform geometric distributions, compact spatial structures, urban infrastructure mapping	k-means	Fixed number of clusters, spherical cluster shapes, sensitive to outliers
Density-based	Irregular point distributions, noisy environments, complex geological surveys	DBSCAN	Adapts to nonlinear shapes, handles noise effectively, captures local density variations
Connectivity-based	Nested spatial relationships, multiscale analysis, architectural and forest canopy mapping	Connected components	Reveals hierarchical structures, no predefined cluster number, preserves spatial semantics
Distribution-based	Probabilistic spatial modeling, medical imaging, biomechanical analysis	Gaussian mixture models	Probabilistic cluster memberships, captures statistical variations, handles complex data distributions

Now that you understand the various ways we can leverage clustering algorithms for 3D unsupervised segmentation, let's create solutions with Python.

> This chapter, as outlined in the introduction, delves into the initial two clustering methodologies: centroid-based and density-based approaches. Indeed, they represent the most fundamental elements and, owing to their inherent flexibility, facilitate the development of solutions. For the sake of thoroughness, I present an in-depth solution to connectivity-based and distribution-based approaches (utilizing image projection and deep learning) in Chapter 13.

First, let's leverage *k*-means clustering.

k-Means Clustering

k-means is a very simple and popular algorithm for computing a scene "clustering." Let's leverage the technique for generalizable approaches in this section (Figure 12-14).

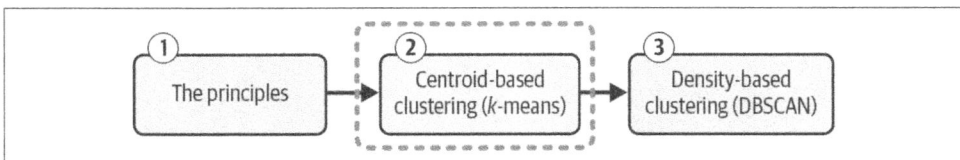

Figure 12-14. Centroid-based clustering with k-means

It is typically an unsupervised process, so we do not need any labels, as in classification problems. But what data points do we want to cluster?

These can be arbitrary points, such as 3D points recorded with a LiDAR scanner. But they can also represent spatial coordinates, color values in your dataset (mesh, point cloud, etc.), or other features such as keypoint descriptors extracted from an image to build a bag of words dictionary. You can see them as arbitrary vectors in space, each holding a set of attributes. Then, we gather many of those vectors in a defined "feature space," and we want to represent them with a small number of representatives.

The only thing we need to know is a distance function that tells us how far two data points are apart. In the simplest form, this is the Euclidean distance. But depending on your application, you may also want to select a different distance function, as seen earlier. Then, we can decide if two data points are similar to one another and if they belong to the same cluster.

This permits abstracting and representing all the data points with k representatives, which gave the algorithm its name. So, k is a user-defined number that we put into the system. For example, take all the data points and represent them with three points in space (see Figure 12-15).

Figure 12-15. k-means clustering, showing how data points are assigned to clusters based on their proximity to cluster centroids

k-means works as follows: first, we have some data points in feature space (x, y, and z in the Euclidean space). Then, we compute the k representative and run k-means to assign data points to a cluster represented by the representative.

In the preceding example above, we set $k = 3$. That means we want to represent all the data points with three representatives. The circled points illustrate those representatives, orienting the corresponding assignments of data points to the "best" representatives. We then obtain three clusters of points.

k-means does it in a way that minimizes the squared distance between the data point and its closest representative. The algorithmic part that achieves this is constructed in two simple steps iteratively remade, namely (1) the initialization and (2) the assignment:

1. We randomly initialize k centroid points as representatives and compute a data association of every data point to the closest centroid. So, we are doing a nearest neighbor query here.

2. Every data point is assigned to its closest centroid, and then we reconfigure each centroid's location in our space. This is done by computing the primary vector out of all the data points assigned to that centroid, which changes its location.

In the next iteration of the algorithm, we get a new assignment and then a new centroid location, and we repeat this process until convergence.

> We should note that k-means is not an optimal algorithm. This means that k-means tries to minimize the distance function, but we are not guaranteed to find a global minimum. So, depending on your starting location, you may end up with a different result for your k-means clustering. Suppose we want to implement k-means quickly. In that case, we typically need to have an approximate nearest-neighbor function in our space because this is the most time-consuming operation with this algorithm.

k-means is thus a relatively simple two-step iterative approach to finding representatives for a potentially large number of data points in high-dimensional spaces.

k-Means: Workflow Definition

You are tasked with monitoring a vast airspace. The challenge lies in efficiently detecting and identifying individual stationary aircraft within a massive LiDAR point cloud. This daunting task is crucial for ensuring airspace security and preventing potential threats. By analyzing the 3D point cloud data, we can identify and locate aircraft, assess their size and type, and monitor their movements over time. This information can be used to detect unauthorized aircraft, identify potential security risks, and respond to emergencies promptly.

By tracking the changes in aircraft positions and configurations, we can detect suspicious activity or potential threats early on, allowing security personnel to take appropriate measures to protect critical infrastructure and personnel. You will build an automated approach that answers these challenges, saving valuable time and resources while minimizing human error risk and allowing for a proactive response to potential security threats.

The first step is to gather a nice dataset! This time, I want to share another excellent place to find nice LiDAR datasets: the Geoservices Catalog of the National Geography Institute of France (*https://oreil.ly/7iXSz*). The LiDAR HD campaign (*https://oreil.ly/BDgQg*) provides an open data portal where you can get crisp 3D point clouds of some regions of France. For this section, I went on the portal, selected one tile, extracted a subtile from it, deleted the georeferencing information, prepared some extra attributes for the LiDAR file, and then made it available in the chapter's folder (see steps 1 to 3 in Figure 12-16). The data you are interested in is *KME_planes.xyz* and *KME_cars.xyz*. I propose to follow a simple procedure that you can replicate to label your point cloud datasets, as illustrated in Figure 12-16.

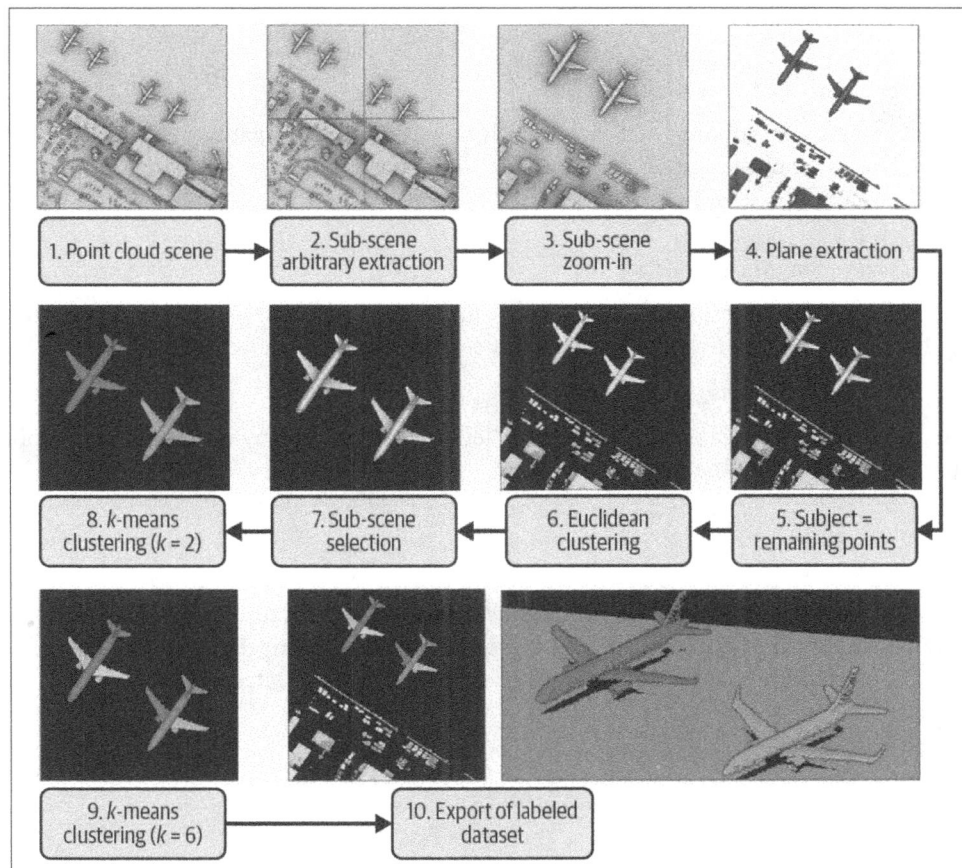

Figure 12-16. Semisupervised workflow to label 3D point cloud datasets

This section covers steps 4 to 10, the others being already part of your new skills based on previous chapters. Now that the theory is over, let's dive into the Python implementation.

3D Python Context Definition

As always, focusing on efficient and minimal library usage is nice. We could do it all with libraries like Open3D or PyTorch3D, but for the sake of mastering 3D Python, let's leverage only NumPy, Matplotlib, and scikit-learn. Six lines of code to start your script:

```
#numpy
import numpy as np

#matplotlib
import matplotlib.pyplot as plt
from mpl_toolkits import mplot3d
# Scikit-learn
from sklearn.cluster import KMeans
from sklearn.cluster import DBSCAN
```

From there, I propose that we relatively express our paths, separating the data_folder containing our datasets from the dataset name to switch easily on the fly:

```
data_folder="../DATA/"
dataset="KME_planes.xyz"
```

Then, let me illustrate a nice trick to load your point cloud with NumPy. The intuitive way would be to load everything in a point cloud variable, such as:

```
pcd=np.loadtxt(data_folder+dataset)
```

But because we will play with the features, let's save some time by unpacking each column in a variable on the fly:

```
x, y, z, illuminance, reflectance, intensity,
nb_of_returns = np.loadtxt(data_folder+dataset,skiprows=1, delimiter=';',
unpack=True)
```

We now use variable names for the downward processes. Because of the essence of k-means, we have to be careful with the ground element omnipresence, which would provide something that we do not specifically expect, as illustrated in Figure 12-17.

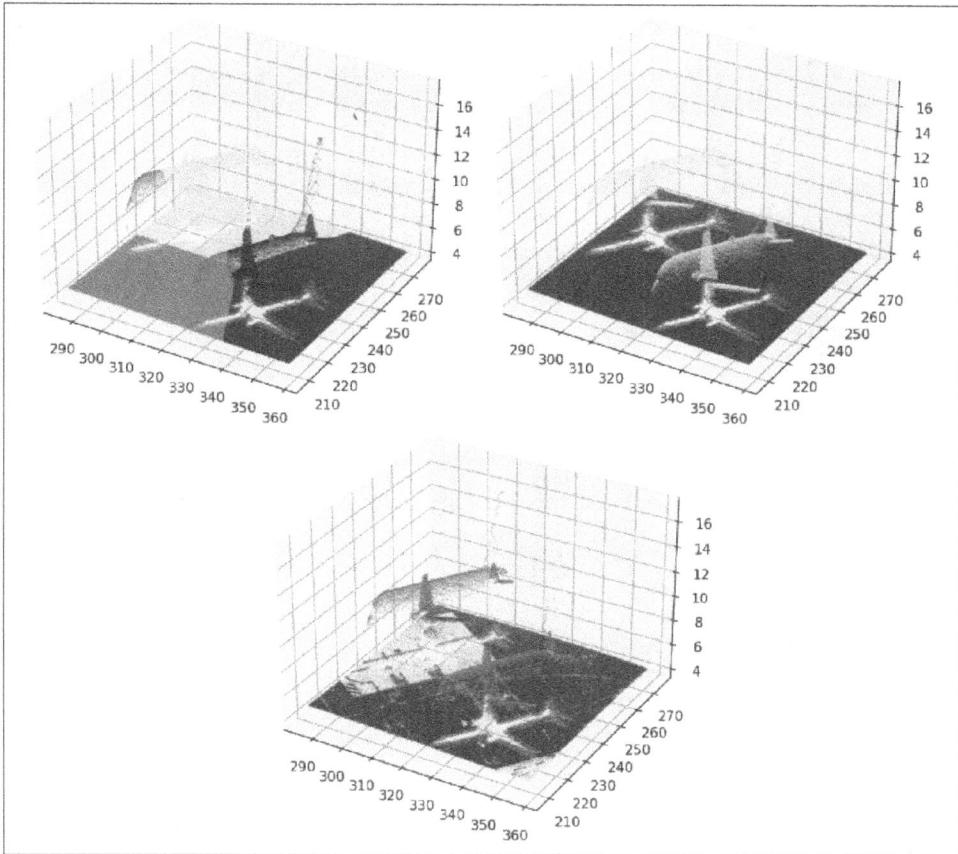

Figure 12-17. Unpacking point cloud data with NumPy for k-means segmentation

In Figure 12-17, notice the regular triangular separation on the first image using only spatial attributes. It's not ideal, as you cannot gather each plane independently. To avoid these results, we should handle what we consider outliers to detect the planes: the ground. At this stage, you should have some clear ideas for getting rid of the ground plane (i.e., you can leverage RANSAC or even *k*-means with a seed of 2). Let me show you an analytical base for achieving this (see Chapter 8 for the complete approach).

LiDAR Data Preprocessing

Let's look at how to handle this with a supervised step using Matplotlib. This also allows me to give you pieces of code that always come in handy with subplot creation and line layering.

We will monitor on a 2D plot where the mean value of our points falls and see if this can be useful to filter out the ground in a later step. First, let's make a subplot element that will hold our points on an x-z view and also plot the mean value of our spatial coordinates:

```
plt.subplot(1, 2, 1) # row 1, col 2 index 1
plt.scatter(x, z, c=intensity, s=0.05)
plt.axhline(y=np.mean(z), color='r', linestyle='-')
plt.title("First view")
plt.xlabel('X-axis')
plt.ylabel('Z-axis')
```

If you look within the code, I use the intensity field as the coloring element for our plot. I can do this because it is already normalized at an [0,1] interval. The s stands for size and permits us to give a size to our points. Then, let's do the same trick, but this time on the y-z axis:

```
plt.subplot(1, 2, 2) # index 2
plt.scatter(y, z, c=intensity, s=0.05)
plt.axhline(y=np.mean(z), color='r', linestyle='-')
plt.title("Second view")
plt.xlabel('Y-axis ')
plt.ylabel('Z-axis ')
```

From there, we can draw the plot using the following command:

```
plt.show()
```

The result is shown in Figure 12-18. The line, which represents the mean value, is positioned in a way such that we can use this information to filter out the ground element with a minimalistic approach. If you want a more robust solution, I encourage you to use the percentile and interquartile distribution to ensure you are robust to outliers.

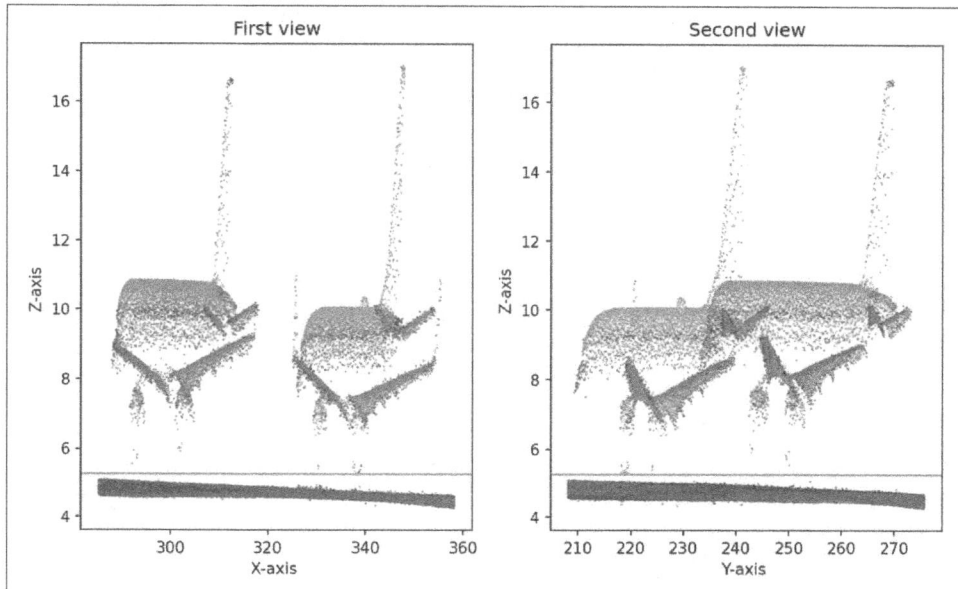

Figure 12-18. Visualizing the y-z axis of a point cloud with ground element filtering

Now, we want to find a mask that permits us to eliminate points that do not satisfy the "non-ground selection" query. The query we are interested in only considers points with a z value above the mean, with z>np.mean(z).

We can store the results in our variable spatial_query:

```
pcd=np.column_stack((x,y,z))
mask=z>np.mean(z)
spatial_query=pcd[z>np.mean(z)]
```

The column_stack function of NumPy is super handy but be careful because it can create overhead if applied to too big of a vector. Nevertheless, playing around with a set of feature vectors makes it very useful.

Now, let's plot the results, this time in 3D, with the following commands:

```
# Plotting the results in 3D
ax = plt.axes(projection='3d')
ax.scatter(x[mask], y[mask], z[mask], c = intensity[mask], s=0.1)
plt.show()
```

And again, if you want a top view well adapted with our LiDAR HD data:

```
# Plotting the results in 2D
plt.scatter(x[mask], y[mask], c=intensity[mask], s=0.1)
plt.show()
```

As you can see in Figure 12-19, we got rid of the outlier points. We can now focus on these two planes and try to attach semantics to each one.

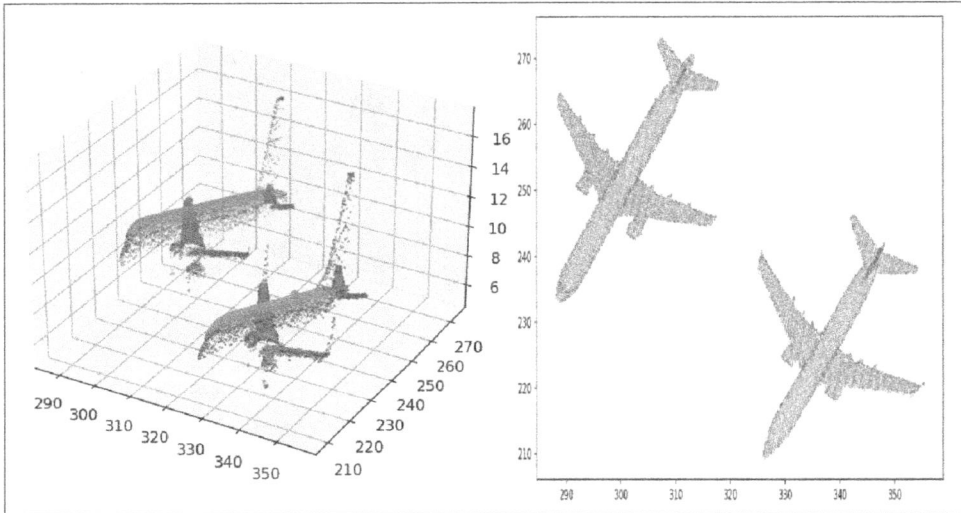

Figure 12-19. 3D and 2D visualization of LiDAR data after filtering outliers

k-Means Implementation

The construction of the high-level scikit-learn library is a great asset for leveraging algorithms efficiently. In as little as one line of code, we can fit the clustering k-means machine learning model. I will emphasize the standard notation, where our dataset is usually denoted "X" to train or fit on. In this first case, let's create a feature space holding only the X and Y features after masking:

```
X=np.column_stack((x[mask], y[mask]))
```

From there, we can run our k-means implementation, with $k = 2$, to see if we can retrieve the two planes automatically:

```
kmeans = KMeans(n_clusters=2).fit(X)
plt.scatter(x[mask], y[mask], c=kmeans.labels_, s=0.1)
plt.show()
```

We retrieve the ordered list of labels from the *k*-means implementation by calling the .labels_ method on the sklearn.cluster._kmeans.KMeans object. This means that we can directly pass the list to the color parameter of the scatter plot. We retrieve the two planes correctly in two clusters. Increasing the number of clusters (*k*) will provide different results that you can experiment on (see Figure 12-20).

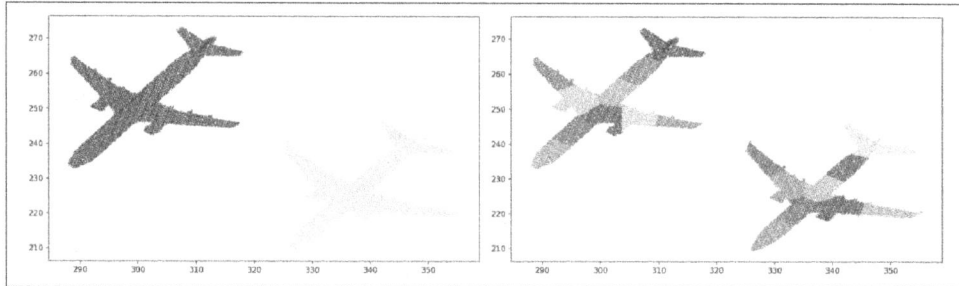

Figure 12-20. 3D point cloud instance segmentation versus part segmentation, given by k-means parametrization

Choosing the correct number of clusters may not be initially so obvious. We can use the elbow method if we want to have some heuristics that help decide this process in an unsupervised fashion. We are playing with parameter *k* in the elbow method, which is the number of clusters we want to extract.

To implement the method, we can loop *k* over a range of a specific range, e.g., 1 to 20, execute our *k*-means clustering with the *k* parameter, and compute a WCSS (within-cluster sum of squares) value that we will store in a list:

```
X=np.column_stack((x[mask], y[mask], z[mask]))
wcss = []
for i in range(1, 20):
  kmeans = KMeans(n_clusters = i, init = 'k-means++', random_state = 42)
  kmeans.fit(X)
  wcss.append(kmeans.inertia_)
```

The init argument is the method for initializing the centroid, which we set to k-means++ for clustering with an emphasis on speeding up convergence. The WCSS (*https://oreil.ly/tBlN6*) value generated from kmeans.inertia_ indicates the total sum of squared distances between each data point and the centroid within a cluster.

> If you paid attention to the details of the *k*-means line, you might have wondered: why 42? Well, for no smart reason. The number 42 is an ongoing joke in the scientific community and is derived from the legendary book *The Hitchhiker's Guide to the Galaxy* (*https://oreil.ly/6IFwJ*), wherein an enormous supercomputer named Deep Thought calculates the answer to the "Ultimate Question to Life...."

Then, once our WCSS list is complete, we can plot the graph `wcss` depending on the k value, which kind of looks like an elbow (maybe this has to do with the name of the method?):

```
plt.plot(range(1, 20), wcss)
plt.xlabel('Number of clusters')
plt.ylabel('WCSS')
plt.show()
```

This looks like magic, as we see that the value that creates the elbow shape is situated where the number of clusters of 2, which makes perfect sense (see Figure 12-21).

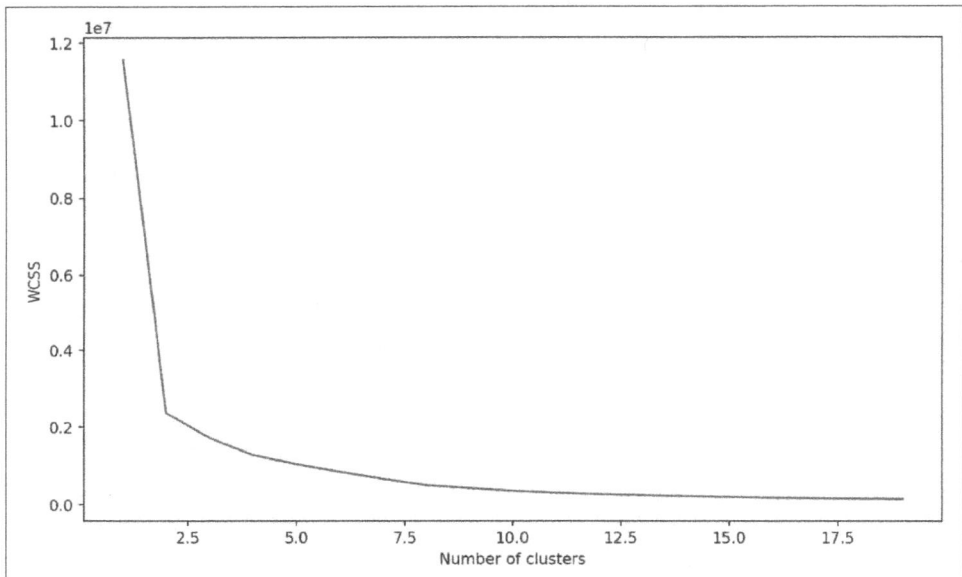

Figure 12-21. WCSS plot for determining optimal k in k-means clustering

Now, you can leverage centroid-based clustering with k-means for 3D point clouds. On top of that, you can automatically leverage the elbow method to find the best parameter for a specific 3D dataset. Finally, we are just left with exporting the data to a coherent structure, for example, an *.xyz* ASCII file, holding only the spatial coordinate with label information that can be read in external software:

```
result_folder="../DATA/RESULTS/"
np.savetxt(result_folder+dataset.split(".")[0]+"_result.xyz",
np.column_stack((x[mask], y[mask], z[mask],kmeans.labels_)), fmt='%1.4f',
delimiter=';')
```

You have learned to create an automated semisupervised segmentation using *k*-means clustering. This technique is valuable in cases where semantic labeling is absent alongside the 3D data. You can deduce semantic information by examining inherent geometrical patterns within the data. Now, let's explore a superb spatial clustering approach: DBSCAN (see Figure 12-22).

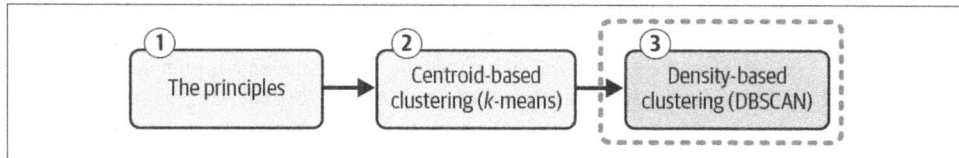

Figure 12-22. Stage 3: Density-based clustering with DBSCAN

DBSCAN for Unsupervised Segmentation

After our case on airport security, we want to develop an autonomous robot that can navigate a complex indoor environment like a kitchen. The robot needs to understand the spatial layout and the objects within the scene to perform tasks like fetching a specific item or cleaning a surface. Detecting individual planar constituents, such as walls, floors, and countertops, is fundamental to this understanding. By identifying these planes, the robot can establish a precise 3D map of the environment, enabling accurate localization, navigation, and object interaction. Without this knowledge, the robot would struggle to avoid obstacles, plan efficient paths, and execute tasks reliably.

By accurately segmenting the scene into planar regions and main constituents, we empower our robots to operate safely and efficiently. This is our goal for this project.

And now that you have experience working with point clouds (or any data), you know the significance of identifying patterns among your observations. Often, we require extracting higher-level knowledge that significantly depends on identifying "objects" produced by data points that exhibit a pattern (see Figure 12-23).

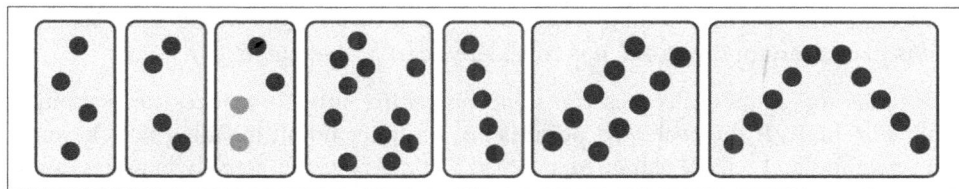

Figure 12-23. The visual patterns progress from left to right: ungrouped, proximity, similarity, common cluster region, linearity, parallelism, and symmetry (reference: Poux et al., 2019)

Our visual cognitive system easily accomplishes this task. Nevertheless, replicating this human talent through computational approaches poses a significant challenge. Essentially, we aim to use the inherent tendency of the human visual system to categorize collections of things.

The primary motivations for point cloud segmentation can be categorized into three distinct factors. Firstly, it allows end users to efficiently access and manipulate specific content using higher-level abstractions called segments. Furthermore, it generates a condensed depiction of the data, enabling subsequent processing to be conducted at a regional level rather than at the level of individual points. This can lead to substantial computing advantages. Ultimately, it provides the ability to derive connections between neighborhoods, graphs, and topologies, a feature absent in unprocessed point-based datasets. Segmentation is primarily a preliminary step in annotating, enriching, analyzing, classifying, categorizing, extracting, and summarizing information from point cloud data.

Which method can we use to accomplish this task? Let me introduce DBSCAN's principles.

DBSCAN Principles

When working with point cloud datasets, it is frequently necessary to group collections of points that seem physically "contiguous." However, what is the most effective way to accomplish this?

The DBSCAN algorithm was initially presented in 1996 to fulfill this particular purpose. This algorithm has gained significant popularity and recognition, leading to a prestigious scientific contribution award in 2014, and has remained highly regarded.

DBSCAN sequentially processes each point in the dataset. For every point it examines, the algorithm creates a list of points that can be reached based on their density: it calculates the neighborhood of each point, and if this neighborhood contains more points than a certain threshold, it is considered part of the region.

Each adjacent point undergoes the same procedure until it reaches its maximum expansion limit. If the evaluated point is not an inlier point, meaning it lacks sufficient neighbors, it will be classified as noise. DBSCAN can effectively handle outliers by isolating them through this approach (see Figure 12-24).

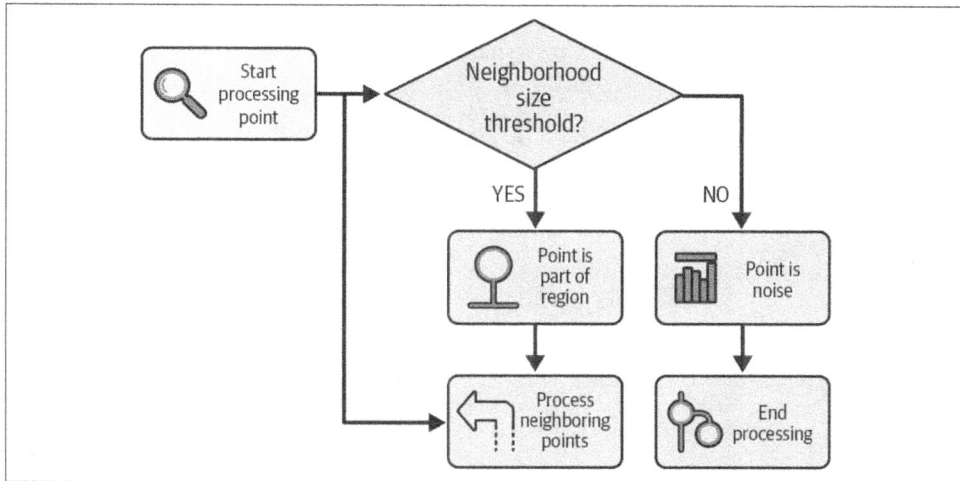

Figure 12-24. Visual representation of Euclidean clustering using DBSCAN: identifying contiguous point groups in a point cloud dataset

However, it is essential to mention the use of parameters. Determining the appropriate parameters (ε for the neighborhood and n_min for the minimum number of points) might also be challenging.

It is crucial to exercise caution while establishing parameters to ensure an adequate number of inlier points are generated. This may not occur if the value of n_min is excessively big or if ε is excessively small. Specifically, DBSCAN will have difficulties in identifying clusters with varying densities. However, unlike algorithms like *k*-means, DBSCAN offers the significant benefit of being computationally efficient without predefining the number of clusters. Lastly, it enables the identification of clusters with any shape.

Now, let's take all of this confusing and complicated information and input it into highly practical "software" using a five-step procedure! Let's proceed to open the box.

The Strategy

Look at the challenge we face when analyzing a 3D point cloud of a kitchen scene. We want to extract the main structural elements, such as walls, countertops, cabinets, and appliances. Manually segmenting these components point by point is incredibly tedious, especially for large datasets.

I see a clear opportunity to streamline this process by leveraging a combination of RANSAC and DBSCAN. My approach involves first utilizing RANSAC to iteratively detect and extract large planar surfaces like walls and countertops. These dominant planar segments are then removed from the point cloud, leaving behind the remaining objects like cabinets and appliances. Next, we can apply DBSCAN on the remaining point cloud to cluster these spatially adjacent points into distinct objects based on their density and proximity.

This strategic workflow illustrated in Figure 12-25 allows us to efficiently segment the main constituents of the kitchen scene in an automated fashion, paving the way for higher-level scene understanding.

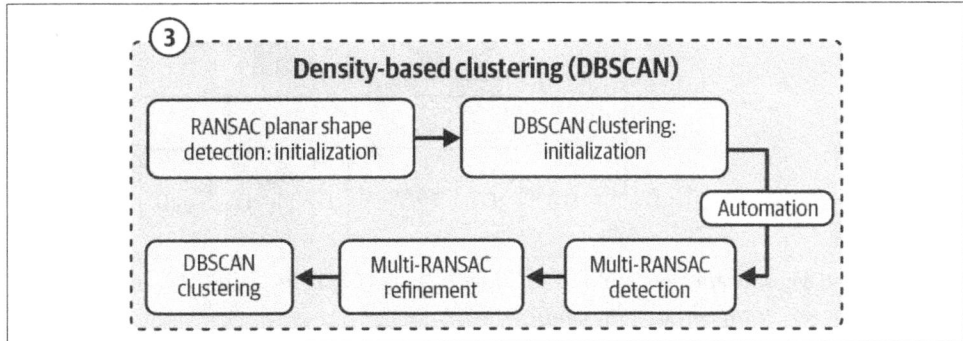

Figure 12-25. The five-step strategic workflow for DBSCAN clustering of a kitchen scene

You should be pretty familiar with random sample consensus (RANSAC) by now. It is a straightforward yet very efficient method that is particularly useful when dealing with data influenced by outliers, as in our situation. When dealing with actual sensors in practical applications, the data obtained will never be flawless. Frequently, your sensor data is influenced by outliers. RANSAC is an iterative method that partitions your data points into two sets: inlier and outlier sets. Subsequently, you can disregard the data points that deviate significantly from the norm and focus on the data points within the expected range. I encourage you to refer to Chapter 9 for more clarity and know-how about RANSAC.

Let's execute on this strategic plan, by first laying down our experimental setup.

Experimental Setup

Let's utilize a dataset that I collected using a terrestrial laser scanner. It is a point cloud of a kitchen that you want your robot to operate in. I will omit the specifics about input/output processes and file formats discussed in the preceding chapters (see Chapter 4 for a reminder). Today, we immediately begin utilizing the widely recognized *.ply* file format.

I shall adopt a decisive approach to obtain results rapidly. We will successfully achieve precise segmentation by adhering to a basic coding strategy. This entails being extremely discerning when it comes to the foundational libraries. We utilize three highly resilient libraries, namely NumPy, Matplotlib, and Open3D.

To install the library packages mentioned here in your environment, you know the drill:

```
pip install numpy matplotlib open3d
```

> We have selected Python over C++ or Julia, and as a result, our system's performance is as it is. Presumably, it will suffice for your application, specifically for what we refer to as "offline" processes, which are not conducted in real time.

It is time to move on to the first two steps of our strategy (see Figure 12-26): defining how to detect planar shapes and verifying that DBSCAN works as expected.

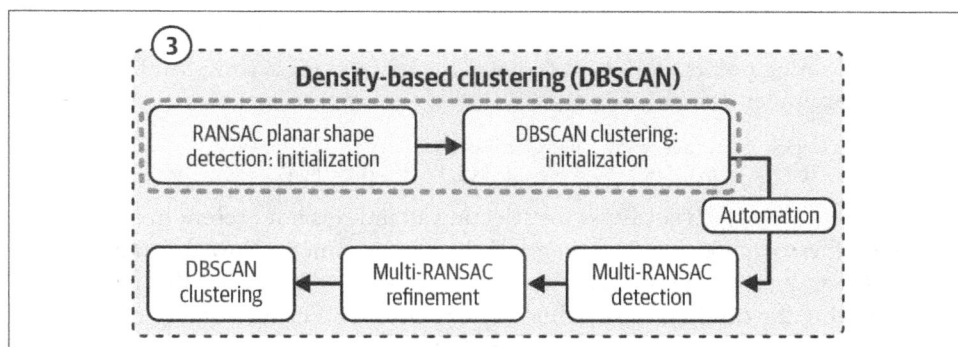

Figure 12-26. Step 1: RANSAC planar shape detection initialization

3D Planar Shape Recognition with RANSAC

Let's first refresh on the use of RANSAC for planar shape recognition. We begin by importing the data into the variable named pcd using the following line of code:

```
pcd = o3d.io.read_point_cloud("../DATA/TLS_kitchen.ply")
```

This time, let's leverage the Open3D library, including a RANSAC implementation specifically designed to detect planar shapes in point clouds. The sole code snippet to compose is as follows:

```
plane_model, inliers = pcd.segment_plane(distance_threshold=0.01, ransac_n=3,
num_iterations=1000)
```

The `segment_plane()` method contains three parameters, as shown. The distance threshold (`distance_threshold`) determines whether a point is considered an inlier or outlier based on its distance from the plane. The number of sampled points drawn, which in this case is 3, is utilized to estimate each plane candidate using the RANSAC algorithm. The number of iterations, denoted as `num_iterations`, specifies the frequency at which the estimation process is performed. These numbers are considered conventional, but it is important to exercise caution as well as verify the `distance_threshold` based on the specific dataset being used. To automate these based on the data characteristics, I encourage you to refer to Chapter 9.

The output of the preceding line consists of the optimal plane candidate parameters a, b, c, and d stored in the variable `plane_model`, as well as the indices of the points identified as inliers, stored in the variable `inliers`.

Now, let's proceed to visualize the outcomes. To accomplish this, we must specifically choose the points based on the indices obtained from the inliers and, if desired, designate all remaining points as outliers. What is the method for accomplishing this task? Here is an example:

```
inlier_cloud = pcd.select_by_index(inliers)
outlier_cloud = pcd.select_by_index(inliers, invert=True)
```

The argument `invert=True` allows for selecting all indexes not present in the `inliers` argument, thereby choosing the inverse of the first argument. Now that the variables contain the points, I suggest that we color the inliers red and the other points gray before viewing the findings. To do this, you can provide a list of floating-point values representing the R, G, and B components as follows:

```
inlier_cloud.paint_uniform_color([1, 0, 0])
outlier_cloud.paint_uniform_color([0.6, 0.6, 0.6])
```

And now, let us visualize the results with the following line:

```
o3d.visualization.draw_geometries([inlier_cloud, outlier_cloud])
```

If you wish to enhance your understanding of the geometric properties revealed by the color, you can calculate the surface normals by executing the following command beforehand:

```
pcd.estimate_normals(search_param=o3d.geometry.KDTreeSearchParamHybrid(radius=0.1,
max_nn=16), fast_normal_computation=True).
```

This will ensure you get a much nicer rendering (see Figure 12-27A).

This displays the outcome of the previously described RANSAC script. The data points that fall within the expected range are in darker shading (top), while those that deviate from the expected range are shown in gray (bottom).

We verified that we can divide a point cloud into two sets: a planar-based inlier point set and an outlier point set. Now, let's examine the process of delineating clusters that are in close proximity to each other with DBSCAN. This will pave the way for our robot's main object delineation. Let's consider a scenario where we have identified large flat surfaces, and now we need to outline things that appear floating. What is the procedure for accomplishing this task?

DBSCAN for 3D Point Cloud Segmentation

Initially, we choose a sample in which we presume all planar sections have been eliminated (see Figure 12-27B). We aim to segment the remaining items using Euclidean clustering. The dataset can be accessed via the chapter's folder. Let's proceed with DBSCAN clustering. We will utilize the DBSCAN method from the Open3D package to streamline the process. However, it is important to note that if you require greater adaptability, the implementation in scikit-learn may be a more viable option in the long run. In terms of computing times, they are almost the same.

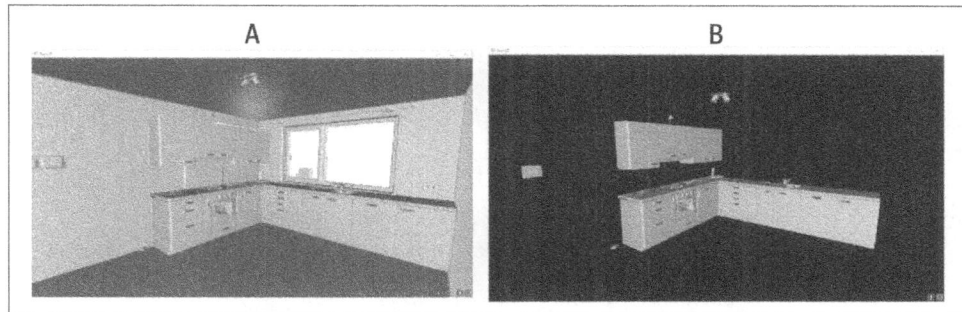

Figure 12-27. (A) Visualization of RANSAC results: inliers in darker shading (top) and outliers in gray (bottom); (B) DBSCAN for 3D point cloud segmentation: preprocessing step with planar sections removed

The `cluster_dbscan` method operates directly on the `pcd` point cloud entity and produces a list of labels after the initial indexing of the point cloud:

```
labels = np.array(pcd.cluster_dbscan(eps=0.05, min_points=10))
```

The labels range from -1 to n, with -1 indicating a "noise" point and values 0 to n being the cluster labels assigned to each associated point. It is important to mention that our objective is to obtain the labels in the form of a NumPy array. Additionally, we employ a radius of 5 cm to expand clusters, and we only consider a cluster if it contains a minimum of 10 points after this expansion. You are welcome to engage in experimentation without any restrictions.

Now that we have established groupings of points with a corresponding label for each point, we can proceed to assign colors to the findings. While not obligatory, it is advantageous to seek for the optimal values of parameters via iterative operations. In order to achieve this objective, I suggest utilizing the Matplotlib package to obtain precise color ranges, such as the `tab20` palette:

```
max_label = labels.max()
colors = plt.get_cmap("tab20")(labels / (max_label
if max_label > 0 else 1))colors[labels < 0] = 0
pcd.colors = o3d.utility.Vector3dVector(colors[:, :3])
o3d.visualization.draw_geometries([pcd])
```

The `max_label` variable should have an apparent purpose: it is used to record the highest value in the list of labels. This allows us to use it as a denominator in the color scheme while addressing the specific scenario where the clustering is imbalanced and produces only noise and one cluster using an if statement. Subsequently, we ensure that these disruptive data points are assigned the label –1, which is represented by the color black (0). Next, we assign the 2D NumPy array of three "columns," representing the colors R, G, B, to the property `colors` of the point cloud `pcd`.

The outcome of our clustering analysis is shown in Figure 12-28. DBSCAN was executed with parameters `eps=0.05` and `min_points=10`.

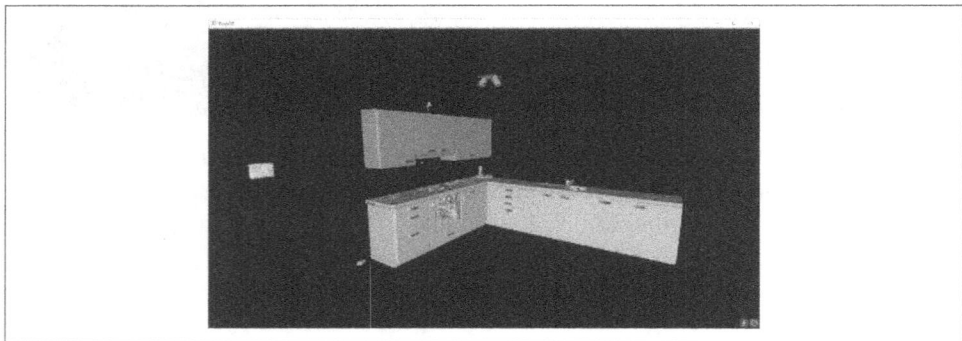

Figure 12-28. DBSCAN point cloud clustering: groupings and visual differentiation of objects

We can denote a clear distinction between the upper and lower cabinets, the heating controls, and the lamp. All these elements can be easily identified.

This validates that our system is functioning smoothly. Now, how can we effectively implement and automate this on a larger scale?

Our philosophy is straightforward and uncomplicated. Initially, we want to execute the RANSAC algorithm several times (denoted as *n* times) to identify and extract the distinct planar sections that make up the scene. Next, we want to address the issue of "floating elements" using density-based clustering (DBSCAN). Consequently, providing a mechanism for storing the outcomes throughout the iterative process is imperative. Let's execute this in the three stages illustrated in Figure 12-29.

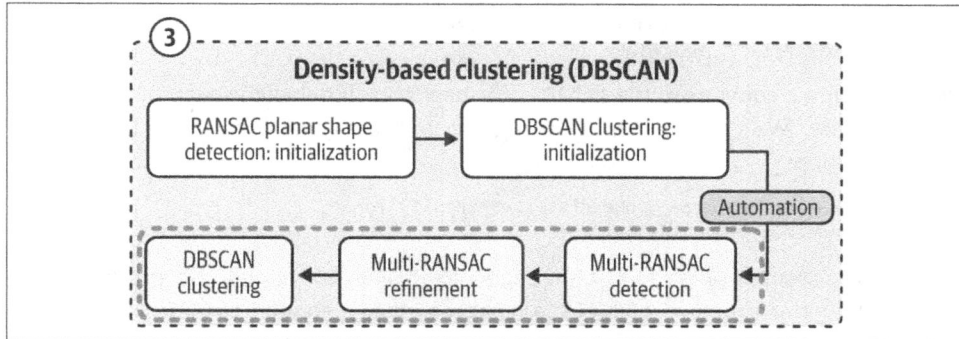

Figure 12-29. The automation loop to cluster the scene efficiently with RANSAC and DBSCAN

The Multi-RANSAC Framework

Let's establish a vacant dictionary to store the iteration results. The dictionary will store the plane parameters in `segment_models` and the planar regions identified from the point cloud in `segments`:

```
segment_models={}
segments={}
```

Next, we aim to ensure that we have control over the number of iterations required for plane detection in the future. To achieve this objective, we will define a variable called `max_plane_idx` that will store the number of iterations:

```
max_plane_idx=20
```

Here, we express the desire to iterate 20 times to locate 20 planes, although more intelligent methods exist to establish such a parameter, as shown in Chapter 9. Let's enter a functional loop, which I will briefly demonstrate. We distinguish the inliers from the outliers during the initial iteration (loop i=0). The inliers are stored in segments, and the remaining points are the focus of interest for the subsequent loop n + 1 (loop i=1).

This implies that we intend to regard the outliers from the previous phase as the fundamental point cloud until we reach the specified threshold of iterations (distinct from RANSAC iterations). This can be expressed as follows:

```
rest=pcd
for i in range(max_plane_idx):
    colors = plt.get_cmap("tab20")(i)    segment_models[i],
    inliers = rest.segment_plane(
    distance_threshold=0.01,ransac_n=3,num_iterations=1000)
    segments[i]=rest.select_by_index(inliers)
    segments[i].paint_uniform_color(list(colors[:3]))
rest = rest.select_by_index(inliers, invert=True)
print("pass",i,"/",max_plane_idx,"done.")
```

We will assign a color from the `tab20` colormap to each detected segment to visualize the ensemble. We can do this by using the `plt.get_cmap("tab20")(i)` function within the loop:

```
o3d.visualization.draw_geometries([segments[i] for
i in range(max_plane_idx)]+[rest])
```

The list `[segments[i] for i in range(max_plane_idx)]` that we provide as an argument to the function `o3d.visualization.draw_geometries()` is essentially a "list comprehension." It is useful to compose a for loop that adds the initial element `segments[i]` to a list. Furthermore, we can include the remaining elements in this list and the `draw.geometries()` method will intelligibly interpret our intention to include a point cloud for drawing (see Figure 12-30).

Figure 12-30. The outcome of the iterative RANSAC point cloud segmentation process

At this stage, we may think we obtained the desired results. However, do you observe anything peculiar in this situation? Upon careful observation, you can discern some artifacts, such as "lines," that effectively sever some planar components. What is the reason?

Indeed, due to our fitting of all the points to RANSAC plane candidates, regardless of the density continuity of the points, we encounter these "lines" artifacts that vary depending on the sequence in which the planes are discovered. Notably, these plane candidates have no limit in Euclidean space. Therefore, the subsequent course of action is to deter such conduct.

To achieve this, I suggest incorporating a condition based on another type of density-based clustering into the iterative process. This will allow for the refinement of inlier point sets located in adjacent clusters. Are you prepared?

Multi-RANSAC Refinement with DBSCAN

To achieve this objective, we depend on the DBSCAN algorithm. Allow me to elucidate the logical process, albeit not in a simplistic manner. Within the previously specified for loop, we executed the DBSCAN algorithm immediately after assigning the inliers (`segments[i]=rest.select_by_index(inliers)`). This is accomplished by inserting the following line directly afterward:

```
labels = np.array(segments[i].cluster_dbscan(eps=d_threshold*10, min_points=10))
```

I set the epsilon in the function of the initial threshold of the RANSAC plane search to a value that is 10 times greater. This is not a complex scientific matter; rather, it is a merely observational decision, but it typically yields favorable results and simplifies matters with parameters. We typically derive this value from the point cloud's characteristics, multiplying the average point spacing (`d_threshold`) by a factor (here, set to 10). This multiplication ensures we capture meaningful local point neighborhoods while accounting for varying point densities.

Next, inside the loop, we tally the number of points in each cluster we have identified, utilizing a list comprehension. The outcome is recorded in the variable `candidates`:

```
candidates=[len(np.where(labels==j)[0]) for j in np.unique(labels)]
```

From there, we want to identify the "optimal candidate," typically the cluster that contains the highest number of points:

```
best_candidate=int(np.unique(labels)[np.where(candidates==
np.max(candidates))[0]])
```

Several complex operations are being performed behind the scenes. Still, fundamentally, we utilize NumPy's expertise to search for and retrieve the index of the points that belong to the largest cluster. Subsequently, it is imperative to include the leftover clusters in every iteration to account for the subsequent RANSAC iterations.

I recommended reading these lines multiple times to fully get their meaning:

```
rest = rest.select_by_index(inliers,
invert=True) + segments[i].select_by_index(list(np.where(labels!=best_candidate)
[0]))
segments[i]=segments[i].select_by_index(list(np.where(labels== best_candidate)
[0]))
```

The rest variable now stores the remaining points obtained from the RANSAC and DBSCAN algorithms. Additionally, the inliers are refined by selecting only the most significant cluster from the original RANSAC inlier set. Upon completion of the loop, you obtain a pristine collection of segments that contain spatially adjacent sets of points that conform to planar geometries as in Figure 12-31.

Figure 12-31. Downhill skiing approach in RANSAC: refining segments with iterative clustering and point selection

We have successfully resolved the issue with the "lines." However, certain elements remain unassigned (see your experiments). Let us cluster these as well if we want to later allow our robot to use the light panel, change the light bulb, or put shades on the window.

DBSCAN Refinement

At this stage, we reason outside the loop, and our new input is constituted by the remaining elements in the rest variable that have not yet been assigned to any segment. To process them, we can use a straightforward application of density-based clustering with the DBSCAN algorithm, as initially presented:

```
labels = np.array(rest.cluster_dbscan(eps=0.05, min_points=5))
max_label = labels.max()
print(f"point cloud has {max_label + 1} clusters")colors = plt.get_cmap("tab10")
(labels / (max_label if max_label > 0 else 1))
colors[labels < 0] = 0
rest.colors = o3d.utility.Vector3dVector(colors[:, :3])
```

I utilize it the same way as previously employed. I ensure the use of consistent parameters to enhance clustering and achieve similar results as shown in Figure 12-32.

Figure 12-32. Final clustering outcome: applying DBSCAN to remaining points for complete point cloud segmentation

This is the conclusive clustering outcome using the present methodology. Congratulations on completing a state-of-the-art detection of items in a planar fashion. It can now be manipulated as the foundation for further advanced visual processes!

DBSCAN Versus k-Means

DBSCAN is a valuable tool for segmenting 3D point clouds, particularly when dealing with clusters of varying shapes and sizes. Its density based approach allows it to identify clusters in noisy data effectively.

However, proper parameter tuning and preprocessing are crucial for achieving optimal results. Additionally, evaluation metrics can help quantitatively assess the performance of the algorithm on a given dataset. While DBSCAN may face challenges with clusters of varying densities or elongated shapes, it remains a powerful and versatile choice for many 3D point cloud segmentation tasks.

Importantly, it offers a distinct advantage over *k*-means clustering when dealing with clusters of varying shapes and sizes. *k*-means assumes spherical clusters and requires specifying the number of clusters beforehand. In contrast, DBSCAN is density-based, identifying clusters based on regions of high density without requiring prior knowledge of the number of clusters. This makes it more suitable for data with complex structures or when the number of clusters is unknown. DBSCAN is also less sensitive to noise and outliers compared to k-means. With that in mind, you should have a clear decision on when to use one or the other.

A Visual Comparison

You may wonder if there are cases where k-means shines over DBSCAN in the case of 3D point clouds. Let me illustrate some cases where you may want to switch. In the following, you can see an aerial LiDAR dataset, which contains three cars close to one another. Let's run k-means with $k = 3$:

```
data_folder="../DATA/"
dataset="KME_cars.xyz"
x,y,z,r,g,b = np.loadtxt(data_folder+dataset,skiprows=1, delimiter=';',
unpack=True)
X=np.column_stack((x,y,z))
kmeans = KMeans(n_clusters=3).fit(X)
```

In this example, we use the k-means algorithm to segment a portion of the aerial LiDAR dataset that contains three cars. When we set $k = 3$, k-means divides the point cloud into three clusters (see Figure 12-33).

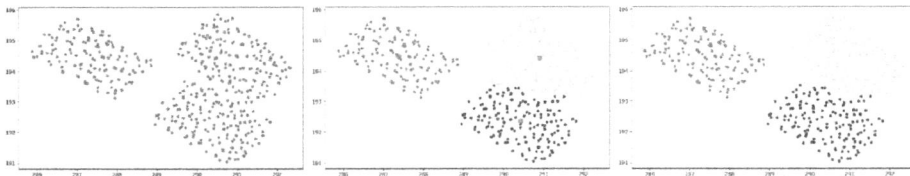

Figure 12-33. 3D point cloud segmentation comparison: the original point cloud (left), clustering with k-means (center), clustering with DBSCAN (right)

As you can see, we get excellent clustering even if we cannot delineate the objects spatially. What does it look like with DBSCAN? Well, let's check with the following code lines:

```
# Analysis on DBSCAN
clustering = DBSCAN(eps=0.5, min_samples=2).fit(X)
plt.scatter(x, y, c=clustering.labels_, s=20)
plt.show()
```

In the DBSCAN approach, the clustering process is more flexible and adaptable to the natural distribution of data (see Figure 12-34). As you can see, in addition to having difficulties setting the epsilon parameter, we cannot delineate, at least with these features, the two cars on the right; 1–0 for k-means in that case.

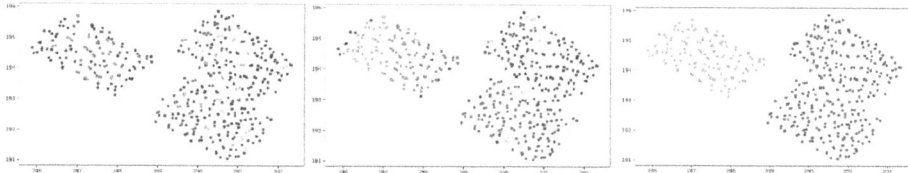

Figure 12-34. DBSCAN clustering of the point cloud: respectively, the epsilon value was set to 0.1, 0.2, and 0.5

At this point, we have only illustrated *k*-means using spatial features. But we can use any combination of features, which makes it super flexible for different applications. For the purposes of the section, you can also experiment with illuminance, intensity, number of returns, and reflectance. The following are two examples using these features (see Figure 12-35):

```
X=np.column_stack((x[mask], y[mask], z[mask], illuminance[mask],
nb_of_returns[mask], intensity[mask]))
kmeans = KMeans(n_clusters=3, random_state=0).fit(X)
plt.scatter(x[mask], y[mask], c=kmeans.labels_, s=0.1)
plt.show()

X=np.column_stack((z[mask] ,z[mask], intensity[mask]))
kmeans = KMeans(n_clusters=4, random_state=0).fit(X)
plt.scatter(x[mask], y[mask], c=kmeans.labels_, s=0.1)
plt.show()
```

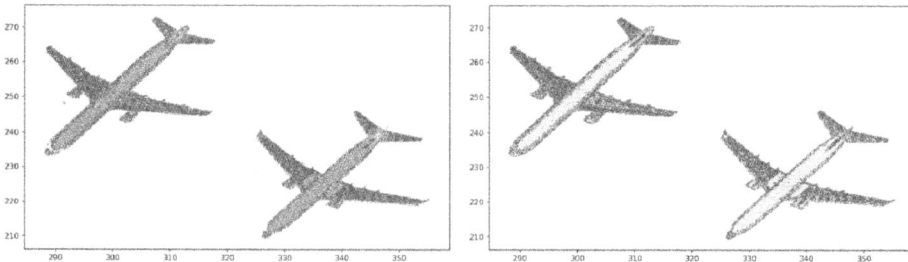

Figure 12-35. k-means results on the 3D point cloud using different feature spaces and k values

To go deeper, we could better describe the local neighborhood around each point, for example, through a principal component analysis (see Chapters 6 and 7). Indeed, this could permit the extraction of a large set of more or less relevant geometric features; feel free to experiment.

Summary

Throughout this chapter, we've worked with 3D machine learning through the lens of unsupervised learning. We focused on the major clustering algorithms to unlock a first step toward 3D scene understanding. 3D clustering methods partition a three-dimensional model, like a point cloud, into distinct clusters or regions, each representing a different object or part. We found that these methods offer a high degree of autonomy, eliminating the need for labeled datasets. This makes them ideal as a seed process to accelerate the creation of labeled datasets for semantic segmentation tasks, such as 3D object recognition. By providing a semantic-aware structure on top of the base unit, clustering empowers us to understand 3D digital environments, potentially leading to more advanced segmentation and semantic injection techniques.

In this chapter, we delved into the art of monitoring these results, ensuring the groupings accurately reflect the underlying structure of the data. We then explored centroid-based clustering, represented by k-means, and density-based clustering, with DBSCAN as a prime example. We hybridized these approaches with RANSAC for robust 3D shape recognition. This means that you can be proud to exhibit your new skills:

- You can group similar data points based on proximity or building on features.
- You can leverage k-means for partition-based clustering with centroids.
- You can leverage DBSCAN for density-based clustering, which is ideal for arbitrary shapes.
- You can use RANSAC for 3D shape detection and binary segmentation.
- You can leverage different distance metrics to measure similarity between groups.

Furthermore, unsupervised learning is rapidly evolving with promising perspectives. As hardware capabilities advance, we can expect to see even more complex and large-scale applications. This drives the development of novel algorithms that can efficiently handle such data. Moreover, integrating deep learning techniques with traditional clustering methods holds excellent promise. Deep learning models can learn complex patterns and features from large datasets, enabling more accurate and robust segmentation results. This is our goal for our next chapter, Chapter 13, focusing on the use of graphs and foundation models to extend your unsupervised learning skillset.

But before jumping from our horse, I would encourage you to test your newfound skills with the chapter's project.

Hands-on Project

Now, let's test your skills. Can you build a 3D point cloud segmentation script that does the following?

1. Load the point cloud using a suitable library (e.g., Open3D).
2. Preprocess the data (e.g., remove outliers, normalize).
3. Apply k-means clustering with an appropriate number of clusters.
4. Apply DBSCAN clustering with different eps and minPts values.
5. Apply RANSAC segmentation for a chosen model (e.g., plane, sphere).
6. Compare results and analyze their strengths and weaknesses.
7. Explore feature engineering for improved performance.
8. Consider hybrid approaches combining different algorithms.

Graphs and Foundation Models for Unsupervised Segmentation

Extracting meaningful information from 3D datasets is challenging. We have massive data with intricate details, but we lack the integrated intelligence needed for high-level tasks. This gap limits the potential for advanced 3D scene understanding: without semantics and topology, we cannot extract individual objects and their relationships, such as chairs and tables, and their arrangement within a room. Let's leverage 3D machine learning to extract these.

Supervised learning, which thrives on labeled data, is our first investigation. However, a major hurdle is the scarcity of labeled datasets for 3D data: without a lot of data, building such a system is limited. The good news is that technological leaps are astonishing, especially when we leverage cutting-edge research in unsupervised segmentation. But to bring human-level reasoning to computers, extracting formalized meanings from the 3D entities we observe is crucial.

This is why we combine 3D point clouds, graph theory, and deep learning in this chapter to unlock new scene-understanding capabilities for interpreting our visual world. Among these advancements, I want to focus on two major solutions:

- Connectivity-based clustering (see Chapter 12), leveraging the power of graph theory (*https://oreil.ly/sTzwc*)

- Image-based 3D segmentation using a foundation model: the Segment Anything Model (*https://oreil.ly/M39Rb*)

Before proceeding, I want to emphasize a key point: there is a critical distinction between semantic segmentation and what we discuss in this chapter. Semantic segmentation aims to assign labels with a predefined meaning, like "chair" or "wheel." In contrast, as in Chapter 12, we focus on models that aim to group points or surfaces without this conceptualized label (group 1, group 2, group n).

This strategy provides numerous benefits. In contrast to semantic segmentation, unsupervised segmentation with clustering does not require training data, resulting in higher efficiency and cost-effectiveness.

On the contrary, semantic segmentation frequently depends on a substantial amount of annotated data, which can be very expensive and usually time-consuming to acquire. Therefore, due to the scarcity of labeled 3D datasets, unsupervised models are central to 3D machine learning. Additionally, unsupervised learning offers a world of possibilities. The model isn't tied to a specific set of labels. It can discover new groupings based on the data itself, potentially revealing previously unknown structures and opening up new avenues of exploration.

Chapter Resources

In this chapter, I introduce two powerful unsupervised segmentation techniques. First, we explore how to represent point clouds as graphs, leveraging the principles of graph theory and connectivity-based clustering to group points based on spatial proximity. We use Python to implement this approach using k-d trees for neighbor searches and connected component analysis to delineate distinct objects. The second solution introduces the Segment Anything Model (SAM),a revolutionary foundation model that enables fully automated and unsupervised segmentation of images. By leveraging 2D projections, we can harness the power of SAM to achieve accurate segmentation results without the need for manual labeling or training data.

As always, you can find all the materials for this chapter on the 3D Data Science Resource Hub (*https://learngeodata.eu/3d-data-science-with-python*). To access the files (code, datasets, articles) and resources, you may need to share an email address, a personal password, or proof that you are the book's owner.

Let's start with a solution that leverages the power of graphs (*https://oreil.ly/if6tC*).

Connectivity-based Clustering

If we step back, our eyes can capture spatial information and process it through our cognition system. And this is where the magic lies: our brain helps us make sense of the scene and its relational decomposition. With internal knowledge representation, you can instantly know that your scene is made of floors and walls, that the floor hosts chairs and tables, and that, in turn, the cup, microwave, and laptop stand on the desks (see Figure 13-1).

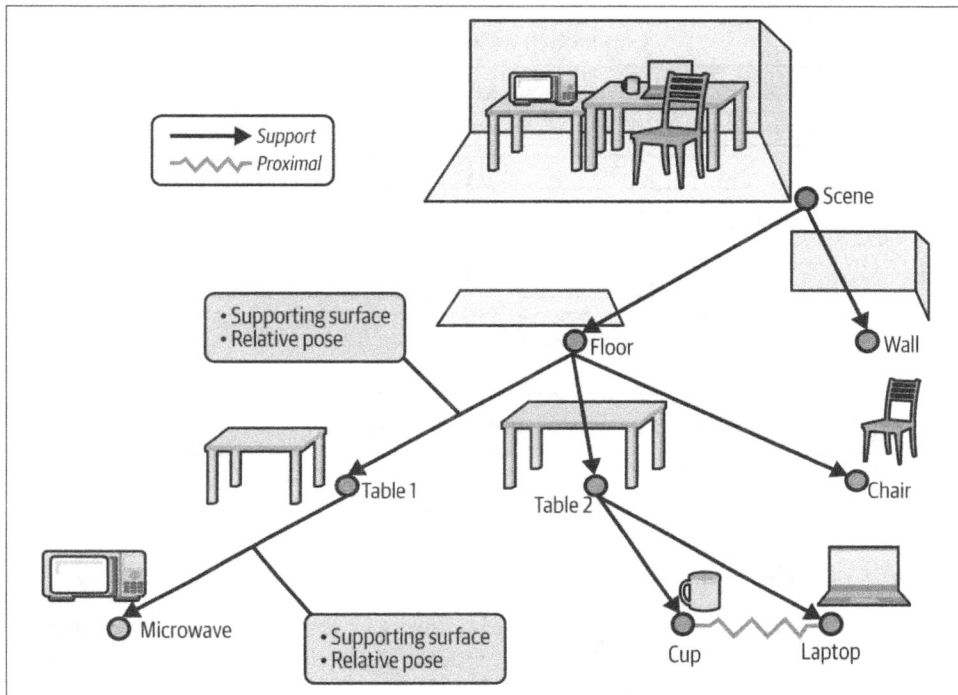

Figure 13-1. Example of the relationships in a scene as described by the authors in "Scene Reconstruction with Functional Objects for Robot Autonomy" (https://oreil.ly/0P8-W) (2022)

Mimicking this with 3D tech means we need to find a way to leverage a 3D dataset that represents the scene and then build a decomposition into the main constituents: a clustering.

From my example, you can understand how "connectivity" plays a major role: it is simple to distinguish "objects" that do not touch (once the support is removed, i.e., the ground). But can we achieve this decomposition efficiently with 3D digital datasets? Let's develop an advanced solution with graph theory for 3D datasets that builds on the workflow illustrated in Figure 13-2. We'll start with the mission brief.

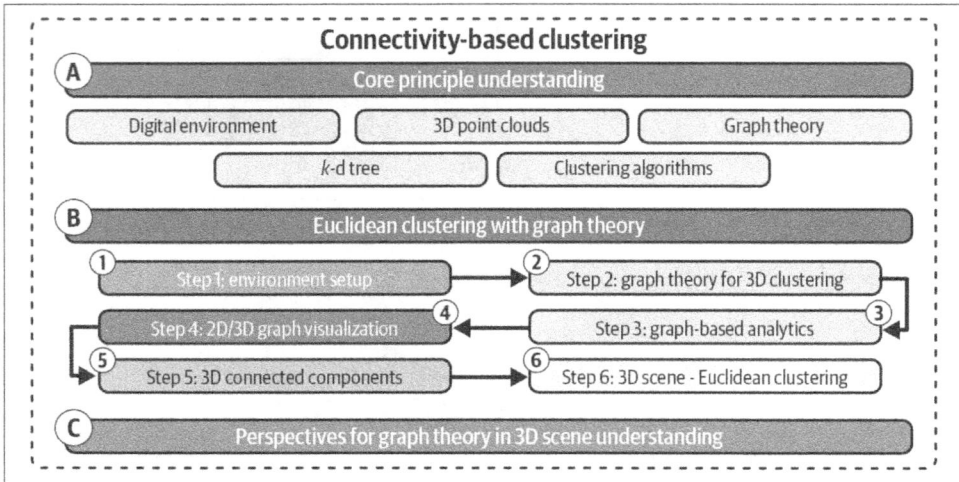

Figure 13-2. Organization of the connectivity-based clustering project.

The Mission Brief

You are organizing a massive library of 3D objects to develop a next-generation AI system that can identify in real time the model of each piece of furniture. The issue: you have three days to do 10 entire buildings.

You load your 3D point clouds and try the first approach: manual segmentation. It becomes a significant bottleneck for identifying individual objects like chairs, tables, and lamps in a scanned environment. You encounter memory limitations and juggle multiple software platforms, each introducing its own constraints, learning curves, and errors.

After losing one day in a tedious cycle of exporting, importing, and converting data across various tools, you feel discouraged by your disjointed workflow, which hampers productivity and scalability. You decide to get a good night of sleep, and while dreaming, you imagine a streamlined solution that harnesses the power of graph theory to redefine how you approach point cloud segmentation.

The little dreambee tells you to automate the process by employing Euclidean clustering in Python, eliminating the reliance on manual intervention. This innovative method organizes point clouds into structured graphs, where each point is a node connected by edges that signify proximity. Then, through algorithms for connected component analysis, you could enable efficient, automated delineation of objects, transforming how you analyze and interpret the 3D data.

You wake up, and that is it. You are going to develop a solution that presents a clear and effective method: Euclidean clustering utilizing graph theory in Python. Initially, you eliminate significant structural components such as walls and floors, which can be segmented using alternative methods like RANSAC, thereby isolating the objects of interest.

Subsequently, you want to create a graph in which each point is represented as a node and connections. This graphical representation and practical algorithms should then be able to leverage connected component analysis to enable the automatic delineation of individual objects, which is significantly more efficient than manual segmentation. It's time to dive into the solution! But first, let's discuss the core principles of what we use.

Core Principles

Clustering describes the process of categorizing a collection of data points so that points within the same category, known as a cluster, are more "alike." While interchangeably used with the term "unsupervised segmentation," we usually refer to the latter when we aim to partition an entire 3D space into meaningful regions or objects. In contrast, clustering is mainly used to denote the algorithms and a "feature level," not necessarily resulting in a complete partitioning of a 3D digital environment. For wielding connectivity-based clustering, it is essential to understand our digital environments, the role of point clouds, graph theory, k-d trees, and clustering algorithms (see Figure 13-3).

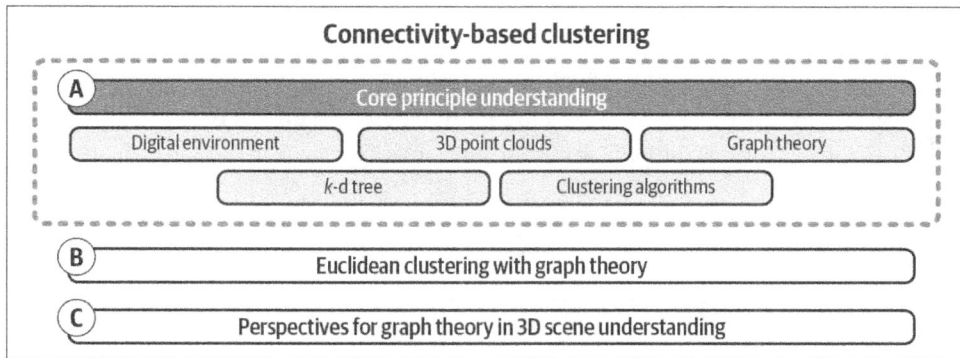

Figure 13-3. Focus on the core principles

Aside from digital environments, let me highlight the core concepts that we leverage and the related resources if you feel like you need a refresher:

3D point cloud (unstructured data)
A point cloud represents a 3D scene as a collection of individual points, each with its own 3D coordinates. Unlike mesh data, which explicitly defines surface connectivity, point cloud data is unstructured. This lack of inherent relationships between points poses a challenge for segmentation.

Graph theory basics

Graph theory provides a powerful framework for representing relationships between entities. A graph consists of *nodes* (vertices) connected by *edges*. In our case, each point in the point cloud becomes a node. Edges connect spatially proximate points, transforming the unstructured point cloud into a structured graph.

k-d tree

A *k*-d tree (*k*-dimensional tree) is a space-partitioning data structure that efficiently organizes points in a *k*-dimensional space (3D in our case). It recursively divides the space with hyperplanes, creating a tree-like structure. This initially enables fast nearest-neighbor searches, crucial for efficiently finding points within a given radius.

(3D) clustering

Our approach centers on clustering, which transforms disconnected points into meaningful groups representing objects or distinct regions within the 3D scene. While other approaches like *k*-means (*https://oreil.ly/Xwp6z*) or DBSCAN (*https://oreil.ly/w1GQ7*) exist (see Chapter 12), we unveil graph-based clustering to leverage connectivity. Our method relies on constructing a graph from the point cloud and then identifying connected components within this graph. Each connected component represents a cluster.

We can now proceed with a comprehensive Python implementation. Let me illustrate the six-step workflow in Figure 13-4 using a synthetic demonstration dataset and a real-world point cloud to highlight its practical application. The solution is designed to provide an effective tool for managing point cloud data and maximizing its potential for unsupervised segmentation scenarios.

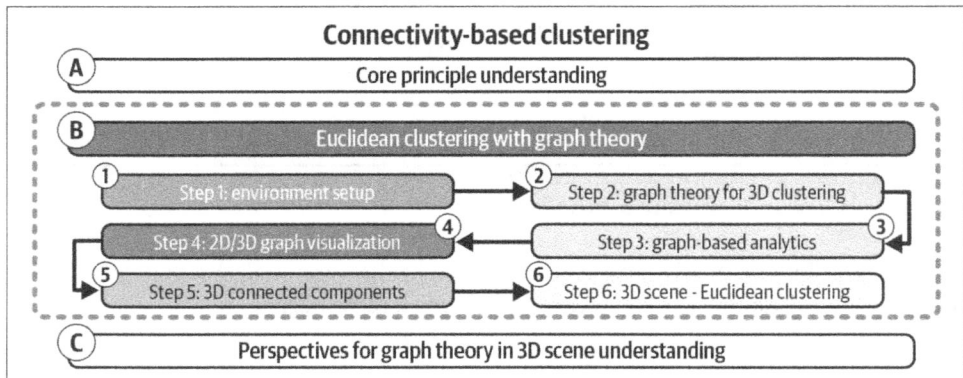

Figure 13-4. Focus on Euclidean clustering with graph theory

Step 1: Environment Setup

The first stage is to set up our environment for a successful project. We have three main constituents: Python setup, data setup, and utilities (Figure 13-5).

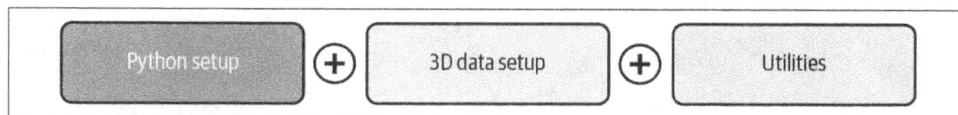

Figure 13-5. The three components of our environment setup

Concerning our Python setup, we leverage Python 3.10 with Anaconda (*https://anaconda.com*). After the creation of a virtual environment using Anaconda, we install the five following libraries:

- NumPy (*https://numpy.org*) for numerical operations
- SciPy (*https://scipy.org*) for scientific computing, including *k*-d tree implementation
- NetworkX (*https://networkx.org*) for graph manipulation and analysis
- Matplotlib (*https://matplotlib.org*) for visualization and color mapping
- Open3D (*http://open3d.org*) for advanced 3D data processing and visualization

To install these libraries, we use the pip package manager as follows:

```
pip install numpy scipy networkx matplotlib open3d
```

If you want to ensure you understand how to set up your environment for local 3D development quickly, I encourage you to follow this video tutorial (*https://oreil.ly/IKOJG*). You may want to use an IDE for easier development; I use Spyder, installed with `pip install spyder`, and launched with `spyder`.

Then, we gather a 3D point cloud of an indoor scene that our method is going to attack (see Figure 13-6). The dataset was obtained using the Naavis VLX2 (*https://navvis.com*) handheld laser scanner. Finally, we are going to leverage CloudCompare (*https://oreil.ly/rGpLw*) as a supplementary tool for visualization and demonstration outside of the Python environment.

Figure 13-6. The source dataset for this project

Now that we are set, let's test our system by loading our point cloud (e.g., in PLY format) into Python:

```python
# Importing the libraries
import numpy as np
from scipy.spatial import KDTree
import networkx as nx
import matplotlib.pyplot as plt
import open3d as o3d

# Load point cloud (PLY format)
pcd = o3d.io.read_point_cloud("../DATA/room_furnitures.ply")
translation = pcd.get_min_bound()
pcd.translate(-translation)

# Visualization with Open3D
o3d.visualization.draw_geometries([pcd])
```

We leverage Open3D (o3d) to load a point cloud dataset representing our room with furniture, as illustrated in Figure 13-7.

The critical operation here is translating the point cloud to its minimum bound, effectively centering the dataset at the origin. This preprocessing step is crucial for subsequent geometric operations, ensuring that our spatial analysis starts from a standardized reference point. By calling `pcd.translate(-translation)`, we shift the entire point cloud, which helps in maintaining consistent spatial relationships during further processing.

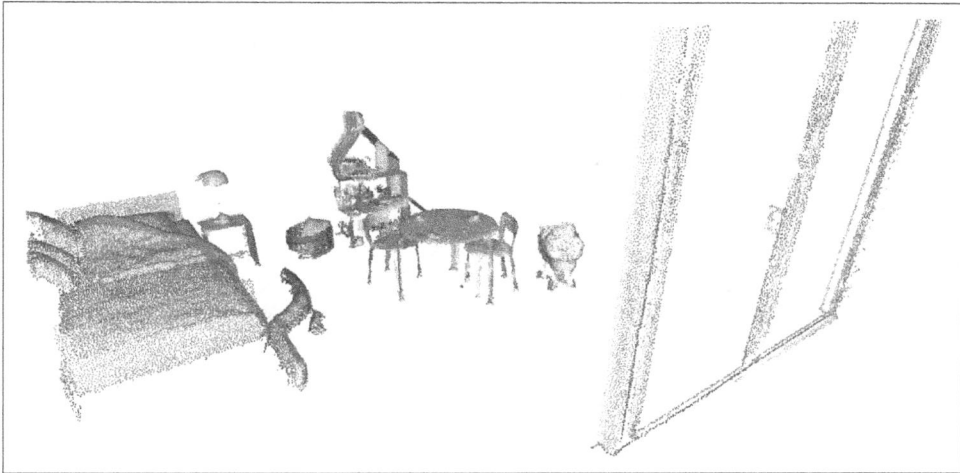

Figure 13-7. The floating elements that need to be clustered

> The visualization using `o3d.visualization.draw_geometries` (`[pcd]`) allows us to inspect the loaded point cloud visually, providing an immediate verification of data integrity and spatial configuration. This step is fundamental in 3D data science, as visual confirmation helps us quickly identify potential issues or characteristics in the dataset before diving into complex analytical procedures.

Great, we can move on to the second stage: constructing our graph.

Step 2: Graph Theory for 3D Clustering

Let's generate graphs with Python. But before moving there, I want to give you the fundamentals of graph theory that we are going to leverage so that what I code makes sense (step 2 of 6 in the workflow process; see Figure 13-8).

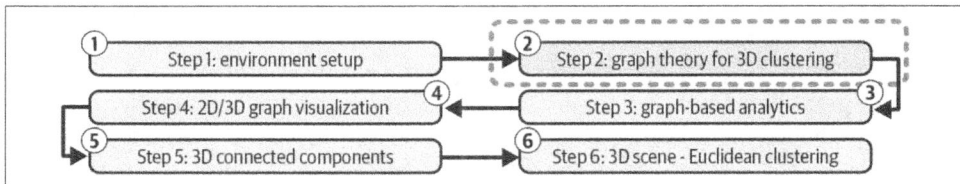

Figure 13-8. Graph theory for 3D clustering

Let's explore the fundamental concepts of graph theory used in our point cloud clustering approach, drawing directly from the interactive explanation presented in the audio tutorial. Visualizing these concepts makes them much more intuitive and engaging.

What is a graph?

A graph is a structure composed of *vertices* (also called nodes) and *edges* (example in Figure 13-9). Think of vertices as individual entities and edges as connections or relationships between these entities. The graph in Figure 13-9 is composed of six vertices and seven edges, which we denote as such:

$$V = \{v_1, v_2, v_3, v_4, v_5, v_6\}$$
$$E = \{v_5v_6, v_5v_4, v_6v_4, v_4v_3, v_6v_3, v_6v_1, v_3v_2\}$$

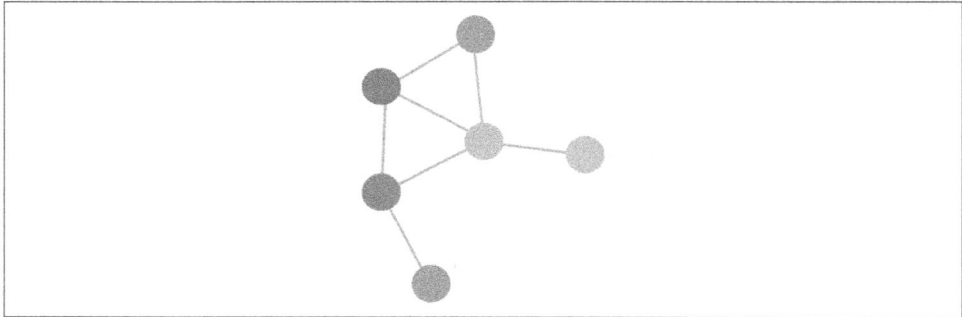

Figure 13-9. A graph composed of six vertices (nodes) and seven edges

The vertices (nodes) hold information relevant to the entity they represent. In our point cloud scenario, each point's 3D coordinates would be associated with a vertex. The edges (connections) are represented as lines connecting vertices. They signify a relationship between the entities represented by the vertices. In our case, an edge between two vertices would indicate that the corresponding points in the point cloud are spatially close to each other, within the defined radius. Now, you have some key graph properties to be aware of. Let's consider the graph in Figure 13-10. First, the *order* of a graph is simply the number of vertices it contains. In the example, you have 10 vertices, hence the order is 10. Adding or removing vertices changes the order of the graph. In our point cloud graph, the order is equal to the number of points in the cloud.

Then, the *size* of a graph refers to the number of edges (in this example, it is 7). As we connect or disconnect vertices, the size of the graph changes accordingly. In our point cloud graph, the size reflects the density of connections based on the chosen radius and the `max_neighbors` parameter. Finally, the *degree* of a vertex is the number of edges connected to it (see Figure 13-11).

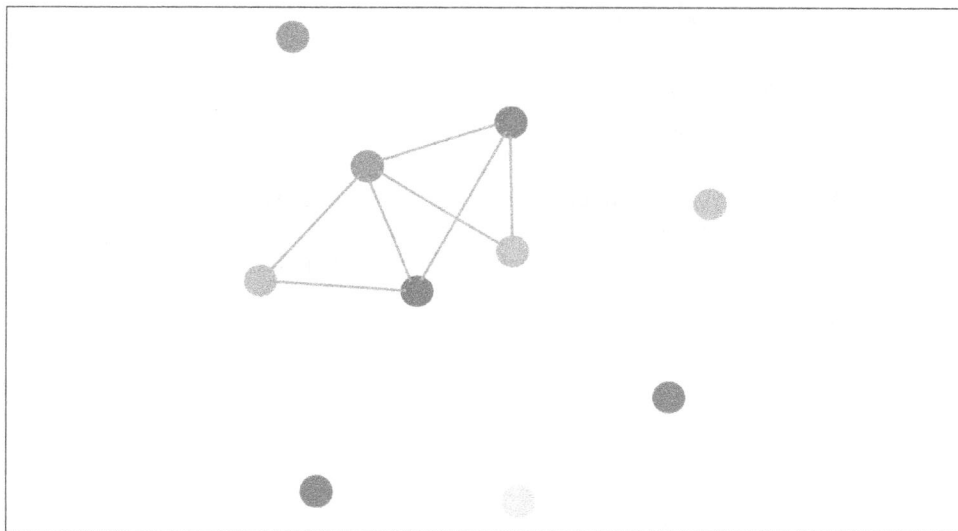

Figure 13-10. A graph of order 10 with 7 edges

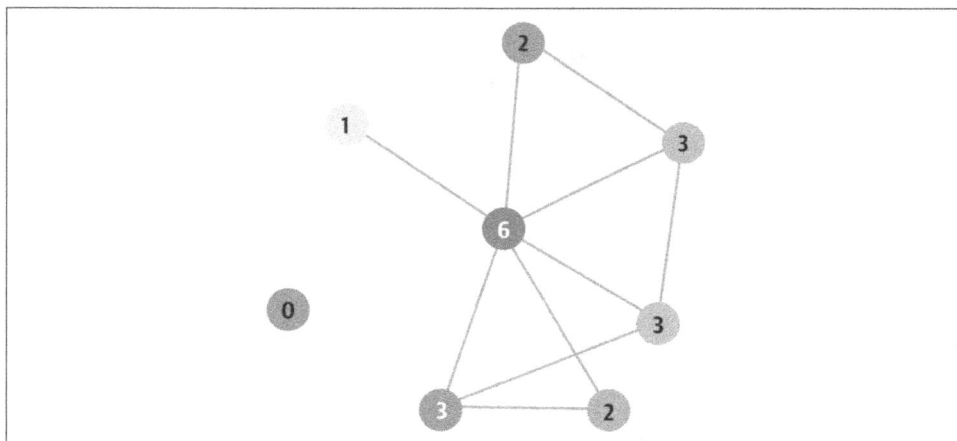

Figure 13-11. The degree of a graph

A vertex with a high degree is connected to many other vertices, signifying a central or highly connected entity. In the context of point clouds, a point with a high degree might indicate that it lies within a dense region or is part of a large object.

Now, imagine a scenario where one vertex represents the ground plane, and other vertices represent objects resting on the ground (i.e., a node is tied to a cluster instead of a single point). The edges connecting the object vertices to the ground vertex signify the "on" relationship. By analyzing the graph structure, we can infer that vertices with a high degree connected to a central "ground" vertex likely represent objects resting on that surface.

The case of connected components for 3D

The concept of *connected components* is crucial for our clustering approach. A connected component is a subgraph where there exists a path between any two vertices within the subgraph, but no path exists between a vertex in the subgraph and any vertex outside of it. Take the graph illustrated in Figure 13-12. In the example, you can see a disconnected graph with seven connected components, three of which are isolated vertices in the sense that there is no vertex in the component that isn't connected to it and no "outside" vertex connected to it.

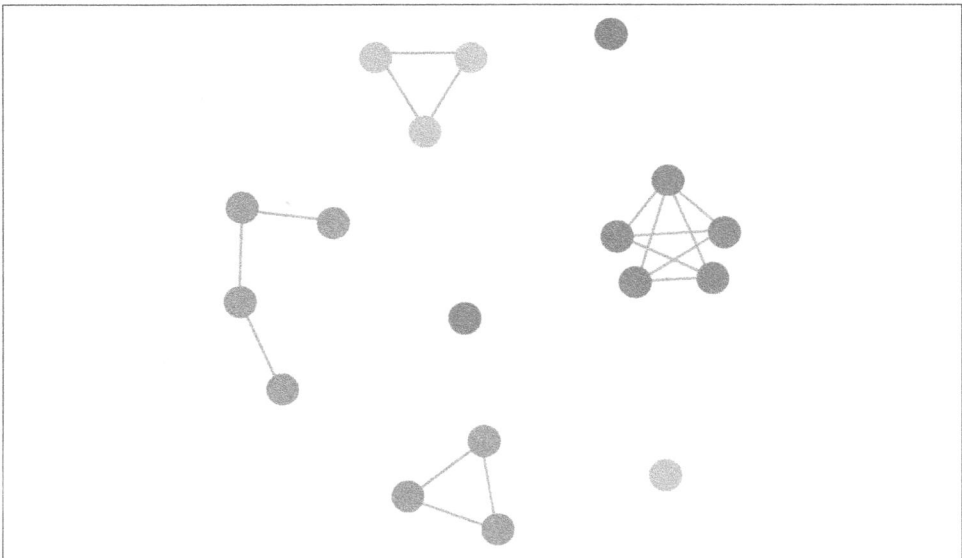

Figure 13-12. A graph with seven connected components, three of which are isolated vertices

In simpler terms, it's a self-contained group of interconnected vertices. In the interactive visualization, we can see how disconnected graphs have multiple connected components. In our point cloud scenario, each connected component corresponds to a potential object. By identifying these connected components, we can effectively segment the point cloud into meaningful clusters.

If you want to explore graph theory interactively, I highly recommend using the online tool D3 Graph Theory (*https://d3gt.com*). It allows for interactive graph manipulation to illustrate many very useful concepts.

Now that you have a solid foundation for understanding how we transform unstructured point cloud data into a structured graph representation let's enable it with Python.

Python function definition

We introduce a sophisticated graph construction method `build_radius_graph()` that transforms point cloud data into a networked representation. The function utilizes a *k*-d tree for efficient spatial querying, connecting points within a specified radius and optionally limiting the number of neighbors per node:

```python
def build_radius_graph(points, radius, max_neighbors):

    # Convert points to numpy array if not already
    points = np.asarray(points)

    # Create k-d tree
    kdtree = KDTree(points)

    # Initialize graph
    graph = nx.Graph()

    # Add nodes with position attributes
    for i in range(len(points)):
        graph.add_node(i, pos=points[i])

    # Query the k-d tree for all points within radius
    pairs = kdtree.query_pairs(radius)

    # Add edges to the graph with distances as weights
    for i, j in pairs:
        dist = np.linalg.norm(points[i] - points[j])
        graph.add_edge(i, j, weight=dist)

    # If max_neighbors is specified, prune the graph
    if max_neighbors is not None:
        prune_to_k_neighbors(graph, max_neighbors)

    return graph
```

By using NetworkX's graph structure, we create a flexible framework for topological analysis that captures local spatial relationships.

The *k*-d tree plays a pivotal role in establishing connections between points in the point cloud. For each point, we query the *k*-d tree to find its neighbors within a defined radius. These neighbors are the candidates for edge creation in our graph. The efficiency of the *k*-d tree allows us to perform these neighbor searches rapidly, even for large point clouds.

The approach shines here: we leverage the *k*-d tree neighbor searches to build our graph. Each point in the point cloud becomes a node in the graph. An edge is created between two nodes if and only if they are identified as neighbors within the specified radius during the *k*-d tree search. Let's put this "parameter" on hold to explore it with the dataset at hand. Finally, I define an accompanying prune_to_k_neighbors() function that implements a critical graph refinement technique. This ensures that each node maintains only its *k*-nearest neighbors (*k*-NN), effectively controlling graph complexity and reducing computational overhead:

```
def prune_to_k_neighbors(graph, k):
    for node in graph.nodes():
        edges = [(node, neighbor, graph[node][neighbor]['weight'])
                for neighbor in graph[node]]
        if len(edges) > k:
            # Sort edges by weight
            edges.sort(key=lambda x: x[2])
            # Remove edges beyond k nearest
            edges_to_remove = edges[k:]
            graph.remove_edges_from([(e[0], e[1]) for e in edges_to_remove])
```

This pruning mechanism is particularly valuable in high-dimensional point clouds where unconstrained graph construction could lead to exponential complexity, making our approach both computationally efficient and topologically meaningful.

Beautiful! We now have a way to generate graphs, so naturally, let's put it to the test! For this, let me first generate some dummy data (a 2D point cloud):

```
np.random.seed(42)
n_points = 300
xyz = np.random.rand(n_points, 2)
```

We can now use the xyz point cloud for some 2D tests. It is an array, given as follows:

```
array([[0.37454012, 0.95071431],
       [0.73199394, 0.59865848],
       [0.15601864, 0.15599452],
       [0.05808361, 0.86617615],

            ...
       [0.60111501, 0.70807258],
       [0.02058449, 0.96990985],
       [0.83244264, 0.21233911],
       [0.18182497, 0.18340451]])
```

Let's use our defined function with a radius of 0.1 and 4 neighbors:

```
simulation_graph = build_radius_graph(xyz, radius=0.1, max_neighbors=4)
```

Now, if you try to call your simulated graph, you get:

```
<networkx.classes.graph.Graph at 0x1e86ec99460>
```

This means our function works! And from the Graph class, you can explore the nodes (simulation_graph.nodes()) or the edges (simulation_graph.edges()), for example:

```
#For the nodes, you would get:
NodeView((0, 1, 2, ...))

#For the edges, you would get:
EdgeView([(0, 205), (0, 197), ...])
```

But no way to extract its key characteristics? Let's solve this by proposing an analytical step.

Step 3: Graph Analytics

We want to use the concepts highlighted in graph theory (connected components) for our analytical tasks (Figure 13-13).

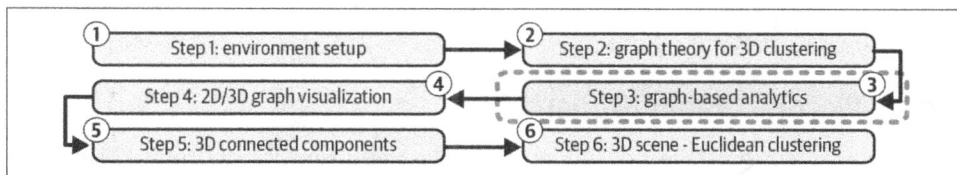

Figure 13-13. Step 3 of graph-based analytics

To this end, we construct an analyze_components() function that provides a comprehensive statistical graph connectivity analysis:

```
def analyze_components(graph):

    components = list(nx.connected_components(graph))

    analysis = {
        'num_components': len(components),
        'component_sizes': [len(c) for c in components],
        'largest_component_size': max(len(c) for c in components),
        'smallest_component_size': min(len(c) for c in components),
        'avg_component_size': np.mean([len(c) for c in components]),
        'isolated_points': sum(1 for c in components if len(c) == 1)
    }
    return analysis
```

By computing metrics such as the number of components, their sizes, and the count of isolated points, we gain deep insights into the topological structure of our point cloud. These metrics are essential for understanding the dataset's spatial distribution and clustering characteristics.

Let's put it to the test:

```
component_analysis = analyze_components(simulation_graph)
print("\nComponent Analysis:")
for metric, value in component_analysis.items():
    print(f"{metric}: {value}")
```

This returns:

```
Component Analysis:
num_components: 9
component_sizes: [169, 10, 32, 43, 2, 23, 11, 5, 5]
largest_component_size: 169
smallest_component_size: 2
avg_component_size: 33.333333333333336
isolated_points: 0
```

As you can see, the analysis reveals critical information about graph topology: how many distinct clusters exist, their relative sizes, and the presence of isolated points. This approach transforms raw geometric data into a rich, quantitative description of spatial relationships, which we can now leverage for further processing of our point cloud. Now, let's plot our graphs.

Step 4: Plotting Graphs (Optional)

We can now focus on stage 4 of our connectivity-based workflow (Figure 13-14). You can define a `plot_components()` function for a sophisticated graph-based point cloud representation visualization technique. At this stage, you should be deep enough into Python code that creating such a function is a great exercise to master Python for 3D. To help you on that front, I encourage you to first get the connected components, create a color iterator, and create a figure that you populate by iterating on the components.

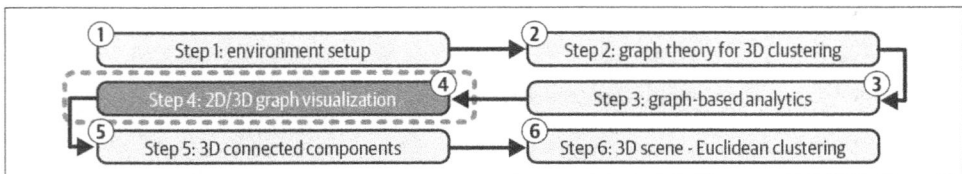

Figure 13-14. Step 4 of the connectivity-based workflow that targets graph visualization

Here is some sample code to help you get started:

```
# Get connected components
components = list(nx.connected_components(graph))
n_components = len(components)

# Create color iterator
colors = plt.colormaps[cmap](np.linspace(0, 1, max(n_components, 1)))
```

By color-coding connected components and rendering their edges and nodes, we can then create an intuitive visual representation of graph topology.

> The best approach is to design a visualization strategy using Matplotlib's color mapping to distinguish between different graph components, providing an immediate visual understanding of spatial clusters. By dynamically generating colors based on the number of components and rendering edges with varying transparencies, we can create a pleasing representation of graph structure.

Step 5: Connected Components for Point Clouds

Let's use our synthetic point cloud data to demonstrate graph construction and analysis techniques (Figure 13-15). The algorithm we use to identify clusters is *connected component analysis*. This fundamental graph algorithm efficiently finds all connected components within a graph.

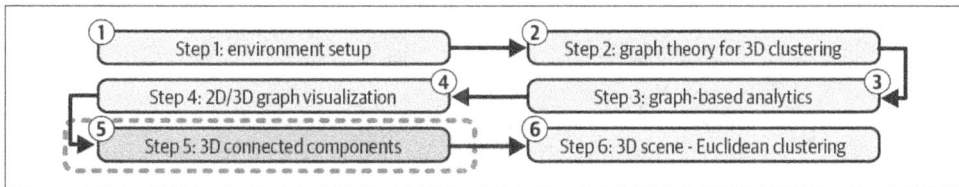

Figure 13-15. Step 5 of the workflow: connected components

As a reminder, a connected component is a maximal subgraph where every pair of nodes is connected by a path. In our context, each connected component represents a potential object in the scene. NetworkX, our chosen Python library for graph manipulation, provides efficient implementations of connected component analysis that we leverage.

We create controlled scenarios to explore graph behavior under different parameters. The code systematically varies radius and neighbor count, allowing us to observe how these parameters influence graph topology.

The simulation approach is a powerful method for understanding graph construction algorithms. By exploring parameter sensitivity through controlled experiments, we can develop robust graph-based clustering strategies that are adaptable to diverse point cloud characteristics.

The radius acts as a crucial parameter, controlling the granularity of the clustering. Let's check and vary the parameter by executing this Python code snippet:

```
for radius_simulation in [0.05,0.1,0.5]:
    simulation_graph = build_radius_graph(xyz, radius=radius_simulation,
    max_neighbors=5)
    plot_components(simulation_graph, xyz, radius_simulation, 5)
    plt.show()
```

A smaller radius will result in more fragmented clusters, while a larger radius will produce fewer, more interconnected clusters (see the following images):

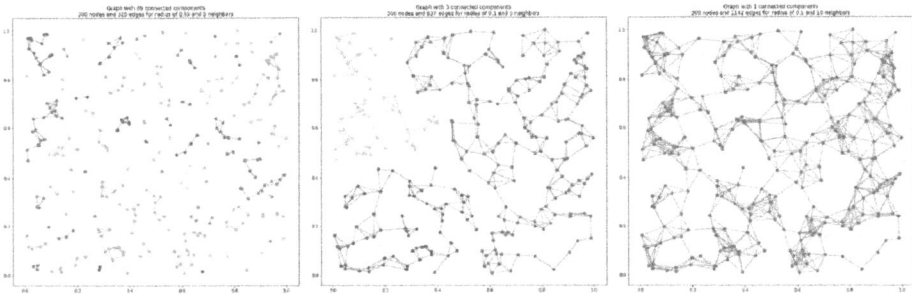

As you can see, by defining a fixed spatial distance, we control the local neighborhood interactions, essentially creating a topological constraint that captures the intrinsic geometric relationships within the dataset. This parameter's sensitivity directly impacts cluster formation, with smaller radii potentially fragmenting the point cloud into numerous small components and larger radii risking overmerging distinct spatial structures.

A small radius results in highly localized connectivity, leading to numerous small, fragmented clusters. This approach can be beneficial for isolating fine details; however, it also raises the likelihood of oversegmentation, which may result in dividing a single object into several clusters due to noise or differences in sampling density. A larger radius leads to wider connections, which may lead to the merging of separate objects, particularly in cluttered environments where objects are closely situated. The optimal radius is significantly influenced by the data available.

We also introduce a *pruning* step, limiting the maximum number of edges connected to each node. This prevents overconnectivity in dense regions, ensuring that the resulting clusters reflect meaningful object boundaries. To check the parameter's influence, let's use this Python code:

```
for neighbors_simulation in [2,5,10]:
    simulation_graph = build_radius_graph(xyz, radius=0.1,
    max_neighbors=neighbors_simulation)
    plot_components(simulation_graph, xyz, 0.1, neighbors_simulation)
    plt.show()
```

Let's explore how the `max_neighbors` parameter influences our results:

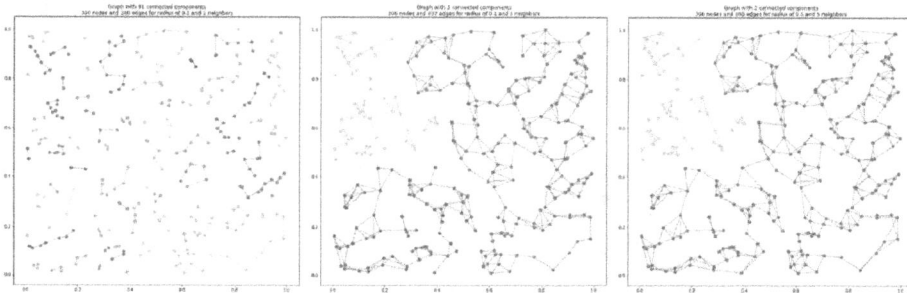

The `max_neighbors` parameter acts as a computational and topological regularization mechanism, preventing excessive graph complexity by limiting the number of connections per node. As we see, this parameter introduces a critical trade-off between local representational fidelity and global graph interpretability.

Too few neighbors might disconnect meaningful spatial relationships, while too many could introduce noise and reduce the graph representation's discriminative power. Selecting optimal values requires iterative experimentation, considering the point cloud's specific geometric characteristics and the intended analytical objectives. This methodology is critical in developing generalizable algorithms for spatial data analysis. Now that we have everything that seems to work on our simulated dataset, let's cluster our 3D point cloud.

Step 6: Euclidean Clustering for 3D Point Clouds

We want to leverage our graph structure for Euclidean clustering (Figure 13-16). We use the distance between points, defined as the Euclidean distance (straight-line distance) between their 3D coordinates. This is a natural choice for many applications, as it reflects the spatial proximity of points in the real world. However, other distance metrics, such as Manhattan distance or geodesic distance, can be used depending on the application's requirements.

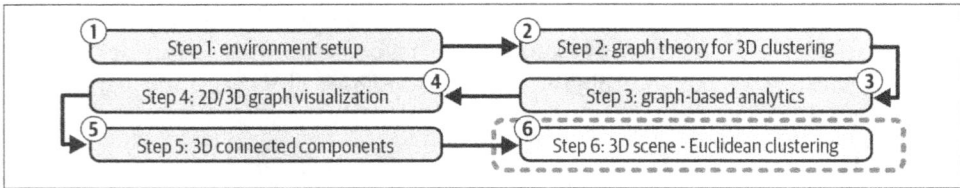

Figure 13-16. Step 6: Euclidean clustering

Transitioning from simulated to real-world data, we construct a graph using the actual point cloud dataset. The innovative approach here is computing the graph's radius based on the point cloud's nearest neighbor distance, creating a data-driven connection strategy:

```
# Visualizing our input point cloud
o3d.visualization.draw_geometries([pcd])

# Get point cloud coordinates
xyz = np.asarray(pcd.points)
nn_d = np.mean(pcd.compute_nearest_neighbor_distance()[0])
```

We adaptively capture local spatial relationships by scaling the radius to three times the average nearest neighbor distance. Then, we can use our graph construction process to demonstrate our approach to point cloud analysis:

```
# Build the graph
graph = build_radius_graph(xyz, radius=nn_d*3, max_neighbors=10)

# Analyze components
component_analysis = analyze_components(graph)
print("\nComponent Analysis:")
for metric, value in component_analysis.items():
    print(f"{metric}: {value}")
```

By analyzing the connected components in the real-world dataset, we can extract meaningful structural information, potentially identifying distinct object segments or spatial clusters:

```
Component Analysis:
num_components: 52
component_sizes: [5836, 743, 214, 6, 1, 2, 1, 1, 1, 16771, 1935, 3567, 1872, 16,
266, 1359, 352, 1648, 8792, 1246, 2710, 81, 74, 47, 5, 18, 8, 13, 1, 3, 14, 4, 8,
1, 2, 8, 2, 1, 7, 2, 8, 1, 1, 1, 2, 2, 1, 1, 1, 1, 1, 1]
largest_component_size: 16771
smallest_component_size: 1
avg_component_size: 916.5192307692307
isolated_points: 16
```

We have 52 main components, 16 of which are isolated points. The average size is 917 points, with the biggest (likely the bed) being 16,771 points.

Finally, let's visualize the results of our clustering. For this, let's create plot_cc_o3d( ), a specialized function for visualizing connected components in 3D point clouds using Open3D:

```
def plot_cc_o3d(graph, points, cmap = 'tab20'):

    # Get connected components
    components = list(nx.connected_components(graph))
    n_components = len(components)

    # Create color iterator
    colors = plt.colormaps[cmap](np.linspace(0, 1, max(n_components, 1)))
    rgb_cluster = np.zeros(np.shape(points))

    # Plot each component with a different color
    for idx, component in enumerate(components):
        if len(component)<=10:
            rgb_cluster[list(component)] = [0, 0, 0]
            idx-=1
        else:
            color = colors[idx][:3]
            rgb_cluster[list(component)] = color

    pcd_clusters = o3d.geometry.PointCloud()
    pcd_clusters.points = o3d.utility.Vector3dVector(xyz)
    pcd_clusters.colors = o3d.utility.Vector3dVector(rgb_cluster)
    return pcd_clusters
```

By color-coding components and handling small clusters differently, we create a nuanced visualization that highlights significant spatial structures while minimizing visual clutter from minor components.

The implementation is particularly clever in its handling of small clusters, assigning them a neutral color to distinguish them from major components. This approach provides a clear, informative visualization that helps researchers quickly understand the spatial organization of complex point cloud datasets, bridging computational analysis with intuitive visual interpretation. Now let's plot our scene:

```
pcd_cluster = plot_cc_o3d(graph, xyz)
pcd_cluster.estimate_normals()
o3d.visualization.draw_geometries([pcd_cluster])
```

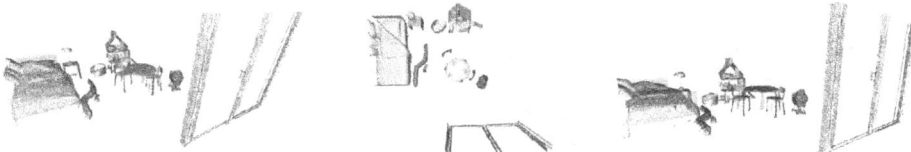

The results of the Euclidean clustering demonstrated on the real-world point cloud are particularly compelling due to their effectiveness in a practical scenario. We can see that the algorithm successfully isolates individual furniture pieces, including the bed, chairs, table, lamps, basket, and even small toys within the dollhouse.

This ability to delineate object boundaries, even in a cluttered scene with varying object sizes and shapes, showcases the approach's robustness. Notably, it handles overlapping objects like the chair and table, separating them into distinct clusters despite the overlap in the top-down view. Now, let's analyze a bit how we could extend the solution.

Discussion and Perspectives

Let's focus on our new approach's benefits, and how it can be improved (Figure 13-17). One of the most significant advantages of this approach is its automation. The entire process, from graph construction to cluster visualization, is automated within a Python script. This eliminates the need for manual segmentation, significantly saving time and effort.

Additionally, the clustering process inherently acts as a noise reduction technique. Small, isolated clusters, often representing noise or spurious points, are filtered out, leaving behind more substantial clusters representing actual objects.

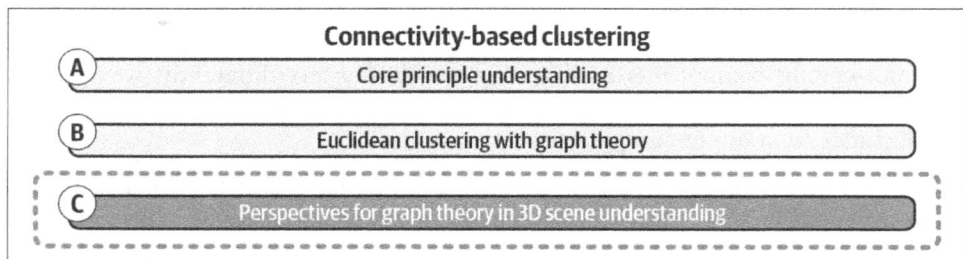

Figure 13-17. The final part of the project: perspectives

Another key advantage of this approach is its scalability. The efficiency of the k-d tree for neighbor searches and the connected component analysis algorithm allows for the processing of large, complex point clouds, opening up possibilities for applications in urban mapping, 3D modeling, and robotics.

The segmented clusters serve as a solid foundation for further analysis. Extracting features like size, shape, position, and orientation from each cluster enables object recognition, scene understanding, and other downstream applications. For example, in a point cloud representing a colored object, integrating color information into the edge weights can assist in distinguishing between spatially close areas that exhibit different colors. Additionally, advanced graph algorithms like spectral clustering and community detection can reveal higher-level relationships among clusters, facilitating scene understanding and object recognition. The next logical step is implementing a classifier on these segmented clusters to automate object identification and categorization.

> By incorporating attributes beyond spatial proximity (e.g., color, intensity, or normal vectors) into the edge weighting scheme, we significantly improve the segmentation's robustness and accuracy. Furthermore, advanced graph algorithms like community detection can be applied to identify higher-level relationships between clusters, uncovering hierarchical structures within the data. This opens up exciting possibilities for 3D scene understanding and object recognition.

The results demonstrate the practical utility of graph-based Euclidean clustering for real-world unsupervised segmentation. Its automated nature, robustness to complex scenes, noise reduction capabilities, and scalability make it a powerful tool for extracting meaningful information from 3D data. From a computational geometry perspective, our graph representations transform point clouds into robust topological networks that capture intrinsic spatial relationships. Encoding geometric information as a graph enables topology extraction and connectivity analysis. For instance, in urban mapping, these graphs can represent not just spatial proximity but semantic relationships between building structures, infrastructure, and terrain. The graph's edge weights and connectivity patterns become informative features for machine learning models to train algorithms with the spatial context on top of the geometry.

The effectiveness of graph-based Euclidean clustering is found in its capacity to convert the unstructured characteristics of point cloud data into a structured and analyzable format. We have another way to achieve this: raster-based representation through the projection mechanisms shown in Chapter 4. To leverage this, I propose using a foundational model: the Segment Anything Model.

The Segment Anything Model

Let's leverage the Segment Anything Model (SAM) for automation without supervision (Figure 13-18). This section explores unsupervised segmentation using the SAM. It comprises an image encoder, image embeddings, and some preprocessing operations to finally pass into the decoder and prompt encoder, giving the results as masks. The goal is to transform a complex 3D point cloud into a segmented representation without manual labeling through image-based mechanisms.

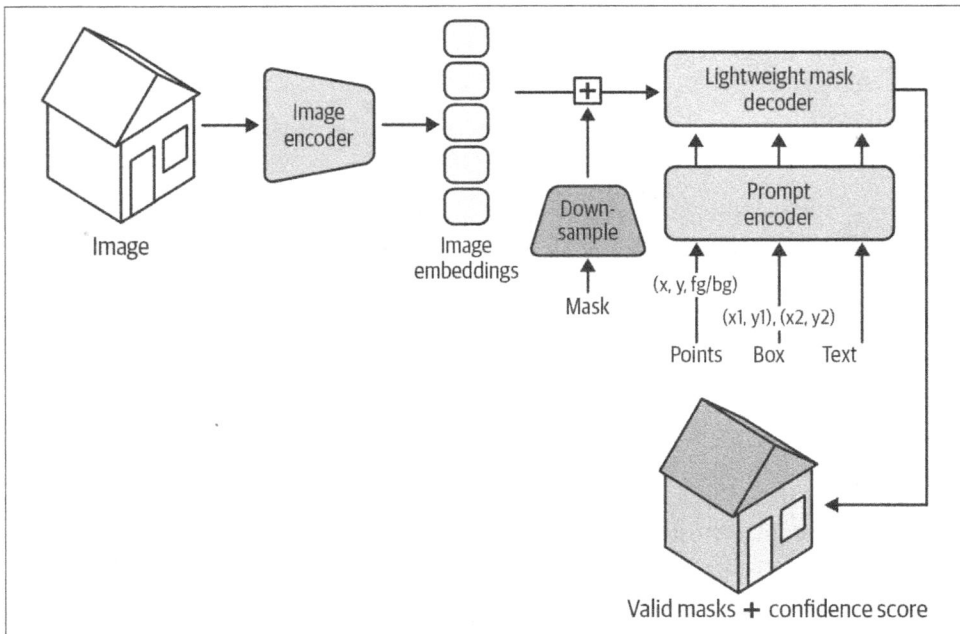

Figure 13-18. The Segment Anything Model architecture we use for 3D data

The Mission

OK, it's time for the mission brief. You are now a multiclass member of your country's special forces, and you must find some dangerous materials hidden inside a specific building without ever being detected. With your superb internet hacking skills, you find the 3D scans for the part of the building you are interested in. You now need to find a way to define the path for your dangerous material recovery team quickly. After that, the team can proceed unnoticed to recover the materials, and you have saved the day!

After careful research and using your various skills, you develop a 3D data processing workflow. This involves setting up a 3D Python code environment to process the 3D point cloud by using the SAM to highlight the composition of the scene (Figure 13-19). We have four main steps (3D project setup, SAM, 3D point cloud projections, and unsupervised segmentation).

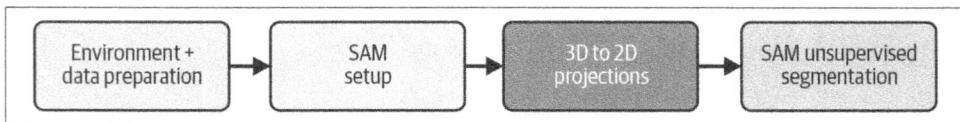

Figure 13-19. The workflow for 3D data processing with the Segment Anything Model

This workflow allows you to produce a 3D semantic map that permits pinpointing the location of the materials at least 90 minutes before the team arrives on-site. Are you ready?

3D Project Setup

Before we dive into the marvels of the SAM, it is crucial to establish a robust foundation. Setting up the appropriate environment ensures smooth sailing throughout our journey, allowing seamless experimentation and exploration. At this stage, we want to ensure that our coding environment is correctly set up with robust libraries.

3D code environment setup

It is time to get our hands dirty! We aim to use the SAM to semantically segment a 3D point cloud. And that is no easy feat. The first idea is to check out the Segment Anything dependencies: SAM GitHub (*https://oreil.ly/2VRFO*). From there, we check out the necessary prerequisites of the package (Figure 13-20).

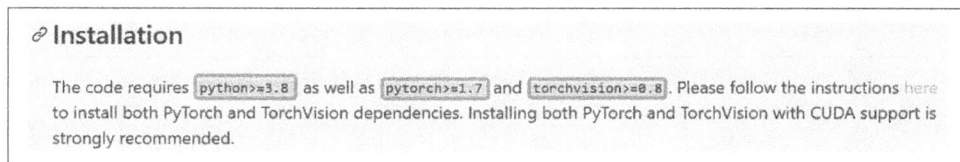

 ⌀ **Installation**

The code requires `python>=3.8` as well as `pytorch>=1.7` and `torchvision>=0.8`. Please follow the instructions here to install both PyTorch and TorchVision dependencies. Installing both PyTorch and TorchVision with CUDA support is strongly recommended.

Figure 13-20. The dependencies are highlighted in the Segment Anything repository

> When dealing with deep learning libraries or new research code, it is essential to check out the dependencies and installation recommendations. Indeed, this will strongly influence the follow-up of your experiments and the time needed for replication.

We need to use the following library versions:

```
python ≥ 3.8
pytorch ≥ 1.7
torchvision ≥ 0.8
```

With this information in place, we will generate a virtual environment to ensure smooth sailing. But, not to keep you high and dry, let's leverage our quick and lightweight setup from Chapter 3 using Miniconda (*https://oreil.ly/XqHHv*).

As a reminder, here are the steps to ensure a coherent setup:

1. Create a new environment: `conda create -n SAMSEG python=3.10`

2. Switch to the newly created environment: `conda activate SAMSEG`

3. Install pip in the new environment: `conda install pip`

We are now ready to move on installing the necessary libraries for playing with SAM. We install the base libraries needed to use SAM: NumPy, laspy, OpenCV, and Matplotlib. To install these libraries, we can use pip:

```
pip install numpy matplotlib laspy opencv-python
```

We've used these libraries in previous chapters, except OpenCV, which provides computer vision and image-processing functionalities. We can now explore the installation of deep-learning libraries, starting with my favorite: PyTorch. Since its launch in 2017, PyTorch has prioritized flexibility with performance. Today, using PyTorch for deep learning applications is an excellent choice if you want high-performance execution, Pythonic internals, and effective abstractions for valuable tasks.

> Since 2017, hardware accelerators (such as GPUs) have become approximately 15 times faster in computing tasks. One can only imagine what advancements are in store for the coming years. Therefore, it is essential to keep an eye out for flexible libraries that can adapt quickly, including those that can refactor their "internals" to languages like C++, as PyTorch does.

To simplify the installation of the appropriate PyTorch distribution and avoid the complexities of installing CUDA, developers created a web app that generates the exact command for you to copy and paste into your terminal. You can jump on this PyTorch Getting Started page (*https://oreil.ly/oR_df*) and select the most relevant way to install your distribution.

> We want to leverage our GPU. Therefore, it is essential to note that we need an installation with CUDA. But this is possible only if you have an NVIDIA GPU at the time of writing. If not, you may want to use the CPU or switch to a cloud computing service such as Google Colab. If you need guidance with CUDA, use the additional 3D Data Science Hub resources (*https://learngeodata.eu/3d-data-science-with-python*).

Therefore, our code is the following:

```
conda install pytorch torchvision torchaudio pytorch-cuda=11.7 -c pytorch
-c nvidia
```

This line triggers the retrieval and installation of the necessary elements for PyTorch to function coherently. The second deep-learning library that we want to use is Segment Anything. While PyTorch is being installed, we can download and install "software" that will make it easier for us to manage versions and access online libraries. This software is *Git* and is accessible from the Git website (*https://oreil.ly/RUCSV*). Once the installation is finished, PyTorch should be installed in your environment. To install Segment Anything, we can write the line below:

```
pip install git+https://github.com/facebookresearch/segment-anything.git
```

At this stage, we have installed the base and deep learning libraries. Before using them, let's install an IDE to ensure a smoother workflow (e.g., for Spyder: `pip install spyder`).

Next, we import our installed packages into our Python session:

```
# The base libraries
import numpy as np
import matplotlib.pyplot as plt
import cv2
import laspy

# The deep learning libraries
import torch
from segment_anything import sam_model_registry
from segment_anything import SamAutomaticMaskGenerator
```

We're all set up. Before proceeding with the next steps of coding our model, let's retrieve data.

3D dataset curation

This chapter uses a dataset acquired through a terrestrial laser scanner. The dataset pertains to an indoor scene. You can download the data from the chapter's folder and put it in the folder that holds your datasets (in my case, "*DATA*"). At this stage of the process, we have a coding setup with all the necessary libraries in a lightweight, isolated conda environment. You can test your CUDA setup:

```
import torch
print('CUDA available -> ', torch.cuda.is_available())
print('CUDA GPU number -> ', torch.cuda.device_count())
print('GPU -> ', torch.cuda.get_device_name())
```

It is now time to set up our model.

Segment Anything Model Core Concepts

The SAM presents an innovative architecture and training process, making it a perfect candidate to be tested on indoor applications. Let's first clarify its core concepts.

Segment Anything fundamentals

Meta AI has delved into the fascinating realm of natural language processing (NLP) and computer vision with their SAM, which enables *zero-shot* and *few-shot learning* on novel datasets and tasks using foundation models.

> For clarity concerns, let me summarize the terms: zero-shot learning refers to the ability to recognize something without having seen it at all (zero times). Similarly, few-shot learning involves using a limited number of labeled examples for each new class, to make predictions for new classes based on just these few examples of labeled data.

Overall, the SAM "AI" algorithm can significantly reduce the human effort required for image segmentation. To do so, you provide the model with foreground/background points, a rough box or mask, some text, or any other input that indicates what you want to segment in an image. The Meta AI team has trained SAM to generate a proper segmentation mask. This mask is the model's output and should be a suitable mask to delineate one of the things that the prompt might refer to. For instance, if you indicate a point on the roof of the house, the output should correctly identify you meant the roof of the house (Figure 13-21).

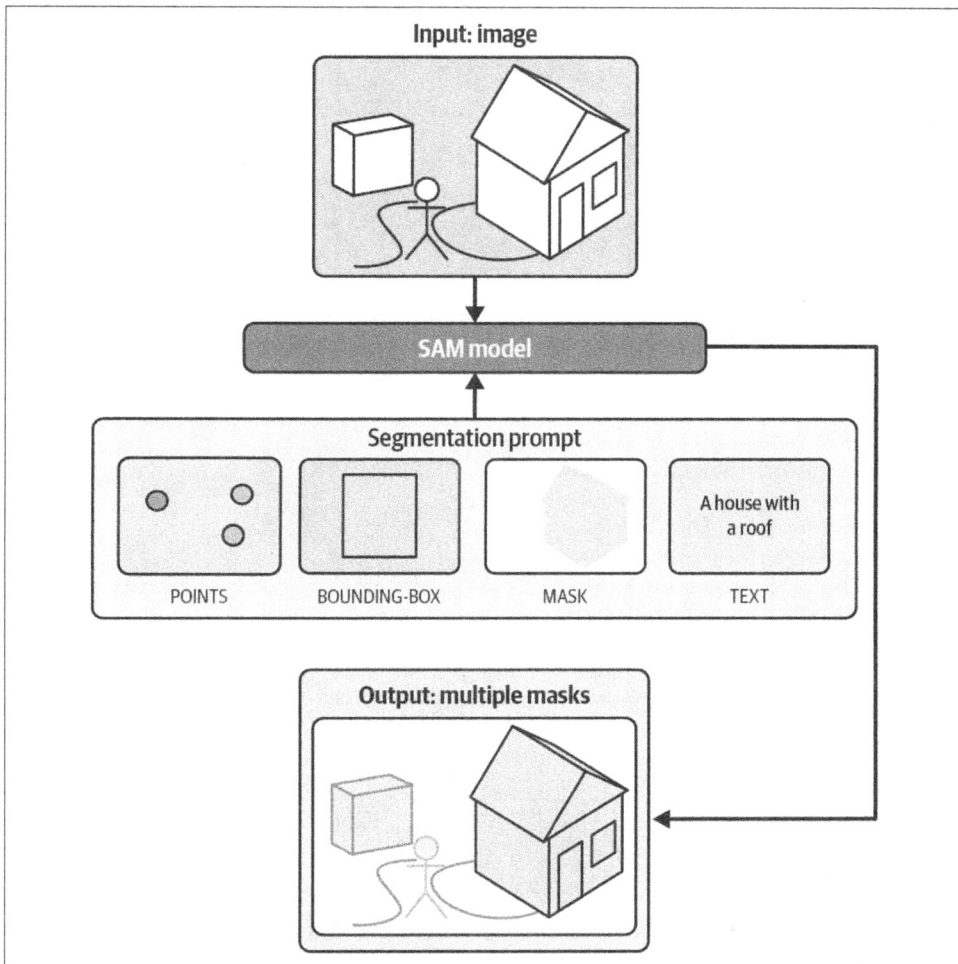

Figure 13-21. The segmentation prompt to generate valid masks (case of a house)

This segmentation task can then serve as model pretraining and guide solutions for various downstream segmentation problems. On the technical side, an image encoder creates a unique embedding (representation) for each image, and a lightweight encoder swiftly transforms any query into an embedding vector. These two sources are merged using a (lightweight) mask decoder to predict segmentation masks (Figure 13-22).

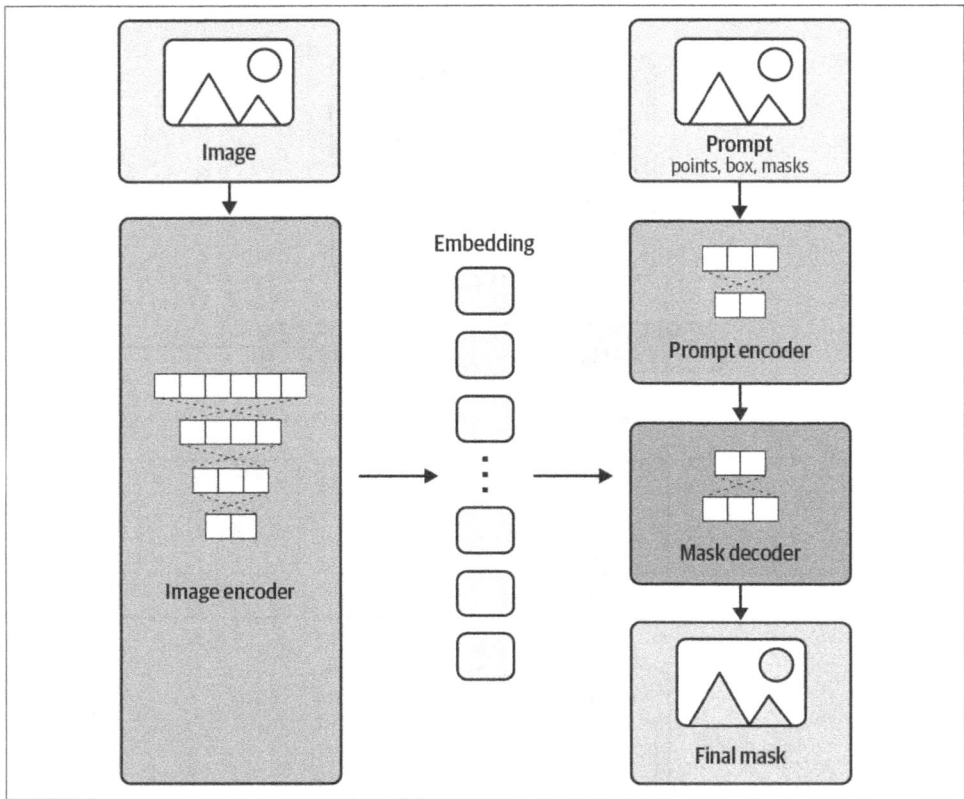

Figure 13-22. Segment Anything Model flowchart

This effective architecture, combined with a massive scale training phase, allows SAM to reach four milestones:

Effortless object segmentation
    With SAM, users can effortlessly segment objects by simply selecting the points they want to include or exclude from the segmentation. You can also use a bounding box as a cue for the model.

Handling uncertainty
    SAM is equipped to handle uncertain situations about the object to be segmented. It can generate multiple valid masks, which is crucial for solving real-world segmentation challenges effectively.

Automatic object detection and masking
    SAM makes automatic object detection and masking a breeze. It simplifies these tasks, saving you time and effort.

Real-time interaction

SAM can instantly provide a segmentation mask for any prompt thanks to pre-computed image embeddings. This means you can interact with the model in real time.

Let's now use it.

SAM parameters

The SAM model can be loaded with three different encoders: ViT-B, ViT-L, and ViT-H. ViT-H gives better results than ViT-B but has only marginal gains over ViT-L:

```
|     Encoder       |  #parameters    |   Speed      |  Quality    |
|-------------------|-----------------|--------------|-------------|
|   ViT-B  (basic)  |      91M        |   Fastest    |  Low        |
|   ViT-L  (large)  |     308M        |   Fast       |  High       |
|   ViT-H  (huge)   |     636M        |   Slow       |  Highest    |
```

These three encoders have different parameter counts that give a bit more freedom to tune an application. ViT-B (the smallest) has 91 million parameters, ViT-L has 308 million parameters, and ViT-H (the biggest) has 636 million parameters.

This difference in size also influences the speed of inference, so this should help you decide the encoder for your specific use case. Following this guide, we will use the heavy artillery: ViT-H, with a Model Checkpoint that you can download from GitHub (*https://oreil.ly/6e6nx*) (2.4 GB) and place in your current parent folder. This is where we can define two variables to make your code a bit more flexible:

```
MODEL = "../../MODELS/sam_vit_h_4b8939.pth"

# You can run the line below to test if you have the ability to leverage CUDA
torch.cuda.is_available()

# Choose between cpu or cuda training. For cpu, input 'cpu' instead 'cuda:0'
USED_D = torch.device('cuda:0')
```

From there, we can initialize our SAM model with the following two lines of code:

```
sam = sam_model_registry["vit_h"](checkpoint = MODEL)

#Cast your model to a specific device (cuda or cpu)
sam.to(device = USED_D)
```

We are all set up. The last step is to see how it performs on a random image on your desktop.

Performance on 2D images

Let's test it and see if everything works as expected on a random image. You can head to Google Earth (*https://earth.google.com*) and zoom in on a spot of interest (Figure 13-23). Next, load the image into memory with OpenCV:

```
# When loading an image with OpenCV, it is in bgr by default
loaded_img = cv2.imread("../DATA/biscarosse.jpg")

# Now we get the R,G,B image
image_rgb = cv2.cvtColor(loaded_img, cv2.COLOR_BGR2RGB)
```

As you can see, by default, OpenCV loads an image by switching to blue, green, and red channels (BGR) that we order as RGB with the second line and store in the image_rgb variable. Now, it is time for us to apply SAM on the image with two lines of code:

```
mask_generator = SamAutomaticMaskGenerator(sam)
result = mask_generator.generate(image_rgb)
```

In around 6 seconds, this returns a list filled with dictionaries, each representing a mask for a specific object automatically extracted, accompanied by its scores and metadata. For a detailed view, the result is a list of dictionaries where each dict holds the following information:

segmentation
    This brings out the mask with (W, H) shape (and Bool type), where W (width) and H (height) target the original image dimensions.

area
    This is the area of the mask expressed in pixels.

bbox
    This is the boundary box detection in xywh format.

predicted_iou
    The model's predicted intersection over union (IoU) metric for the quality of the mask.

point_coords
    This is a list of the sampled input points that were used to generate the mask.

stability_score
    The stability score is an additional measure of the mask quality.

crop_box
    This is a list of the crop_box coordinates used to generate this mask in xywh format (it may differ from the bounding box).

Now that you have a better idea about what we are dealing with, to check out the results, we can plot the masks on top of the image with the following function:

```
def sam_masks(anns):
    if len(anns) == 0:
        return
    sorted_anns = sorted(anns, key=(lambda x: x['area']), reverse=True)
    ax = plt.gca()
    ax.set_autoscale_on(False)
    c_mask=[]
    for ann in sorted_anns:
        m = ann['segmentation']
        img = np.ones((m.shape[0], m.shape[1], 3))
        color_mask = np.random.random((1, 3)).tolist()[0]
        for i in range(3):
            img[:,:,i] = color_mask[i]
        ax.imshow(np.dstack((img, m*0.8)))
        c_mask.append(img)
    return c_mask
```

I admit that this is a bit blunt. But in this function, I sort out the masks by their area and plot them with a random color on top of the image with a transparency parameter:

```
print('Mem allocated by other programs: ', torch.cuda.memory_allocated(),
'reserved:', torch.cuda.memory_reserved())

import os
os.environ["PYTORCH_CUDA_ALLOC_CONF"] = "max_split_size_mb:512"
import gc
gc.collect()
torch.cuda.empty_cache()
```

If not enough GPU memory is free, try rebooting your (Windows) computer. You can also use the following line if memory problems persist:

```
mask_generator = SamAutomaticMaskGenerator(sam, points_per_batch=16)
```

Now, to plot and export the image, we write the following:

```
fig = plt.figure(figsize=(np.shape(image_rgb)[1]/72, np.shape(image_rgb)[0]/72))
fig.add_axes([0,0,1,1])
plt.imshow(image_rgb)
color_mask = sam_masks(result)
plt.axis('off')
plt.savefig("../test_result.jpg")
```

This results in Figure 13-23.

Figure 13-23. Before and after the Segment Anything Model

We have exciting results at this stage, and SAM is working really nicely! For example, you can see that almost all roofs are segments and that the three pools (two blue and one green) are also segments. Therefore, this could well be a starting point for complete automatic detection. If you run into memory errors depending on your computer setup while plotting the masks, loading a lighter SAM model should solve your problem. Now that we have a working SAM setup, let's apply all this hard-earned know-how to 3D point clouds.

3D Point Cloud to Image Projections

To understand the complex 3D world, we reuse the projections teachings from Chapter 4, especially the `generate_spherical_image` function. Now, to use this handy function, let's load and prepare the indoor point cloud first:

```
#Loading the las file from the disk
las = laspy.read("../DATA/indoorpcd.las")

#Transforming to a numpy array
coords = np.vstack((las.x, las.y, las.z))
point_cloud = coords.transpose()

#Gathering the colors
r=(las.red/65535*255).astype(int)
g=(las.green/65535*255).astype(int)
b=(las.blue/65535*255).astype(int)
colors = np.vstack((r,g,b)).transpose()
```

Once prepared, we can define the necessary parameters for projection. These are the center of projection (basically the position from which we want a virtual scan station) and the resolution of the final image (expressed in pixels, as the smallest side of the image):

```
resolution = 500

# Defining the position in the point cloud to generate a panorama
center_coordinates = [189, 60, 2]
```

Finally, we can call the new function to plot and export the results as an image:

```
# Function execution
spherical_image, mapping = generate_spherical_image(center_coordinates,
point_cloud, colors, resolution)

# Plotting with matplotlib
fig = plt.figure(figsize=(np.shape(spherical_image)[1]/72,
np.shape(spherical_image)[0]/72))
fig.add_axes([0,0,1,1])
plt.imshow(spherical_image)
plt.axis('off')

# Saving to the disk
plt.savefig("../DATA/ITC_BUILDING_spherical_projection.jpg")
```

This process results in Figure 13-24.

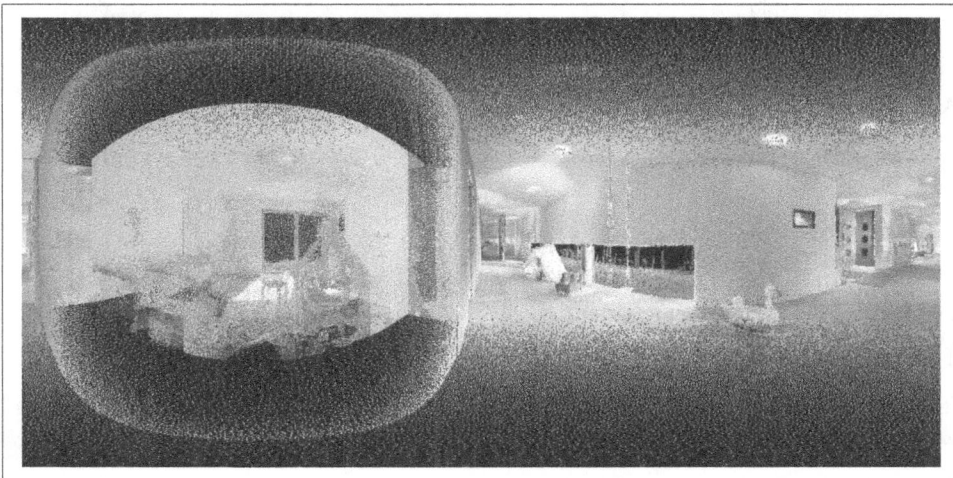

Figure 13-24. The 3D point cloud transformed as an equirectangular image from the projection

You can play around with the various parameters, such as the resolution or the center of projection, to ensure that you get a nice balance between "no data" pixels and relevant panorama. Now let's apply SAM.

Unsupervised Segmentation with SAM

In the case of nonlabeled outputs, we are utilizing SAM's segmentation architecture, which falls under clustering applications. As a reminder, this is opposed to supervised learning approaches that provide labeled outputs. Therefore, the transfer of pixel predictions, coupled with seamless point cloud export, showcases the potential for revolutionizing applications like object detection and scene understanding.

SAM segmentation

To execute the program, we can reexecute the code snippets that we used to test our SAM functionalities on 2D images, which are:

```
sam = sam_model_registry["vit_h"](checkpoint = MODEL)
sam.to(device = USED_D)

mask_generator = SamAutomaticMaskGenerator(sam)

temp_img = cv2.imread("../DATA/ITC_BUILDING_spherical_projection.jpg")
image_rgb = cv2.cvtColor(temp_img, cv2.COLOR_BGR2RGB)

t0 = time.time()
result = mask_generator.generate(image_rgb)
t1 = time.time()
```

And later, we can plot the results on the image itself:

```
fig = plt.figure(figsize=(np.shape(image_rgb)[1]/72, np.shape(image_rgb)[0]/72))
fig.add_axes([0,0,1,1])

plt.imshow(image_rgb)
color_mask = sam_masks(result)
plt.axis('off')
plt.savefig("../DATA/ITC_BUILDING_spherical_projection_segmented.jpg")
```

This results in Figure 13-25.

It already looks like we are delineating significant parts of the image. Let's move forward with the next steps.

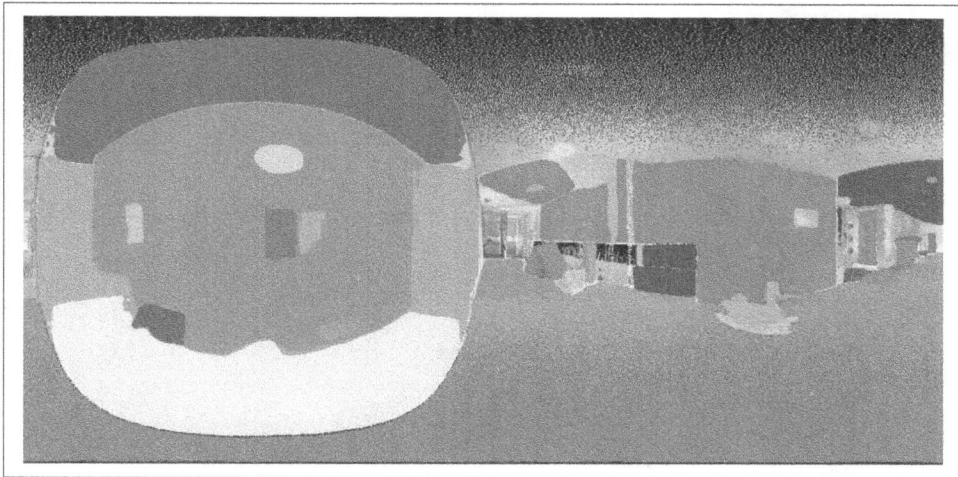

Figure 13-25. Results of the Segment Anything Model on the 3D point cloud projection

Point prediction transfer

We can color the point cloud with this image by defining a coloring function as follows:

```
def color_point_cloud(image_path, point_cloud, mapping):
    image = cv2.imread(image_path)
    h, w = image.shape[:2]
    modified_point_cloud = np.zeros((point_cloud.shape[0],
    point_cloud.shape[1]+3), dtype=np.float32)
    modified_point_cloud[:, :3] = point_cloud
    for iy in range(h):
        for ix in range(w):
            point_index = mapping[iy, ix]
            if point_index != -1:
                color = image[iy, ix]
                modified_point_cloud[point_index, 3:] = color
    return modified_point_cloud
```

This means that to color our point cloud, we can use the following code line that calls our new function:

```
modified_point_cloud = color_point_cloud(image_path, point_cloud, mapping)
```

This returns a NumPy array that holds the point cloud.

Point cloud export

To export the point cloud, you can use NumPy or laspy to extract a *.las* file directly. We proceed with the second solution:

```
def export_point_cloud(cloud_path, modified_point_cloud):
    # 1. Create a new header
    header = laspy.LasHeader(point_format=3, version="1.2")
    header.add_extra_dim(laspy.ExtraBytesParams(name="random", type=np.int32))

    # 2. Create a Las
    las_o = laspy.LasData(header)
    las_o.x = modified_point_cloud[:,0]
    las_o.y = modified_point_cloud[:,1]
    las_o.z = modified_point_cloud[:,2]
    las_o.red = modified_point_cloud[:,3]
    las_o.green = modified_point_cloud[:,4]
    las_o.blue = modified_point_cloud[:,5]
    las_o.write(cloud_path)

    print("Export successful at: ", cloud_path)
    return
```

And with this, we can export our `modified_point_cloud` variable:

```
export_point_cloud("../DATA/pcd_results.las", modified_point_cloud)
```

After this stage, we successfully took our various 2D images resulting from the 3D point cloud projection process. We applied the SAM algorithm, colorized it based on its prediction, and exported a colored point cloud. We can now move on to gaining insights about what we've obtained. As you can see, even with the uneven point distribution and "black zones," SAM is able to pick up what the main parts of the point cloud are about. Specifically, it likely highlights the duck-like furniture on the left, where the "mission's dangerous materials are located," as well as the doors and windows to provide the extraction team with the most direct route.

As we navigate through the 3D point cloud, SAM's segmented predictions bring clarity to the classical "mess of points," showcasing its potential in real-world applications (Figure 13-26). We see the great distinction of the major elements that compose the scene. We have a direct link between the points and pixels with a 100% automated segmentation process.

Acknowledging the rough patches along the way would complete our expedition. While impressive, SAM is not exempt from limitations. By recognizing these shortcomings, we pave the way for refinement and growth. First, all the "unseen" points remain unlabeled. This could limit complete processing; if you use the basic or large model, you will notice more unlabeled points than the Vit-H model. At this stage, we used the automatic prompting engine, which triggered around 50 points of interest, the seeds of the segmentation task. While this is great for getting a direct result, it would be ideal to have the possibility to tune it. Finally, the mapping is somewhat simple at this stage; it would largely benefit from occlusion culling and point selection for a specific pixel of interest.

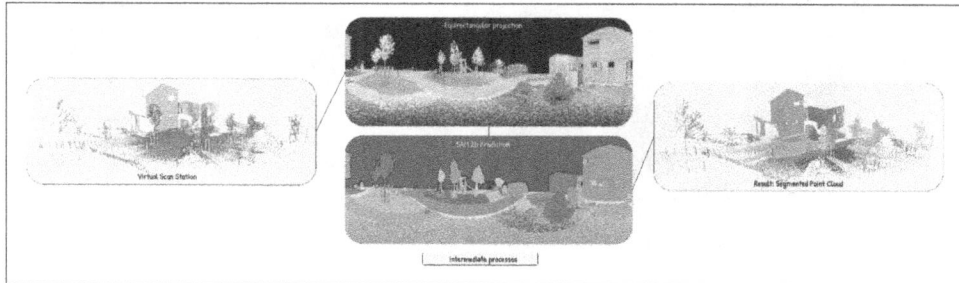

Figure 13-26. 3D point cloud unsupervised segmentation results using Segment Anything

Perspectives

As it stands, our implementation should work pretty well for any application where you can have some distinctive initial features for SAM. As you can see in Figure 13-27, it also works for top-down aerial point clouds.

Figure 13-27. The results of Segment Anything 3D for aerial point clouds

Extending to other indoor scenarios, you can also get some decent and interesting results (Figure 13-28). This may even be useful for changing the light bulb in the hall, automatically by a robot, of course (how many robots does it take to change a light bulb?).

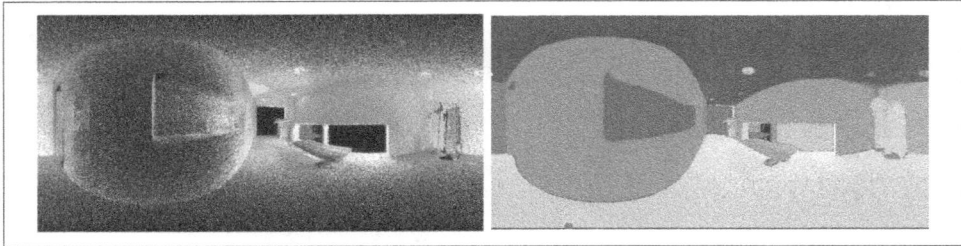

Figure 13-28. The results of Segment Anything 3D for another indoor scenario

Therefore, in addition to generalization, one perspective is to find a way to generate panoramas and fuse predictions from different viewpoints. Another approach is to expand to custom prompts and address the challenge of improving point-to-pixel accuracy in 2D to 3D mapping.

Summary

Throughout this chapter, we've worked with geometry and semantics through the lens of unsupervised learning with segmentation models. We harnessed the power of graph theory and the SAM, a versatile tool for grouping points based on geometric and semantic features.

This opens up large avenues to mix-and-match graphs and 2D and 3D domains through the art of projections. Of course, this suggests that if you want to extend the work, exploring how to fuse predictions from multiple viewpoints would be beneficial for more comprehensive segmentation. Additionally, we could develop custom prompts to enhance segmentation accuracy or improve point-to-pixel mapping for better 2D-3D correspondence. Finally, extending to complex indoor and outdoor scenarios and integrating other AI models are potential extensions for an enhanced semantic understanding. This means that you are now able to provide a solution for:

- Setting up a Python environment for 3D point cloud processing and deep learning
- Using graph theory for Euclidean clustering of 3D datasets
- Converting 3D point clouds to 2D projections and vice versa
- Applying the SAM to both 2D and 3D data
- Exporting and visualizing segmented point clouds
- Understanding the strengths and limitations of unsupervised segmentation techniques

Equipped with these segmentation skills, we're now prepared for our next chapter, which explores supervised learning. Here, we transition from unsupervised learning to systems where human guidance is key.

Hands-on Project

I have an exercise for you. Take a different 3D point cloud dataset (e.g., an outdoor scene or a different indoor environment) and apply the Segment Anything Model pipeline we have learned. Experiment with varying methods of projection, SAM model sizes, and segmentation parameters. Compare the results and reflect on the following questions:

1. How well does SAM perform on your chosen dataset?
2. What challenges did you encounter, and how did you overcome them?
3. Can you think of any specific applications for the segmentation results you obtained?
4. How might you improve the segmentation results further?

Completing this exercise will give you hands-on experience with real-world data and deepen your understanding of unsupervised 3D point cloud segmentation using cutting-edge AI models.

Supervised 3D Machine Learning Fundamentals

We aim to design machines that can perceive space precisely as humans do. This means understanding depth, predicting spatial relationships, recognizing objects, and making intelligent decisions in digital environments. This is a complex goal but more manageable when broken into chunks. In previous chapters, we solved the first two chunks: understanding depth and finding spatial relationships. But what about recognizing objects with a clear concept attached to them?

This could allow us to build systems that analyze medical images to detect anomalies like tumors, helping physicians formulate diagnostic and treatment plans. Beyond healthcare, self-driving cars rely heavily on 3D perception to navigate complex environments safely. We can create autonomous driving assistance by identifying obstacles, pedestrians, and traffic signs using LiDAR sensors and cameras. Again, every 3D data science application detailed in Chapter 1 relies on semantic extraction.

However, transforming complex 3D data into meaningful insights is the fundamental challenge. 3D machine learning, with supervised learning models, brings a critical paradigm shift in this context.

Point clouds, depth maps, meshes, and volumetric representations all contain intricate information that classical algorithms struggle to interpret. We previously saw that unsupervised learning can help us "decode" the overall meaning of our 3D environments. This is a first step toward a full semantic description of objects. But we need to move one extra stride, i.e., give meaning to our main constituents for a complete understanding of 3D scenes.

This chapter aims to find a supervised learning solution that helps "teach" computational systems to comprehend spatial data. We want to leverage sophisticated features (like those extracted in Chapters 6 to 9) and combine them with "ground-truth labels" to design a 3D machine learning model capable of predicting semantic labels in unseen 3D datasets.

> Let me highlight the three high-level tasks where you may recognize a supervised learning AI footprint: 3D reconstruction, 3D scene understanding, and 3D visualization. Within these buckets, more specialized tracks leverage generative AI, differentiated rendering, and large language models discussed in the last chapter. While these are exciting avenues, it is essential to understand how to build on some core concepts that are part of this chapter's teachings.

By the end of this chapter, you will have the skills to create, train, and deploy your 3D scene understanding model, by going through three main sections of the workflow shown in Figure 14-1.

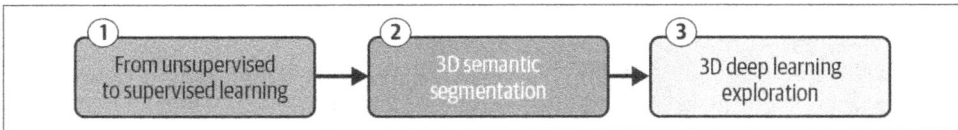

Figure 14-1. 3D machine learning fundamentals workflow

The first section compares supervised learning to unsupervised learning, highlighting how supervised learning enables a deeper understanding of 3D scenes by assigning meaning to objects within a scene. The chapter then focuses on 3D semantic segmentation, a specific type of supervised learning where every data unit (like a point, voxel, or triangle) is classified, leading to a comprehensive understanding of the 3D environment. Finally, we transition onto the difference with 3D deep learning architectures, introducing our future chapters.

In this chapter, we set out to create systems that could perceive the 3D world and truly comprehend it. It's not enough to simply capture the geometry; we need to imbue our systems with the ability to understand what those shapes and points represent. This leads us to the core of this chapter: supervised learning, specifically in the context of 3D semantic segmentation. By providing our algorithms with labeled data and carefully selected features, we demonstrate how to train models that could distinguish between different classes of objects within a 3D point cloud, achieving impressive levels of accuracy. We leverage the pandas (*https://oreil.ly/zJmjj*), scikit-learn (*https://scikit-learn.org*), Matplotlib (*https://matplotlib.org*), Pickle (*https://oreil.ly/yxAJg*) (to save the trained machine learning model to a file for later use), and NumPy (*https://numpy.org*) Python libraries. As always, you can find all the materials for this chapter on the 3D Data Science Resource Hub (*https://learngeodata.eu/3d-data-science-with-python*). To access the files (code, datasets, articles) and resources, you may need to share an email address, a personal password, or proof that you are the book's owner.

From Unsupervised to Supervised Learning

Most modern applications build on the ability to recognize objects of interest with a defined delineation strategy. The fundamentals of these approaches usually depend on the system's capacity to inject/retrieve semantics from a scene.

By training algorithms on large 3D datasets, we can develop systems capable of performing tasks that were once the exclusive domain of human experts. In Chapters 12 and 13, we saw how to accelerate the generation of labels for 3D datasets through unsupervised learning techniques. This chapter focuses on the second main track: supervised learning, as illustrated in Figure 14-2. Let me share some preliminary words on supervised learning in the context of 3D scene understanding before moving into hands-on work with Python development.

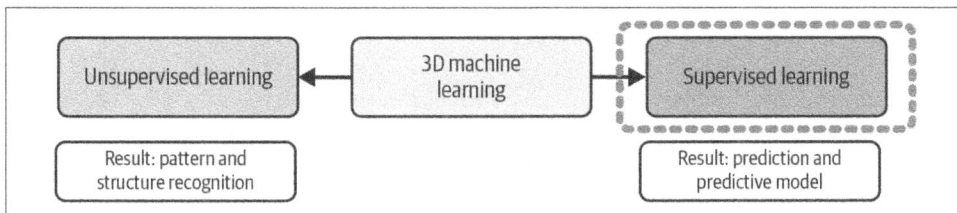

Figure 14-2. 3D machine learning main tracks: unsupervised and supervised learning

You have some other tracks aside from unsupervised and supervised learning, which are beyond the scope of this chapter. The first is reinforcement learning (*https://oreil.ly/f0N6n*), which develops optimal decision-making strategies through trial-and-error interactions with an environment, maximizing cumulative rewards. Then, you have hybrid learning methods such as self-supervised learning (*https://oreil.ly/7kmtf*) (generates supervisory signals from the data, creating proxy tasks that enable learning representations without explicit human annotations), semisupervised learning (*https://oreil.ly/nS9fQ*) (leverages a small set of labeled data combined with a larger pool of unlabeled data to improve learning performance), or active learning (*https://oreil.ly/hO8Lu*) (strategically selects the most informative data points for manual labeling, minimizing annotation effort while maximizing model performance).

Supervised Learning Concepts

Supervised machine learning entails training a model using labeled data to discern patterns and relationships (phase 1 in Figure 14-3). The logic is then captured and subsequently applied to generate accurate predictions on unseen data (phase 2 in Figure 14-3).

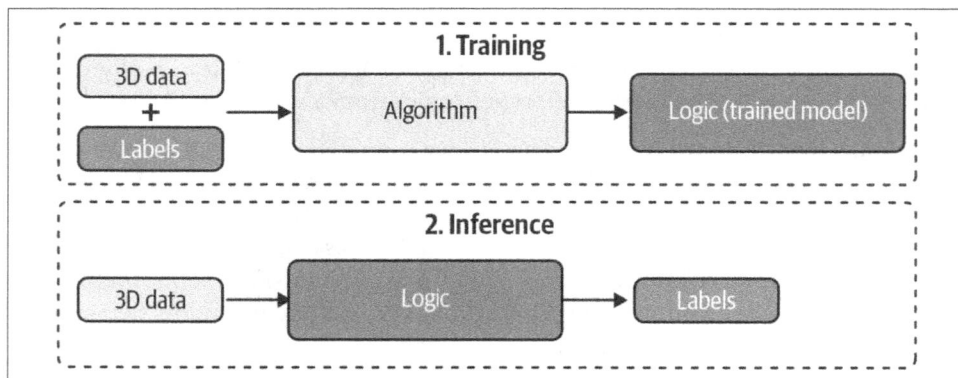

Figure 14-3. Training versus inference in 3D machine learning

Therefore, training involves "teaching" a model internal logic on labeled data, meaning each input is paired with a *ground truth* label. The model can then improve the quality of its predictions by iteratively checking its guesses against this ground truth label.

The overall objective of supervised learning is achieved in stage 2: inference. Once we have a coherent trained model, we want to test its logic on new data it hasn't seen before and monitor the quality of the predictions. Take the case illustrated in Figure 14-4, which shows the constitution of a trained 3D machine learning model to classify 3D shapes as ground or buildings.

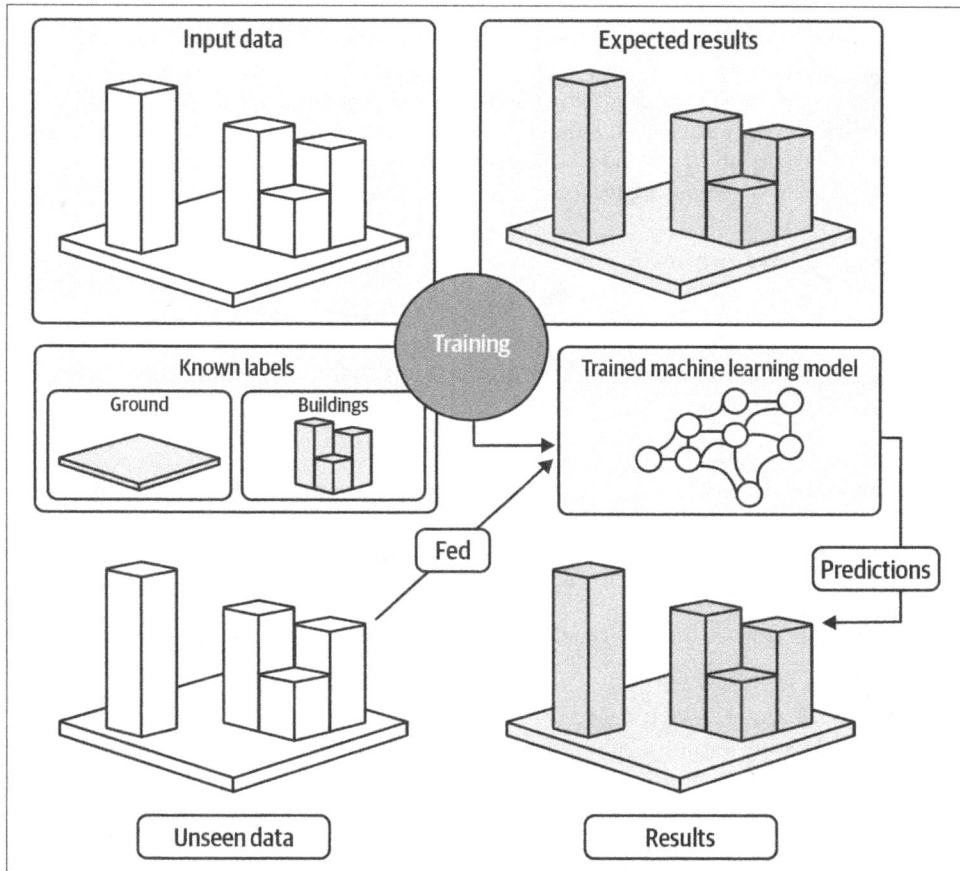

Figure 14-4. The supervised learning workflow: the training phase leads to a machine learning model that predicts labels over unseen data

Our input data (features; see Chapter 6) is associated with known labels (ground and building) and constitutes our training data. We then leave the task of defining a specific model's logic for use with unseen data to the training process. The model is ready when it proposes good predictions for unseen datasets where we want to detect buildings and the ground.

> I know, at this stage, you have many questions. How can we quantitatively know when a prediction is good? What if the unseen data presents a widely different statistical distribution? What if we get good results on the training data but not on the testing data? Let's keep these questions and answer them with practical Python code in the following sections.

Therefore, we have two main phases in the training stage to constitute a sound 3D machine learning model:

Training data preparation
    You give the model a prepared training dataset that has relevant input data (features) and output data that goes with it (labels).

Learning
    By working with the training data and some parameters, the algorithm learns how to connect the features it receives with the labels it gives. One way to do this is to change the model's settings so that the difference between what it says and finds is as tiny as possible.

Once locked in a state, our 3D machine learning model uses new input features and output labels.

In Chapters 12 and 13, we focused on unsupervised learning. This can help you create labeled datasets (among other things) more efficiently, which solves one main challenge for 3D data. But usually, when we need to push our system to get highly accurate/exhaustive/precise predictions, we have to rely on the second aspect of 3D machine learning: supervised learning. On this "secondary" trunk, we have two major branches: classification and regression, as illustrated in Figure 14-5.

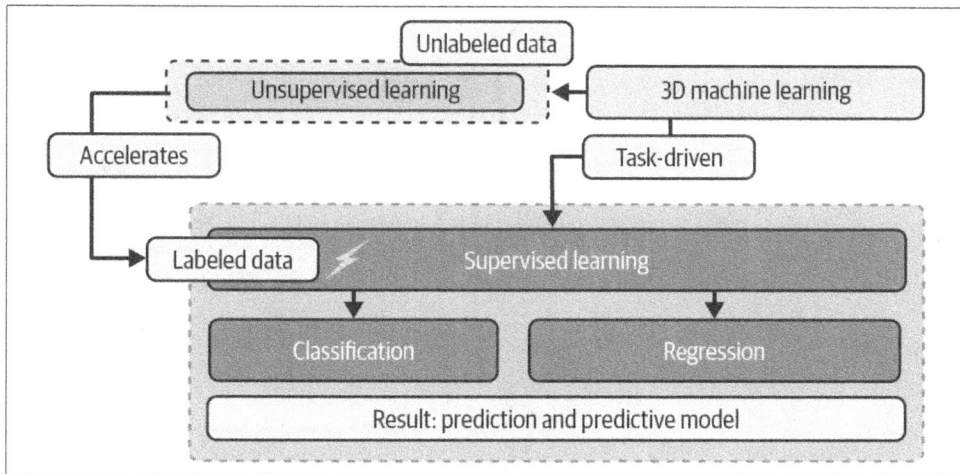

Figure 14-5. 3D machine learning: supervised learning for classification tasks

In a regression scenario, we want to predict a continuous variable, e.g., estimating the mathematical best fit of a set of points. In contrast, for classification scenarios, we want to assign a discrete categorical label (label 1, label 2, label *n*), e.g., this data point is a car (label 1), and this other one is a tree (label 2). I illustrate the differences in Figure 14-6.

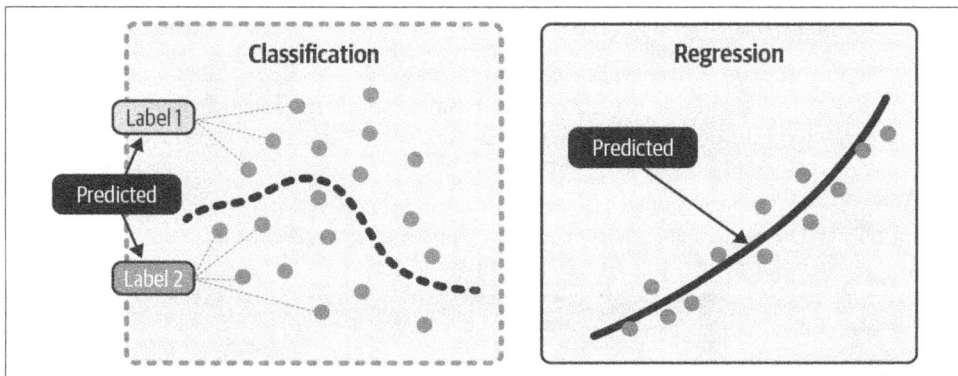

Figure 14-6. Classification versus regression: the predicted element is a discrete categorical value (classification) or a continuous variable (regression)

I compiled in Table 14-1 the main objective, output, characteristics, and performance metrics that we usually use for each task-driven task.

Table 14-1. Regression versus classification tasks

Model type	Primary objective	Output type	Key characteristics	Performance metrics
Regression	Predict continuous numerical values	Continuous scalar / vector	Predicts exact numeric values Handles continuous data distributions	Mean squared error (MSE) R-squared (R^2) Mean absolute error (MAE)
Classification	Assign discrete categorical labels	Discrete class labels	Determines class membership Probabilistic output possible	Accuracy Precision Recall F1-score, mAP

As you can see, to assess the quality of our model's performance, we have various metrics explained in depth in the Resource Hub (*https://learngeodata.eu/3d-data-science-with-python*).

Previous chapters, especially Chapters 9 and 10, addressed regression (linear and nonlinear) cases to generate 3D shapes from 3D point clouds. Therefore, in this chapter, we focus on classification models. Let's quickly discuss the strategy we can employ to detect and delineate objects with some predefined granularity (at the point level, at the group level, at the object level, at the dataset level).

Supervised Learning Classification

If you remember from Chapter 2, we can add semantics to 3D scenes through various strategies. Let me highlight the three main classification tasks you can encounter in your journey: semantic segmentation, instance segmentation, and 3D object detection, as summarized in Table 14-2.

Table 14-2. The three subcategories of classification strategies: semantic segmentation, instance segmentation, and 3D object detection

Model type	Primary objective	Output type	Key characteristics
Semantic segmentation	Assign semantic labels to every spatial data unit (point, triangle, segment, etc.)	Unit/point-wise class labels	Fine-grained scene understanding Per-point classification Preserves spatial relationships
Instance segmentation	Identify and differentiate individual object instances	Unique instance labels per object	Distinguishes between objects of the same class Assigns unique identifiers Combines detection and segmentation
3D object detection	Locate and classify 3D objects with spatial information	3D bounding boxes with class labels	Provides 3D spatial coordinates Orientation and dimensions High spatial awareness

The choice is usually task driven and highly dependent on the analytic granularity that you need for your application. 3D semantic segmentation provides a more comprehensive understanding of spatial environments by classifying every data unit (point, voxel, triangle), which enables holistic scene comprehension.

Unlike instance segmentation or object detection, which focuses on individual object boundaries or specific instances, semantic segmentation captures the intrinsic relationships and contextual information within the entire 3D space, which usually reveals nuanced structural and categorical insights that are essential for intelligent spatial reasoning and decision making. With the methods provided in Chapters 12 and 13, we can detect the instances with clustering approaches using each class independently as input. Therefore, let's focus on developing a 3D semantic segmentation system.

3D Semantic Segmentation Example

We are in the context of supervised learning: we want to develop an algorithm that can associate input data with specified output values. This means training a model using a labeled dataset, where outputs are supplied for each input.

Upon completion of training, the model should be capable of generating predictions on novel, unobserved data. This means that the model learns to map input data (features) to corresponding output values (labels) during training.

Let me share some examples with scikit-learn:

```
import numpy as np
from sklearn.neighbors import KNeighborsClassifier
from sklearn.metrics import accuracy_score
```

Features are the specific characteristics or attributes extracted from the input data. In the context of 3D data, features can include geometric properties like shape, size, and orientation; statistical measures like mean and variance; or handcrafted descriptors. For simplicity, let's assume we'll use the nearest-neighbor distance as a feature. This can be calculated using scikit-learn's `NearestNeighbors` class:

```
from sklearn.neighbors import NearestNeighbors

# Find the k-nearest neighbors for each point
knn = NearestNeighbors(n_neighbors=5)
knn.fit(point_cloud)
distances, _ = knn.kneighbors(point_cloud)

# Calculate the mean distance to neighbors as a feature
features = np.mean(distances, axis=1)
```

Now, we are almost ready to train a supervised model, such as a *k*-NN classifier (*https://oreil.ly/zurTd*), on these features. All we need is labels. Labels are the desired output values associated with each data point. These are categorical, such as "car," "pedestrian," or "tree." (For regression tasks, these labels would be continuous, like object size or depth.) Now, we can train our classification model to learn the relationship between features and labels.

> Supervised learning can leverage different algorithms (or models), each with its own unique characteristics and applications. I compiled for your reference the most common types of supervised learning algorithms: linear regression (*https://oreil.ly/J09Yh*), logistic regression (*https://oreil.ly/pj8bS*), support vector machines (*https://oreil.ly/Wdmlj*), decision trees (*https://oreil.ly/ocgil*), random forests (*https://oreil.ly/rGl1Y*), gradient boosting (*https://oreil.ly/F2Q0d*), naïve Bayes algorithm (*https://oreil.ly/KrPzE*), *k*-NN (*https://oreil.ly/C5JfV*), and neural networks (*https://oreil.ly/yuZr2*). In this chapter, we will leverage some of them for classification purposes.

Let me showcase the *k*-NN classifier on our dataset:

```
# Create a KNN classifier
knn_classifier = KNeighborsClassifier(n_neighbors=3)

# Train the classifier
knn_classifier.fit(features.reshape(-1, 1), labels)
```

Once a trained model is ready, it can be deployed to make predictions on any data that follows the same preprocessing steps to prepare the training data. Here, we try our model on our training data:

```
# Make predictions
predictions = knn_classifier.predict(features.reshape(-1, 1))
```

Now that the demystification process has been successfully applied, let's develop a complete 3D machine learning system.

3D Point Cloud Semantic Segmentation

We aim to extract semantic information and integrate it into our 3D data as point clouds. To achieve this, we will explore a specific strategy that helps us derive this information from the sensor. We focus on one algorithm family, supervised learning methods—as opposed to the unsupervised methods shown in the previous chapter.

A well-structured 3D machine learning system is essential for achieving optimal performance and scalability. The system design involves a specific linkage of phases and various factors, including data preprocessing, feature engineering, model selection, training methodologies, and inference optimization (Figure 14-7).

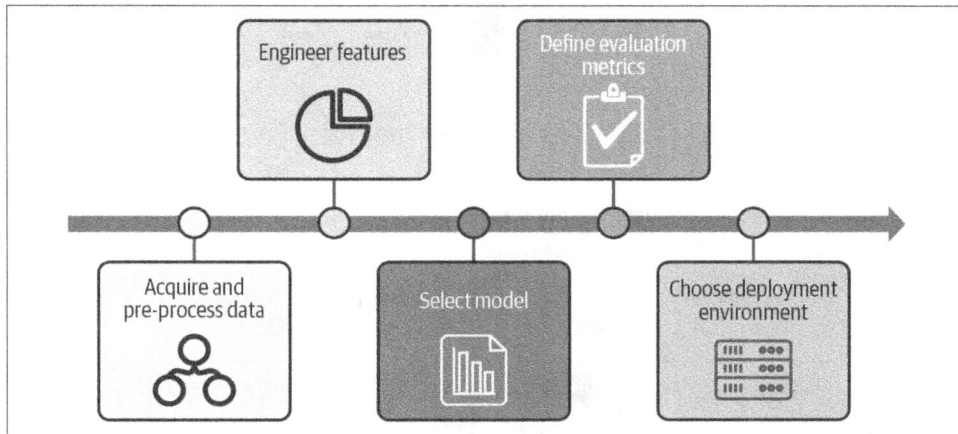

Figure 14-7. The various stages of a 3D machine learning development cycle

---

3D Machine Learning Checklist

As a simple guide, it's good to keep these points in mind:

- Data acquisition and preprocessing
  - Identifying reliable and diverse data sources
  - Data cleaning to remove noise, outliers, and inconsistencies
  - Data augmentation to increase data variability
  - Data format conversion (e.g., point clouds to voxel grids)
  - Data normalization and standardization
- Feature engineering (only for non–deep learning designs)
  - Extracting relevant information from raw data
  - Handcrafted features (e.g., geometric, statistical)
  - Learned features (e.g., deep learning embeddings)
  - Feature selection and dimensionality reduction
- Model selection
  - Choosing appropriate algorithms based on problem type and dataset characteristics
  - Considering trade-offs between accuracy, speed, and interpretability
  - Exploring different model architectures
- Evaluation metrics
  - Defining relevant performance metrics (accuracy, precision, recall)

> — Considering domain-specific metrics (intersection over union for segmentation)
- Deployment environment
  > — Selecting hardware and software infrastructure for training and inference
  > — Considering scalability, performance, and cost-efficiency
  > — Designing for real-time or batch processing

I propose a straightforward procedure that you can easily replicate to train a 3D machine learning model and apply it to real-world scenarios (Figure 14-8).

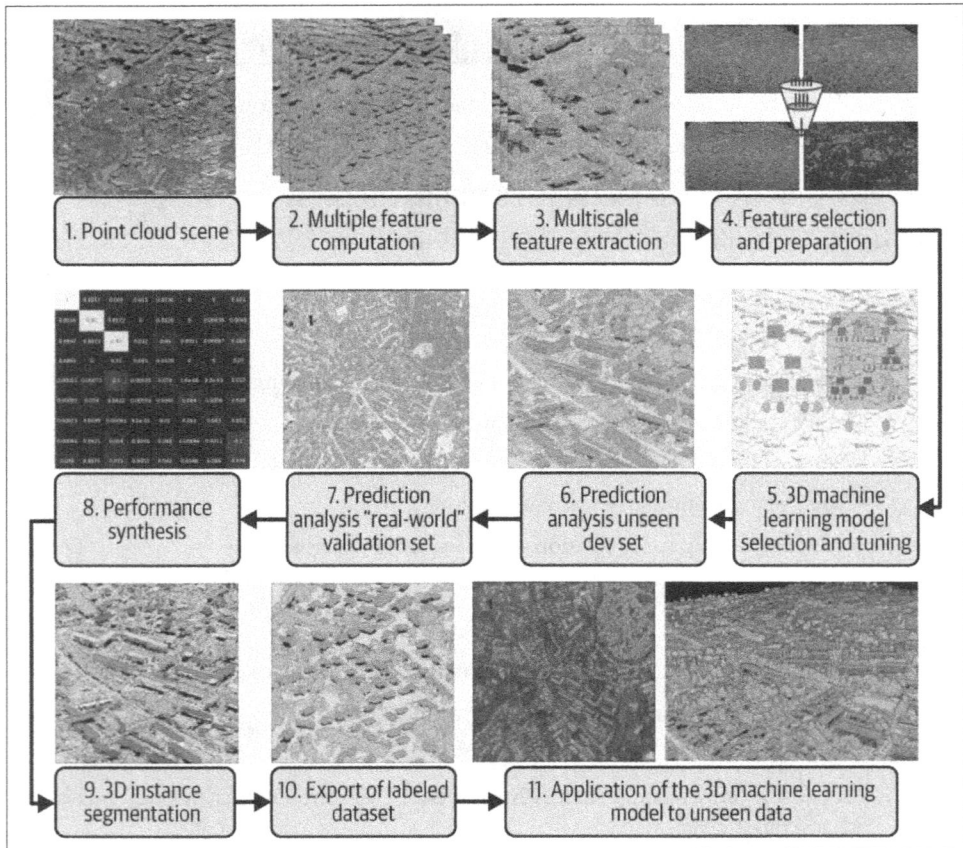

Figure 14-8. 3D Machine learning workflow: point cloud semantic segmentation

Since we've already covered steps 1 to 3 in previous chapters, let's move on to the next stages of our journey. But first, let's set up our favorite Python environment.

3D Python and Data Setup

We want to start with preparing our dataset and our Python environment.

3D LiDAR data

The first step we take is to dive into the web and source some 3D data. This time, I want to explore a French source for chilled LiDAR datasets: the National Geography Institute (IGN) of France. Through their LiDAR HD campaign, France has initiated an OpenData effort, offering crisp 3D point clouds of various regions. Even better, some of these datasets include labels, making it easier to start without building everything from scratch, as you'll find here: LiDAR HD (*https://oreil.ly/xfyCq*).

To keep things straightforward, I accessed the portal, selected data covering part of the city of Louhans, removed the georeferencing information, computed some additional attributes (which are explained in Chapter 6), and made it available. In Figure 14-9, on the left side, we see the point cloud with RGB colors, and on the right are the three classes of interest (ground, vegetation, and buildings) we will study. The data you're interested in is *3DML_urban_point_cloud.xyz* and *3DML_validation.xyz*.

Figure 14-9. A view of the aerial LiDAR point cloud data from the city of Louhans provided by IGN

3D Python environment setup

Let's focus on efficient and minimal library usage with only two libraries: pandas and scikit-learn. With these, we'll achieve great results using just five lines of code to start your script:

```
import pandas as pd

# scikit-learn imports
from sklearn.model_selection import train_test_split
from sklearn.metrics import classification_report
from sklearn.ensemble import RandomForestClassifier
from sklearn.preprocessing import MinMaxScaler
```

As shown, I import functions and modules from the libraries in different ways. For pandas, I use the `import` module, which demands less maintenance of the import statements. However, when you want more control over which items of a module can be accessed, I recommend using `from module import foo`.

From there, I propose that we relatively express our paths, separating the `data_folder` containing our datasets from the dataset name to switch easily on the fly:

```
data_folder="../DATA/"
dataset="3DML_urban_point_cloud.xyz"
```

Now, we can quickly load the dataset into the `pcd` variable using pandas. And because the original file is not clean and contains NaN values, we will use the very handy `dropna` method to ensure we start with a filtered DataFrame with only complete rows. This means we will drop some points along the way (<1% of records), but this time we are OK with that:

```
pcd=pd.read_csv(data_folder+dataset,delimiter=' ')
pcd.dropna(inplace=True)
```

The `inplace` argument set to `True` allows us to directly replace the Python object instead of making a DataFrame copy. We can now move on to feature selection and preparation.

Feature Selection and Preparation

To grasp what we do when using machine learning frameworks, it's important to understand that we rely on feature sets or vectors of features that are variably representative. In our approach, the key is to thoroughly understand the context in which we operate and to be creative in engineering features that we believe will be effective descriptors of the variance in our data, or at least help us distinguish between the classes of interest.

By following the Chapter 7 feature extraction process, we are going to use:

- Spatial features: X, Y, Z coordinates
- Radiometric features: R, G, B values
- Planarity (from covariance matrix)
- Verticality (from covariance matrix)
- Omnivariance (from covariance matrix)
- Normals and normal change rate

These are illustrated in Figure 14-10.

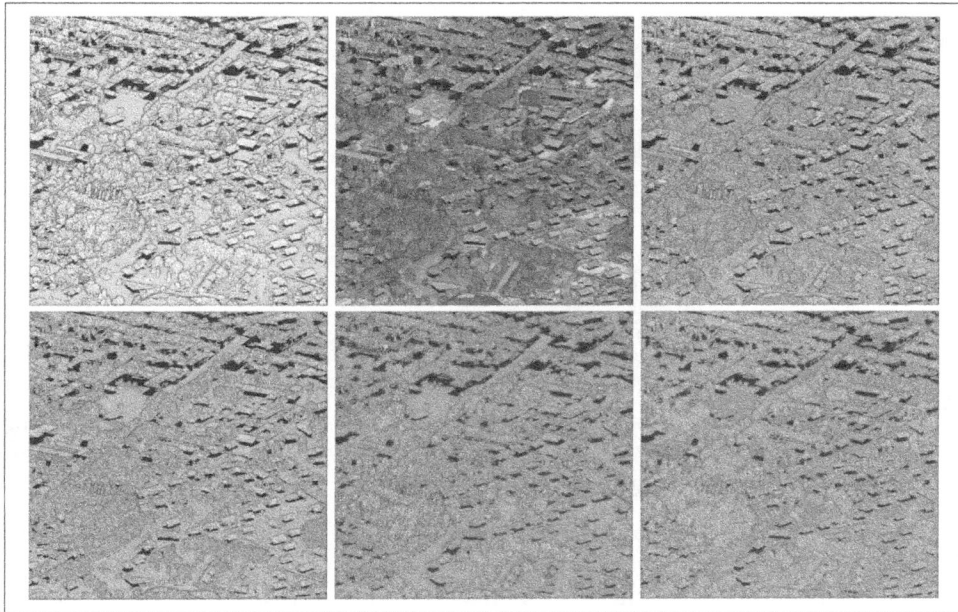

Figure 14-10. Aerial LiDAR point cloud training feature set from top left to bottom right: (0) raw point cloud, (1) RGB colors, (2) planarity, (3) verticality, (4) omnivariance, and (5) normal change rate

For the sake of simplicity, I computed several of them for you, filtered with the primary intention that they are pertinent for subsequent semantic segmentation tasks (Figure 14-11).

Figure 14-11. DataFrame extract

To establish a solid foundation, we organize our data into labels—what we aim to predict—and features—what we will use to make those predictions. With pandas, we can easily accomplish this in just two lines of code:

```
labels=pcd['Classification']
features=pcd[['X','Y','Z','R','G','B']]
```

This DataFrame structure permits quickly switching to a specific set of pertinent features without using numeric indexes. Thus, feel free to return to this step and change the features vector set.

Selecting the features

Feature selection is the process of minimizing the input variables utilized by a model by retaining just the most pertinent ones and discarding extraneous data. It is the procedure of selecting characteristics relevant to your machine learning model according to the specific problem you aim to address. If executed mechanically, this pertains to the AutoML technique (*https://oreil.ly/cjXff*) of automating the application of machine learning to practical issues.

You have two options: either a fully unsupervised approach (e.g., reducing correlations among your features) or a supervised approach (e.g., aiming to improve the model's final score by adjusting features and parameters). To ensure clarity, we manually modify the selection of our existing feature vector in a supervised manner: we will conduct experiments and adjust if the outcomes are unsatisfactory. Ready?

Preparing the features

Once our initial feature vector is ready, we can rush into processing. Or can we?

Be wary! We may encounter some surprises depending on the machine learning model we use. Indeed, let's look at a simple scenario. Let's imagine that our selected feature vector is the following:

```
features=pcd[['X','Y']]
```

If we use that to train our algorithm, we would be stuck to the observed range, e.g., X varying between 0 and 100. If, after training on this range, the model is fed future data with a similar distribution but a different range, e.g., X from 1100 to 1200, then we could get catastrophic results, even if it is the same dataset with just a translation. Indeed, for some models, an X value above 100 can make the model predict an erroneous value, whereas if we ensured that we translated the data to the same range observed in training, the predictions would have a better chance of making sense.

Let's turn to the concept of feature scaling and normalization. It is a crucial part of the data preprocessing stage, but I have seen many beginners overlook it (to the detriment of their machine learning model). This is a mistake we do not want to make.

Normalization and standardization are critical preprocessing techniques in 3D data science that transform raw spatial data into mathematically consistent and comparable representations. These approaches address fundamental challenges by preparing datasets for optimal model performance and interpretation.

Min-max scaling represents the most intuitive normalization method, linearly mapping feature values to a predefined range, typically between zero and one. By preserving the original data distribution's shape while constraining values, this technique ensures that each feature contributes proportionally without being dominated by extreme magnitudes. Because we are in a spatial-heavy context, this is a good way to avoid generalization problems. For this, we can use the `MinMaxScaler` function:

```
from sklearn.preprocessing import MinMaxScaler
features_scaled = MinMaxScaler().fit_transform(features)
```

Other approaches are worth mentioning. Z-score normalization (*https://oreil.ly/xMZuL*), or standardization, centers data around zero with unit variance, making it particularly powerful for normally distributed datasets. With 3D point clouds, z-score normalization helps mitigate variations caused by sensor noise, calibration differences, or measurement inconsistencies. Robust scaling (*https://oreil.ly/YVX2G*) is a great alternative, utilizing median and interquartile range to minimize outlier impacts. This technique is especially valuable with data presenting significant measurement variations, such as point clouds from terrestrial laser scanning or mobile mapping systems. By reducing sensitivity to extreme values, robust scaling provides a more stable normalization approach for complex spatial datasets.

Now that our feature vector is ready, we can move on to the next stage: setting up our training workbench.

3D machine learning training setup

We have a labels vector and a proper features vector. Now, we need to prepare a setup for the training phase. Training a 3D machine learning model involves iteratively improving its performance on our dataset. If I had to summarize the key steps, I would include:

Data splitting (https://oreil.ly/HyTnc)
Dividing the dataset into training, validation, and testing sets

Hyperparameter tuning (https://oreil.ly/rue3u)
Optimizing model parameters through techniques like grid search, random search, or Bayesian optimization

Model training (https://oreil.ly/-dCwr)
  Iteratively updating model parameters using optimization algorithms to minimize the loss function

Regularization (https://oreil.ly/Q36k0)
  Preventing overfitting by adding constraints or penalties to the model

Early stopping (https://oreil.ly/jkbZq)
  Terminating training to avoid overfitting by monitoring performance on a validation set

To start, we split both vectors—while keeping a proper index match between labels and features—to use a portion for training the machine learning model and another portion only for examining performance.

We use 60% of the data for training, and 40% for looking at performance, both taken from the same distribution randomly. We make the split using the `train_test_split` function from scikit-learn:

```
X_train, X_test, y_train, y_test = train_test_split(features_scaled, labels,
test_size=0.4)
```

We apply naming conventions when dealing with data for machine learning tasks. X denotes the features (or data) fed to a model and y denotes the labels. Each is decomposed into `_train` or `_test` depending on their finality.

We then create a "classifier object" through:

```
rf_classifier = RandomForestClassifier()
```

The aforementioned classifier is a random forest classifier. It employs multiple decision tree classifiers on diverse feature subsamples and uses averaging to enhance forecast accuracy and mitigate overfitting.

After classifier initialization, we fit the classifier to the training data to adjust its core parameters. This phase is the training phase, which can take some minutes depending on the hyperparameters (i.e., the parameters that define the machine learning model architecture) that we used beforehand (number of trees, depth):

```
rf_classifier.fit(X_train, y_train)
```

And at the end, we have a trained model. Yes, it is that easy! It's also easy to take shortcuts. During the prediction phase, regardless of whether you have labels or not, you need to do the following:

```
rf_predictions = rf_classifier.predict(X_test)
```

You can then visualize results and differences with the following code block that will create three subplots: the 3D point cloud data ground truth, the predictions, and the difference between both:

```
fig, axs = plt.subplots(1, 3, figsize=(20,5))
axs[0].scatter(X_test['X'], X_test['Y'], c =y_test, s=0.05)
axs[0].set_title('3D Point Cloud Ground Truth')
axs[1].scatter(X_test['X'], X_test['Y'], c = rf_predictions, s=0.05)
axs[1].set_title('3D Point Cloud Predictions')
axs[2].scatter(X_test['X'], X_test['Y'], c = y_test-rf_predictions,
cmap = plt.cm.rainbow, s=0.5*(y_test-rf_predictions))
axs[2].set_title('Differences')
```

If you want to check out some metrics, we can print a classification report with a bunch of numbers using the `classification_report` function of scikit-learn:

```
print(classification_report(y_test, rf_predictions))
```

Let's now analyze the metrics to assess the quality of our trained 3D machine learning model.

Metrics and Models

Evaluating model performance is crucial for understanding its capabilities and limitations. Let's dive into the most important metrics.

Performance and metrics

We can use several quantitative metrics for assessing 3D semantic segmentation and classification outcomes. I'll introduce four metrics that are useful for 3D point cloud semantic segmentation assessment: precision, recall, F1-score, and overall accuracy. They all depend on what we call true positive and true negative:

True positive (TP)
    Observation is positive and is predicted to be positive.

False negative (FN)
    Observation is positive but is predicted negative.

True negative (TN)
    Observation is negative and is predicted to be negative.

False positive (FP)
    Observation is negative but is predicted positive.

These are extracted from the confusion matrix as illustrated in Figure 14-12.

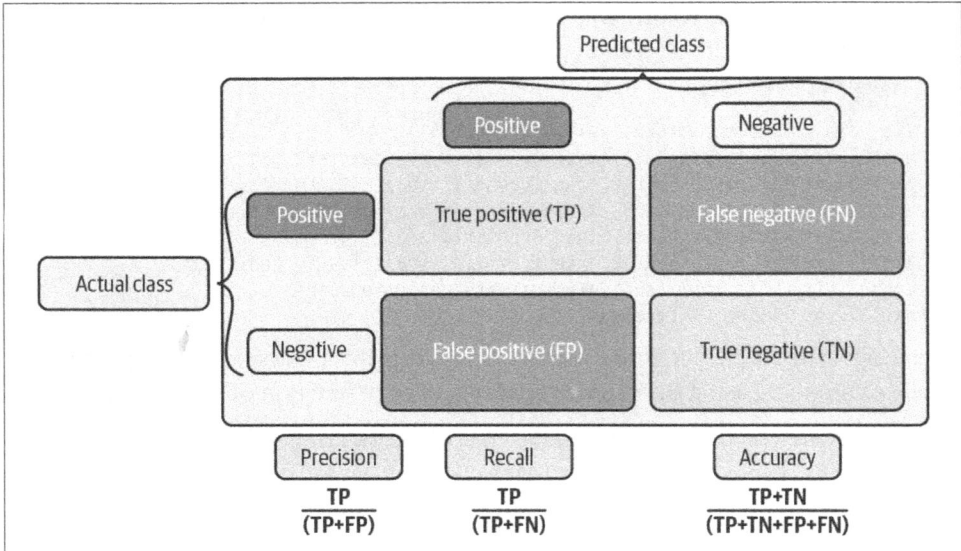

Figure 14-12. Metrics for 3D machine learning: classification cases

The *overall accuracy* is a comprehensive metric assessing the classifier's efficacy in accurately predicting labels across all observations. *Precision* refers to the classifier's capability to avoid incorrectly labeling negative samples as positive, whereas *recall* intuitively denotes the classifier's capacity to identify all positive examples. The F1-score serves as a weighted harmonic mean of precision and recall, providing a singular metric for evaluating classifier performance.

Alternative global accuracy metrics are unsuitable evaluation measures when class frequencies are imbalanced, a common occurrence in both natural indoor and outdoor environments, as the predominant classes skew the results. Henceforth, the F1-score in our experiments indicates the average performance of a proposed classifier. You can also explore mean average precision (mAP), intersection over union (IoU), or the Dice coefficient for classification metrics. As for regression metrics, these are the mean squared error (MSE), mean absolute error (MAE), or R-squared.

Now, let's select a model.

Model selection

It is time to select a specific 3D machine learning model. For this chapter, I limited the choice to three machine learning models: random forests, *k*-NN, and a multilayer perceptron that falls within the deep learning category. To use them, we will first import the necessary functions with the following:

```
from sklearn.neighbors import RandomForestClassifier
rf_classifier = RandomForestClassifier()
```

```
from sklearn.neighbors import KNeighborsClassifier
knn_classifier = KNeighborsClassifier()
from sklearn.neural_network import MLPClassifier
mlp_classifier = MLPClassifier(solver='lbfgs',
alpha=1e-5,hidden_layer_sizes=(15, 2), random_state=1)
```

Then, you just have to replace XXXClassifier() in the following code block with the wanted algorithm stack:

```
XXX_classifier = XXXClassifier()
XXX_classifier.fit(X_train, y_train)
XXX_predictions = XXXclassifier.predict(X_test)
print(classification_report(y_test, XXX_predictions, target_names=['ground',
'vegetation','buildings']))
```

For simplicity, I passed the list of the tree classes that correspond to the ground, vegetation, and buildings present in our datasets to the classification_report. And now on to the test phase using the three classifiers with the following parameters:

```
Train / Test Data: 60%/40%
Number of Point in the test set: 1 351 791 / 3 379 477 pts
Features selected: ['X','Y','Z','R','G','B'] - With Normalization
```

Let's first explore the results with random forests.

Random forests

We start with random forests. Some sort of magical tree conjuration through an ensemble algorithm that combines multiple decision trees to give us the final result: an overall accuracy of 98%, based on a support of 1.3 million points:

classes	precision	recall	f1-score	support
ground	0.99	1.00	1.00	690670
vegetation	0.97	0.98	0.98	428324
buildings	0.97	0.94	0.96	232797

It is further decomposed in Figure 14-13.

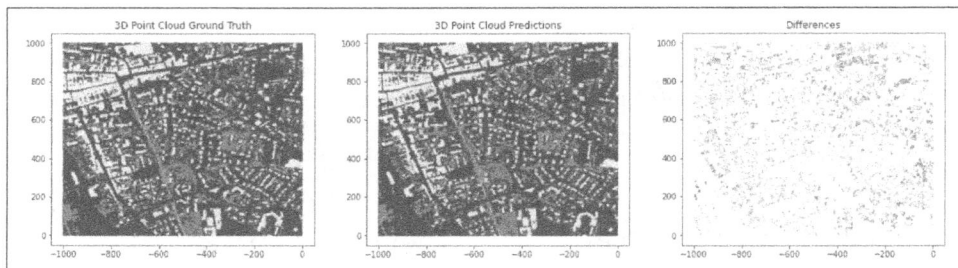

Figure 14-13. Results of the supervised machine learning random forest approach on the 3D aerial LiDAR point cloud

The ground points are almost perfectly classified: the 1.00 recall means that all points that belong to the ground were found, and the 0.99 precision means that there is still a tiny margin of improvement to ensure no false positives. We can see that the errors are distributed a bit everywhere, which could be problematic if this would have to be corrected manually.

Now, what can *k*-NN propose?

k-nearest neighbors

The *k*-NN classifier uses proximity to make predictions about an individual data point grouping. We obtain 91% global accuracy:

classes	precision	recall	f1-score	support
ground	0.92	0.90	0.91	690670
vegetation	0.88	0.91	0.90	428324
buildings	0.92	0.92	0.92	232797

This is further decomposed in Figure 14-14.

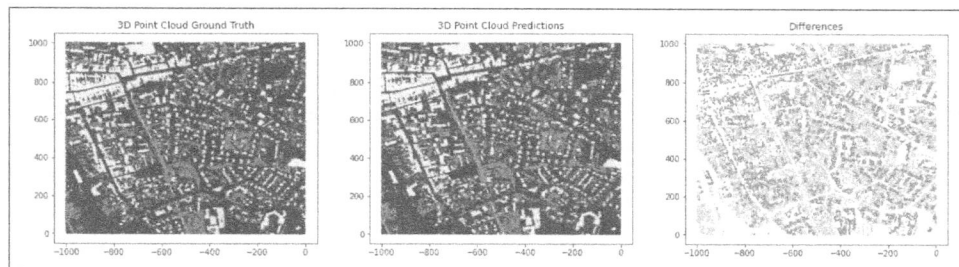

Figure 14-14. Results of the supervised machine learning k-nearest neighbors approach on the 3D Aerial LiDAR point cloud

The results are worse than random forests, which is expected because we are more subject to local noise in the current vector space. We have a homogeneous precision/recall balance over all classes, which is a good sign that we avoid overfitting problems (at least in the current distribution).

Now, let's explore the case of a specific artificial neural network: a multilayer neural perceptron.

Multilayer perceptron

The multilayer perceptron (MLP) is a neural network technique that learns linear and nonlinear data correlations. The MLP necessitates the adjustment of multiple hyper-parameters, including the quantity of hidden neurons, layers, and iterations, complicating the achievement of optimal performance immediately. For example, with the hyperparameters set, we have a global accuracy of 64%:

classes	precision	recall	f1-score	support
ground	0.63	0.76	0.69	690670
vegetation	0.69	0.74	0.71	428324
buildings	0.50	0.13	0.20	232797

This is illustrated in Figure 14-15.

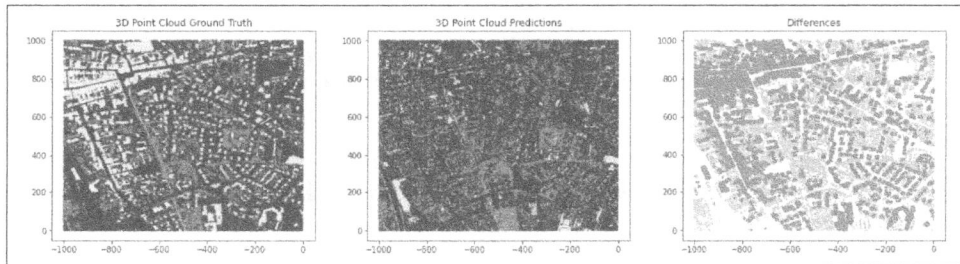

Figure 14-15. Results of supervised deep learning multilayer perceptron approach results on 3D aerial LiDAR point cloud

Intentionally, the MLP metrics illustrate what constitutes poor performance. We have an accuracy score lower than 75%, which is often the primary metric to target. Additionally, there are significant differences between intra- and interclass performance. Notably, the building class is far from robust, suggesting a potential overfitting problem. This is also evident visually, as it represents the primary source of confusion for the deep learning model.

At this stage, we won't improve through feature selection, but we always have the option available. We will choose the best-performing model: the random forest approach. Let's investigate if the currently trained model performs well under challenging unseen scenarios.

Inference and Generalization

Now, it gets tricky. Considering what we have discussed, scaling the current model to real-world applications that extend beyond the scope of the existing sample dataset could present significant challenges. Therefore, let's proceed with a full deployment of the model.

The test dataset

Instead of using only a training dataset and a validation dataset from the same distribution, it is vital to have another unseen dataset with different characteristics to measure real-world performances. This is a critical concept to help avoid issues related to overfitting. As such, we have:

Training data
> The data sample used to fit the model

Validation data
> The dataset employed to deliver an impartial assessment of a model trained on the training data while simultaneously adjusting the model's hyperparameters and feature vector. The evaluation is consequently somewhat skewed due to the utilization of this data to adjust input parameters.

Test data
> This uncorrelated data sample is utilized to deliver an impartial assessment of a final model calibrated on the training data.

The validation dataset may also contribute to other aspects of model preparation, including feature selection. The final model might have been fitted using the combined training and test data; however, we opted against it.

The chosen validation data is from the city of Manosque, which presents a different urban context with a different topography and a widely different urban context (Figure 14-16). This way, we increase the challenge of coping with generalization. On the left of Figure 14-16, we have the dataset from which we trained and evaluated (40% unseen). On the right, we have a different validation dataset, which will be a significant mark of the real-world possibilities given by our model.

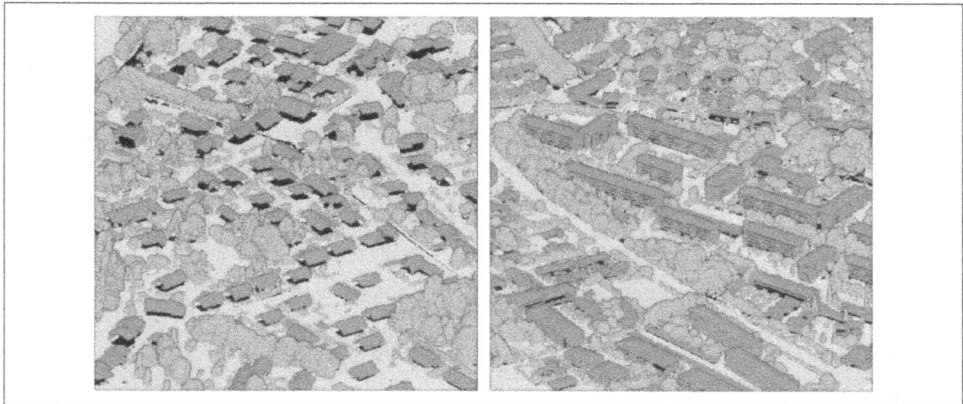

Figure 14-16. Training dataset (left) and validation data from the city of Manosque (right)

If you haven't already, you can download the *3DML_validation.xyz* dataset from the data given. As explained in the following, you will also find the labels to study metrics and potential gains on the different iterations.

Improving the generalization results

The goal is to check the test dataset results and see if we bypassed some possibilities. First, we import the test data in our script with the following three lines of code:

```
val_dataset="3DML_validation.xyz"
val_pcd=pd.read_csv(data_folder+dataset,delimiter=' ')
val_pcd.dropna(inplace=True)
```

Then, we prepare the feature vector to have the same features as the one used to train the model: no less, no more. We further normalize our feature vector to be in the same condition as our training data:

```
val_labels=val_pcd['Classification']
val_features=val_pcd[['X','Y','Z','R','G','B']]
val_features_scaled = MinMaxScaler().fit_transform(val_features)
```

Then we apply the already trained model to the test data, and we print the results:

```
val_predictions = rf_classifier.predict(val_features_scaled)
print(classification_report(val_labels, val_predictions, target_names=['ground',
'vegetation','buildings']))
```

That leaves us with a final accuracy of 54% for the 3.1 million points (against 98% for the test data containing 1.3 million points) present in the validation dataset:

classes	precision	recall	f1-score	support
ground	0.65	0.16	0.25	1188768
vegetation	0.59	0.85	0.70	1315231
buildings	0.43	0.67	0.53	613317

This is decomposed in Figure 14-17.

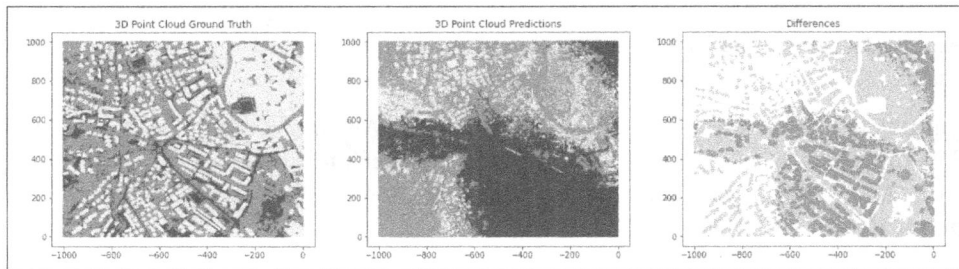

Figure 14-17. Qualitative results of the random forest classifier over the validation point cloud dataset: only spatial coordinates and R, G, B channels were used as input features

You just witnessed the true dark side power of machine learning: overfitting a model to a sample distribution and having massive trouble generalizing. Because we have already ensured that we normalized our data, we can investigate the possibility that this low-performance behavior may be due to features not being distinctive enough.

We used some of the most common/basic features. So let's improve this with a better feature selection, for example, the one in the following:

```
features=pcd[['Z','R','G','B','omnivariance_2','normal_cr_2','NumberOfReturns',
'planarity_2','omnivariance_1','verticality_1']]

val_features=val_pcd[['Z','R','G','B','omnivariance_2','normal_cr_2',
'NumberOfReturns','planarity_2','omnivariance_1','verticality_1']]
```

We now restart the training phase on the validation data. We check out the performance of the model and then we check out how it behaves on the validation dataset:

```
features_scaled = MinMaxScaler().fit_transform(features)
X_train, X_test, y_train, y_test = train_test_split(features_scaled, labels,
test_size=0.3)
rf_classifier = RandomForestClassifier(n_estimators = 10)
rf_classifier.fit(X_train, y_train)
rf_predictions = rf_classifier.predict(X_test)
print(classification_report(y_test, rf_predictions, target_names=['ground',
'vegetation','buildings']))
val_features_scaled = MinMaxScaler().fit_transform(val_features)
val_rf_predictions = rf_classifier.predict(val_features_scaled)
print(classification_report(val_labels, val_rf_predictions,
target_names=['ground','vegetation','buildings']))
```

Let's study the results. We now have 97% accuracy on the validation data, further decomposed as follows:

classes	precision	recall	f1-score	support
ground	0.97	0.98	0.98	518973
vegetation	0.97	0.98	0.97	319808
buildings	0.95	0.91	0.93	175063

Adding features introduced a slight drop in performance compared to using only the base X, Y, Z, R, G, B set, which shows that we added some noise.

But this was worth it for the sake of generalization. We now have a global accuracy of 85% on the test set, so an increase of 31%, only through feature selection. This is massive. And as you can see, the buildings are the central aspect that hurt the performance:

classes	precision	recall	f1-score	support
ground	0.89	0.81	0.85	1188768
vegetation	0.92	0.92	0.92	1315231
buildings	0.68	0.80	0.73	613317

This is explained mainly by the fact that they are very different from the ones in the validation set and the feature set cannot truly represent them in an uncorrelated context (Figure 14-18).

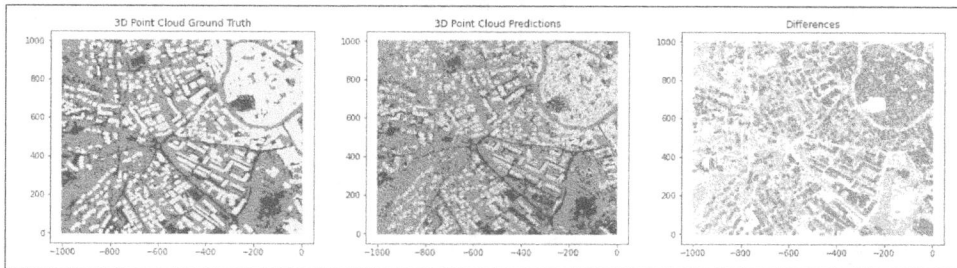

Figure 14-18. Qualitative results of the random forest classifier over the validation point cloud dataset after a feature set was selected

This is very good. We now have a model that outperforms most of what you can find, even using deep learning architectures.

Suppose we would like to scale even more. In that case, it may be interesting to inject some data from the test distribution to check if that is what is needed in the model, at the cost that our test loses its stature and becomes part of the validation set. We take 10% of the test dataset and 60% of the initial dataset to train a random forest model. We then use it and check the results on the remaining 40% constituting the validation data, and 90% of the test data:

```
val_labels=val_pcd['Classification']
val_features=val_pcd[['Z','R','G','B','omnivariance_2','normal_cr_2',
'NumberOfReturns','planarity_2','omnivariance_1','verticality_1']]
val_features_sampled, val_features_test, val_labels_sampled,
val_labels_test = train_test_split(val_features, val_labels, test_size=0.9)
val_features_scaled_sample = MinMaxScaler().fit_transform(val_features_test)
labels=pd.concat([pcd['Classification'],val_labels_sampled])
features=pd.concat([pcd[['Z','R','G','B','omnivariance_2','normal_cr_2',
'NumberOfReturns','planarity_2','omnivariance_1','verticality_1']],
val_features_sampled])
features_scaled = MinMaxScaler().fit_transform(features)
X_train, X_test, y_train, y_test = train_test_split(features_scaled, labels,
test_size=0.4)

rf_classifier = RandomForestClassifier(n_estimators = 10)
rf_classifier.fit(X_train, y_train)
rf_predictions = rf_classifier.predict(X_test)
print(classification_report(y_test, rf_predictions,

target_names=['ground','vegetation','buildings']))

val_rf_predictions_90 = rf_classifier.predict(val_features_scaled_sample)
```

```
print(classification_report(val_labels_test, val_rf_predictions_90,
    target_names=['ground','vegetation','buildings']))
```

And to our great pleasure, we see that our metrics are bumped by at least 5% (second table) while losing only 1% on the validation set (first table), at the expense of minimal feature noise, as shown in Figure 14-19:

Validation metrics				
classes	precision	recall	f1-score	support
ground	0.97	0.98	0.97	737270
vegetation	0.97	0.97	0.97	481408
buildings	0.94	0.90	0.95	257806

Test metrics				
classes	precision	recall	f1-score	support
ground	0.88	0.92	0.90	237194
vegetation	0.93	0.94	0.94	263364
buildings	0.87	0.79	0.83	122906

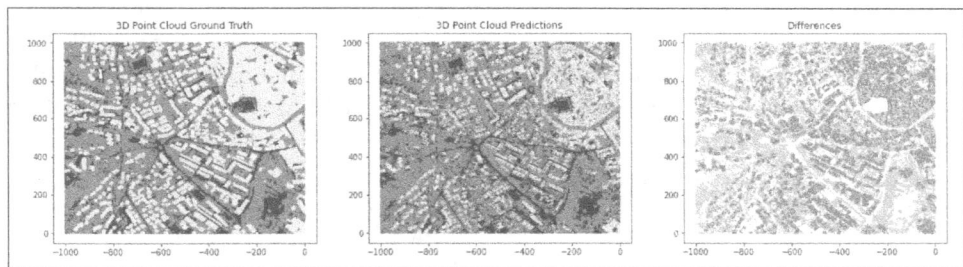

Figure 14-19. Qualitative results of the random forest classifier over the validation point cloud dataset where some data from the validation was used for training

What is usually interesting is checking the final results with different models and the same parameters and then going through a final phase of hyperparameter tuning. But that is for another time.

Exporting the labeled dataset

The title says it all: it is time to export the results to use them in another application (Figure 14-20). Let's export it as an ASCII file with the following lines:

```
val_pcd['predictions']=val_rf_predictions
result_folder="../DATA/RESULTS/"
val_pcd[['X','Y','Z','R','G','B','predictions']].to_csv(result_folder+
dataset.split(".")[0]+"_result_final.xyz", index=None, sep=';')
```

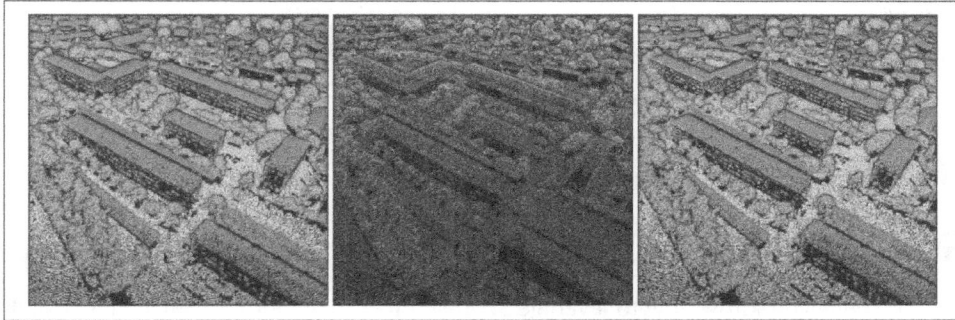

Figure 14-20. Zoomed in final results of the point cloud semantic segmentation machine learning workflow: ground truth (left), the selected differences (center), the predictions (right)

Exporting the 3D machine learning model

Of course, if you are satisfied with your model, you can save it permanently for use in production with unseen or unlabeled datasets. We can use the `pickle` module to accomplish this in just two lines of code:

```
import pickle
pickle.dump(rf_classifier, open(result_folder+"urban_classifier.poux", 'wb'))
```

And when you need to reuse the model:

```
model_name="urban_classifier.poux"
loaded_model = pickle.load(open(result_folder+model_name, 'rb'))
predictions = loaded_model.predict(data_to_predict)
print(classification_report(y_test, loaded_predictions,
target_names=['ground','vegetation','buildings']))
```

And here you are—your first AI model, trained from scratch, evaluated, benchmarked, and tuned to this application.

While this may now seem simple, the underlying principles we apply are actually essential, especially before moving onto the more "black-box" aspects associated with deep learning solutions.

In some advanced cases, you may want to integrate a nice feedback loop to ensure your system stays relevant and ensure ongoing performance. This usually means that you have to add another layer to monitor, identify, and update a model in production by considering these four components:

Data drift
Tracking changes in data distribution over time

Model degradation
Monitoring performance metrics for signs of deterioration

Retraining
    Scheduling periodic retraining to adapt to changes in the data

Explainability
    Understanding the reasons behind model predictions

When you reach this point, it usually indicates that your project is a success.

Specializing 3D Machine Learning with 3D Deep Learning

At this stage, you know how to leverage 3D machine learning for advanced tasks such as 3D point cloud semantic segmentation. The process is clear: we prepare our dataset, compute and select relevant features, establish a training setup (model, hyperparameters, dataset split), and generate a 3D machine learning semantic segmentation model. Depending on the performance of the training round that we closely monitor, we select the best candidate and ensure that generalization tests are successful.

However, this system depends on the critical task of extracting meaningful features from 3D data. These features should capture the data's geometric, topological, and statistical properties. This process often requires domain expertise and careful consideration.

Some promising directions have been explored to overcome these challenges. Specifically from the 2D world, deep learning, particularly convolutional neural networks (CNNs), has shown remarkable success in handling complex datasets. CNNs can automatically learn features from raw data, reducing the need for extensive feature engineering. This is worth investigating, and, more importantly, it is worth developing some hands-on experience with deep learning.

However, deep learning is very different from 3D deep learning. In most cases, it is not just a matter of adding one dimension but rather changing the lens through which we look at our algorithms. But this is for the next chapter.

AI Learning Paradigms

As researchers, we are investigating diverse methodologies for developing AI systems that more accurately emulate human intellect. One promising avenue is unsupervised and semisupervised learning. These methods aim to reduce reliance on labeled data by allowing models to learn patterns directly from raw data. This is especially beneficial when labeled data is limited or costly to acquire. We have already explored these in previous chapters, but it is good to know that new approaches are being developed within the AI world.

Reinforcement learning holds great potential. By learning through trial and error, reinforcement learning agents can acquire complex behaviors without explicit programming. This can be applied to tasks that require decision making and problem solving, such as in 3D games or robotics. In 3D data science, its use is still marginal, often in cases where we can clearly formulate some objectives, as exemplified.

Transfer learning is another valuable technique that can help address data scarcity. By leveraging knowledge gained from one task to improve performance on a related task, transfer learning can enable AI systems to learn more efficiently and effectively. For example, an AI system trained on a large dataset of 3D models could transfer its knowledge to a new task, such as recognizing objects in a different domain.

Neuro-symbolic AI is another promising methodology. It integrates the advantages of neural networks with symbolic reasoning. By combining these two paradigms, researchers hope to create AI systems that can reason and explain their decisions more effectively. This could enhance the transparency and trustworthiness of AI systems and is a current research track through explainable AI.

Finally, embodied AI is a relatively new field that seeks to develop AI systems that interact with the physical world. By grounding AI in the real world, researchers hope to create more human-like learning capabilities. Embodied AI has applications in robotics, autonomous vehicles, and other areas where AI systems need to interact with the physical environment. Naturally, this is the kind of field that 3D data science can greatly benefit from, and that should continue to provide new solutions to workflows, systems, and challenges.

Now that I've given you something to think about, I hope this newfound understanding of the fundamental differences between human and machine learning will help you develop effective AI systems. By overcoming the shortcomings of existing methodologies and emulating human cognition, we can develop AI that learns and reasons more akin to humans.

As mentioned, AI learning currently delivers the most efficient results when supervised learning approaches are applied to narrow tasks. Let's dive into the specifics of supervised learning and its application in the 3D domain.

As 3D AI systems become increasingly sophisticated, ethical considerations become paramount. We can outline four main guidelines:

Bias and fairness
  Ensuring that 3D AI systems do not perpetuate or amplify existing biases in data or society

Privacy
  Protecting sensitive information contained within 3D data

Transparency
  Making AI systems understandable and accountable

Safety
> Ensuring 3D AI systems operate safely and reliably, especially in critical applications

With that in mind, innovations have a high chance of positively impacting our lives.

Summary

As we conclude this chapter, I can't help but feel excited about the new innovations you can create with sophisticated semantic segmentation models capable of interpreting complex 3D environments. AI does not only revolve around deep learning approaches. The power of machine learning, particularly random forests, has been demonstrated in its ability to accurately classify urban elements. However, we've also uncovered the challenges of generalization and the importance of thoughtful feature selection and model evaluation.

Now, it is only logical to explore 3D deep learning approaches to extend greater capabilities in feature extraction and complex pattern recognition. As datasets grow larger and computational resources become more accessible, we can anticipate more sophisticated models that can handle increasingly complex scenes and finer-grained classifications. But before we move forward, I think it's vital for you to feel confident with the key learning points from this chapter:

- Understanding the importance of data preprocessing and normalization in 3D machine learning
- Techniques for splitting data into training, testing, and validation sets
- Methods for evaluating model performance using metrics like precision, recall, and F1-score
- Developing 3D machine learning solutions such as random forest
- Strategies for improving model generalization to unseen data
- The process of exporting and deploying trained models for real-world use

As we move forward, it's important to remember that this is just the beginning of our exploration into the world of 3D AI. Supervised learning, while powerful, is only one piece of the puzzle. In the next chapter, we will explore the exciting realm of 3D deep learning, where even greater capabilities await us in automatic feature extraction and the ability to handle increasingly complex patterns. But of course, I encourage you first to ensure that you can replicate and extend the current teachings of this chapter with the provided exercise.

Hands-on Project

To reinforce your learning and expand your skills, try the following "Urban Object Classification Challenge":

1. Download a new urban LiDAR dataset from an open source repository (e.g., ISPRS benchmark datasets).

2. Preprocess the data and extract relevant features, including at least two new features not used in this chapter (e.g., local point density, height above ground).

3. Implement three different machine learning classifiers (e.g., fandom forest, support vector machine (SVM), gradient boosting) to perform semantic segmentation, aiming to classify at least four urban object classes.

4. Assess the efficacy of each classifier with suitable measures and illustrate the outcomes.

5. Experiment with feature selection techniques to improve model performance.

6. Write a brief report comparing the performance of different classifiers and discussing the impact of feature selection on the results.

This activity will facilitate the application of concepts acquired in a novel situation, promoting critical thinking about feature engineering, model selection, and performance evaluation in 3D scene understanding tasks.

3D Deep Learning with PyTorch

This is where things get a bit more complex. Handling 3D data in deep learning architectures is quite different from working with text or images. But don't worry—we'll tackle it step by step at a comfortable pace. One of the primary hurdles lies in data representation. 3D data can be conveyed in various formats, such as point clouds, 3D meshes, or voxel grids as shown in Chapter 4.

Do you remember our previous experiments and writings on 3D data structures and representations? You'll be glad to know that these are crucial factors when working with 3D deep learning. In fact, the choice of 3D data representation significantly impacts the architecture and paradigms of your 3D deep learning solution. At this stage, we can distinguish between four data representations supported by 3D deep learning approaches: 3D point clouds, 3D voxel grids, 3D meshes, and multiview image datasets.

Each representation has its strengths and challenges, influencing the choice of which deep learning architecture is the right fit. Jumping into the specifics at this stage would be like throwing you into the deep end before you've learned to swim. You might feel overwhelmed by layers of complexity and end up completely lost. Jokes aside, let me first share some key concepts and tools that will be helpful before we dive into 3D deep learning architectures. I structured this chapter to guide you through the fundamentals of 3D deep learning with PyTorch (Figure 15-1).

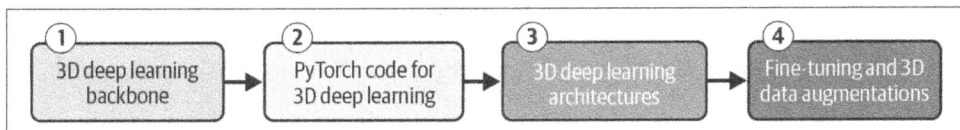

Figure 15-1. The four main components of a successful 3D deep learning project with PyTorch

This chapter is quite dense, with fewer illustrations and complex topics. I recommend a high-level readthrough before diving into the code on a second pass.

We'll begin by exploring the core concepts and components that form the backbone of 3D deep learning. Then, we delve into practical implementation using PyTorch, providing you with hands-on experience in building 3D deep learning models. The heart of our discussion focuses on various 3D deep learning architectures, including 3D convolutional neural networks (3D CNNs) for voxel data, graph neural networks (GNNs) for mesh processing, point-based architectures like PointNet for point cloud analysis, and multiview CNNs (MVCNNs) for 2D projections of 3D objects.

We'll then compare 3D machine learning with 3D deep learning, helping you understand when to apply each approach. Finally, we explore advanced techniques such as fine-tuning, transfer learning, and 3D data augmentation, equipping you with strategies to enhance model performance and generalization.

This chapter comprehensively introduces 3D deep learning with PyTorch (*https://pytorch.org*), outlining key concepts and practical implementations for experts in cutting-edge applications. We explore the foundational elements of 3D deep learning, including the distinct challenges of data representation and model architecture. We discuss the importance of selecting the appropriate 3D deep learning architecture (3D CNNs (*https://oreil.ly/3o99b*), GNNs (*https://oreil.ly/6wV4G*), PointNet (*https://oreil.ly/zksbt*), MVCNNs) based on the type of 3D data used. We focus on practical implementation using PyTorch, covering topics such as tensors, neural network modules, and data loaders. You can find step-by-step guidance on building a simple 3D deep learning model with code examples to guide your work. Finally, we explore how transfer learning (*https://oreil.ly/vvS-Y*), fine-tuning (*https://oreil.ly/92y3k*), and 3D data augmentation (*https://oreil.ly/Qt-G9*) can help you overcome generalization constraints and enhance your model's performance. As always, you can find all the materials for this chapter on the 3D Data Science Resource Hub (*https://learngeo data.eu/3d-data-science-with-python*). To access the files (code, datasets, articles) and resources, you may need to share an email address, a personal password, or proof that you are the book's owner.

3D Deep Learning Backbone

The fundamentals of 3D deep learning are linked with deep learning. The end goal is to create a 3D deep learning model for a specific objective (classification, segmentation, regression). But to make such a model, we must first design the architecture, i.e., the global structure on which you want to build your neural network (Figure 15-2).

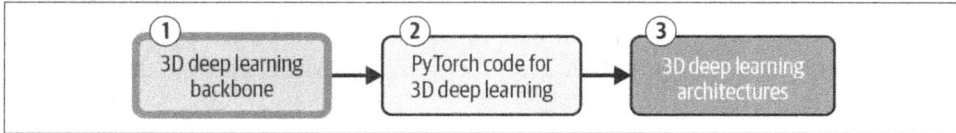

Figure 15-2. The 3D deep learning backbone

For example, you want a few convolutional layers to extract visual features from a voxel grid, some fully connected layers, and a softmax layer to make predictions. This is a great moment to make sure that you spot the difference with the deep learning model, that is, the neural network itself, with specific kernel sizes, activation functions, number of nodes, etc. It is a virtual object that you can save, load, train, and deploy to production to make predictions on new data.

I know many terms may sound unfamiliar, and this is what we will ensure you understand. At this stage, it is good to know that we have three main components that are very important to designing 3D deep learning models: the network architecture, how data should be prepared for this architecture, and how the training process should be developed. Let's start by covering the terms you will encounter concerning network architecture.

Network Architecture

Your neural network's architecture refers to its overall structure, determining how neurons in different layers are interconnected and how information flows through the network. You've already encountered layers in our discussions. In a standard neural network, we can distinguish our layers based on their position relative to the input data. The first layer is called the "input layer." Then we'll move onto one or more hidden layers, to finally make sure to share predictions through an output layer.

Each layer comprises one or more neurons. It is a mathematical function that takes numerical data and multiplies it by weight coefficients to produce a new value (Figure 15-3).

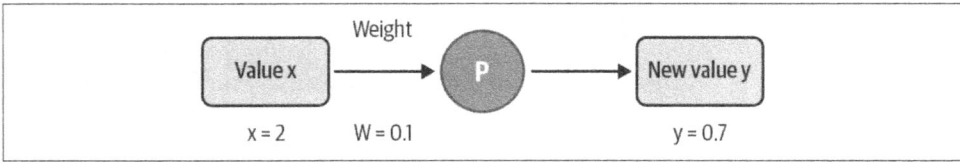

Figure 15-3. A neuron, called a perceptron, and its function

Some neurons are "on" (meaning they are activating or firing) when a particular input comes in; other neurons are "off." A neuron that is on sends its signal to the next layer (Figure 15-4).

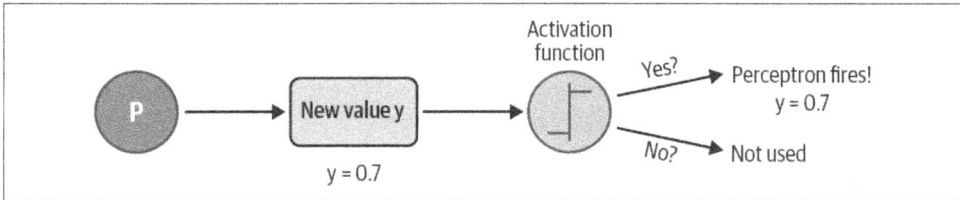

Figure 15-4. The role of the activation function that activates the neuron

A neuron that is off does not. Once a signal has made it through to the output layer, the network has done its "computation." Then, you can use a cost function to measure how accurate the output is.

All of this revolves around the most fundamental component of any neural network model: layers. A neural network can be viewed as a stack of layers that comprise its structure (Figure 15-5).

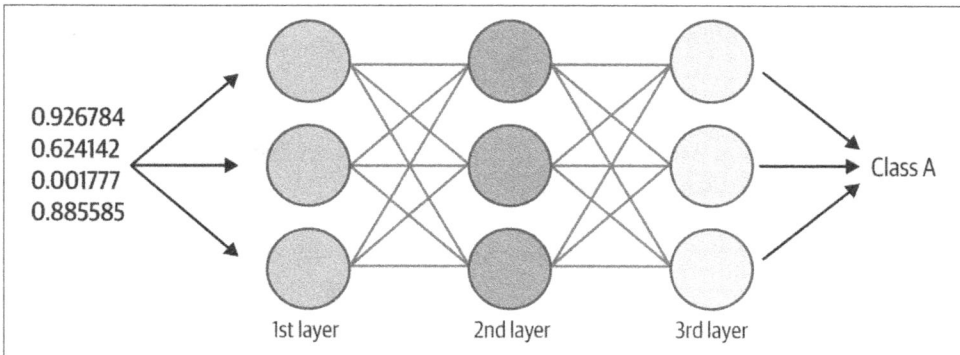

Figure 15-5. Three layers in a neural network

Each layer contains several neurons that perform a "transformation" on each element of the input tensor. This "transformation" consists of a weighted sum of the inputs plus a bias term acquired during the training process. The primary distinction among the many types of layers resides in the behavior of the neurons.

We typically categorize three primary layers: the input layer, which receives the raw data; the hidden layers, which analyze complicated patterns and correlations; and the output layer, which generates the output or prediction.

As a good starting point, below are the five most used layers that you are likely to encounter:

Convolutional layers

These layers extract features from the input data by applying filters, or kernels, to an input "image." The filters detect specific patterns, such as edges, corners, or textures.

Pooling layers

Pooling layers downsample the feature maps generated by convolutional layers, diminishing the spatial dimensionality while retaining the most critical information. This mitigates computational complexity.

Fully connected layers

In a completely linked layer, each neuron is interconnected with every neuron in the preceding layer. This means that every feature extracted by the earlier layers can influence the final output. Unlike the convolutional layers that focus on local patterns, fully connected layers are responsible for combining extracted features into a global representation and performing high-level reasoning.

Normalization layers

Normalization layers help to stabilize the training process by normalizing the activations of neurons in a layer. This can improve convergence speed and prevent what is known as "vanishing" or "exploding" gradients.

Regularization layers

These usually help prevent overfitting. The regularization layer is used to improve the model's learning. With normalization, we simplify our model's task, whereas with regularization, we make it more complex. Indeed, regularization (for example, the dropout layer) transforms some values of our tensor to 0. This strategy will enhance the complexity of our model, but paradoxically, it will improve its efficiency.

Additional layers are also accessible. Recurrent layers, known as RNNs, are mostly utilized in NLP for text analysis. This layer is designed to enable the deep learning model to possess memory capabilities. The recurrent layer will repeatedly evaluate the incoming data.

The attention layer is undoubtedly the most sophisticated method. The attention layer's principle is to enhance RNNs and potentially exceed their performance. In a recurrent neural network, each previously evaluated word is retained in memory to facilitate the comprehension of subsequent words. The issue is in our sentence analysis, which proceeds from beginning to conclusion rather than in reverse order.

Through attention, we can examine the interconnections of each word in a sentence to comprehend the relationships that bind them together. Consequently, it represents an enhancement of RNNs, as attention facilitates the extraction of greater information from the same subject under examination (e.g., the phrase).

You may have seen me use the word "activation." This relates to the use of "activation functions." These play a crucial role in enabling neural networks to learn in a nonlinear way. They do this by being applied after each layer of the neural network, taking the weighted sum of the inputs to that layer and turning it into the output of that layer.

The activation function (*https://oreil.ly/htIxE*) "activates" the layer's output by making the weighted sum of the inputs nonlinear. Without that nonlinearity, the neural network doesn't have its shapes—or isn't able to learn its shapes—and therefore doesn't have what we think of as "intelligence."

The sigmoid function (*https://oreil.ly/qYrpq*) transforms inputs into a scaled output that fits into the binary classification tasks but suffers from the vanishing gradient problem when neurons produce large negative inputs. The tanh function is similar to sigmoid in that it also produces outputs in the bounded range but is centered around zero and, therefore, does not seem to possess the same shortcoming as sigmoid when visualizing the function as a neural network.

The rectified linear unit (ReLU) function (*https://oreil.ly/tVN0U*) is frequently the favored activation function for deep feedforward neural networks because of its processing efficiency and its ability to circumvent the vanishing gradient problem. However, in certain circumstances, other functions, such as sigmoid or tanh, can be more appropriate choices.

You now have a more coherent picture in broad strokes of an architecture for deep learning. Let's impart some of the basics of how we convey information to that kind of setup.

Data Preparation

Before diving in, we typically undergo a data preparation phase. After data collection, there are three main steps you're already familiar with:

1. Data cleaning to remove or correct inconsistencies, noise, and errors in your dataset.

2. Data normalization to scale your data to a common range (e.g., [0,1] or [-1,1]) to ensure consistent processing.

3. Data augmentation to increase your dataset size and improve model generalization by creating data variations. This usually includes 3D rotations, translations, and scaling.

These phases are implemented by creating a custom "dataset class" that proposes some custom processing steps on the three aspects. However, while some processes are optional (e.g., data augmentation), the dataset class should bear these three components:

- Load data from disk or memory.
- Apply necessary transformations.
- Return data in the format expected by your model.

In order to clarify even more, this means that for 3D data, the input might be:

- $N \times 3$ for point clouds (N points with x, y, z coordinates)
- $H \times W \times D \times C$ for volumetric data (height, width, depth, and channels for voxels)

The output depends on the task (e.g., a single class label for classification or another 3D structure for reconstruction tasks). When you have successfully prepared your dataset, we'll move on to the second phase: training our 3D deep learning model.

AI Model Training

The training process of a 3D neural network involves, in essence, a cyclical process of learning from its mistakes. The neural network systematically modifies its internal parameters (weights and biases) throughout the training. This is known to reduce the discrepancy between its predicted output and what we actually want it to predict: the actual target output for a specified collection of training instances.

This learning process is controlled through some hyperparameters, which are configuration variables. You will encounter the following:

Learning rate
    This controls the step size during what we call "optimization."

Batch size
    This is the number of data samples processed before updating the model. Data is typically processed in batches to improve computational efficiency. Batch size is a hyperparameter that specifies the number of samples processed prior to updating the model's internal parameters.

Number of epochs
    How many times the model processes the entire dataset.

This learning process also involves the following steps: initialization, forward propagation, loss calculation, backpropagation, and optimization through several iterations, as illustrated in Figure 15-6.

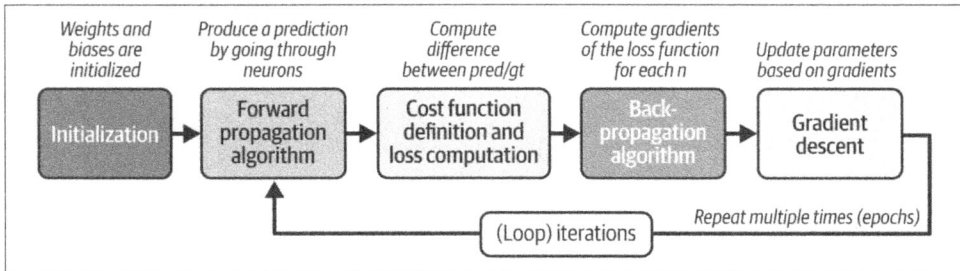

Figure 15-6. The learning process that occurs during the training phase of the neural network

In the initialization phase, the neural network's weights and biases are randomly initialized. These initial values establish the foundation for the learning process. Subsequently, we input data into the neural network. The activations of every neuron in each layer are calculated using a sequence of matrix multiplications and activation functions. This process is known as forward propagation, producing the predicted output of the neural network. The difference between the predicted output and the actual target output (ground truth) is computed using a chosen loss function. This loss function quantifies how well the model performs.

Common loss functions (*https://oreil.ly/Lv45j*) include:

- MSE for regression tasks
- Cross-entropy for classification tasks
- Custom loss functions for specific 3D tasks (e.g., Chamfer distance for point clouds)

At this stage, we have made an entire forward pass with a loss metric. This means we can use this metric through a back propagation strategy to optimize the network's weights and biases. This is possible through the use of an optimizer algorithm (*https://oreil.ly/7MdIY*), whose goal is to minimize the loss function.

An optimizer initially computes the gradients of the loss function concerning the network's weights and biases. It updates them based on the computed gradient; this process is repeated for multiple iterations until the loss function converges to a minimum. The most common optimizers are the gradient descent (*https://oreil.ly/BL7lU*) (updates the weights in the direction of the steepest descent of the loss function; see Figure 15-7), stochastic gradient descent (*https://oreil.ly/cXGE8*) (a version of gradient descent that uses a single training example to update weights in each iteration, enhancing efficiency for big datasets), Adam (*https://oreil.ly/FhUGA*) (a popular optimizer that combines the best aspects of adaptive moment estimation and RMSprop), RMSprop (*https://oreil.ly/MeNaf*) (an adaptive learning rate algorithm that modifies the learning rate for each parameter according to past gradient data), and AdaMax (*https://oreil.ly/Qavqi*) (a variant of Adam that uses a different infinity norm for the learning rate update).

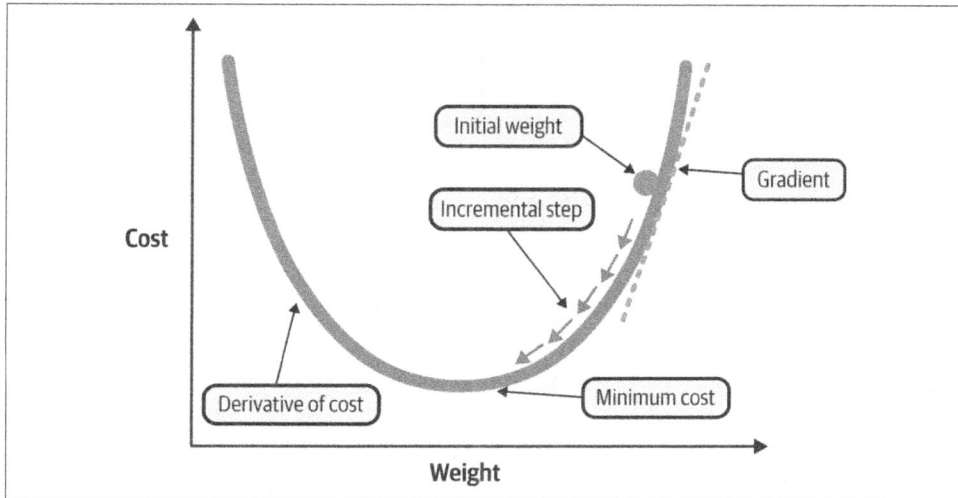

Figure 15-7. Example of how gradient descent works to find a local minima

The learning rate is the hyperparameter that controls the magnitude of this update. Once all the weights and biases are modified, we repeat this for each batch of training data until all batches are processed to complete the first epoch of training. All that is left is to repeat that multiple times (epochs) until the neural network's performance on the training data reaches a satisfactory level, converges to a solution, or has reached a certain number of iterations. The network is then evaluated on a separate set of data called the validation set to check for overfitting (the model exhibits strong performance on the training data but has inadequate performance on novel data).

Of course, we have a hand in how the model learns. Neural networks acquire knowledge by progressively modifying their parameters with regard to one specific feature: the gradients of the loss function. This is very important to keep in mind because this process is often computationally intensive, especially for large and deep networks. Hence, training neural networks is typically performed on powerful hardware or specialized hardware accelerators (GPUs or TPUs) to speed up the process.

Upon completing the training phase, the neural network should normally have assimilated fundamental descriptors. These should relate the patterns and relationships within the training data, enabling it to generate precise predictions on novel, unseen data. This is where we can fix our model and move on to serving it and putting it to the test.

Serving a Trained Model

Upon successful training, our model can be deployed to provide predictions on novel, unobserved data. This procedure is referred to as model serving. The initial stage in deploying a model is to preserve it in a designated format.

Popular formats include TensorFlow's SavedModel (*https://oreil.ly/aZeOZ*), PyTorch's TorchScript (*https://oreil.ly/zOaWR*), and ONNX (*https://oreil.ly/qxMfv*). The choice of format depends on the deployment environment and the tools being used.

Once the model is saved, it needs to be deployed to a suitable platform. Standard options include cloud platforms (e.g., AWS, GCP, Azure), edge devices (e.g., IoT devices, smartphones), or on-premise servers. The choice of a specific platform will depend on factors such as the latency requirements, the scalability needs, and security considerations.

To serve the model, it must first be loaded into the deployment environment. This involves using the appropriate tools and libraries to load the saved model and prepare it for inference. This may include setting up input and output tensors, loading necessary data preprocessing steps, and initializing any required resources.

Before passing data to the model, it must be preprocessed to match the format and scaling expected by the model. This plays on the same "data preparation" phase that involves tasks such as normalization and resizing. Once the data is preprocessed, it can be fed into the model to obtain a prediction.

The predicted output may need to be postprocessed to make it more meaningful. For example, if the model is predicting probabilities, the output may need to be converted to class labels.

Finally, a serving infrastructure must be set up to handle incoming requests, process the data, and return the predictions. This serving is possible through the use of a web server, REST API, or other suitable infrastructure. To ensure the model continues to perform well, it is important to monitor its performance and retrain it as needed. This involves continuously evaluating the model's accuracy and making adjustments as necessary.

But enough about theoretical concepts. Let's move on to how to actually implement a 3D deep learning model.

Implementation with PyTorch

Let's move on to the implementation with PyTorch. When we want to dive into 3D deep learning, we depend on deep learning frameworks (Figure 15-8). Currently, I recommend three general-purpose libraries in use:

TensorFlow
> Developed by Google, TensorFlow (*https://tensorflow.org*) is a versatile platform for machine learning, including deep learning. It offers a flexible architecture and supports a variety of tasks.

PyTorch
> Created by Meta, PyTorch (*https://pytorch.org*) is known for its dynamic computational graph and ease of use. It's popular among researchers and for prototyping.

Keras
> An advanced API (*https://keras.io*) capable of operating over TensorFlow or Theano, Keras streamlines the construction and training of deep learning models.

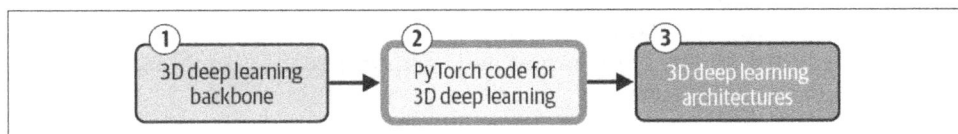

Figure 15-8. Stage 2: PyTorch for deep learning

In addition to these, other libraries have emerged, such as Caffe (*https://oreil.ly/x8O4h*), MXNet (*https://oreil.ly/7ZfuZ*), Chainer (*https://chainer.org*), and FastAI (*https://fast.ai*).

These libraries provide an extensive array of features and functionalities to accommodate diverse requirements and preferences. PyTorch is currently the most used and promising framework and the one I will be using in this book. It has become a popular choice for 3D data science due to its dynamic computational graph, ease of use, and strong community support. Its ability to integrate with other tools and its efficient GPU utilization make it well suited for the computationally intensive tasks associated with 3D data. PyTorch's built-in functions for 3D data operations and its support for geometric deep learning make it a versatile tool for a variety of 3D data science projects.

Let's install PyTorch first.

Installing PyTorch (with CUDA)

When dealing with 3D data, we should prioritize training on the GPU. This means installing CUDA. The first stage is to check CUDA compatibility. Ensure your GPU is compatible with the latest CUDA version. You can visit NVIDIA's website (*https://oreil.ly/FEkjN*) for details. From there, download the appropriate CUDA Toolkit, and follow the installer on-screen instructions to install it.

I suggest creating a new conda environment with Python 3.9+ and using the PyTorch installation command, specifying the CUDA version as proposed on the PyTorch Start Locally website (*https://oreil.ly/DtumA*). In my case (GPU NVIDIA 3090 GTX, CUDA 12.1), this translates to:

```
conda install pytorch torchvision torchaudio pytorch-cuda=12.1 -c pytorch
-c nvidia
```

Once the process is complete, you can then check if everything went smoothly by launching a Python session and typing:

```
import torch
torch.cuda.is_available()
```

If the result is true, you can leverage your CUDA GPU for computations! If you don't have success, I advise you to investigate this troubleshooting guide (*https://oreil.ly/r1KuA*). Now, what are the key concepts and components of PyTorch that are essential for working with 3D data?

Tensors: The Building Blocks

At the core of PyTorch are tensors, which are multidimensional arrays similar to NumPy's ndarrays but with the added benefit of GPU acceleration (Figure 15-9).

In 3D deep learning, we may work with 5D tensors:

```
import torch

# Create a 5D tensor: (batch_size, channels, depth, height, width)
x = torch.randn(32, 1, 64, 64, 64)
```

This tensor could represent a batch of 32 features (1 channel) with dimensions of $64 \times 64 \times 64$ voxels. Due to the memory requirements of 3D data, you may need to use small batch sizes. Also, adjusting the learning rate accordingly is a great practice. Now, let's develop the neural network modules.

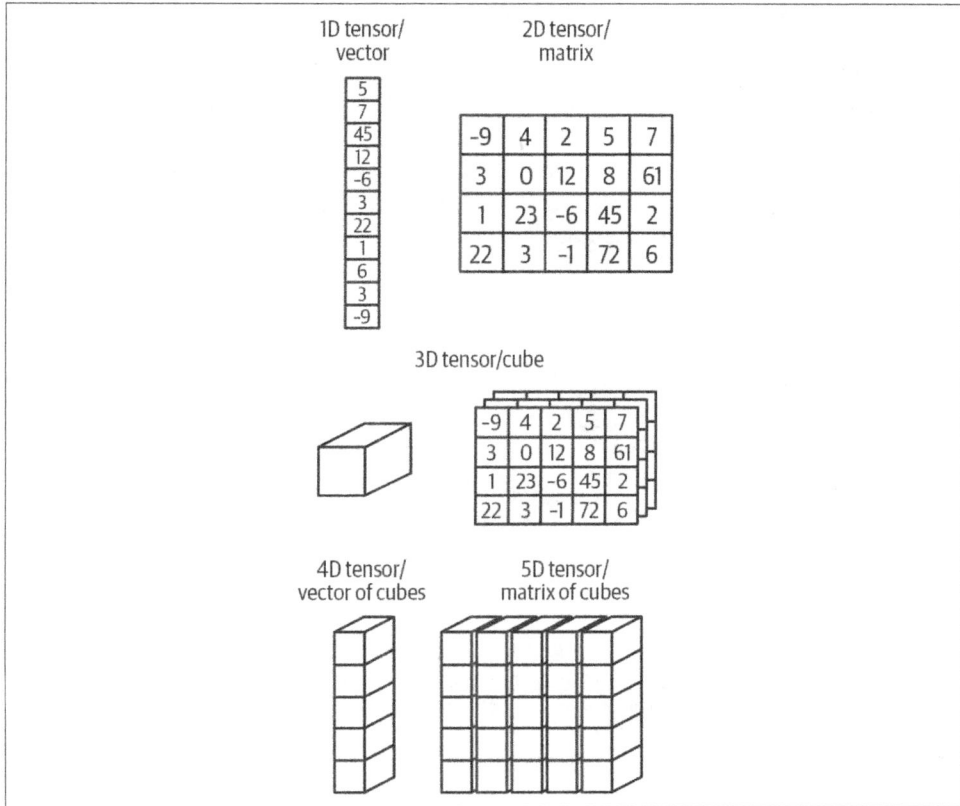

Figure 15-9. Example of tensors based on the data

Neural Network Modules

PyTorch provides the nn module, which contains building blocks for creating neural networks. For 3D operations, we have specialized layers such as:

```
import torch.nn as nn

# 3D convolutional layer
conv3d = nn.Conv3d(in_channels=1, out_channels=16, kernel_size=3, stride=1,
padding=1)

# 3D max pooling layer
maxpool3d = nn.MaxPool3d(kernel_size=2, stride=2)
```

Defining a 3D Neural Network

Let's create a simple 3D deep learning model using PyTorch that takes in 3D voxels and predicts their class. We'll build a basic 3D CNN for this task. The first thing that we want to do is to create a `VoxelClassifier` class that defines the model architecture. We will create three convolutional layers followed by two fully connected layers, a typical pattern in CNNs, adapted here for 3D data. For this, let me show you the full class definition, which I will then break down to make sure you understand what the main constituents are:

```python
class VoxelClassifier(nn.Module):
    def __init__(self, num_classes):
        super(VoxelClassifier, self).__init__()
        self.conv1 = nn.Conv3d(1, 32, kernel_size=3, padding=1)
        self.conv2 = nn.Conv3d(32, 64, kernel_size=3, padding=1)
        self.conv3 = nn.Conv3d(64, 128, kernel_size=3, padding=1)
        self.pool = nn.MaxPool3d(kernel_size=2, stride=2)
        self.fc1 = nn.Linear(128 * 4 * 4 * 4, 512)
        self.fc2 = nn.Linear(512, num_classes)
        self.relu = nn.ReLU()

    def forward(self, x):
        x = self.pool(self.relu(self.conv1(x)))
        x = self.pool(self.relu(self.conv2(x)))
        x = self.pool(self.relu(self.conv3(x)))
        x = x.view(-1, 128 * 4 * 4 * 4)
        x = self.relu(self.fc1(x))
        x = self.fc2(x)
        return x
```

Now, let me break down this mumbo jumbo. The first thing is to understand that in PyTorch, you define the layers in `__init__` and specify how they connect in `forward`. The `forward` method is called automatically when you pass data through your model. PyTorch uses dynamic computation graphs, meaning the `forward` method can contain arbitrary Python code, not just a fixed sequence of operations.

The second thing to note is that the dimensions of the data change as it passes through the network. Convolutional layers maintain spatial structure but change the number of channels, while pooling layers reduce spatial dimensions. The `view` operation reshapes the data into a 1D vector for the fully connected layers.

First, let's look at the class definition:

```python
class VoxelClassifier(nn.Module):
```

This line defines a new class called `VoxelClassifier` that inherits from `nn.Module`. In PyTorch, `nn.Module` is what I would consider the base class for all neural network modules. By inheriting from it, our class gains several useful methods and properties for building neural networks.

Now, let's go through the methods of this class. First, the initialization method:

```
def __init__(self, num_classes):
    super(VoxelClassifier, self).__init__()
    self.conv1 = nn.Conv3d(1, 32, kernel_size=3, padding=1)
    self.conv2 = nn.Conv3d(32, 64, kernel_size=3, padding=1)
    self.conv3 = nn.Conv3d(64, 128, kernel_size=3, padding=1)
    self.pool = nn.MaxPool3d(kernel_size=2, stride=2)
    self.fc1 = nn.Linear(128 * 4 * 4 * 4, 512)
    self.fc2 = nn.Linear(512, num_classes)
    self.relu = nn.ReLU()
```

This is the constructor method. It's called when you create an instance of the class. `super(VoxelClassifier, self).__init__()` calls the constructor of the parent class (`nn.Module`), which then allows us to leverage `nn.Module` directly.

`super` gives you the base class that a class inherits from. So if you have a class `cat` that inherits from a class `animals`, then calling `super` on `cat` will give you `animals`. Most of the PyTorch models inherit from `nn.Module`, so calling `super` on your neural network class will allow you to call methods from the base `nn.Module` class. `__init__()` is then just a call to the constructor of the base class. The nice thing about object-oriented programming is that you don't really need to worry about what constructors do for predefined classes. It's convention to call them, but otherwise, you assume that they will do some required stuff and make all the methods from the base class usable, initiate variables, and so on.

This `__init__` function, called a method (as it is part of a class), defines several layers of the neural network:

- Three 3D convolutional layers (`nn.Conv3d`) with `self.conv1`, `self.conv2`, `self.conv3`
- A 3D max pooling layer (`self.pool`), which reduces the spatial dimensions by half in each direction
- Two fully connected (linear) layers (`self.fc1` and `self.fc2`)
- The ReLU activation function (`self.relu`)

Now, if I take the first 3D convolutional layer, this is what we have:

```
self.conv1 = nn.Conv3d(1, 32, kernel_size=3, padding=1)
```

If I had to specify its definition a bit more, then the first argument is the number of input channels (1); the second is the number of output channels (32); the `kernel_size` of 3 means the convolutional filter is $3 \times 3 \times 3$; and `padding=1` adds a one-voxel border around the input, which helps maintain spatial dimensions.

Our network thus presents three consecutive 3D convolutional layers (conv1, conv2, conv3) that progressively increase the number of feature maps from 1 to 32, then 64, and finally 128, each followed by a ReLU activation function and a max pooling layer that reduces spatial dimensions, allowing the network to extract increasingly complex and abstract features from the input voxel data. After the convolutional layers, the tensor is flattened and passed through two fully connected layers (fc1 and fc2), with the first producing a 512-dimensional representation and the second generating the final classification output, with the number of output classes determined by the num_classes parameter passed during initialization.

Now, let's focus on the forward method, which defines the forward pass of the network—how data flows through the layers:

```python
def forward(self, x):
    x = self.pool(self.relu(self.conv1(x)))
    x = self.pool(self.relu(self.conv2(x)))
    x = self.pool(self.relu(self.conv3(x)))
    x = x.view(-1, 128 * 4 * 4 * 4)
    x = self.relu(self.fc1(x))
    x = self.fc2(x)
    return x
```

In the method, the input x is passed through each convolutional layer, followed by ReLU activation and max pooling. After the convolutional layers, x.view() reshapes the tensor to flatten it for the fully connected layers. The data then passes through the fully connected layers, with ReLU activation between them. The final output, which will be the class predictions (probability based on the number of classes), is returned.

Now, let's set our number of classes to 10:

```python
num_classes = 10
```

We can initialize the model:

```python
model = VoxelClassifier(num_classes)
```

This brings:

```
VoxelClassifier(
  (conv1): Conv3d(1, 32, kernel_size=(3, 3, 3), stride=(1, 1, 1),
  padding=(1, 1, 1))
  (conv2): Conv3d(32, 64, kernel_size=(3, 3, 3), stride=(1, 1, 1),
  padding=(1, 1, 1))
  (conv3): Conv3d(64, 128, kernel_size=(3, 3, 3), stride=(1, 1, 1),
  padding=(1, 1, 1))
  (pool): MaxPool3d(kernel_size=2, stride=2, padding=0, dilation=1,
  ceil_mode=False)
  (fc1): Linear(in_features=8192, out_features=512, bias=True)
  (fc2): Linear(in_features=512, out_features=10, bias=True)
  (relu): ReLU()
)
```

Beautiful, we are ready for the next stage.

Hyperparameter Definition

Once our architecture is set, we can define the hyperparameters that will guide the training process and can impact the model's performance:

```
# Hyperparameters
learning_rate = 0.001
batch_size = 32
num_epochs = 5
```

 A smaller learning rate (like 0.001) typically leads to more gradual updates, which can help avoid overfitting. However, too small a rate might result in slow convergence.

Our training process will learn at a rate of 0.001, which means that the model's weights are updated by 0.1% of the gradient during each iteration. By considering our data in batches of 32, we try to find a compromise between training speed and stability. Indeed, a smaller `batch size` can introduce more noise into the updates, which can sometimes help prevent overfitting. Finally, we will iterate for 5 epochs. Note that more epochs generally lead to better model performance. Still, there's a risk of overfitting if the model learns the training data too well and struggles to generalize to new data.

Optimizer and Loss Functions

PyTorch provides various optimizers and loss functions. Optimizers update the model parameters based on the computed gradients, and loss functions quantify the difference between the predicted output and the ground truth.

We first have to call a specific module:

```
import torch.optim as optim
```

Then we can define our loss criterion (e.g., cross-entropy loss, commonly used for multiclass classification) and our optimizer, Adam (Adaptive Moment Estimation):

```
criterion = nn.CrossEntropyLoss()
optimizer = optim.Adam(model.parameters(), lr=0.learning_rate)
```

Now that you know how to handle the various components with PyTorch, how do we pass data to our network?

PyTorch DataLoader

Training a deep learning model necessitates data. Data is often accessible in the form of a dataset. The dataset contains several data samples. One may request the model to process a single sample at a time; however, it is typically more efficient to allow the model to handle a batch of multiple samples simultaneously.

A batch can be generated by extracting a segment from the dataset, using the tensor's slicing syntax. To enhance training quality, we usually shuffle the entire dataset at each epoch to ensure that no two batches are identical throughout the training process. Occasionally, we may implement data augmentation to increase the dataset's volatility artificially. One might envision that a substantial amount of code is required to accomplish all of these tasks. However, utilizing the `DataLoader` significantly simplifies the process.

A `DataLoader` in PyTorch is a utility class that iterates over a dataset, creating batches of data and applying transformations to them. It's a crucial component of the PyTorch framework, providing a convenient way to feed data into your neural network models. To prevent overfitting, `DataLoader`s can randomly shuffle the data within each epoch. For faster training, multiple worker processes can be used to load data in parallel.

`DataLoader`s also provide flexibility in handling different data types and formats. The collation function defines how data samples are combined into batches, particularly useful for variable-length sequences or tensors. `DataLoader`s can apply various transformations to the data, such as normalization, augmentation, and tensor conversion, to enhance model performance and robustness.

I will create a custom `Dataset` and `DataLoader` that generates synthetic 3D voxel data. This approach is useful for testing and developing models when actual data is not available or when you want to verify your model architecture.

First, let's import some utilities:

```
from torch.utils.data import Dataset, DataLoader
```

Then, we create our `SyntheticVoxelDataset` class:

```
class SyntheticVoxelDataset(Dataset):
    def __init__(self, num_samples, num_classes, voxel_size=32):
        self.num_samples = num_samples
        self.num_classes = num_classes
        self.voxel_size = voxel_size

    def __len__(self):
        return self.num_samples

    def __getitem__(self, idx):
        # Generate a random voxel grid
```

```
voxel = torch.randn(1, self.voxel_size, self.voxel_size, self.voxel_size)

# Generate a random label
label = torch.randint(0, self.num_classes, (1,)).item()

return voxel, label
```

This custom `Dataset` class generates `num_samples` of synthetic voxel data and `num_classes` labels on the fly. I create two methods: `__len__`, which returns the total number of samples we want to generate, and `__getitem_`, which creates a random voxel grid and a random label each time it is called.

We can then create an instance of `SyntheticVoxelDataset` with our desired number of voxels (1,600, a fixed dataset size to give us a fixed number of batches) and classes (`num_classes`):

```
dataset = SyntheticVoxelDataset(num_samples=1600, num_classes=num_classes)
```

The `DataLoader` is created with this dataset, using the specified batch size and enabling shuffling:

```
# Create the DataLoader
dataloader = DataLoader(dataset, batch_size=batch_size, shuffle=True)
```

This approach mimics real-world scenarios, where you prepare a fixed dataset size and choose a batch size for training. The `DataLoader` then handles creating batches from this dataset.

By structuring it this way, we maintain flexibility. For example, if you wanted to change the batch size to 64, you could do so without changing the total dataset size, and you'd automatically get 25 batches per epoch instead of 50.

The `DataLoader` is created with a batch size of 32 and shuffling enabled. It iterates over the dataset, yielding batches of input and label tensors. Now, what does it look like in a training loop?

PyTorch Training Loop

The training loop is the core of the learning process. In the following, I designed a PyTorch training loop for our synthetic case that we then examine in detail:

```
for epoch in range(num_epochs):
    for batch_idx, (inputs, labels) in enumerate(dataloader):
        # Forward pass
        outputs = model(inputs)
        loss = criterion(outputs, labels)

        # Backward pass and optimize
        optimizer.zero_grad()
        loss.backward()
        optimizer.step()
```

```
        if batch_idx % 10 == 0:  # Print every 10 batches
            print(f'Epoch [{epoch+1}/{num_epochs}],
            Batch [{batch_idx+1}/{len(dataloader)}], Loss: {loss.item():.4f}')
```

```
print("Training finished!")
```

The outer loop iterates over epochs, defined as a complete run over the entire dataset:

```
for epoch in range(num_epochs):
```

Inside each epoch, we iterate over batches of data provided by `dataloader`, where `inputs` is a batch of voxel data, and `labels` is the corresponding class labels. This means that for each single batch:

1. We execute a forward pass through the model (`outputs = model(inputs)`).

2. We compute the loss (`loss = criterion(outputs, labels)`).

3. We perform a backward pass and optimization by clearing the gradients from the previous iteration (`optimizer.zero_grad()`), computing the gradients of the loss with respect to the model's parameters (`loss.backward()`) and updating the model's parameters based on the computed gradients (`optimizer.step()`).

4. The loss for each batch is accumulated to calculate the average loss for the epoch.

5. We update the model parameters (`loss.backward()`).

At the end of each batch of 10, we print the current loss. Once the model is finalized, with our weights set up, we can move on to inference.

PyTorch Inference

To demonstrate the inference mode of our model, let's create a single random voxel grid as a test input:

```
test_input = torch.randn(1, 1, 32, 32, 32)
```

Then, we build our inference code:

```
with torch.no_grad():
    prediction = model(test_input)
    predicted_class = torch.argmax(prediction).item()
    print(f"Predicted class: {predicted_class}")
```

`torch.no_grad()` is a context manager that disables gradient computation, which is not needed during inference and saves memory. Then, we apply our model on the `test_input` to get predictions, and we use `torch.argmax()` to find the index of the highest value in the prediction, which corresponds to the predicted class. Note that `.item()` converts the tensor to a Python scalar.

Finally, we print the predicted class:

```
Predicted class: 5
```

And there you have it: a fully functional 3D deep learning model with PyTorch! All you have to do is adjust your dataset class to your custom data and put it to the test. But this may not be so straightforward, as your dataset may not be a voxel dataset.

So, what kind of 3D deep learning architectures can we leverage?

3D Deep Learning: The Architectures

Choosing the suitable neural network design is essential for optimal performance while handling 3D data. Here's a breakdown of some common choices based on the type of 3D data: voxel, point clouds, meshes, or multiview images:

3D CNNs
Voxel data, which represents 3D space as a grid of voxels, is well suited for 3D CNNs (*https://oreil.ly/xTljI*). These networks extend traditional 2D CNNs to three dimensions, allowing them to capture spatial relationships in 3D data.

PointNet-based architectures
Point clouds are unstructured sets of points representing 3D objects. PointNet (*https://oreil.ly/ckJ4n*) and its variants, such as PointNet++ and DGCNN, are specifically designed to process point clouds. These architectures learn features directly from the raw point coordinates, making them efficient for point cloud tasks like classification and segmentation.

Graph neural networks (GNNs)
Meshes are represented as graphs, where nodes correspond to vertices and edges represent connections between them. GNNs (*https://oreil.ly/92h3-*) are well suited for processing mesh data. They can learn from the graph structure and node features to extract meaningful information from the mesh.

Multiview CNNs
Multiview data consists of multiple 2D views of a 3D object. Multiview CNNs (*https://oreil.ly/CZo_H*) combine information from these views to learn 3D features. They can be used for tasks like 3D object recognition and reconstruction.

The choice of neural network architecture for 3D data depends on several factors, including the type of 3D data (voxel, point cloud, mesh, or multiview), the specific task to be performed (e.g., classification, segmentation, reconstruction), and the available computational resources. These factors will surely guide you in picking the most suitable architecture for your 3D data science project. What may still elude you is how these architectures are defined.

3D Convolutional Neural Networks: Voxels

Before diving into the specifics of 3D CNNs, it is important to understand the basics of CNNs and their extension to 3D space (Figure 15-10). CNNs have been highly successful in processing 2D image data. They use convolutional layers to apply filters across the input, capturing local patterns and features.

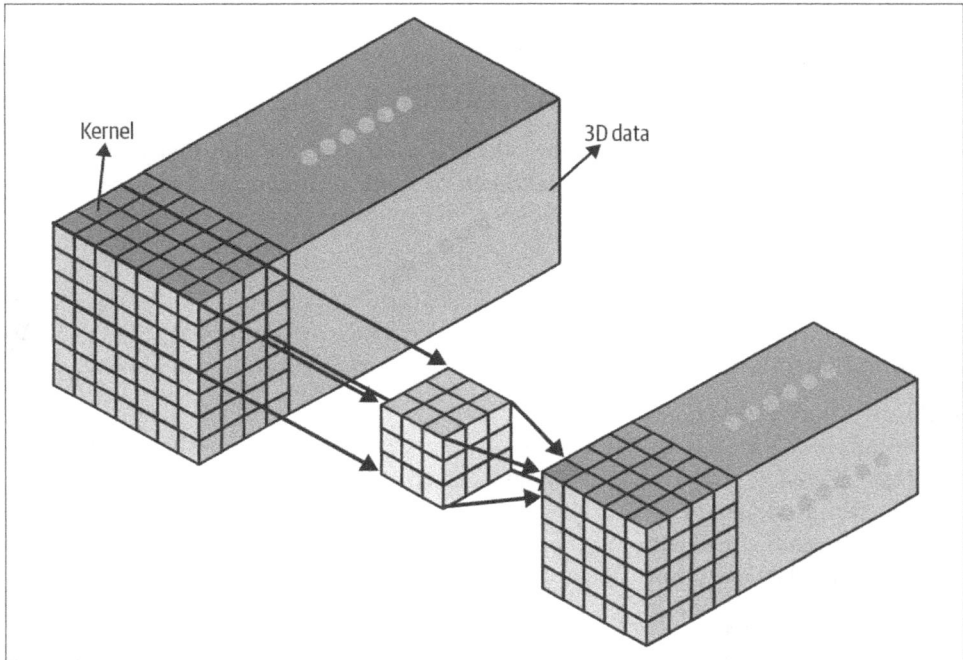

Figure 15-10. Example of a single (3 × 3 ×3) 3D convolution

The key idea is to share weights across the spatial dimensions, which reduces the number of parameters and makes the network translation invariant. When dealing with 3D data, we need to extend these concepts to three dimensions.

This leads us to 3D CNNs for regular, structured datasets. And if you remember, voxels are the analog of pixels, but in 3D. Thus, 3D CNNs are the right fit to process 3D voxel grids. A voxel grid can be considered a 3D tensor with dimensions (depth, height, width, channels). There's typically one channel for binary voxel occupancy grids, while multichannel voxel grids can represent additional properties like color or density. A 3D CNN extends the idea of 2D convolutions to three dimensions.

As we saw before, the main components are:

3D convolutional layers
These layers use 3D filters (e.g., 3 × 3 × 3 or 5 × 5 × 5) that slide across the depth, width, and volume of our input.

3D pooling layers
These layers reduce the spatial dimensions of the feature maps, typically using max pooling or average pooling in all three dimensions.

Fully connected layers
These are used at the end of the network for classification or regression tasks.

Activation functions
ReLU is commonly used, but other functions like LeakyReLU or ELU can also be effective.

Batch normalization
This helps stabilize training and can be applied after convolutional layers.

Here's a basic implementation of a 3D CNN using PyTorch:

```python
import torch
import torch.nn as nn

class Simple3DCNN(nn.Module):
    def __init__(self, in_channels, num_classes):
        super(Simple3DCNN, self).__init__()

        self.conv1 = nn.Conv3d(in_channels, 32, kernel_size=3, padding=1)
        self.bn1 = nn.BatchNorm3d(32)
        self.pool1 = nn.MaxPool3d(kernel_size=2, stride=2)

        self.conv2 = nn.Conv3d(32, 64, kernel_size=3, padding=1)
        self.bn2 = nn.BatchNorm3d(64)
        self.pool2 = nn.MaxPool3d(kernel_size=2, stride=2)

        self.conv3 = nn.Conv3d(64, 64, kernel_size=3, padding=1)
        self.bn3 = nn.BatchNorm3d(64)
        self.pool3 = nn.MaxPool3d(kernel_size=2, stride=2)

        self.fc1 = nn.Linear(64 * 4 * 4 * 4, 512)
        self.fc2 = nn.Linear(512, num_classes)

    def forward(self, x):
        x = self.pool1(torch.relu(self.bn1(self.conv1(x))))
        x = self.pool2(torch.relu(self.bn2(self.conv2(x))))
        x = self.pool3(torch.relu(self.bn3(self.conv3(x))))

        x = x.view(-1, 64 * 4 * 4 * 4)
        x = torch.relu(self.fc1(x))
```

```
        x = self.fc2(x)

        return x
```

Now, to use such a model, we can define the `model` variable, initialize, create an input tensor, and apply the model to the variable:

```
model = Simple3DCNN(in_channels=1, num_classes=10)
input_tensor = torch.randn(1, 1, 32, 32, 32)  # (batch_size, channels, depth,
height, width)

output = model(input_tensor)
```

At this stage, we use a randomly initialized model. Thus, the predictions would likely be very, very poor. This is where the training phase comes in, as seen in the previous section. After this, you are likely to use your newfound model on prepared datasets to see how it behaves. What is important is to understand that 3D CNNs extend the concept of 2D CNNs to volumetric data and can use 3D kernels to convolve over width, height, and depth. Also, they can preserve spatial relationships in all three dimensions and are an effective approach for dense 3D data representations. They usually follow an approach like this:

Input (N × D × H × W × C) → 3D Conv → 3D Pooling → ... → Fully Connected → Output

Here N is the batch size, D is depth, H is the height, W is what we consider the width, and C stands for channels. This approach remains limited by its computationally intensive approach and memory constraints for high-resolution inputs (Table 15-1).

Table 15-1. Advantages and limitations of 3D CNNs

Architecture	Advantages	Limitations	Best use cases
3D CNNs	Regular grid-based data, local spatial dependencies, leverages 2D CNNs	Inefficient for irregular point clouds, requires voxelization	Volumetric images, point clouds converted to voxels

3D Graph Neural Networks

GNNs are a powerful tool for processing graph-structured data. Unlike CNN models that primarily work on grid-like data, GNNs are designed to capture the complex relationships and dependencies inherent in graph-structured data (Figure 15-11).

GNNs consist of four key components: a graph representation, a message-passing mechanism, an aggregation function, and an update function. If you need to refresh yourself on graph theory, I recommend that you go back to Chapter 13. If you remember, a graph is defined by a set of nodes (vertices) and edges connecting them. Each node and edge can have associated features. Message passing is the fundamental activity of GNNs, wherein nodes disseminate information to their neighboring nodes.

The aggregation function determines how messages are combined from neighboring nodes. After aggregating messages, each node updates its state based on the aggregated information and its original features.

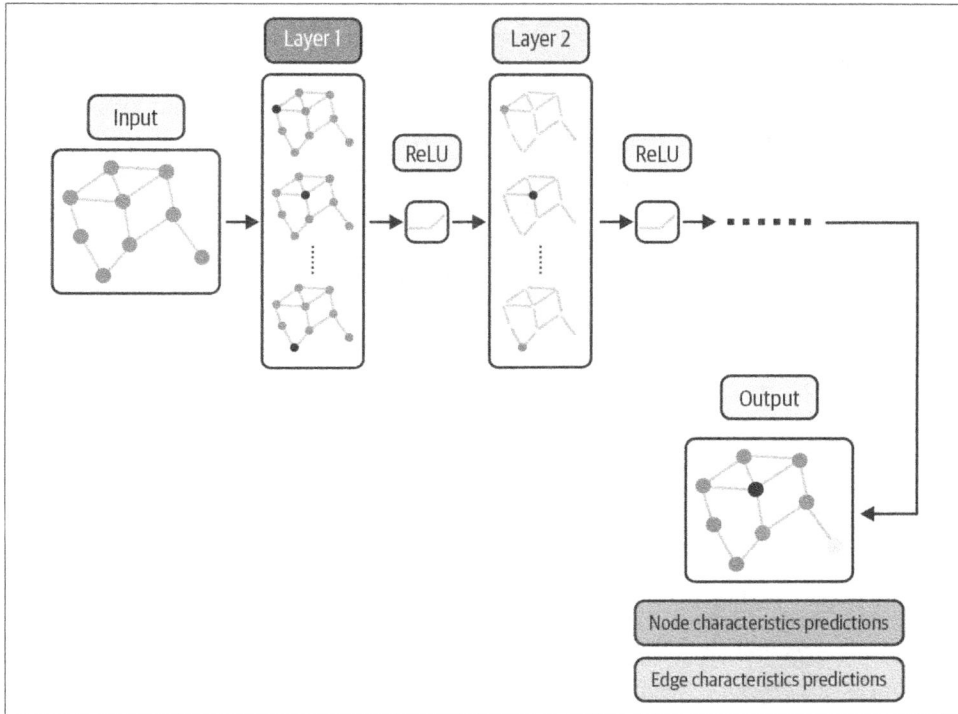

Figure 15-11. Graph neural network architecture example

To adapt GNNs for 3D datasets such as point clouds or 3D meshes, we need to prepare the datasets in a graph representation. *k*-NN graphs or radius graphs can be used for point clouds, while the mesh itself can be directly used as a graph. In message passing, points in point clouds can aggregate features from their neighbors, considering geometric information. In 3D meshes, messages can be passed along edges or faces, evaluating their properties. The aggregation functions can be adapted to the specific characteristics of point clouds and meshes. For example, weighted sums or geometric means can be used for 3D meshes. Update functions can also be tailored to the data type. For point clouds, neural networks or geometric transformations can be used, while for 3D meshes, vertex updates or edge/face updates can be applied.

Now, how does this play out with PyTorch? First, we need to load PyTorch with the geometric extension:

```
import torch
import torch_geometric
from torch_geometric.data import Data, DataLoader
from torch_geometric.nn import GCNConv, global_mean_pool
```

Then, let's imagine you face a 3D shape classification problem. You want to leverage a database of IKEA furniture to create a 3D GNN model that can predict the specific class of the mesh computed from a depth scanner. We can define a MeshGCN class that uses graph convolutional layers (GCNConv) from PyTorch Geometric:

```
class MeshGCN(nn.Module):
    def __init__(self, num_features, num_classes):
        super(MeshGCN, self).__init__()
        self.conv1 = GCNConv(num_features, 64)
        self.conv2 = GCNConv(64, 128)
        self.conv3 = GCNConv(128, 256)
        self.fc = nn.Linear(256, num_classes)

    def forward(self, data):
        x, edge_index, batch = data.x, data.edge_index, data.batch
        x = torch.relu(self.conv1(x, edge_index))
        x = torch.relu(self.conv2(x, edge_index))
        x = torch.relu(self.conv3(x, edge_index))
        x = global_mean_pool(x, batch)
        x = self.fc(x)
        return x
```

We could then define a mesh loading function that reads *.ply* files and converts them into PyTorch Geometric data objects. This function could then be passed to a data loader function that can load all meshes from a directory, organizing them by class. From there, we could train our model and test it on this newfound mesh.

The choice of node features can significantly influence the performance of a GNN. Common features include vertex coordinates, normals, curvature, and other geometric properties. These features provide valuable information about the mesh's structure and geometry. Also, different aggregation functions can be used to combine messages from neighboring nodes. For example, mean, sum, or max aggregation can be employed. The choice of aggregation function can affect the GNN's ability to capture different types of information from the graph. Various update functions can be applied to the aggregated messages to compute new node features. ReLU, LeakyReLU, and tanh are some common choices. These functions introduce nonlinearity and help the GNN learn complex patterns.

By carefully considering these factors and customizing the GNN architecture, you can effectively process 3D mesh data for classification, segmentation, and reconstruction tasks. Assuming you have loaded your mesh data into a `DataLoader`, this is the code sample that you can tune to leverage your model:

```
model = MeshGCN(num_features, num_classes).to(device)
optimizer = torch.optim.Adam(model.parameters(), lr=0.01)
criterion = torch.nn.CrossEntropyLoss()

for epoch in range(num_epochs):
    for data in data_loader:
        optimizer.zero_grad()
        output = model(data)
        loss = criterion(output, data.y)
        loss.backward()
        optimizer.step()
```

When you have a 3D mesh that you want to process as it stands, 3D GNNs can be decisive, but they are often prone to highly expensive computing costs (Table 15-2).

Table 15-2. Advantages and limitations of graph neural networks

Architecture	Advantages	Limitations	Best use cases
GNNs for 3D	Irregular point clouds, complex relationships, 3D shape tasks	Computational expense, graph construction	3D shape classification, segmentation, generation

We saw ways to process voxels or meshes directly, but what about point clouds? What if we do not want to abstract their representation into voxels or meshes?

Point-based Architectures: PointNet and Point Clouds

Point-based architectures have emerged as a powerful approach for processing 3D point clouds. These architectures directly operate on the raw point cloud data, avoiding the need for intermediate representations like voxels or meshes. Indeed, these "traditional" computer vision techniques, centered around 2D convolutions and MLPs, prove inadequate when confronted with point clouds. These unstructured datasets, devoid of a fixed grid or pixel-based representation, present a formidable challenge. To address this, the groundbreaking PointNet architecture emerged, pioneering a novel approach to processing point clouds.

PointNet consists of a symmetric function that aggregates information from all points in the cloud, followed by an MLP to extract features. It is invariant to the order of points, making it suitable for tasks like classification and segmentation. PointNet++ is an extension of PointNet that addresses its limitations by introducing hierarchical feature learning. It uses a grouping mechanism to partition the point cloud into local regions, applies PointNet to each area, and then aggregates the results. This hierarchical approach captures both local and global features of the point cloud.

Since PointNet (2016) and PointNet++ (*https://oreil.ly/Jn1bh*) (2017), many other extractors and architectures have followed, such as PointCNN (*https://oreil.ly/XFUAR*) (2018), DGCNN (*https://oreil.ly/muUHP*) (2019), PointNeXt (*https://oreil.ly/Jg8O7*) (2022), PointMLP (*https://oreil.ly/wfZea*) (2022), and even—what's today considered state-of-the-art—Point Transformer v3 (*https://oreil.ly/lIvoW*) (2023/2024). Point Transformer is a recent architecture that leverages transformer architectures for point cloud processing. It uses self-attention mechanisms to capture global dependencies between points, allowing it to learn complex relationships and achieve state-of-the-art performance on various 3D tasks.

Table 15-3. Advantages and limitations of PointNet architectures

Architecture	Advantages	Limitations	Best use cases
PointNet	Directly process point clouds, efficient, 3D shape classification	Limited spatial relationships, noise sensitivity	3D shape classification, part segmentation
Point Transformer	Global dependencies, large-scale point clouds	Computational expense, hyperparameter tuning	3D shape tasks, large point clouds

This line of 3D deep learning architectures, with its ability to process "raw data," gives the ability to transcend simple applications and aim at very advanced innovations. As PointNet is a core component of point-based supervised learning, and as we saw how to go from 3D datasets to point clouds, having hands-on experience with these approaches is a great feat. I dedicate a complete chapter to PointNet and how to build supervised learning segmentation and classification systems in Chapter 16.

Multiview CNNs

Multiview CNNs (MVCNNs) are another powerful approach for processing 3D data (Figure 15-12). They leverage the strengths of 2D CNNs by analyzing multiple 2D views of a 3D object.

This enables MVCNNs to effectively learn 3D features and perform classification, detection, and segmentation tasks. The core components of an MVCNN include view generation, feature extraction, feature aggregation, and classification.

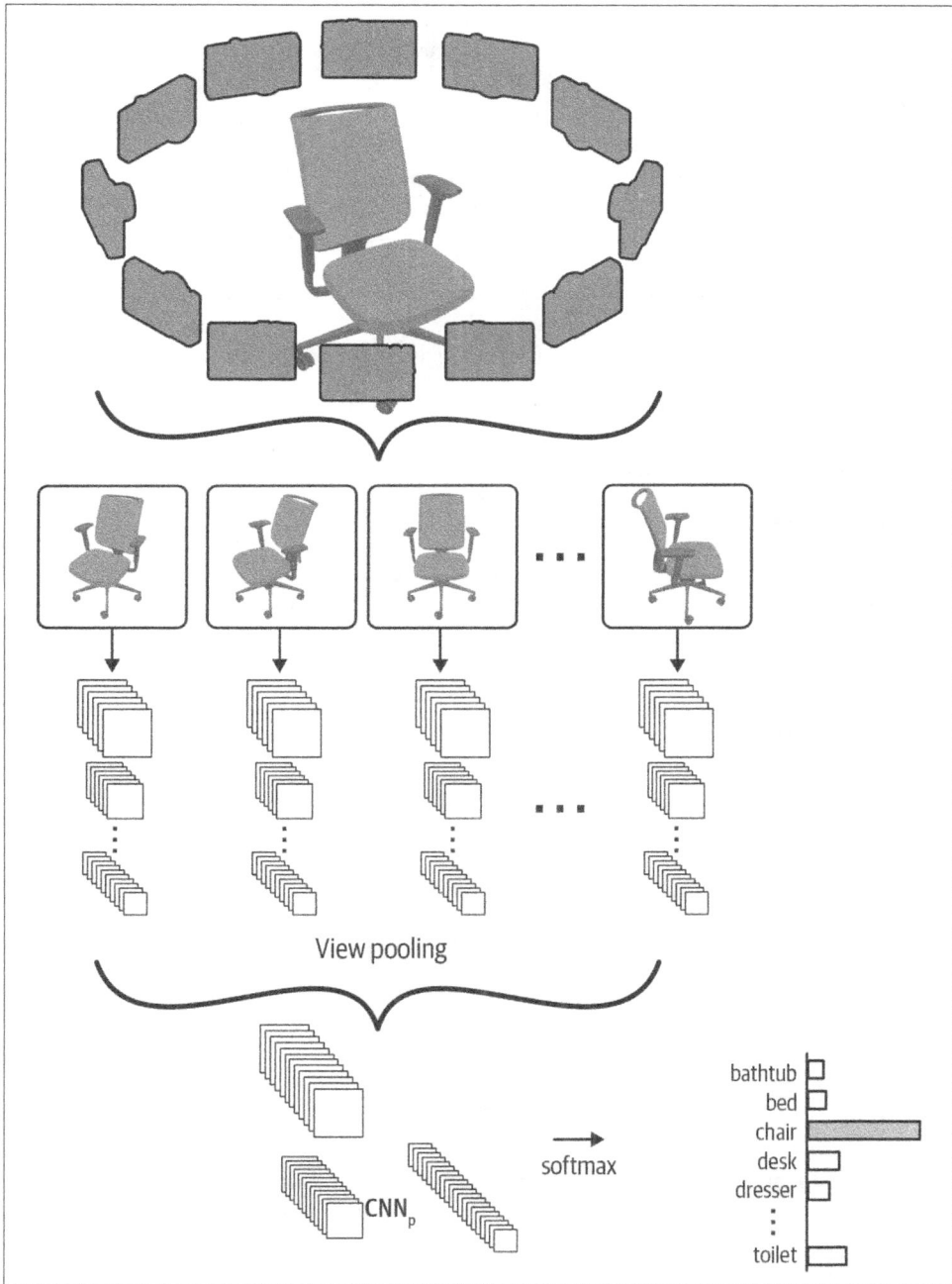

Figure 15-12. Example of a multiview CNN architecture

View generation involves creating multiple 2D images of the 3D object from different angles. A 2D CNN then processes each 2D image to extract features. These features are aggregated into a single 3D representation, which is finally fed into a classification or regression layer. One of the challenges in MVCNNs is selecting the optimal number of viewpoints for a given object. Various techniques, such as random sampling, canonical views, or learned view selection, can be employed to address this. Another challenge is effectively combining features from multiple views. Techniques like averaging, concatenation, or attention mechanisms can be used to achieve this.

Advanced MVCNN architectures incorporate techniques like residual connections, attention mechanisms, and hybrid approaches with other 3D deep learning methods. These developments improve the performance and functionality of MVCNNs.

The following is the creation of such a model:

```python
import torch
import torch.optim as optim

import torch.nn as nn

import torchvision.transforms as transforms
import torchvision.datasets as datasets

from torchvision.models import resnet18

class MVCNN(nn.Module):
    def __init__(self, num_views, feature_dim, num_classes):
        super(MVCNN, self).__init__()
        self.num_views = num_views
        self.feature_dim = feature_dim
        self.num_classes = num_classes

        # Load pretrained 2D CNN (e.g., ResNet18)
        self.backbone = resnet18(pretrained=True)
        self.backbone.fc = nn.Linear(self.backbone.fc.in_features, feature_dim)

        # Feature aggregation layer
        self.aggregation = nn.AdaptiveAvgPool2d((1, 1))

        # Classification layer
        self.classifier = nn.Linear(feature_dim * num_views, num_classes)

    def forward(self, x):
        batch_size, num_views, channels, height, width = x.size()

        features = []
        for i in range(num_views):
            view = x[:, i, :, :, :]
            view_features = self.backbone(view)
            features.append(view_features)
```

```
features = torch.cat(features, dim=1)
features = self.aggregation(features).view(batch_size, -1)
output = self.classifier(features)
return output
```

As you can see, we leverage an existing pretrained 2D CNN such as ResNet18 as part of the backbone (Table 15-4).

Table 15-4. Advantages and limitations of multiview CNNs

Architecture	Advantages	Limitations	Best use cases
Multiview CNNs	Leverage 2D CNNs, 3D object recognition	Preprocessing required, limited 3D structure	3D object recognition, pose estimation

3D Machine Learning Versus 3D Deep Learning

The choice between 3D (nondeep) machine learning and 3D deep learning is a critical one, influenced by a variety of factors, some of which are shown in Figure 15-13.

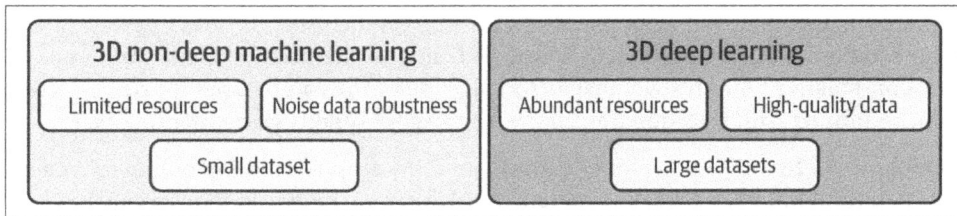

Figure 15-13. 3D machine learning versus 3D deep learning: factors to weigh in deciding which paradigm to use

For small datasets, traditional machine learning algorithms often perform better due to their simpler models and fewer parameters. In contrast, deep learning models can leverage large datasets to learn complex patterns and achieve high performance. High-quality labeled data benefits deep learning models in extracting meaningful features, while traditional machine learning algorithms may be more robust to noisy or incomplete data.

Computational resources are another consideration: traditional machine learning models typically have lower requirements, while deep learning models can be computationally intensive. Traditional machine learning algorithms might adequately address simple problems with well-defined features and clear patterns. In contrast, complex problems with nonlinear relationships in the data are better suited for deep learning models.

Interpretability is also a factor: traditional machine learning models are generally more interpretable, allowing for understanding of decision-making processes, while deep learning models often lack interpretability but can achieve high performance.

Ultimately, temporal limitations are significant: conventional machine learning models may be taught and implemented more rapidly, whereas deep learning algorithms sometimes necessitate extended periods for experimentation and optimization. Table 15-5 summarizes the considerations for using 3D machine learning versus 3D deep learning.

Table 15-5. 3D machine learning versus 3D deep learning

When to use 3D machine learning	When to consider 3D deep learning
Small datasets: limited data availability necessitates feature engineering and careful model selection.	Large datasets: abundant data is available to train complex models.
Interpretability: understanding the model's decision-making process is crucial.	High performance: achieving state-of-the-art results is the primary goal.
Fast deployment: quick turnaround time is required.	Complex patterns: the problem involves intricate relationships and patterns in the data.
Simple problems: the problem can be solved with well-defined features and linear relationships.	Feature learning: automatic feature extraction is desired.

In many cases, a combination of traditional machine learning and deep learning can yield optimal results. For example, deep learning can be used for object detection and segmentation, while traditional machine learning can be used for path planning and decision making. In industrial inspection, deep learning can be used for defect detection, combined with traditional machine learning for quality assessment.

The choice between 3D machine learning and 3D deep learning is not always clear-cut. A careful evaluation of the problem, available data, computational resources, and desired outcomes is essential. A hybrid strategy that integrates both techniques often yields optimal results.

Fine-Tuning, Transfer Learning, and 3D Data Augmentation

Pretrained models trained on massive datasets have become a cornerstone of modern deep learning. These models capture rich feature representations that could be very beneficial for out-of-domain tasks or when we have limited data access. But how can we leverage such models? This typically relies on two prevalent methodologies: fine-tuning and transfer learning. These methodologies enable us to leverage the information and expertise encapsulated in established models.

Transfer Learning

Transfer learning entails utilizing a pretrained model as a foundation for a novel task or domain. The objective is to utilize the information gained from the pretrained model on an extensive dataset and implement it in a related task with a more limited dataset. By doing so, we can leverage the overarching characteristics and patterns acquired by the pretrained model, conserving time and computational resources. Transfer learning often comprises two primary stages:

Feature extraction
In this phase, we utilize the pretrained model as a static feature extractor. We eliminate the last levels designated for classification and substitute them with new layers tailored to our specific goal. The weights of the pretrained model are fixed, and only the weights of the newly incorporated layers are trained on the limited dataset.

Refinement
Fine-tuning advances the process by unfreezing certain layers of the pretrained model, enabling their modification with the fresh dataset. This phase allows the model to adjust and acquire more specialized characteristics pertinent to the current job or domain.

In certain cases, transfer learning is relevant. In other cases, fine-tuning may be best.

Fine-Tuning

Transfer learning entails freezing the weights of the pretrained model and training only the new layers, but fine-tuning advances this process by permitting updates to the pretrained layers. This supplementary step is advantageous when the new dataset is sufficiently large and analogous to the original dataset used for training the pretrained model.

This means that we can leverage:

Feature extraction
Employing the pretrained model as a feature extractor. The output of the penultimate layer may serve as input for a new classifier or regressor.

Fine-tuning the entire model
Unfreezing all layers and retraining the entire model can lead to better performance but requires more computational resources.

Fine-tuning specific layers
Unfreezing only the top layers while keeping the lower layers frozen strikes a balance between performance and computational efficiency.

However, it is essential to understand that the performance of fine-tuned models can be affected by the domain gap between the pretrained model and the target task. Sometimes, we need to reduce this by working on the statistical distribution of the data. Also, fine-tuning with limited data can lead to overfitting, requiring regularization techniques. Finally, fine-tuning large pretrained models can be computationally demanding, which also demands that we balance the effort. These challenges can be partially solved with 3D data augmentation techniques.

3D Data Augmentation: Expanding the Dataset

Data augmentation is an effective method for enhancing the diversity and volume of a training dataset, resulting in increased model generalization and resilience. We are essentially relying on geometric transformations when it comes to 3D data:

Rotation
> Rotating the 3D object around random axes

Scaling
> Uniformly or nonuniformly scaling the object

Translation
> Shifting the object along the x, y, and z axes

Shearing
> Applying a shear transformation to the object

You can see some of the effects in Figure 15-14.

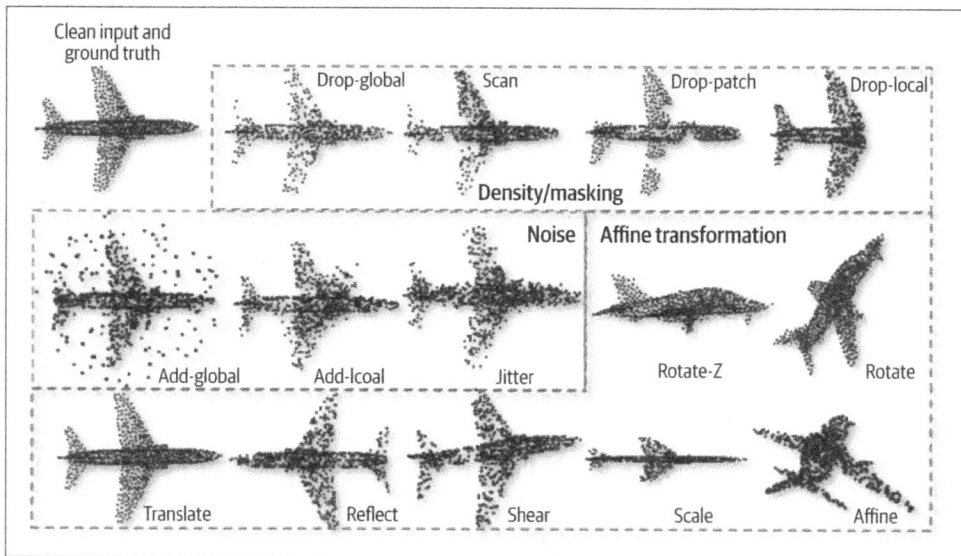

Figure 15-14. Example of some augmentations of a 3D point cloud for deep learning purposes

We can also work with a more local augmentation to add noise and play with the local data distribution through:

Random point dropping
  Removing a random subset of points

Point jittering
  Adding random noise to point coordinates

Feature perturbation
  Adding random noise to point features

Finally, for cases such as 3D object classification, we can always rely on data synthesis, which generates synthetic 3D data using generative models or rule-based approaches.

In the next chapter, we will use specific 3D data augmentation techniques to increase the size of our training dataset for 3D deep learning applications with PointNet. In the meantime, combining fine-tuning with data augmentation can significantly boost model performance. By augmenting the limited target dataset, you can provide the fine-tuned model with a richer and more diverse training experience.

Summary

While we have explored the foundations, techniques, and applications of 3D deep learning with PyTorch it may be helpful to step back and remind ourselves of the implications of all this information.

First off, 3D deep learning requires careful consideration of data representation and model architecture. As we have learned, PyTorch provides powerful tools for implementing 3D deep learning models. But of course, this means that we need to use different architectures (3D CNNs, GNNs, PointNet, MVCNNs), each suited for different 3D data representations. As we touch on data, understand that the "sample" we have in the 3D domain is nowhere near the massive amount of 2D labeled data. Thus, transfer learning and fine-tuning can significantly boost performance, especially in this context of limited data. In addition, data augmentation is crucial for improving model generalization in 3D deep learning.

So, at this stage, what newfound skills can you count on?

- Comprehending the advantages and constraints of various 3D deep learning architectures
- Implementing custom datasets and data loaders in PyTorch for 3D data
- Mastering the PyTorch training loop for 3D deep learning models
- Recognizing when to use traditional machine learning versus deep learning for 3D tasks
- Applying transfer learning and fine-tuning techniques to 3D models

Remember that 3D deep learning is a great way to develop the perception and reasoning capabilities of spatial AI systems. If you want to deepen your skills, you can follow the chapter's hands-on exercise.

Hands-on Project

As always, to solidify your understanding of 3D deep learning with PyTorch, try to create a PyTorch implementation of a simple 3D CNN for voxel classification. Use the `SyntheticVoxelDataset` class provided in the chapter as a starting point. Implement the following steps:

1. Define a 3D CNN architecture with a minimum of two convolutional layers and two fully connected layers if you have the heart.

2. Create a custom dataset class that generates synthetic voxel data (you can use the provided `SyntheticVoxelDataset` class as a reference).

3. Set up a `DataLoader` to efficiently batch and shuffle your data.

4. Implement a training loop that trains your model for a specified number of epochs.

5. Add a simple inference function that takes a single voxel grid and returns a predicted class.

As a bonus, you can try to experiment with data augmentation techniques specific to 3D voxel data, such as random rotations or adding noise to the voxel grid. This exercise will help you practice implementing a complete 3D deep learning pipeline in PyTorch, from data preparation to model training and inference.

PointNet for 3D Object Classification

3D scene understanding, a crucial aspect of spatial AI systems, depends heavily on the effective semantic extraction of 3D data. Chapters 12 through 14 have allowed us to leverage both unsupervised and supervised 3D machine learning for this goal, when we have limited labeled datasets. However, when you benefit from large-scale data repositories, 3D deep learning shows promise, which we highlighted in Chapter 15 with 3D CNNs.

However, 3D CNNs are inadequate for handling the complexities of point clouds, which are unstructured datasets without a fixed grid or pixel-based representation. This limitation highlights the critical need for innovative approaches to directly processing and interpreting point cloud data.

Indeed, various methods exist for representing and processing 3D data, such as voxels, meshes, and multiview images (see Chapter 4). However, each of these representations has its drawbacks. While suitable for 3D CNNs, voxels can be computationally intensive and memory-demanding, especially for high-resolution inputs. Meshes or B-reps, processed using GNNs, present challenges in graph construction and computational expense. Multiview CNNs, which leverage 2D CNNs on multiple 2D views, require extensive preprocessing and may not fully capture the inherent 3D structure. These limitations underscore the need for a more efficient and direct approach to 3D data processing (see Chapter 15).

PointNet emerges as a pivotal solution to address these limitations and enable practical 3D scene understanding. This groundbreaking architecture directly processes raw point cloud data, circumventing the need for intermediate representations like voxels or meshes. By operating directly on unstructured (tabular) point data, PointNet simplifies the processing pipeline and avoids information loss associated with converting point clouds to other representations. This characteristic makes PointNet a compelling area of investigation to advance 3D deep learning approaches for 3D scene understanding.

This chapter delves into the intricacies of this architecture and highlights the possibilities for 3D object classification and 3D semantic segmentation. We explore the fundamental concepts of point cloud data preparation for PointNet, discuss its architecture and components, and implement solutions for object classification.

> In this chapter, we leverage several Python libraries: NumPy (*https://numpy.org*), PyTorch (*https://pytorch.org*), Matplotlib (*https://matplotlib.org*), Plotly (*https://plotly.com/python*), Open3D (*http://open3d.org*), urllib (*https://oreil.ly/ctGL-*) (to download files from the internet, specifically the ModelNet10 dataset (*https://oreil.ly/e_ZRM*)), zipfile (*https://oreil.ly/1TBh-*) (to extract files from a ZIP archive, specifically for unpacking the downloaded ModelNet10 dataset), pathlib (*https://oreil.ly/lfYZX*) (for handling filepaths in a more object-oriented way), and scikit-learn (*https://scikit-learn.org*) (provides the `confusion_matrix` function to evaluate the performance of the trained PointNet). As always, on the 3D Data Science Resource Hub (*https://learngeodata.eu/3d-data-science-with-python*) you can find all the materials for this chapter. To access the files (code, datasets, articles) and resources, you may need to share an email address, a personal password, or proof that you are the book's owner.

By understanding and implementing PointNet, we unlock a powerful tool that will enhance your ability to tackle digital environment generation.

PointNet: A Point-based 3D Deep Learning Architecture

Let's take a step back to understand better how the PointNet architecture works. PointNet is one of the pioneers in neural networks for 3D deep learning (Figure 16-1). If you understand PointNet, you can use all the other advanced models. But, of course, understanding is only a part of the equation.

The main challenge is getting the daunting system to work and adapting it to use it with your data! This is a challenging feat, even for seasoned coders. Therefore, we divide the process into several parts, including data preparation for the architecture to work in real-world conditions.

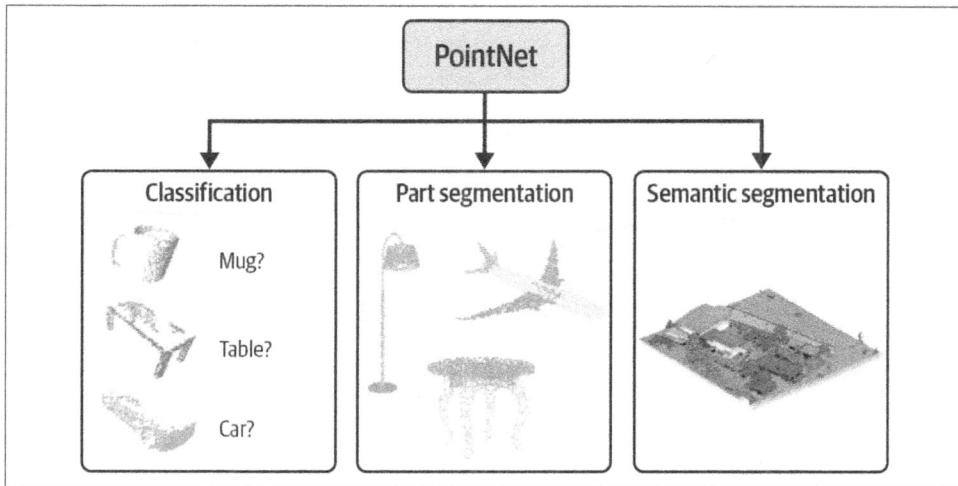

Figure 16-1. The PointNet architecture has the ability to attack three semantization applications: classification, part segmentation, and semantic segmentation

Understanding the network's building blocks is essential to correctly preparing the data. Figure 16-2 shows the critical aspects to consider when preparing your data with the network.

The PointNet architecture consists of several layers of neural networks that process the point cloud data. The input to PointNet is a simple set of points, each represented by its 3D coordinates and additional features such as color or intensity. These points are fed into successive shared MLP networks that learn to extract local features from each point (Figure 16-3).

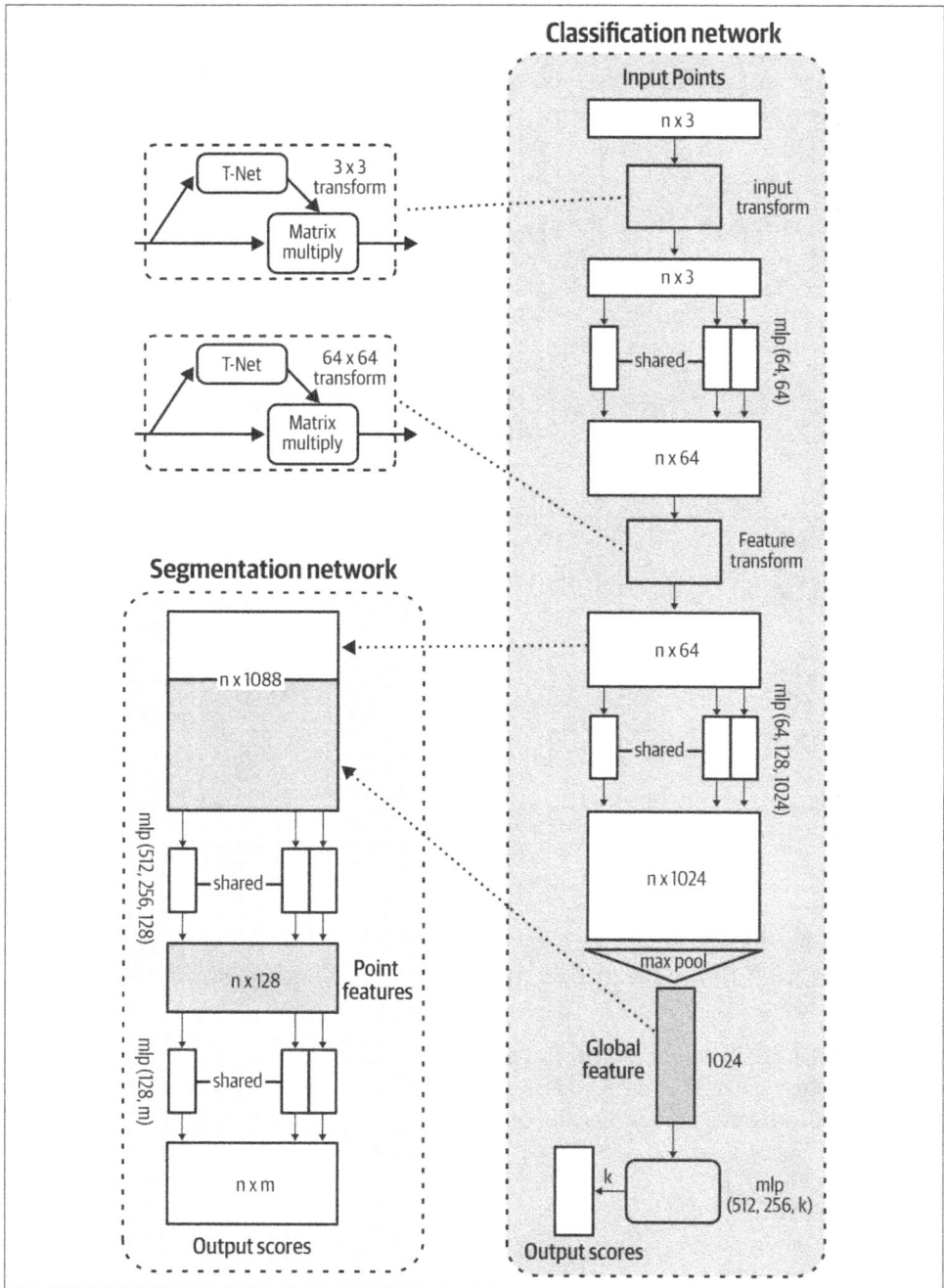

Figure 16-2. The PointNet model architecture as described by its authors

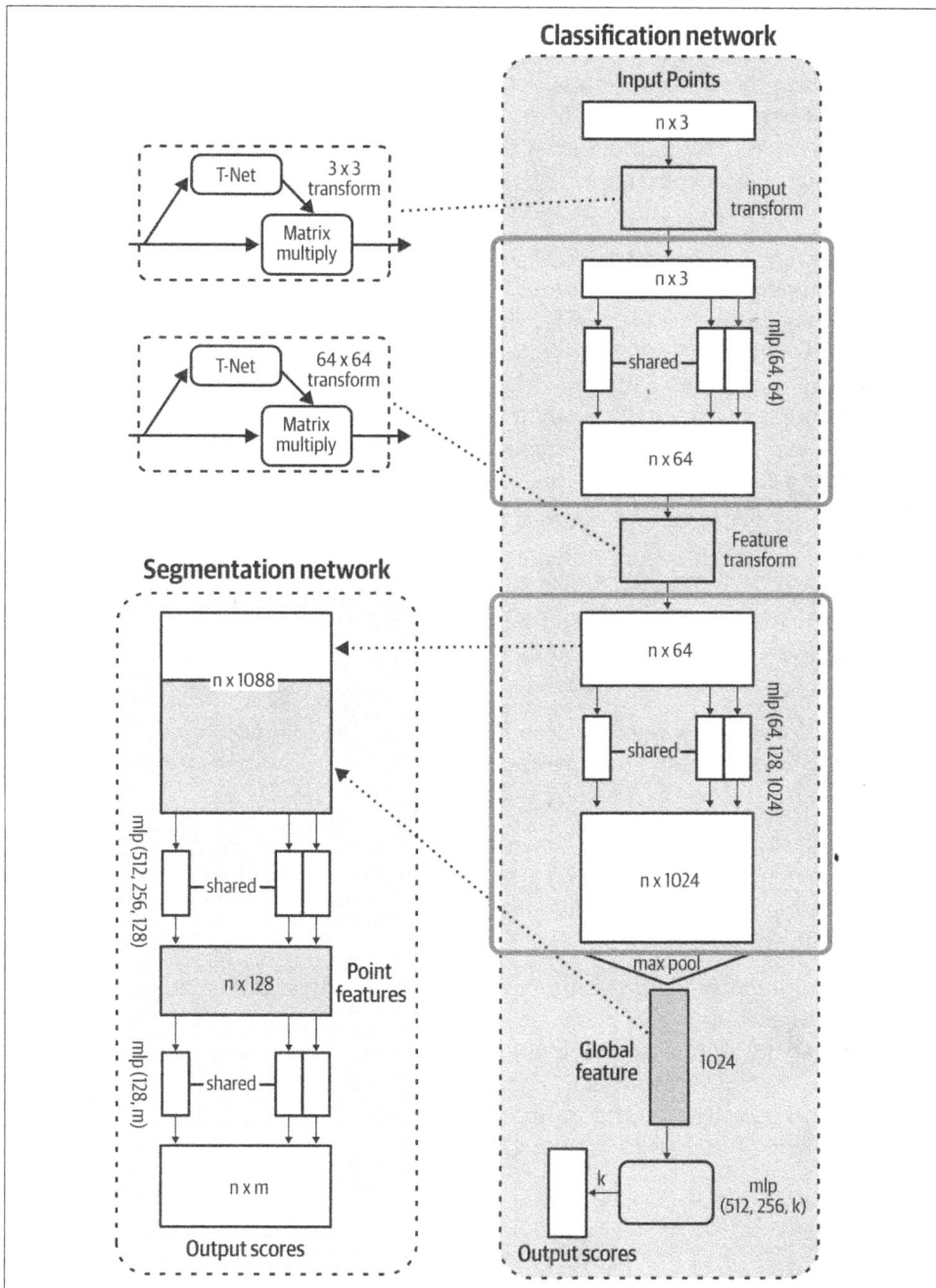

Figure 16-3. MLP networks within the PointNet architecture (ArXiv paper (https://oreil.ly/2oHMt))

An MLP is a neural network of multiple layers of connected nodes or neurons. Each neuron in the MLP first receives an input from all neurons that constitute the previous layer. From there, it applies a transformation to this input using weights and biases and then passes the result to the neurons in the next layer. What is interesting to note is that the weights and biases in the MLP are the two variables learned during training using backpropagation. They are adjusted to minimize the difference between the network's predictions and the true output.

These MLPs are fully connected layers, each followed by what we call "a nonlinear activation function" (such as the ReLU function). The number of neurons in each layer (e.g., 64) and the number of layers themselves (e.g., 2) can be adjusted depending on the specific task and the complexity of the input point cloud data. As you can guess, the more neurons and layers, the more complex the targeted problem can be because of the combinatorial possibilities given by the architecture's plasticity. If we continue to explore the PointNet architecture, we see that we describe the original n input points with 1,024 features that emerge from the initial ones provided (x, y, and z).

This is where the architecture provides a global description of the input point cloud using a max-pooling operation on the locally learned features to obtain a global feature vector that summarizes the entire point cloud. This global feature vector is then fed through several fully connected layers to produce the final output of the classification head, i.e., the score for k classes (Figure 16-4).

Notice that the semantic segmentation head in PointNet is a fully connected network that concatenates the global feature vector and the local feature vectors to produce a per-point score or label for each point in the input point cloud data.

The semantic segmentation head consists of several fully connected layers with ReLU activation functions and a final softmax layer. The output of the final softmax layer represents the per-point probability distribution over the different semantic labels or classes (Figure 16-5).

The PointNet architecture can capture important geometric and contextual information for tasks such as object classification and segmentation in 3D data by learning local and global features from each point in the input point cloud. One of the critical innovations of PointNet is using a symmetric function in the max-pooling operation, which ensures that the output is invariant to the order of the input points. This makes PointNet robust to variations in the ordering of the input points, which is essential in 3D data analysis.

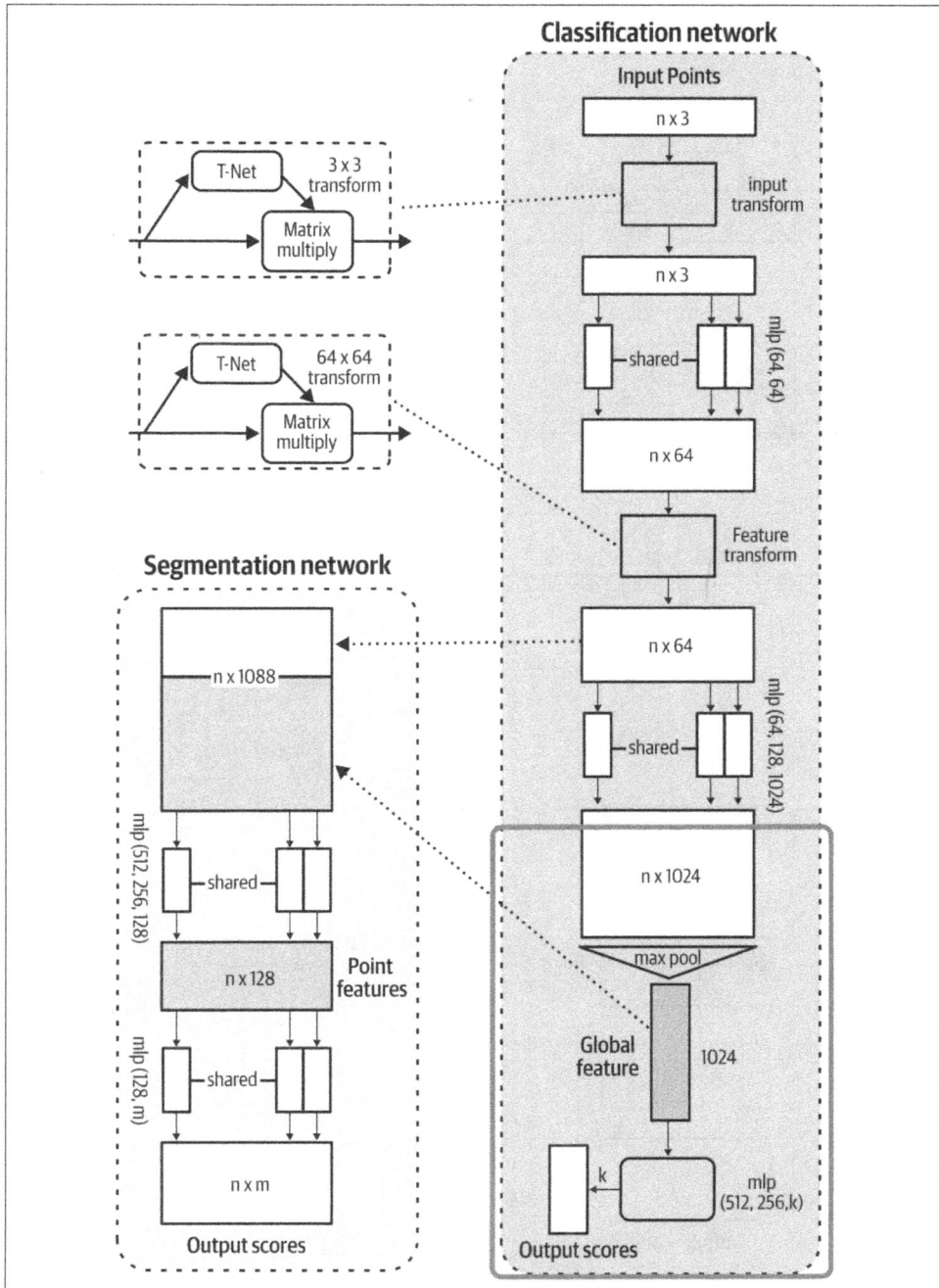

Figure 16-4. Using max pooling to obtain a global feature vector in the PointNet architecture (ArXiv paper (https://oreil.ly/b0iPQ))

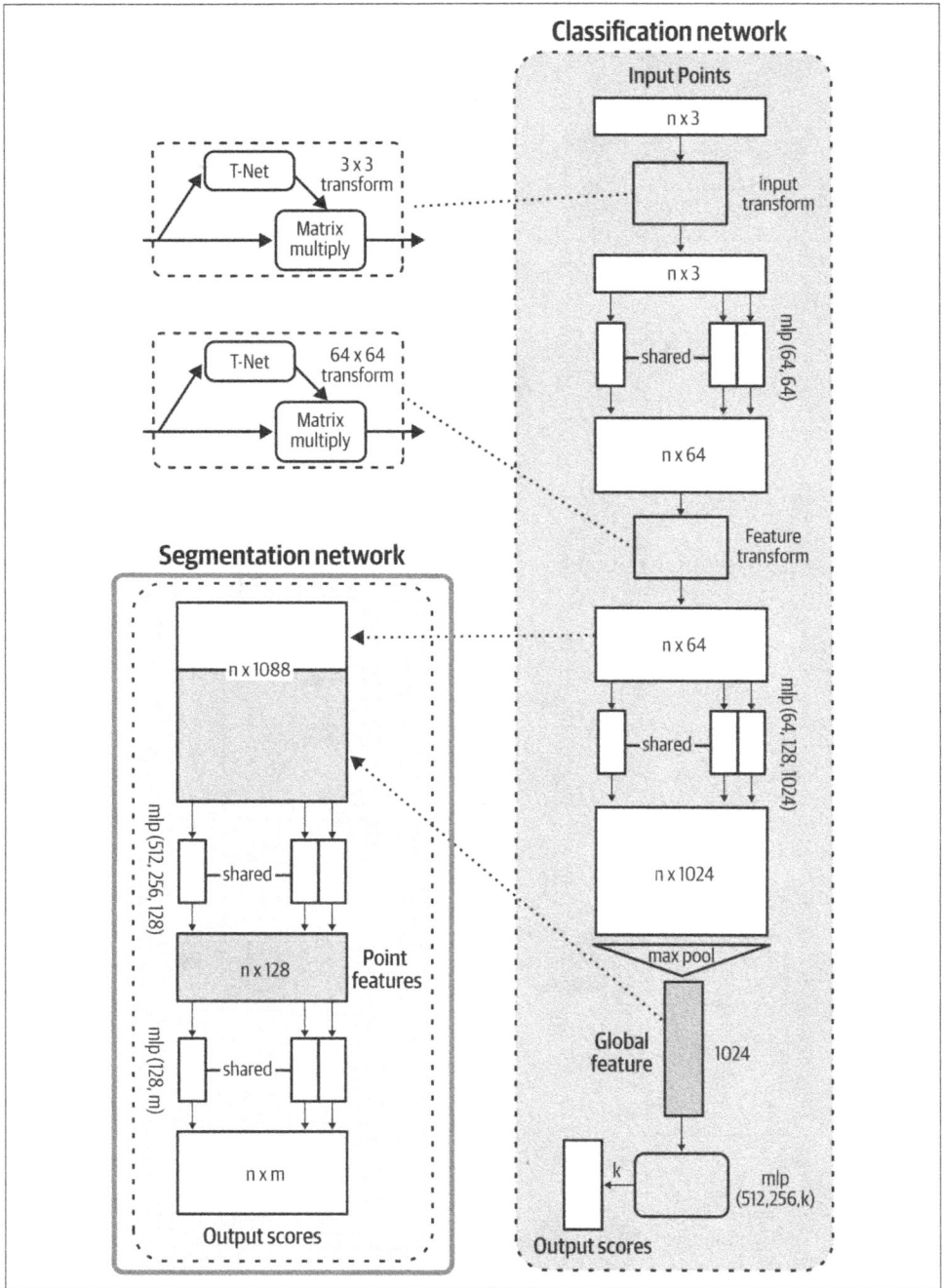

Figure 16-5. The segmentation head of the PointNet architecture (ArXiv paper (https://oreil.ly/wBesf))

At a high-level view, if we study the original paper (*https://oreil.ly/r_zzS*), we see that PointNet functions in a very straightforward manner (see Figure 16-6):

1. We take a point cloud and normalize the data to a canonical space.

2. We compute a bunch of features (without ingesting our expert knowledge but by leveraging network capabilities to create relevant ones).

3. We aggregate these features into a global signature for the considered cloud.

4. *Option 1*: We use this global signature to classify the point cloud.

5. *Option 2*: We combine this global signature with the local signature and build even sharper features for semantic segmentation.

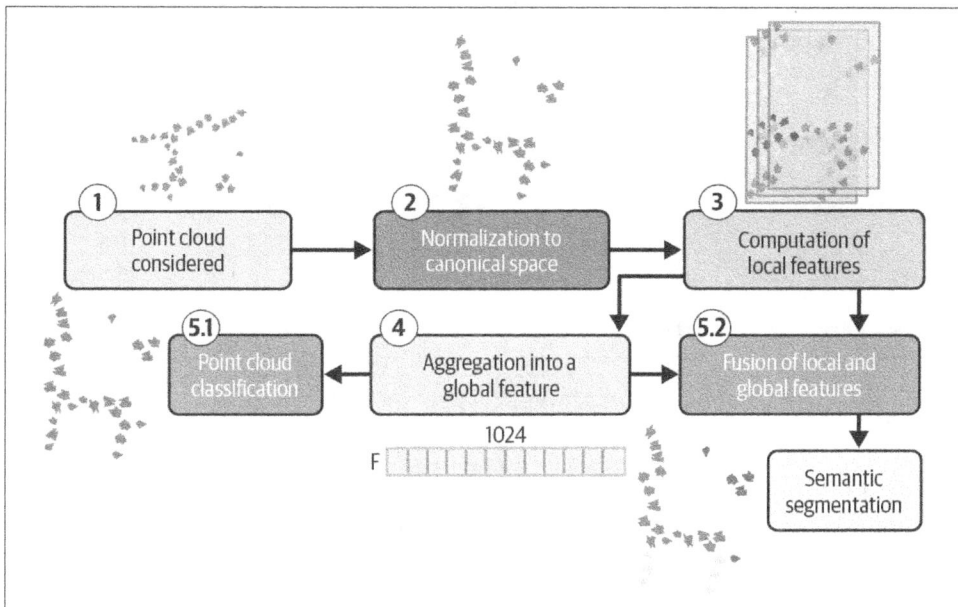

Figure 16-6. The five steps of PointNet, leading to either semantic segmentation or classification tasks

It is all about features, meaning the chunk we provide in the network should be very relevant. So, how do we approach this?

3D Object Classification

Let's contextualize your mission. You are a brave engineer and you want to improve citizens' lives. You have the idea to equip your robotic companion with the power to perceive and understand its surroundings. Using LiDAR, your robot must learn to detect and classify various types of furniture. Imagine the possibilities! With this newfound ability, your robot will be able to identify beds, chairs, tables, and more and then dispatch other robots to carry out specific tasks like making beds or rearranging furniture.

To achieve this, you develop a sophisticated 3D deep learning system. This system analyzes the LiDAR data, extracting meaningful features from the point cloud. It then learns to categorize these features into different furniture types, enabling your robot to make informed decisions. Naturally, with your first-hand experience with PointNet, this is what you will explore (Figure 16-7).

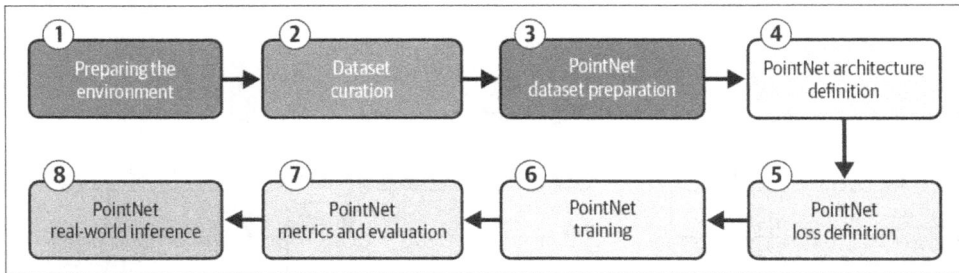

Figure 16-7. The 3D object classification workflow with PointNet

The success of this quest will not only enhance your robot's capabilities but also pave the way for a future where robots seamlessly integrate into our daily lives, assisting us with mundane tasks and making our homes more efficient and comfortable. Let's get started.

3D Object Classification Fundamentals

If you remember, PointNet allows us to predict labels to achieve the 3D semantic segmentation goal. But PointNet delineates two main granularities for this: either at the point level (3D semantic segmentation, to assign a label to each point in the point cloud) or at the point cloud level (3D object classification, which seeks to assign a single label to the entire cloud), illustrated in Figure 16-8.

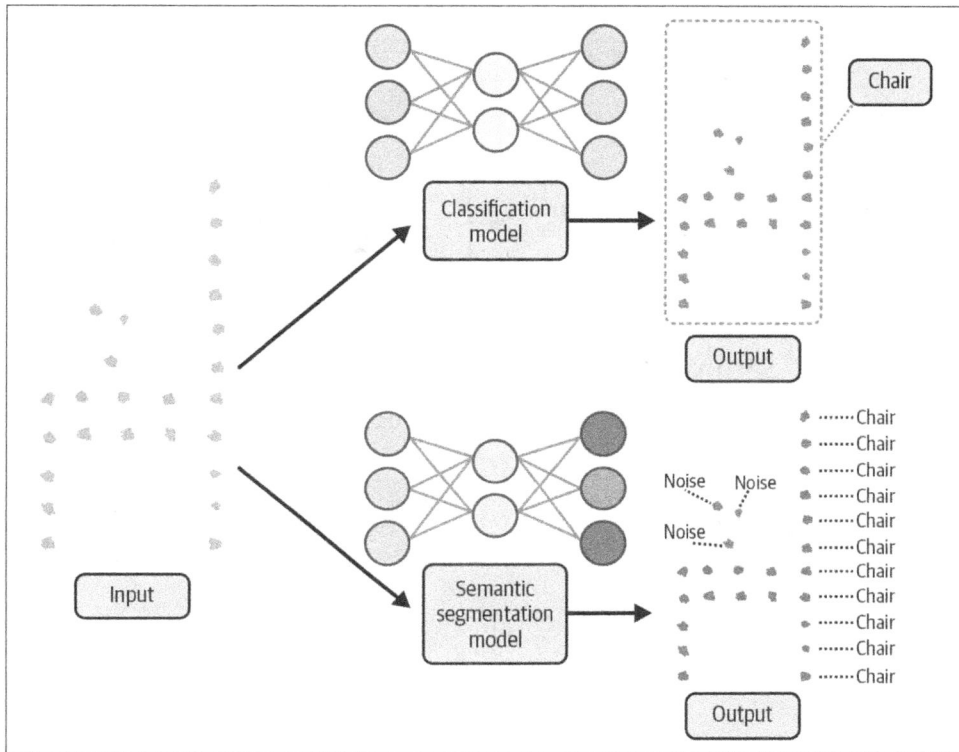

Figure 16-8. The difference between the classification model and the semantic segmentation model

In both cases, we pass a point cloud, but for the classification task, the whole point cloud is the entity, whereas, in the semantic segmentation case, each point is an entity to classify. For example, using the PointNet architecture, 3D point cloud classification involves passing the entire point cloud through the network and outputting a single label representing the entire cloud. In contrast, the semantic segmentation "header" will assign a label to each point in the cloud. The difference in approach is because segmentation requires a more detailed understanding of the 3D space represented, as it seeks to identify and label individual objects or regions within the point cloud. In contrast, classification only requires a high-level understanding of the overall shape or composition of the point cloud.

Overall, while 3D semantic segmentation and classification are essential tasks for analyzing 3D point cloud data, the main difference is the level of detail and granularity required in the labeling process. With this in mind, let's develop a complete solution for the 3D object classification goal, consisting of eight stages as illustrated in Figure 16-9. Let's start by setting up our environment.

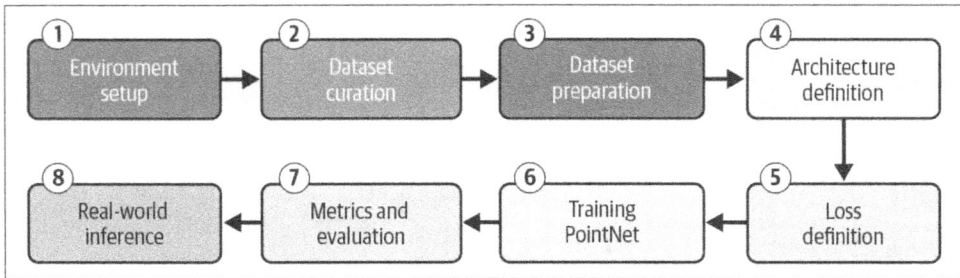

Figure 16-9. Complete PointNet implementation workflow for 3D object classification

Environment Setup

Let's explore a full 100% Python-based solution, including the dataset sourcing. We first load data science base libraries:

```
# Data science base libraries
import numpy as np
import random
import math
import scipy.spatial.distance
```

Followed by utility libraries:

```
import os
import copy
import urllib.request
import zipfile
from pathlib import Path
```

Finally, we import PyTorch and associated methods and utilities:

```
# PyTorch and dependencies
import torch
from torch.utils.data import Dataset, DataLoader
from torchvision import transforms, utils
import torch.nn as nn
import torch.nn.functional as F
```

From there, we can retrieve our dataset.

Dataset Curation

I want to showcase the power of research and open initiative. In some cases, you do not have to constitute your own dataset, and you can use existing datasets. In our case, we leverage *ModelNet10,* a widely used benchmark dataset, specifically for object classification tasks. It consists of 10 object categories, each containing 984 CAD models. These models are carefully cleaned and aligned, making them suitable for training and testing deep learning models for 3D object recognition. ModelNet10 includes 10 categories (airplane, bathtub, bed, bookshelf, bottle, chair, cup, lamp, laptop, table) ensuring a comprehensive evaluation of 3D object classification models.

First, let's get the dataset and download it in the local *DATA* folder:

```
file_path_zip = "../DATA/ModelNet10.zip"

urllib.request.urlretrieve("http://3dvision.princeton.edu/projects/2014/
3DShapeNets/ModelNet10.zip", file_path_zip)
```

This leverages the `urllib` library to download the specified file to the specified path directly. As you can see, this is a *.zip* file that we need to extract:

```
with zipfile.ZipFile(file_path_zip,"r") as zip_ref:
    zip_ref.extractall("../DATA/")
```

Once this is done, we create three variables, namely `path`, `folders`, and `classes`, that we can leverage later in our script:

```
path = Path("../DATA/ModelNet10/")

folders = [dir for dir in sorted(os.listdir(path)) if os.path.isdir(path/dir)]
classes = {folder: i for i, folder in enumerate(folders)};
```

Now, it is time to read our files, which are in the *.off* file format (see Figure 16-10). A simple solution would be to leverage Open3D to do just that:

```
import open3d as o3d
mesh = o3d.io.read_triangle_mesh(str(path / "bed/train/bed_0001.off"))
mesh.compute_vertex_normals()
o3d.visualization.draw_geometries([mesh], mesh_show_back_face = True)
```

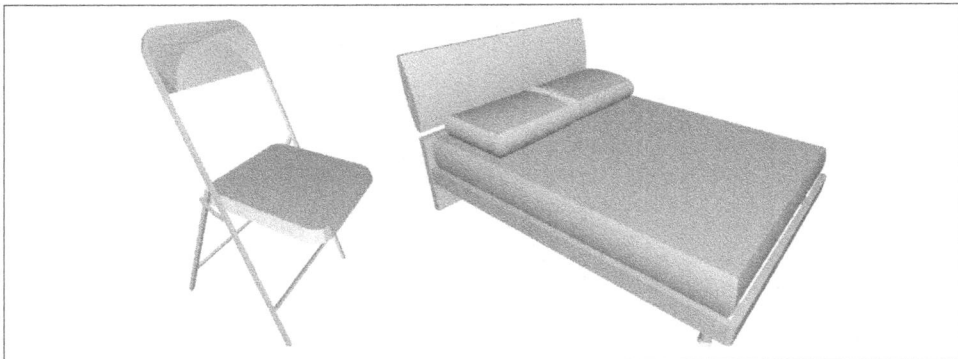

Figure 16-10. Example of the 3D objects, visualized with Open3D

However, for the sake of customization, what if you face an unknown format? Well, it means we need to develop a data reader. If we check the file format of the ModelNet10 dataset, we see that it is an *.off* file. Let us develop a reader to read *.off* files:

```
def read_off(file):
    if 'OFF' != file.readline().strip():
        raise('Not a valid OFF header')
    n_verts, n_faces,
    __ = tuple([int(s) for s in file.readline().strip().split(' ')])
    verts = [[float(s) for s in file.readline().strip().split(' ')] for i_vert
    in range(n_verts)]
    faces = [[int(s) for s in file.readline().strip().split(' ')][1:] for i_face
    in range(n_faces)]
    return verts, faces
```

You can now leverage the reader, which returns the geometry as two variables: a list of vertices and a list of faces. Now, we can start creating our dataset class.

PointNet: Dataset Preparation

The first stage is to define a function that transforms the mesh into a 3D point cloud. You can leverage the teachings of Chapter 4, which are incorporated in a PyTorch Dataset class for handling point cloud data for 3D object classification:

```
class PointCloudData(Dataset):
    def __init__(self, root_dir, valid=False, folder="train",
    transform=default_transforms()):
        self.root_dir = root_dir
        folders = [dir for dir in sorted(os.listdir(root_dir)) if
        os.path.isdir(root_dir/dir)]
        self.classes = {folder: i for i, folder in enumerate(folders)}
        self.transforms = transform if not valid else default_transforms()
        self.valid = valid
        self.files = []
        for category in self.classes.keys():
            new_dir = root_dir/Path(category)/folder
            for file in os.listdir(new_dir):
                if file.endswith('.off'):
                    sample = {}
                    sample['pcd_path'] = new_dir/file
                    sample['category'] = category
                    self.files.append(sample)

    def __len__(self):
        return len(self.files)

    def __preproc__(self, file):
        verts, faces = read_off(file)
        if self.transforms:
            pointcloud = self.transforms((verts, faces))
        return pointcloud

    def __getitem__(self, idx):
        pcd_path = self.files[idx]['pcd_path']
        category = self.files[idx]['category']
        with open(pcd_path, 'r') as f:
```

```
        pointcloud = self.__preproc__(f)
     return {'pointcloud': pointcloud,
             'category': self.classes[category]]}
```

Let's break it down. First, we initialize everything with the __init__ method (the dataset with a root directory, validation flag, folder name, transforms, dictionary mapping folder, and list of files). Then, the __len__ method returns the total number of samples in the dataset.

The most useful method is the __preproc__ method, which preprocesses a single point cloud file by reading the mesh and applying transforms as specified. Finally, the __getitem__ method retrieves a single sample from the dataset, reads, preprocesses, and returns a dictionary with the processed point cloud and its category index.

So at this stage, our dataset loader class is prepared. Now, following the recommendations of Chapter 15, let's tackle the definition of the 3D deep learning architecture: the PointNet architecture.

PointNet Architecture Definition

The architecture comprises three main classes: Tnet, Transform, and PointNet. The Transformer network class, or Tnet, is a transformation network used to learn input transformations. It's a crucial component of PointNet that helps make the network invariant to certain geometric transformations. This is our Tnet class:

```python
class Tnet(nn.Module):
    def __init__(self, k=3):
        super().__init__()
        self.k=k
        self.conv1 = nn.Conv1d(k,64,1)
        self.conv2 = nn.Conv1d(64,128,1)
        self.conv3 = nn.Conv1d(128,1024,1)
        self.fc1 = nn.Linear(1024,512)
        self.fc2 = nn.Linear(512,256)
        self.fc3 = nn.Linear(256,k*k)

        self.bn1 = nn.BatchNorm1d(64)
        self.bn2 = nn.BatchNorm1d(128)
        self.bn3 = nn.BatchNorm1d(1024)
        self.bn4 = nn.BatchNorm1d(512)
        self.bn5 = nn.BatchNorm1d(256)


    def forward(self, input):
        # input.shape == (bs,n,3)
        bs = input.size(0)
        xb = F.relu(self.bn1(self.conv1(input)))
        xb = F.relu(self.bn2(self.conv2(xb)))
        xb = F.relu(self.bn3(self.conv3(xb)))
        pool = nn.MaxPool1d(xb.size(-1))(xb)
```

```
flat = nn.Flatten(1)(pool)
xb = F.relu(self.bn4(self.fc1(flat)))
xb = F.relu(self.bn5(self.fc2(xb)))

#initialize as identity
init = torch.eye(self.k, requires_grad=True).repeat(bs,1,1)
if xb.is_cuda:
  init=init.cuda()
matrix = self.fc3(xb).view(-1,self.k,self.k) + init
return matrix
```

The network is initialized with a default input dimension of 3 (for 3D point clouds) and consists of three 1D convolutional layers and three fully connected layers. The architecture progressively increases the number of channels from the input dimension to 64, then 128, and finally 1,024, along with corresponding batch normalization layers to stabilize and accelerate the training process. After this, the convolutional layers are followed by three fully connected layers that reduce the feature representation from 1024 to 512, then to 256, and finally to a transformation matrix with dimensions $k \times k$.

The forward method implements the actual transformation learning process, where the input point cloud goes through a series of operations: first, it applies ReLU-activated 1D convolutions with batch normalization, then uses max pooling to aggregate features across all points and flattens the result. The network then passes this representation through two more fully connected layers with ReLU and batch normalization.

The unique aspect of this implementation is the final transformation matrix generation: it creates an identity matrix as an initial transformation, adds the learned transformation from the network, and returns a matrix that can be used to spatially transform the input point cloud, effectively learning an alignment that makes the point cloud features more invariant to geometric variations.

Now, we can move onto the Transform class, which is responsible for applying two Tnets, one for input transformation and another for feature transformation:

```
class Transform(nn.Module):
    def __init__(self):
        super().__init__()
        self.input_transform = Tnet(k=3)
        self.feature_transform = Tnet(k=64)
        self.conv1 = nn.Conv1d(3,64,1)

        self.conv2 = nn.Conv1d(64,128,1)
        self.conv3 = nn.Conv1d(128,1024,1)


        self.bn1 = nn.BatchNorm1d(64)
        self.bn2 = nn.BatchNorm1d(128)
```

```
        self.bn3 = nn.BatchNorm1d(1024)

    def forward(self, input):
        matrix3x3 = self.input_transform(input)
        # batch matrix multiplication
        xb = torch.bmm(torch.transpose(input,1,2), matrix3x3).transpose(1,2)

        xb = F.relu(self.bn1(self.conv1(xb)))

        matrix64x64 = self.feature_transform(xb)
        xb = torch.bmm(torch.transpose(xb,1,2), matrix64x64).transpose(1,2)

        xb = F.relu(self.bn2(self.conv2(xb)))
        xb = self.bn3(self.conv3(xb))
        xb = nn.MaxPool1d(xb.size(-1))(xb)
        output = nn.Flatten(1)(xb)
        return output, matrix3x3, matrix64x64
```

It also includes three 1D convolutional layers with batch normalization. In its forward pass, the `Transform` class first applies the input transformation using the first `Tnet`, then processes the data through the convolutional layers, and finally applies the feature transformation using the second `Tnet`. The output of this class is the transformed features along with both transformation matrices (3×3 and 64×64).

Finally, the `PointNet` class is the main architecture that connects everything. It starts by using the `Transform` class to process the point cloud data:

```
class PointNet(nn.Module):
    def __init__(self, classes = 10):
        super().__init__()
        self.transform = Transform()
        self.fc1 = nn.Linear(1024, 512)
        self.fc2 = nn.Linear(512, 256)
        self.fc3 = nn.Linear(256, classes)

        self.bn1 = nn.BatchNorm1d(512)
        self.bn2 = nn.BatchNorm1d(256)
        self.dropout = nn.Dropout(p=0.3)
        self.logsoftmax = nn.LogSoftmax(dim=1)

    def forward(self, input):
        xb, matrix3x3, matrix64x64 = self.transform(input)
        xb = F.relu(self.bn1(self.fc1(xb)))
        xb = F.relu(self.bn2(self.dropout(self.fc2(xb))))
        output = self.fc3(xb)
        return self.logsoftmax(output), matrix3x3, matrix64x64
```

In the `__init__` method, the network is constructed with several key components: a transformation module that learns spatial alignments of the input point cloud; three fully connected layers that progressively reduce the feature dimensionality from 1,024 to 512, then to 256, and finally to the number of classification classes (defaulting to

10); and batch normalization layers (to stabilize the learning process). The choice of a dropout rate of 0.3 is a regularization technique to prevent overfitting by randomly setting a fraction of input units to zero during training, while the LogSoftmax activation at the end ensures that the output represents log-probability distributions across the classification classes.

The forward method implements the network's inference process, where the input point cloud first undergoes a spatial transformation using the Transform module, which returns the transformed point cloud features along with two transformation matrices (3 × 3 and 64 × 64). These matrices are crucial for learning point cloud alignment and feature extraction. The features then pass through two fully connected layers with ReLU activations and batch normalization, with the dropout layer applied between the second and third layers to introduce additional regularization.

The final fully connected layer produces the class logits, which are then converted to log probabilities using LogSoftmax. The method returns three elements: the log-probability predictions and the two transformation matrices, which can be used for visualization, further analysis, or as part of a loss function to encourage spatial invariance in the point cloud representation.

A key aspect of this architecture is its ability to process point clouds directly, without the need for voxelization or rendering. The use of Tnets makes the network invariant to certain geometric transformations, which is crucial for handling 3D data. Another important feature is the use of max pooling to create a global feature vector, which makes the network invariant to the order of points in the input (permutation invariance).

The dataset class is created, and the PointNet model is defined; now, what about the loss?

PointNet Loss Definition

Let's define a function, named pointnetloss. It uses several important parameters: the network's predictions (outputs), the true labels (labels), the 3 × 3 transformation matrix from the input transform (m3×3), the 64 × 64 transformation matrix from the feature transform (m64×64), and a regularization parameter alpha, which defaults to 0.0001:

```
def pointnetloss(outputs, labels, m3x3, m64x64, alpha = 0.0001):
    criterion = torch.nn.NLLLoss()
    bs=outputs.size(0)
    id3x3 = torch.eye(3, requires_grad=True).repeat(bs,1,1)
    id64x64 = torch.eye(64, requires_grad=True).repeat(bs,1,1)
    if outputs.is_cuda:
        id3x3=id3x3.cuda()
        id64x64=id64x64.cuda()
    diff3x3 = id3x3-torch.bmm(m3x3,m3x3.transpose(1,2))
```

```
diff64x64 = id64x64-torch.bmm(m64x64,m64x64.transpose(1,2))
return criterion(outputs, labels) + alpha * (torch.norm(diff3x3)+
torch.norm(diff64x64)) / float(bs)
```

The loss function is designed to combine two essential components: a classification loss and a regularization loss. For the classification loss, it employs the negative log-likelihood loss (NLLLoss), which is particularly well suited for the PointNet architecture because the network uses LogSoftmax as its final layer.

This loss measures how well the network's predictions match the true labels. The regularization loss is a more intricate component that plays a vital role in PointNet's ability to handle 3D point cloud data effectively. Its purpose is to encourage the learned transformation matrices (both 3×3 and 64×64) to be as close as possible to orthogonal matrices. This is achieved by penalizing the difference between the identity matrix and the product of each transformation matrix with its transpose.

To compute this regularization loss, the function first creates identity matrices of sizes 3×3 and 64×64, repeating them for each sample in the batch. It then calculates the difference between these identity matrices and the product of each transformation matrix with its transpose. The Frobenius norm of these differences is then computed, which essentially measures the magnitude of the difference matrices.

The total loss is calculated by summing the classification loss and the regularization loss. The regularization loss is scaled by the alpha parameter and normalized by the batch size to balance its contribution to the overall loss. This normalization ensures that the regularization term doesn't dominate the loss as the batch size changes.

The inclusion of this regularization term in the loss function is a key innovation of the PointNet architecture. By encouraging the learned transformations to be close to orthogonal, it helps preserve the structure of the input point cloud throughout the network's processing. This is crucial for maintaining the geometric properties of the 3D data and contributes to the network's ability to achieve invariance to certain geometric transformations.

This custom loss function illustrates the integration of domain knowledge on 3D data characteristics and preferred geometric transformation qualities into the neural network's learning process. It showcases how carefully designed loss functions can guide the network to learn not just to classify correctly, but also to learn meaningful and geometrically sensible internal representations of the input data.

It is time to train our model!

PointNet Training

Let's build a typical training loop for a deep learning model but with some specific adaptations for PointNet. Let's go through it in detail:

```python
def train(model, train_loader, val_loader=None, epochs=15, save=True):
    for epoch in range(epochs):
        pointnet.train()
        running_loss = 0.0
        for i, data in enumerate(train_loader, 0):
            inputs, labels = data['pointcloud'].to(device).float(),
            data['category'].to(device)
            optimizer.zero_grad()
            outputs, m3x3, m64x64 = pointnet(inputs.transpose(1,2))

            loss = pointnetloss(outputs, labels, m3x3, m64x64)
            loss.backward()
            optimizer.step()

            # print statistics
            running_loss += loss.item()
            if i % 10 == 9:    # print every 10 mini-batches
                    print('[Epoch: %d, Batch: %4d / %4d], loss: %.3f' %
                        (epoch + 1, i + 1, len(train_loader), running_loss / 10))
                    running_loss = 0.0

        pointnet.eval()
        correct = total = 0

        # validation
        if val_loader:
            with torch.no_grad():
                for data in val_loader:
                    inputs, labels = data['pointcloud'].to(device).float(),
                    data['category'].to(device)
                    outputs, __, __ = pointnet(inputs.transpose(1,2))
                    _, predicted = torch.max(outputs.data, 1)
                    total += labels.size(0)
                    correct += (predicted == labels).sum().item()
            val_acc = 100. * correct / total
            print('Valid accuracy: %d %%' % val_acc)

        # save the model
        if save:
            torch.save(pointnet.state_dict(), "save_"+str(epoch)+".pth")
```

As you can see, the function takes several parameters: the model (pointnet), training data loader, validation data loader (optional), number of epochs, and a flag to save the model.

For each epoch, the function first sets the model to training mode using `point net.train()`. This ensures that layers like dropout are active during training. It then iterates over the training data. For each batch, it moves the input point clouds and their labels to the specified device (likely a GPU). The optimizer's gradients are zeroed out to prevent accumulation from previous batches.

The point cloud data is passed through the PointNet model. Note the `trans pose(1,2)` operation, which is likely adjusting the input shape to match what PointNet expects. The model returns the class predictions, along with the 3×3 and 64×64 transformation matrices.

The loss is then calculated using the custom `pointnetloss` function we discussed earlier. The network backpropagates this loss, and the optimizer updates the model's parameters.

The function keeps track of the running loss and prints statistics every 10 mini-batches. This gives a real-time indication of how the training is progressing. After processing all batches in an epoch, if a validation loader is provided, the function evaluates the model on the validation set. It switches the model to evaluation mode with `pointnet.eval()` and disables gradient computation with `torch.no_grad()` for efficiency.

During validation, each batch is passed through the model and compared with the predicted class (obtained by taking the maximum of the output) with the true label. It keeps a count of correct predictions and total samples to calculate the accuracy of the validation.

Finally, if the save flag is true, it saves the model's state dictionary after each epoch. This allows for checkpointing and potentially using the best model based on validation performance.

This training function encapsulates the entire training process for PointNet, including forward passes, loss computation, backpropagation, optimization, and validation. It's designed to work with PointNet's specific requirements, such as the custom loss function and the need to transpose the input data. Let's now train our model, which should take approximately 3 hours for 15 epochs with my configuration:

```
train(pointnet, train_loader, valid_loader, save=False)
```

The periodic printing of loss and accuracy statistics allows for monitoring training progress, which is crucial for diagnosing issues and tuning hyperparameters. Also, you can see progress with regard to the number of batches:

```
Batch [   3 /   15]
Batch [   4 /   15]
Batch [   5 /   15]
...
```

Once the model is trained, you can save its weights using this command:

```
torch.save(pointnet.state_dict(), "../MODEL/pointnet_classification_15.torch")
```

Once our model is trained, we can check its performance by running in inference mode on the validation set.

PointNet Metrics and Evaluation

Let's evaluate a trained PointNet model on a validation dataset. First, let's import the `confusion_matrix` function from scikit-learn, which will be used to assess the model's performance:

```
from sklearn.metrics import confusion_matrix
```

Then, we create a new instance of the PointNet model and load its weights from our previously saved state. The `.eval()` method is called to set the model in evaluation mode, which is important for layers like `BatchNorm` and `Dropout`:

```
pointnet = PointNet()
pointnet.load_state_dict(torch.load("../MODEL/pointnet_classification_15.torch"))
pointnet.eval();
```

From there, let's define two empty lists: `all_preds` and `all_labels`. These are initialized to store all predictions and true labels, respectively:

```
all_preds = []
all_labels = []
```

We can then enter a loop that iterates over the validation data loader. For each batch:

1. It prints the current batch number for progress tracking.
2. The point cloud data and labels are extracted from the batch.
3. The point cloud data is passed through the PointNet model. Note the transpose operation, which is adjusting the input shape to what PointNet expects.
4. The model's output is processed to get the predicted class (using `torch.max`).
5. Both the predictions and true labels are added to their respective lists.

This is implemented with this Python code:

```
with torch.no_grad():
    for i, data in enumerate(valid_loader):
        print('Batch [%4d / %4d]' % (i+1, len(valid_loader)))

        inputs, labels = data['pointcloud'].float(), data['category']
        outputs, __, __ = pointnet(inputs.transpose(1,2))
        _, preds = torch.max(outputs.data, 1)
        all_preds += list(preds.numpy())
        all_labels += list(labels.numpy())
```

All of this is done within a `torch.no_grad()` context, which disables gradient computation for efficiency during inference. Let's first sample our validation set to explore some of the predictions. The sample yields an equivalence between the ground truth label and the prediction (Figure 16-11). This is even more impressive when looking at the complex capacity to identify a bed and a dresser, for example.

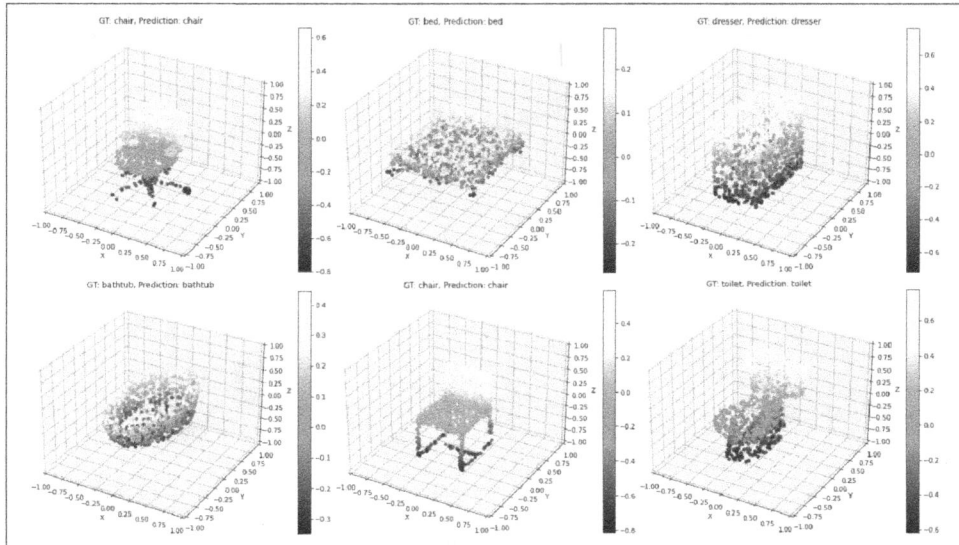

Figure 16-11. The various predictions alongside the plot of our point clouds with Matplotlib

Aside from this qualitative analysis, let's move to quantitative exploration. After processing all batches, we can create a confusion matrix using the accumulated predictions and true labels to show the model's performance for each class. This allows us to see not just overall accuracy but also which classes the model might be confusing with each other (Figure 16-12):

```
cm = confusion_matrix(all_labels, all_preds);
cm
```

This is crucial for understanding how well the trained PointNet model performs on unseen data and provides a detailed view of the model's strengths and weaknesses across different classes, which can be invaluable for further improving the model or understanding its limitations.

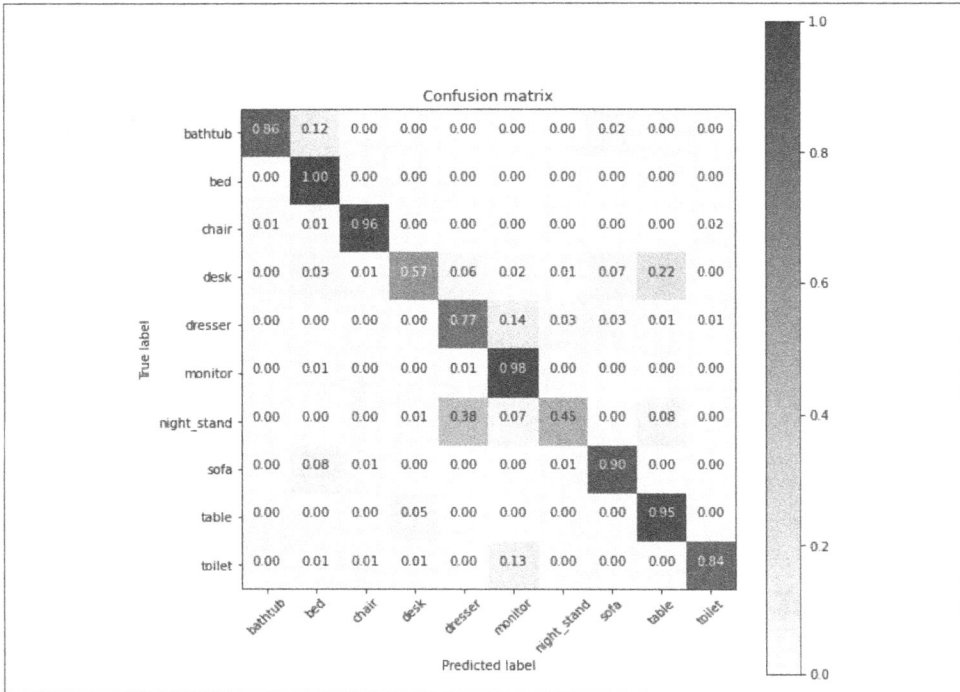

Figure 16-12. The confusion matrix of our 3D object detection case predictions

Looking at this confusion matrix, we can clearly see a diagonal, with some exceptions such as the nightstand and the desk that look to pose some extra challenges. If we push a step further, we can generate a classification report:

```
from sklearn.metrics import classification_report

target_names = ['bathtub', 'bed', 'chair', 'desk', 'dresser', 'monitor',
 'night_stand', 'sofa', 'table', 'toilet']

print(classification_report(all_labels, all_preds, target_names=target_names))
```

This provides us with the following results:

```
              precision    recall  f1-score   support

     bathtub       0.93      0.82      0.87        50
         bed       0.82      0.97      0.89       100
       chair       0.96      0.97      0.97       100
        desk       0.83      0.58      0.68        86
     dresser       0.62      0.77      0.69        86
     monitor       0.76      0.97      0.85       100
 night_stand       0.91      0.48      0.63        86
        sofa       0.90      0.94      0.92       100
       table       0.80      0.92      0.86       100
```

toilet	0.97	0.84	0.90	100
accuracy			0.84	908
macro avg	0.85	0.83	0.83	908
weighted avg	0.85	0.84	0.83	908

We can see that we have a nicely balanced prediction score, with an F1-score that ranges between 63 and 90 points. The first-hand goal, if we were to push our model, could be to try to squeeze this gap. Then, we could also try to push out above 90%, but this is usually a difficult challenge. If we check the overall accuracy, we see we are at 84% with the current mode, which is already a massive feat!

> It's worth noting that this code assumes the model and data are on the CPU, as it's using .numpy() to convert tensors to NumPy arrays. If you're using a GPU, you might need to add .cpu() before .numpy().

Now, what about real-world performance?

PointNet Real-World Inference

The moment you have been waiting for is upon us. Let's leverage the model for pure inference work. For our PointNet model to run in inference mode, we have to make sure that we reuse the same components:

- The PointNet model architecture definition, to create a PointNet instance
- The model weights (which we can load)
- The class definitions
- A preprocessing stage that allows us to prepare the data in the same way as training

As we try and streamline this process, we can cut out the bulk of the libraries. We can work our way up with just these:

```
import numpy as np
import matplotlib.pyplot as plt

import open3d as o3d

import torch
import torch.nn as nn
import torch.nn.functional as F
```

From there, we can make sure our `PointNet` class is usable. After that, we can instantiate a `PointNet` model, initialize its weights with the trained ones, and define our classes:

```
pointnet = PointNet()
pointnet.load_state_dict(torch.load("../MODEL/pointnet_classification_15.torch"))

classes = {'bathtub': 0, 'bed': 1, 'chair': 2, 'desk': 3, 'dresser': 4,
'monitor': 5, 'night_stand': 6, 'sofa': 7, 'table': 8, 'toilet': 9}
```

We can now take a 3D point cloud of an object on which we want to make an inference. This will be the chair from the previous chapter, which was reconstructed via a hybrid approach (LiDAR + photogrammetry):

```
pcd = o3d.io.read_point_cloud('../DATA/FURNITURES/verviers_chair.ply')
```

And we now prepare it in the same way it was prepared for the training process:

```
pcd = pcd.farthest_point_down_sample(1024)
center = pcd.get_center()
pcd.translate(-center)
scale = 1 / np.max(np.abs(np.asarray(pcd.points)))
pcd.scale(scale, center=(0, 0, 0))
o3d.visualization.draw_geometries([pcd])
```

This results in the point cloud illustrated in Figure 16-13.

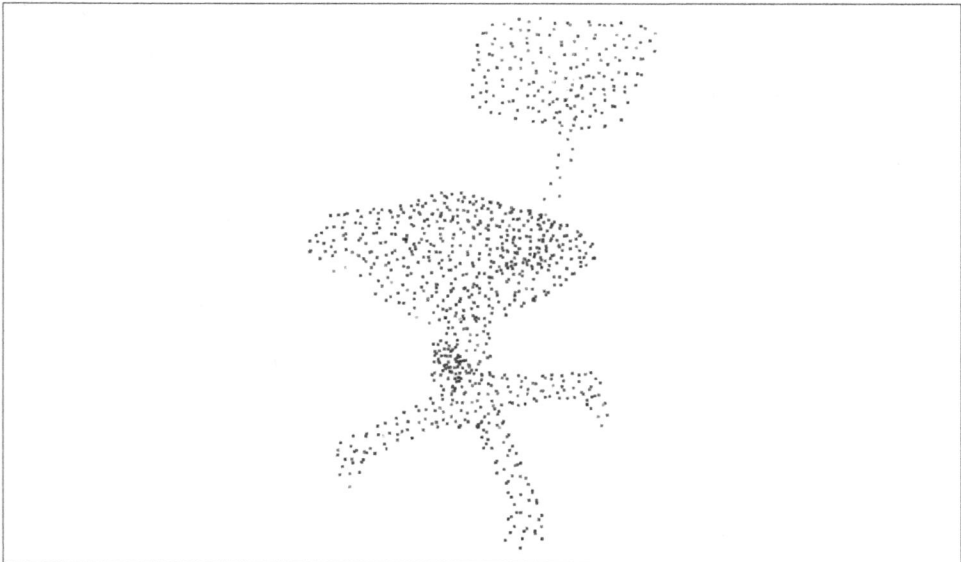

Figure 16-13. The chair used for inference

The last stage is to "tensorize" our point cloud to make sure it fits the input definition:

```
pcd_in = torch.from_numpy(np.asarray(pcd.points)).float()
```

We can now perform inference on a point cloud using our preptrained PointNet model, get the class predictions, and then visualize these predictions. But as I say visualize the predictions, it would also be good to create a tiny, optional function to visualize the point cloud with its ground truth (when accessible) and prediction. For this, you can use this function:

```
def plotPrediction(inputs, i, classes, labels, preds):
    xs, ys, zs = inputs[i].numpy().T

    fig = plt.figure(figsize=(12,7))
    ax = fig.add_subplot(projection='3d')
    img = ax.scatter(xs, ys, zs, c=zs, cmap=plt.hot())
    fig.colorbar(img)

    ax.set_xlabel('X')
    ax.set_ylabel('Y')
    ax.set_zlabel('Z')

    ax.set_xlim([-1, 1])
    ax.set_ylim([-1, 1])
    ax.set_zlim([-1, 1])

    try:
        GTName = list(classes.keys())[list(classes.values()).index(labels[i]
        .numpy())]
    except:
        GTName = 'Unknown'

    PRED = list(classes.keys())[list(classes.values()).index(preds[i].numpy())]

    plt.title(f'GT: {GTName}, Prediction: {PRED}')

    plt.show()
```

Let's first create a context where PyTorch doesn't calculate gradients to save memory and speed up computation; with inference (prediction) we don't need gradients for backpropagation as we're not training. We should make sure we use our model in evaluation mode:

```
pointnet.eval()

with torch.no_grad():
```

Then, we can add the following four code lines:

```
        input_data = pcd_in.unsqueeze(0)
        output, __, __ = pointnet(input_data.transpose(1,2))
        _, preds = torch.max(output.data, 1)

        plotPrediction(input_data, 0, classes, 'None', preds)
```

The `unsqueeze(0)` operation adds a dimension at index 0 to create a batch dimension. This is necessary because our neural networks expect input in batches. Then, we run the PointNet model on the input data. The `transpose(1,2)` operation rearranges the dimensions of the input to match what PointNet expects. The __ in the unpacking suggests that the PointNet model returns additional outputs that are being ignored in this context.

We get predictions from the model output. `torch.max(output.data, 1)` returns two values: the maximum values and their indices along dimension 1. The _ is used to discard the maximum values, keeping only the indices in `preds`. These indices correspond to the batch's predicted class for each point cloud. This results in Figure 16-14.

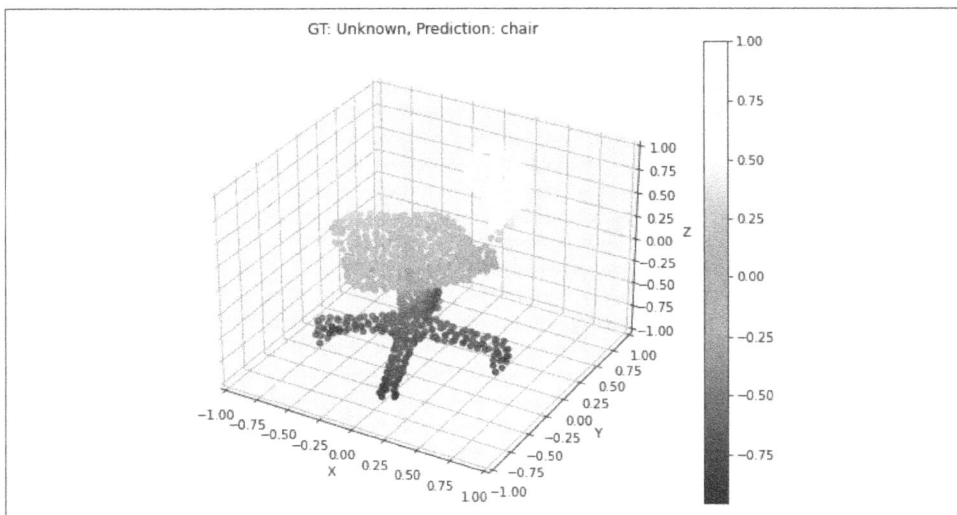

Figure 16-14. The prediction of our network

And there you have it. All that would be left is to push this service as an internal cog in a system, for example, following a clustering phase, to predict what each cluster is based on the preptrained PointNet model. Of course, if you pass an out-of-domain point cloud, you will still get a prediction (see Figure 16-15). We have some predictions for these objects that do not necessarily agree with the ground truth (1 = monitor, 2 = theodolite, 3 = mug, 4 = noise, 5 = box, 6 = table)

At this stage, you have successfully obtained a complete method for predicting the class of a point cloud using a preptrained model. You can fine-tune the model following the strategies highlighted in Chapter 15 if this makes sense for your specific case. Before closing the chapter, I want to highlight a specific aspect: how can we prepare large datasets for semantic segmentation scenarios?

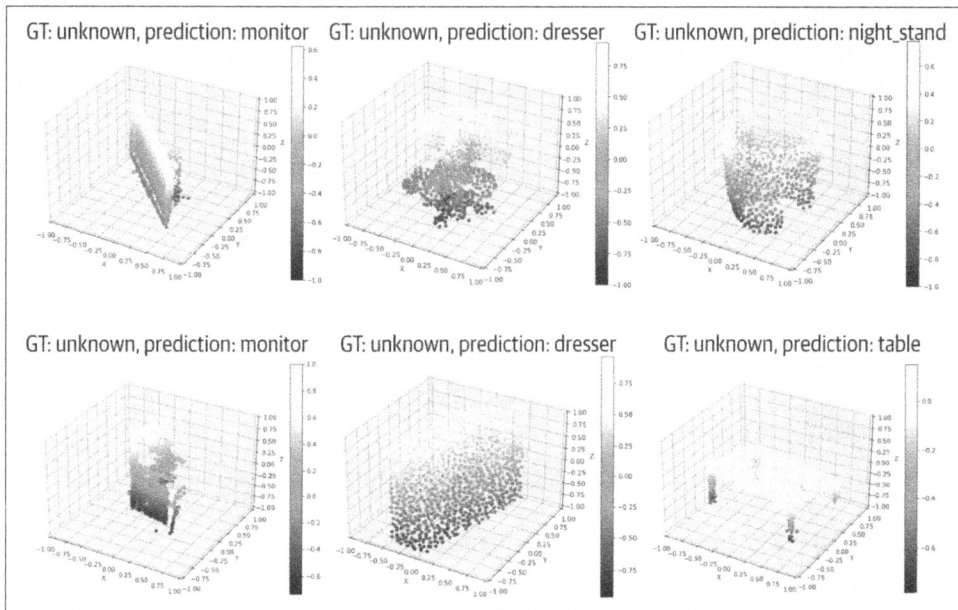

GT: unknown, prediction: monitor — GT: unknown, prediction: dresser — GT: unknown, prediction: night_stand

GT: unknown, prediction: monitor — GT: unknown, prediction: dresser — GT: unknown, prediction: table

Figure 16-15. Various out-of-domain predictions using PointNet

Large-Scale Semantic Segmentation Considerations

One of the most overlooked pain points in 3D deep learning frameworks is preparing the data for use by a selected learning architecture. I'm not referring to a well-curated research dataset but rather a real-world (messy) data silo that you want to use to develop an application.

This is even more pronounced in the case of large and complex 3D point cloud datasets. Therefore, let me provide you with a clear and straightforward series of steps to help you in this context.

You can use any point cloud from previous chapters and prepare it for using PointNet with a semantic segmentation goal in mind. I will illustrate the method by leveraging an aerial LiDAR point cloud of an urban scene.

When using the PointNet architecture for 3D point cloud semantic segmentation, feature selection is essential in preparing the data for training. In the 3D machine learning methods seen in Chapters 13 and 14, feature engineering was required to select and extract relevant features from the data. This step can be circumvented using deep learning techniques such as PointNet, where the model autonomously learns to extract features from the data.

However, ensuring that the input data contains the necessary information for the model to learn relevant and deducted features is still essential. The recommendation is to leverage any features with very low covariance and leave the network the capacity to leverage these.

In the case of a LiDAR point cloud, you could use seven features: X, Y, Z (spatial attributes), R, G, B (radiometric attributes), and the intensity I (LiDAR-derived):

```
X               Y               Z          R   G   B   INTENSITY
205980.49800000 465875.02398682 7.10500002 90  110 98  1175.000000
205980.20100001 465875.09802246 7.13500023 87  107 95  1115.000000
205982.29800010 465875.00000000 7.10799980 90  110 98  1112.000000
```

This is our reference. It means that we build our model with this input, and any other dataset we would like to process with the trained PointNet model must contain these same features.

Before proceeding with the tasks we reviewed before, we must structure the data according to the architecture specifications. Structuring a 3D point cloud into square tiles is essential when processing it with the neural network PointNet architecture.

Indeed, PointNet requires the input data to be of fixed size, meaning that all input samples should have the same number of points. By dividing a 3D point cloud into square tiles, we can ensure that each tile has a more homogeneous number of points, allowing PointNet to process them consistently and effectively without the extra overhead or unreversible loss when sampling to the final fixed-point number (Figure 16-16).

Figure 16-16. Tile definition to train the PointNet 3D deep learning architecture for aerial LiDAR point cloud

With PointNet, we need to set the input tile to a fixed number of points, recommended to be 4,096 points by the original paper's authors. This means that a sampling strategy is needed. As you can see from Figure 16-17, sampling the point cloud with different strategies will yield different results and object identification capabilities (e.g., the electrical pole on the right), which could impact the model's overall performance.

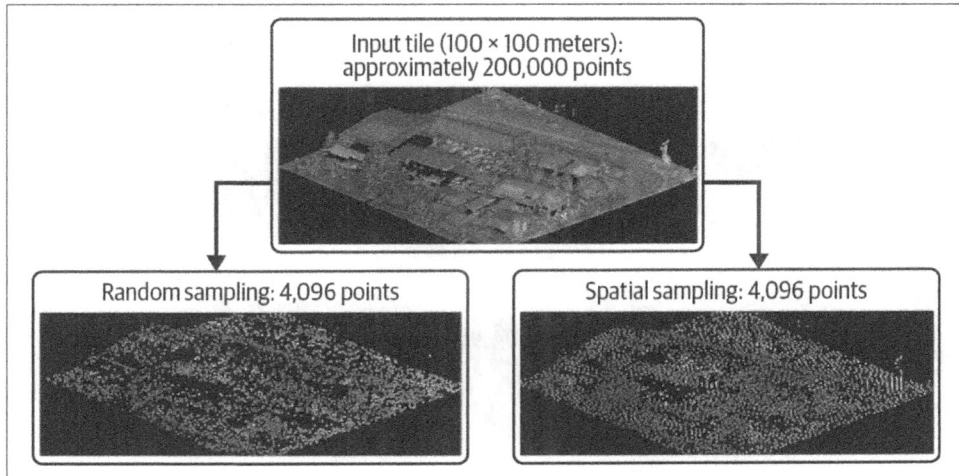

Figure 16-17. Example of the impact of sampling strategies on the 3D point cloud dataset

Secondly, PointNet's architecture involves a shared MLP applied to each point independently, which means that the network processes each point in isolation from its neighbors. By structuring the point cloud into tiles, we can preserve the local context of each point within its tile while still allowing the network to process points independently, enabling it to extract some meaningful features from the data.

Ultimately, organizing the 3D point cloud into tiles enhances the computational efficiency of the neural network by facilitating parallel processing of the tiles, hence diminishing the total processing time needed to analyze the complete point cloud on a GPU.

Structuring a 3D point cloud into square tiles is an essential preprocessing step when using PointNet for semantic segmentation tasks. It allows for consistent input data size, preserves a "local" context, and improves computational efficiency, all of which contribute to more accuracy and efficient processing of the data.

Normalizing a 3D point cloud tile before feeding it to the PointNet architecture is crucial for three main reasons (Figure 16-18). First, normalization ensures that the input data is centered around the origin, which is essential for PointNet's architecture, which applies an MLP to each point independently. The MLP is more effective when the input data is centered around the origin, which allows for more meaningful feature extraction and better overall performance.

> Having some intuition beforehand is helpful, rather than applying normalization blindly. For example, we predominantly use gravity-based scenes, meaning that the z-axis is almost always colinear to the gravity-based axis. Therefore, how would you approach this normalization?

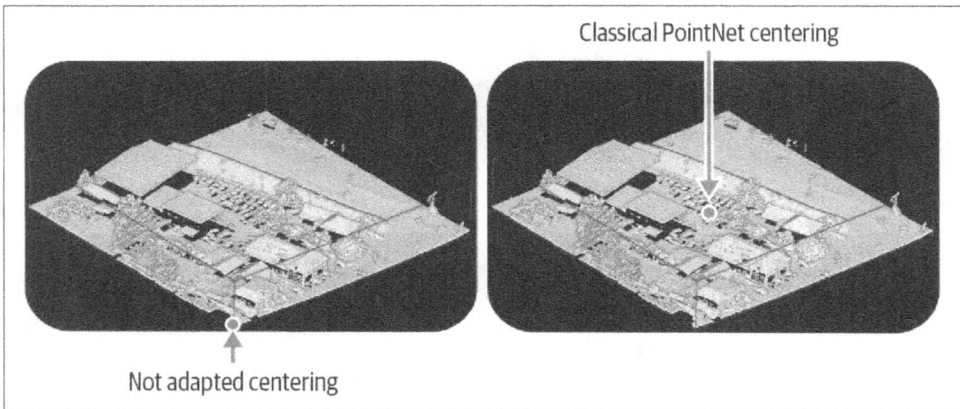

Figure 16-18. Illustration of the impact of normalization on the results of training 3D deep learning models

For our experiments, we can capture the minimum value of the features in `min_f`, and the average in `mean_f`:

```
cloud_data=temp.transpose()
min_f=np.min(cloud_data,axis=1)
mean_f=np.mean(cloud_data,axis=1)
```

We can then normalize the different features to use in our PointNet networks. First, we work with the spatial coordinates X, Y, and Z. We center our data on the planimetric axis (X and Y) and ensure that we subtract the minimal value of Z to account for discrimination between roofs and ground, for example:

```
n_coords = cloud_data[0:3]
n_coords[0] -= mean_f[0]
n_coords[1] -= mean_f[1]
n_coords[2] -= min_f[2]
print(n_coords)
```

Now we can scale our colors by ensuring we are in a [0,1] range. This is done by dividing by the max value (255) for all our colors:

```
colors = cloud_data[3:6]/255
```

This normalization scales the point cloud data to a consistent range, which helps prevent saturation of the activation functions within the MLP. This allows the network to learn from the entire range of input values, improving its ability to accurately classify or segment the data. It also helps reduce the impact of different scales in the point cloud data, which can be caused by differences in sensor resolutions or distances from the scanned objects (which is a bit flattened in the case of aerial LiDAR data). This improves the consistency of the data and the network's ability to extract meaningful features from it. Take the example of the intensity in Figure 16-19.

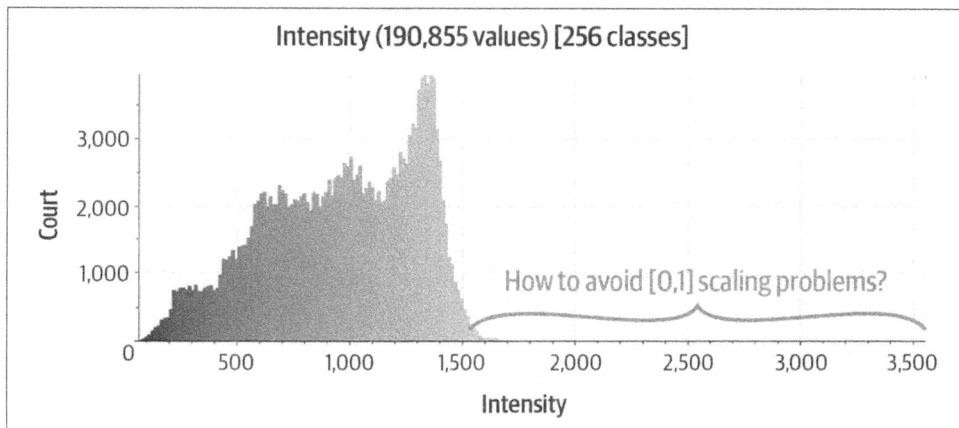

Figure 16-19. The problem with [0,1] scaling illustrated by the intensity field of a 3D point cloud

Here, we work with quantiles to obtain a normalization robust to outliers, as we saw when exploring our data. This is done in a three-stage process. First, we compute the interquartile range (IQR), which is the difference between the 75th and 25th quantile. Then we subtract the median from all the observations and divide by the interquartile difference. Finally, we subtract the minimum value of the intensity to achieve significant normalization:

```
# The interquartile difference is the difference between the 75th and 25th
# quantile
IQR = np.quantile(cloud_data[-2],0.75)-np.quantile(cloud_data[-2],0.25)
# We subtract the median to all the observations and then divide by the
# interquartile difference
n_intensity = ((cloud_data[-2] - np.median(cloud_data[-2])) / IQR)
# This gives us scaling robust to outliers (which is often needed)
n_intensity -= np.min(n_intensity)
print(n_intensity)
```

At this stage, we have a point cloud normalized and ready to be fed to a PointNet architecture. You can follow the complimentary resources on the 3D Resource Hub (*https://learngeodata.eu/3d-data-science-with-python*) if you want to explore the complete method for 3D semantic segmentation. It specifically includes the code for 3D semantic segmentation experiments (or you can just leverage the code for 3D object classification and activate the semantic segmentation header).

Summary

This chapter explored using PointNet for 3D object classification and semantic segmentation, explicitly focusing on preparing real-world data for 3D scene understanding objectives. I want to emphasize the importance of having correctly formatted data, which is crucial for successful training and inference in 3D deep learning tasks. At this stage, you should feel confident with the following list of learning points (to help you self-assess your newfound skills):

- You can understand and build a PointNet architecture, with key components like the `Tnet` and the core `PointNet` class.
- You can build a data preparation pipeline including:
    - 3D data curation: downloading and understanding the structure and metadata of 3D datasets
    - 3D data analysis: visually inspecting the data, identifying outliers, and understanding feature distributions before addressing them
    - 3D data labeling: refining existing labels, concatenating point clouds, and establishing a clear labeling scheme
    - 3D feature selection: selecting relevant features like spatial coordinates (X, Y, Z), color (R, G, B), and intensity
    - Data structuration (tiling): dividing the point cloud into fixed-size tiles to ensure consistent input for PointNet and preserve local context
    - 3D Python data loading and normalization: implementing Python code to load, normalize, and prepare the data for PointNet using libraries like NumPy and PyTorch
- You can develop a normalization strategy to ensure that the input data is centered and scaled appropriately, which is essential for the effective functioning of PointNet's MLP layers.
- You have a trained PointNet model ready for real-world inference on a new point cloud.

Now, what about actually extending your expertise? Well, in this chapter, we focused on using PointNet for 3D object classification. Now, you can use this newfound knowledge to explore techniques to address imbalanced datasets, a common issue, particularly with real-world data like LiDAR point clouds. Additionally, you could investigate the impact of different point cloud sampling strategies and tile sizes on

PointNet's performance further and explore the use of additional features, such as the near-infrared (NIR) channel, to improve vegetation segmentation. For a deeper dive into the field, exploring more advanced PointNet variations like PointNet++ (*https://oreil.ly/ljbHK*) and other point-based deep learning models is the next logical step in broadening your perspective on the evolution of 3D perception techniques.

Hands-on Project

As experts in cutting-edge applications, you understand that 3D deep learning offers powerful techniques to extract meaningful information from 3D data. This project focuses on applying PointNet to a real-world scenario involving aerial LiDAR data for 3D semantic segmentation. You will work with raw aerial LiDAR data and learn how to prepare this data for PointNet to create a model capable of segmenting LiDAR point clouds into distinct categories: vegetation, buildings, and ground:

Step 1: Data acquisition and exploration
- Download a dataset of aerial LiDAR data from an open source repository. This data should include point coordinates (X, Y, Z), intensity, and color (R, G, B) information.
- Analyze the data using software like CloudCompare. This will allow you to visualize the point cloud, understand the distribution of features, and identify any outliers or potential issues with the data.

Step 2: Data preprocessing
- Refine the labels by isolating and potentially cleaning existing labels to create distinct classes (ground, vegetation, buildings, water, etc.).
- Structure the point cloud into tiles of a fixed size (e.g., 100 m × 100 m) using a tool like CloudCompare. This step ensures consistent input for PointNet and facilitates parallel processing.

Step 3: Feature engineering and normalization
- Select relevant features from the LiDAR data. As a starting point, use X, Y, Z coordinates, intensity, and color (R, G, B).
- Develop a normalization strategy for the features.
- Implement the data loading and normalization process in Python.

Step 4: PointNet implementation and training
- Implement the PointNet architecture in PyTorch, including the TNet modules and the classification head.
- Define a custom loss function that combines the classification loss with a regularization term that encourages the learned transformation matrices to be orthogonal.

- Train the PointNet model using your preprocessed LiDAR data. Experiment with different hyperparameters and training strategies to optimize performance.

Step 5: Model evaluation and deployment
- Evaluate the trained PointNet model on a separate test set of LiDAR tiles.
- Visualize the segmentation results by overlaying the predicted labels on the point cloud.
- Deploy the trained PointNet model for semantic segmentation of aerial LiDAR data.

This project provides a hands-on opportunity to apply PointNet to a challenging real-world problem involving 3D point cloud data.

It is now time to ensure the complete 3D data science workflow from acquisition to the definition of a digital environment is robust and extendable.

CHAPTER 17

The 3D Data Science Workflow

Our exploration of 3D data science together is nearing its end. You should be proud of all the modular aspects brought to life with the power of Python. We developed end-to-end workflows for any 3D dataset and ran advanced scientific experiments that unleashed critical outputs. Now, let's take a step back and look at the bigger picture.

Right from the start, we learned that 3D data science enables the most capable spatial AI systems based on 3D data—systems that try to mimic human behavior. This essentially means grasping the complexity of reality and scheming agents that act human-like in a 3D digital world. So, where do we stand in Chapter 17?

Well, in this chapter, we will adjust our focal length to focus on what 3D data science has enabled us to do: design 3D digital environments, which are the candidates for reasoning agents to act on. This means reviewing the main techniques that you are now proud to master, making sure the learnings are clear, and laying out a blueprint for creating any 3D data science system. To be super pragmatic, we have a complete 3D data science modular workflow at this stage. This means you have an input/output workflow to take any modality and turn it into 3D output or analytical insights.

In addition, we can automate complex 3D processing pipelines, which means having many different modules that address a specific part of a pipeline. This pipeline is usually defined to answer an application challenge. It can compute the number of chairs in a room, detect obstacles for a self-driving car, infer the brain's structure from an MRI, and so on. But now, if we dive a bit more, let's ensure that you feel confident about your capacity to address the various challenges in the pipeline to define digital environments, as illustrated in Figure 17-1. Let's synthesize these concepts while highlighting their real-world impact and interconnections.

587

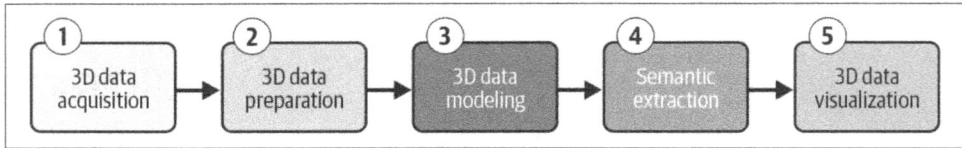

Figure 17-1. The main workflow modules developed with 3D data science

3D Data Acquisition

The first identified module is 3D data acquisition (Figure 17-2). While we moved quickly here to avoid overflowing the whole book, you have solutions to create 3D datasets from multiview images. At this stage, these are 3D point clouds, with the possibility of deriving 3D meshes with texture.

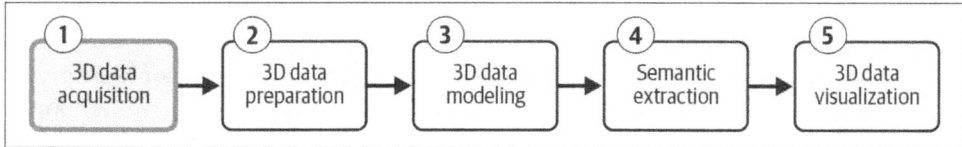

Figure 17-2. Phase 1: 3D data acquisition

With new technological advancements, we can capture that information using inference processes, new spatial AI systems, and deep learning architectures. For example, we saw how to leverage single (monocular) image-based reconstruction like MiDaS (*https://oreil.ly/IljzU*) or Depth Anything (*https://oreil.ly/DF4is*) to obtain depth information and generate 3D point clouds. This provides other exciting ways to retrieve depth information from the environment.

However, multiview reconstruction offers a more complete way to generate a 3D dataset by combining different camera perspectives. It uses several pictures of the same scene or object to make a 3D model. This method usually produces more accurate or exhaustive results than only single-image 3D reconstruction processes (Figure 17-3).

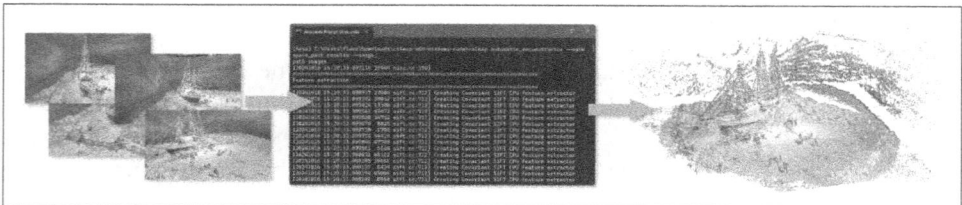

Figure 17-3. The 3D reconstruction process from multiview images

Systems in large-scale production rely on structure from motion (SfM) (*https://oreil.ly/dWJyA*) to obtain camera poses and reconstruct sparse 3D points from multiple images. This allows us to estimate the structure (i.e., the intrinsic and extrinsic parameters) from camera motion through feature detection and feature matching. Multiview stereo (MVS) (*https://oreil.ly/ca3l5*) through dense matching (*https://oreil.ly/e35Qy*) then allows the creation of dense 3D reconstructions with known camera poses. In Chapter 2 and through the 3D Resource Hub (*https://learngeo data.eu/3d-data-science-with-python*), we saw how to leverage these photogrammetry concepts through software and tools that combine SfM and MVS techniques to create accurate 3D models.

With Python, COLMAP (*https://oreil.ly/V5K8W*) is the most popular solution for aligning and generating a dense point cloud from an image set. You can use PyCOL-MAP (*https://oreil.ly/mgJqn*) or the command-line interface wrapped in Python to achieve a proper 3D reconstruction from a set of images. You can use the dataset provided in the 3D Resource Hub and the following Python script to automate the 3D reconstruction of a 3D point cloud from the input images:

```python
import subprocess

def run_colmap(image_dir, output_dir):
    # Feature extraction
    subprocess.run(["colmap", "feature_extractor",
                    "--database_path", f"{output_dir}/database.db",
                    "--image_path", image_dir])

    # Feature matching
    subprocess.run(["colmap", "exhaustive_matcher",
                    "--database_path", f"{output_dir}/database.db"])

    # Sparse reconstruction
    subprocess.run(["colmap", "mapper",
                    "--database_path", f"{output_dir}/database.db",
                    "--image_path", image_dir,
                    "--output_path", f"{output_dir}/sparse"])

    # Dense reconstruction
    subprocess.run(["colmap", "image_undistorter",
                    "--image_path", image_dir,
                    "--input_path", f"{output_dir}/sparse/0",
                    "--output_path", f"{output_dir}/dense"])

    subprocess.run(["colmap", "patch_match_stereo",
                    "--workspace_path", f"{output_dir}/dense"])

    subprocess.run(["colmap", "stereo_fusion",
                    "--workspace_path", f"{output_dir}/dense",
                    "--output_path", f"{output_dir}/dense/fused.ply"])

# Usage
run_colmap("path/to/images", "path/to/output")
```

Of course, after reading Chapter 10, you can also generate 3D datasets from a monocular image. Still, we face some direct challenges in generalizing 3D reconstruction for any scenario, such as expanding applications that can deal with surfaces with low light reflectance, transparency, or holes and optimizing cost, efficiency, simplicity, and robustness.

Current industrial trends point toward increased automation, interconnectivity, and real-time processing. This enables collaborative work between multiple systems and robots equipped with machine vision, leading to more efficient and informed decision making. With that in mind, let's approach 3D data preparation and engineering.

3D Data Preparation and Engineering

The next logical high-level module prepares our 3D geometric base for high-level tasks (Figure 17-4).

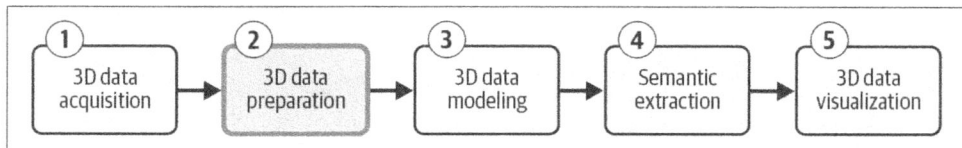

Figure 17-4. Phase 2: 3D data preparation

Here, we can find the preprocessing algorithms (trying to clean, filter out outliers, and see where we have occlusions). I also grouped the solutions that leverage data engineering, like feature extraction. If this is not yet part of your DNA, let me point out the old adage "garbage in, garbage out" one more time. This holds especially true for 3D data science tasks.

The quality and structure of your 3D data can make or break your entire pipeline and the compatibility of your digital environments with spatial AI systems. Let me walk you through, one last time, the essential steps you must master to prepare the best-quality 3D base.

Noise Removal

Noise in 3D data is a persistent beast, often arising from sensor imperfections, environmental factors, or registration errors. Remember the project where we worked with LiDAR scans of an urban environment? The raw point cloud was riddled with noise—floating points in the air, warped surfaces, or density-based noise.

Our go-to approach for noise removal typically involves a combination of statistical outlier removal and radius-based filtering. Here is a snippet for reference using the Open3D library in Python:

```
import open3d as o3d

def remove_noise(pcd, nb_neighbors=20, std_ratio=2.0):
    cl, ind = pcd.remove_statistical_outlier(nb_neighbors, std_ratio)
    return cl, ind

# Load point cloud
pcd = o3d.io.read_point_cloud("DATA/noisy_point_cloud.ply")

# Remove noise and color
clean_pcd, noise_idx = remove_noise(pcd)
clean_pcd.paint_uniform_color([1, 0.706, 0])

# Select the noise
noise = pcd.select_by_index(noise_idx, invert=True)
noise.paint_uniform_color([1, 0.0, 0])

#visualize
o3d.visualization.draw_geometries([noise, clean_pcd])
```

The results of this generated point cloud are shown in Figure 17-5.

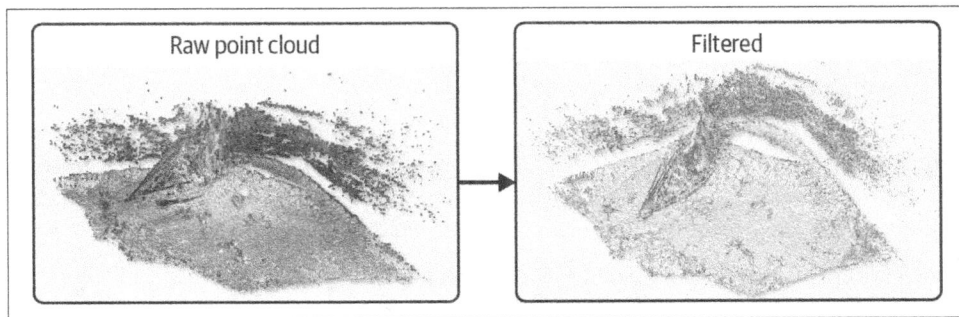

Figure 17-5. From raw point cloud to filtered point cloud

However, our fundamental learning point is that different types of 3D data require different noise removal strategies. For structured data like depth maps, edge-preserving filters like bilateral filters work wonders. Machine learning–based approaches and training models to distinguish between noise and valid data points are also possible for unstructured point clouds, especially in outdoor environments.

Subsampling

Subsampling is crucial for computational feasibility when dealing with massive data-sets, such as point clouds with hundreds of millions of points. But it is not just about randomly throwing away points. Depending on your application (semantic segmentation, visualization, modeling), adapting your sampling strategy is often key to optimized results.

You can directly extend our learning by developing a multiscale subsampling approach that preserves detail in complex areas while aggressively simplifying flat or uniform regions. Here is a simplified version of how you might implement this:

```
def adaptive_downsample(pcd, voxel_size=0.05, detail_size=0.02):
    # Compute normals
    pcd.estimate_normals()

    # Identify high-detail areas
    normals = np.asarray(pcd.normals)
    curvature = np.var(normals, axis=0)
    high_detail = curvature > np.percentile(curvature, 75)

    # Separate high and low detail points
    points = np.asarray(pcd.points)
    high_detail_pcd = o3d.geometry.PointCloud()
    high_detail_pcd.points = o3d.utility.Vector3dVector(points[high_detail])
    low_detail_pcd = o3d.geometry.PointCloud()
    low_detail_pcd.points = o3d.utility.Vector3dVector(points[~high_detail])

    # Downsample
    high_detail_down = high_detail_pcd.voxel_down_sample(detail_size)
    low_detail_down = low_detail_pcd.voxel_down_sample(voxel_size)

    # Combine
    return high_detail_down + low_detail_down
```

This approach has saved me a lot of time, especially when working with large-scale environmental scans where you must preserve fine details (like building facades) while simplifying large uniform areas (like open fields or water bodies).

Now, onto the next data preparation candidate: feature extraction.

Feature Extraction

As we discussed in Chapters 6 and 7, feature extraction is paramount. It is where the rubber meets the road in 3D data engineering. It is about transforming raw geometric data into meaningful descriptors that our down processes and AI models can use.

We have successfully combined local and global features for point cloud classification, segmentation tasks, and registration purposes. The quality of your features should not be overlooked! We also saw that there is usually a fundamental need to sort out the quality and pertinence of our features; for deep learning applications especially, "less is more" is usually a great adage to follow.

Here is another example to guide you on the extraction of some 3D world-related features:

```
import open3d
import numpy

def extract_features(pcd, radius=0.1):
    # Compute normals
    pcd.estimate_normals(search_param=o3d.geometry.KDTreeSearchParamHybrid
    (radius=radius, max_nn=30))

    # Compute local features
    fpfh = o3d.pipelines.registration.compute_fpfh_feature(pcd,
    o3d.geometry.KDTreeSearchParamHybrid(radius=radius*5, max_nn=100))

    # Compute global features
    points = np.asarray(pcd.points)
    centroid = np.mean(points, axis=0)
    distances = np.linalg.norm(points - centroid, axis=1)

    features = {
        'fpfh': np.asarray(fpfh.data).T,
        'distance_to_centroid': distances,
        'height': points[:, 2]  # Assuming Z is up
    }

    return features
```

To visualize the feature, you can follow the solution in Chapter 7 using PyVista. But here is where it gets interesting—different tasks require different features.

For object detection, we have found that geometric primitives (planes, cylinders, spheres) can be incredibly powerful. Topological features that capture the relationships between different scene parts are also crucial for understanding, as seen in Chapter 13.

Finally, high-resolution point clouds, meshes, and geometric models demand significant computational resources and storage. Research into efficient processing, compression, and storage solutions is essential to managing the large datasets generated by 3D spatial measurements (or other approaches). So, let's dive into 3D data modeling.

3D Data Modeling

Molding 3D shapes is fascinating (Figure 17-6). We dived into 3D data modeling and structuration, and you can now move between 3D data representations like a third sense. You know how to go from 3D point clouds to voxel-based assembly, CAD elements, or even a 3D mesh. These are various approaches, algorithms, and techniques that you now have so that you can take on 3D data representation challenges.

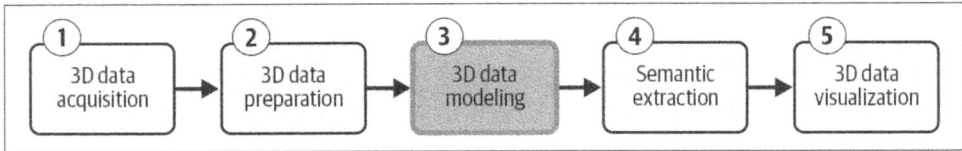

Figure 17-6. Phase 3: 3D data modeling

Also, you can always unify your processes on the 3D point cloud to back-project predictions/processes on the original input data representation. This also plays into the power of structuration. Whenever we need to optimize the computation, we often need to go through a *k*-d tree, an octree data structure, or a bounding volume hierarchy (BVH) that allows us to analyze and process the data more efficiently.

You now have an excellent understanding of how voxels are critical in many processes, especially for 3D data modeling, whether in a processing stage or as input to a structure-aware deep learning approach.

Also, we have realized that how we model and structure our data is just as crucial as the data itself. It is not just about storing points in space but about creating representations that enable efficient querying, analysis, and manipulation.

We walked through all the approaches that could be useful for 99% of your workflow in Chapters 4 and 10. I selected these highly powerful new 3D techniques to summarize your skills.

3D Mesh Reconstruction

Converting point clouds to meshes is often crucial in creating more manageable and visually appealing 3D models (Figure 17-7). However, as we have seen, it is far from trivial, especially when dealing with noisy or incomplete data.

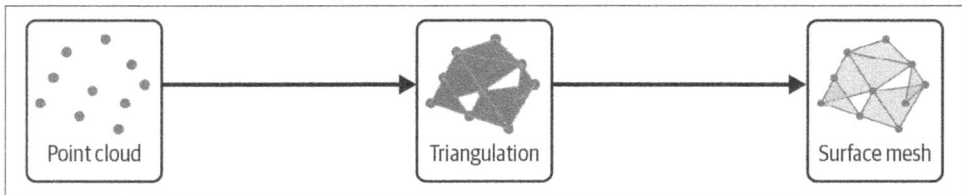

Figure 17-7. From point cloud to surface mesh

We succeeded with several surface reconstruction methods, such as Poisson reconstruction, the Ball Pivoting Algorithm, and Delaunay triangulation. Here is a snippet of how you might implement surface reconstruction using Open3D:

```python
import open3d as o3d
import numpy as np

def point_cloud_to_mesh(pcd, depth=8):
    # Estimate normals if not already present
    if not pcd.has_normals():
        pcd.estimate_normals(search_param=o3d.geometry.KDTreeSearchParamHybrid
        (radius=0.1, max_nn=30))

    # Poisson surface reconstruction
    mesh, densities = o3d.geometry.TriangleMesh.create_from_point_cloud_poisson
    (pcd, depth=depth)

    # Remove low-density vertices
    vertices_to_remove = densities < np.quantile(densities, 0.1)
    mesh.remove_vertices_by_mask(vertices_to_remove)

    return mesh

mesh = point_cloud_to_mesh(pcd, depth=8)
mesh.compute_vertex_normals()
o3d.visualization.draw_geometries([mesh])
```

Figure 17-8 shows the results.

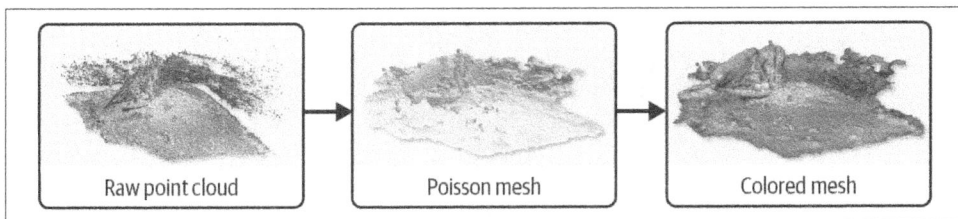

Figure 17-8. Raw point cloud to Poisson mesh to colored mesh

But here's the catch—Poisson reconstruction works wonders for organic shapes but can struggle with sharp edges and planar surfaces. We often use a hybrid approach for architectural or mechanical models, combining planar segmentation for large flat surfaces with Poisson reconstruction for more complex areas. Remember the project where we modeled an entire city from aerial LiDAR data? We developed a custom pipeline that extracted building footprints, reconstructed shapes using primitive fitting, and then used alpha shape reconstruction for other organic elements (see Chapter 11).

The key was to adapt our approach to the nature of the data—a great reminder. But now, let me move to one of my all-time favorites: voxels and voxelization techniques.

Voxelization of 3D Digital Environments

Voxels are cubes, much like 2D pixels, but in 3D. Voxelization is the technique for discretizing 3D space with voxels, making it particularly useful for volumetric analysis and certain types of machine-learning tasks. This is also the go-to representation for medical imaging and datasets that meet the needs of volumetric representations.

I love using voxels to structure digital environments with a regular grid, i.e., trying to get a 3D representation that can be leveraged similarly to 2D pixels (e.g., for convolution algorithms). However, voxels are also vital in extracting topological features and analyzing the connectivity graph, which you can define based on the voxel data structure.

For reference, here is a basic implementation of voxelization that is a bit different from solutions we reviewed in previous chapters:

```python
import open3d as o3d
import numpy as np

def voxelize(pcd, voxel_size):
    # Compute bounds
    points = np.asarray(pcd.points)
    min_bound = np.min(points, axis=0)
    max_bound = np.max(points, axis=0)

    # Compute voxel indices
    voxel_indices = np.floor((points - min_bound) / voxel_size).astype(int)

    # Create voxel grid
    grid_size = np.ceil((max_bound - min_bound) / voxel_size).astype(int)
    voxel_grid = np.zeros(grid_size, dtype=bool)

    # Fill voxels
    voxel_grid[voxel_indices[:, 0], voxel_indices[:, 1],
    voxel_indices[:, 2]] = True

    # Make it with open3d
    voxel_grid_o3d = o3d.geometry.VoxelGrid.create_from_point_cloud(pcd,
    voxel_size)

    return voxel_grid_o3d

voxels = voxelize(pcd, 0.1)
o3d.visualization.draw_geometries([voxels])
```

This results in Figure 17-9. Note that you can automatically adapt the voxel size by leveraging heuristics as shown in Chapters 8 and 9.

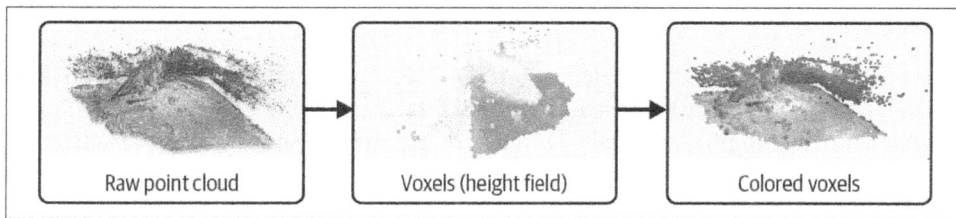

Figure 17-9. The raw point cloud voxelized, with color

But voxelization is not just about creating a binary occupancy grid. We have found that storing additional information in each voxel can be incredibly powerful. For instance, for a recent project analyzing forest structures, I stored occupancy and point density, mean intensity, and derived features like local point distribution entropy in each voxel. This rich voxel representation allows complex analyses of forest structure and health.

All right, when talking about data structures, we cannot escape using *k*-d trees in our experiments.

k-d Trees

Let me share some additional words on *k*-d trees, discussed in Chapter 4. These are your go-to data structures for efficient nearest-neighbor searches in 3D space. They are particularly useful for tasks like normal estimation, feature computation, and local neighborhood analysis.

Here is how you can use a *k*-d tree for efficient radius search:

```
from sklearn.neighbors import KDTree

def radius_search(points, query_point, radius):
    tree = KDTree(points)
    indices = tree.query_radius([query_point], r=radius)[0]
    return points[indices]
```

But here's a pro tip: building a *k*-d tree for the entire dataset can be memory-intensive when working with massive point clouds. I often employ a tiling strategy, building separate *k*-d trees for spatial tiles and only loading the relevant tiles into memory as needed. This approach has allowed me to work with datasets of billions of points on modest hardware.

However, in cases where your data is homogeneously sampled, the ability to leverage octrees is also a great asset.

Octrees

Octrees are another key data structure for 3D datasets you can leverage. They are fantastic for multiresolution representation of 3D data, especially for level-of-detail rendering and hierarchical spatial queries, as we have seen in our visualization experiments.

To complement your knowledge on the matter, I provide you with a condensed octree implementation, and I encourage you to refer to Chapter 4 for an extensive view:

```
class OctreeNode:
    def __init__(self, center, size, points):
        self.center = center
        self.size = size
        self.points = points
        self.children = []

    def subdivide(self):
        if len(self.points) <= 100:  # Arbitrary threshold
            return

        new_size = self.size / 2
        for i in range(8):
            new_center = self.center + new_size * np.array([
                (i & 1) - 0.5,
                ((i >> 1) & 1) - 0.5,
                ((i >> 2) & 1) - 0.5
            ])
            mask = np.all(np.abs(self.points - new_center) <= new_size, axis=1)
            self.children.append(OctreeNode(new_center, new_size,
            self.points[mask]))

def build_octree(points):
    center = np.mean(points, axis=0)
    size = np.max(np.abs(points - center)) * 2
    root = OctreeNode(center, size, points)
    root.subdivide()
    return root
```

Essentially, octrees are incredibly powerful for large-scale 3D visualization. In our project of visualizing city-scale 3D models, we used an octree structure to dynamically load and unload model parts based on the viewer's position and zoom level (Figure 17-10). This allowed us to render massive datasets smoothly.

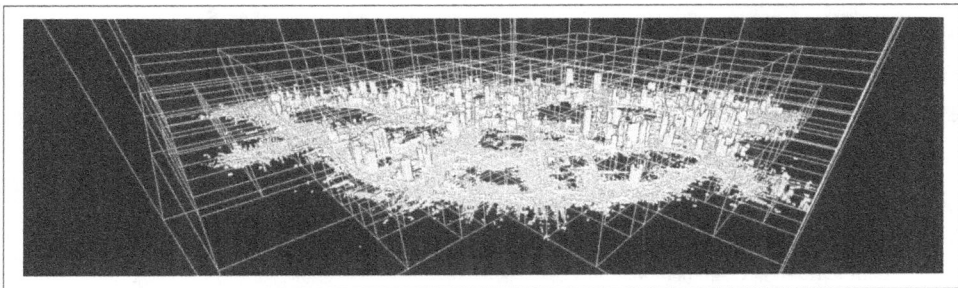

Figure 17-10. Visualization of an octree data structure on a large point cloud

One advanced technique you can experiment with is adaptive octrees, where the subdivision criteria vary based on local data characteristics. For instance, in areas with high detail or rapid change, we might subdivide more aggressively, while in uniform regions, we might use larger nodes. This approach allows us to allocate our computational resources more efficiently, focusing on the areas of the data that matter most.

But let me digress a bit. With real-world applications, I often combine these techniques. If you remember the project for a multimodal 3D viewer experience (Chapter 5) analyzing urban environments, we used a voxel-based representation for broad-scale analysis (like computing building volumes), an octree for multiresolution querying and visualization, and local k-d trees for fine-grained point cloud processing.

The key is understanding each approach's strengths and weaknesses and choosing the right tool (or combination of tools) for the task. It is about applying algorithms and crafting data structures that align with your analytical needs and computational constraints.

Ultimately, 3D data modeling and structuration are critical steps that will make or break your spatial analysis pipeline. It is about finding the right balance between fidelity, efficiency, and usability. As the scale and complexity of 3D data continue to grow, innovative approaches to data structuration will become increasingly important.

I'm particularly excited about the potential of learned data structures, where we use machine learning techniques to adaptively structure our data based on the specific characteristics of the dataset and the intended analyses. Remember that the goal is not just to store your data, but also to organize it in a way that lets you find new ways to analyze it and gain new insights. Get this right, and you've laid the groundwork for compelling spatial AI applications.

Semantic Extraction

Over my years of working with spatial data, I've seen the field of semantic extraction evolve dramatically (Figure 17-11). In our 3D data science workflow, it represents one of the most fascinating challenges.

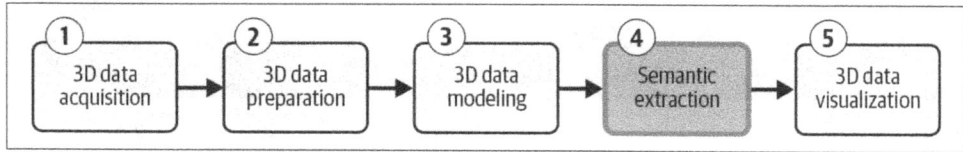

Figure 17-11. Phase 4: Semantic extraction

First, we examined clustering and segmentation scenarios with unsupervised learning techniques (Chapters 12 and 13). From an unlabeled 3D dataset, our goal is to delineate the major components with a granular level that suits a task-driven scenario. Then, we deep-dived into supervised learning approaches, playing on both non–deep learning architectures and 3D deep learning.

We have seen that some semantic extraction can fail when dealing with complex elements (e.g., unique architectural elements). This led us to develop a more nuanced understanding and hands-on knowledge of three fundamental "segmentation tracks":

1. Clustering (Chapter 12) and unsupervised segmentation (Chapter 13)
2. Semantic segmentation (Chapter 14) and instance segmentation (Chapter 15)
3. 3D object detection (Chapter 16)

Let us move on to the first segmentation track: clustering and unsupervised segmentation.

Clustering and Unsupervised Segmentation

We usually do not need to reason precisely at the unit level, which is the point level when it is a point cloud or the triangle level if it is a mesh. If we follow the Gestalt theory (*https://oreil.ly/m1XmJ*), we need to group sets of elements to understand the relationships between them better. So, working on the segment level is a great way to form a meaningful graph-like structure of relationships.

These are not necessarily tied to the object of our application (e.g., the chair) but a little patch of this object (a group of geometrical entities that belong to that chair). This is similar to the SuperPixel (*https://oreil.ly/5osgM*), SuperPoint (*https://oreil.ly/FY4JG*), or SuperVoxel frameworks, where we try to group base units with a common feature space into some granular group that brings more meaning. This is key in many processes and essential, especially as we want to generalize some approaches.

First, we examined the major clustering approaches: *k*-means and DBSCAN (Chapter 12). We found that geometric-based clustering using these methods is remarkably efficient when dealing with unstructured point clouds. We then explored the power of foundation models and graphs to push unsupervised learning capabilities to very advanced solutions (Chapter 13). This allows us to extract segments by working on derived modalities, and thus latent spaces. As I did before, let me share some pseudocode to guide the mind map of your newfound skills:

```
def dbscan_3d(points, eps, min_points):
    # k-d tree for efficient neighbor search
    tree = KDTree(points)
    clusters = []
    visited = set()

    def expand_cluster(point_idx, neighbors):
        cluster = [point_idx]
        For neighbor in neighbors:
            if neighbor not in visited:
                visited.add(neighbor)
                new_neighbors = tree.query_ball_point(points[neighbor], eps)
                if len(new_neighbors) >= min_points:
                    cluster.extend(expand_cluster(neighbor, new_neighbors))
        return cluster

    for i in range(len(points)):
        if i not in visited:
            neighbors = tree.query_ball_point(points[i], eps)
            if len(neighbors) >= min_points:
                visited.add(i)
                cluster = expand_cluster(i, neighbors)
                clusters.append(cluster)

    return clusters
```

This approach is very beneficial and particularly efficient for our experiments with industrial facilities. These algorithms identify pipes and structural elements based on their spatial distribution and local geometric features (Figure 17-12).

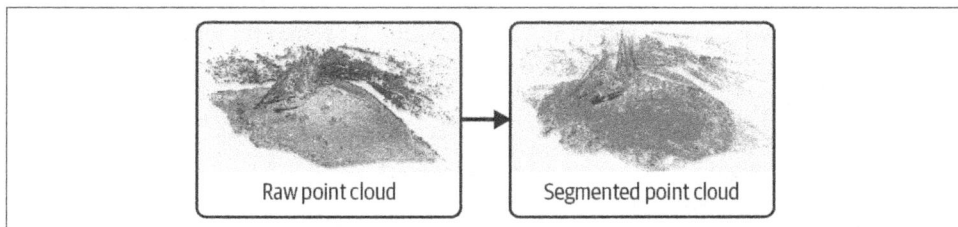

Figure 17-12. From raw point cloud to segmented point cloud

However, the real breakthrough comes from incorporating local geometric descriptors like eigenvalue-based features and normal vectors. As seen in our projects (Chapters 7 and 9), we have consistently achieved superior results by combining multiple descriptors—surface variation, linearity, and planarity—rather than relying on single geometric properties.

Therefore, the key to our unsupervised approaches lies in defining the most powerful features. Usually, this means creating a latent space with the best balance between a low number of features and a synergy that allows the creation of groups of base units that best segment the scene.

This means we must be cautious with undersegmentation and oversegmentation pitfalls. Indeed, the best segment-based structuration never bleeds over the natural borders of shapes. This also highlights that we deal with many geometric shape combinations, especially within artificial environments.

After obtaining such a segment-based decomposition, the labeling process is also accelerated. Instead of working at the unit level, we label at the segment level, which speeds up and allows for robust labeling pipelines. This is key to supervised learning scenarios leveraging 3D semantic segmentation.

Semantic Segmentation

Semantic segmentation is at the heart of intelligent reasoning services. Indeed, identifying the various entities that compose a scene at a specified granular level is critical for reasoning AI agents. In our 3D data science journey, we naturally stopped on this aspect for several chapters: Chapters 14, 15, and 16.

First, we used 3D machine learning through nondeep architectures with handcrafted features. At least in the first round, the choice of features was guided by our prior knowledge of the data, algorithm, and application domains.

The idea is to be able to assign each point (with our point clouds) a specific class from a predefined set of classes. This process was carried out in a supervised learning fashion, which is essential, at least, to get a first-level working methodology. As a guide for your experiments, you can find in the following 3D semantic segmentation pseudocode that takes in a 3D mesh and leverages random forests to predict semantic labels per vertex:

```
import numpy as np
from sklearn.ensemble import RandomForestClassifier
from sklearn.preprocessing import StandardScaler
from sklearn.model_selection import train_test_split

def train_rf(point_cloud, labels=None, test_size=0.2, n_estimators=100):

    # Normalize features
    scaler = StandardScaler()
```

```python
    X_scaled = scaler.fit_transform(point_cloud)

    # Split the data
    X_train, X_test, y_train, y_test = train_test_split(X_scaled, labels,
    test_size=test_size, random_state=42)

    # Train random forest classifier
    rf_classifier = RandomForestClassifier(n_estimators=n_estimators,
    random_state=42)
    rf_classifier.fit(X_train, y_train)

    # Optional: Print model performance
    train_accuracy = rf_classifier.score(X_train, y_train)
    test_accuracy = rf_classifier.score(X_test, y_test)
    print(f"Training Accuracy: {train_accuracy:.2f}")
    print(f"Testing Accuracy: {test_accuracy:.2f}")

    return rf_classifier

# Loading the training dataset and preparing our feature vector
x, y, z, r, g, b, label, nx, ny, nz = np.loadtxt("DATA/training_data.txt",
delimiter=' ', skiprows=1, unpack=True)
pcd_features = np.column_stack((x, y, z, r, g, b, nx, ny, nz))

# Training our classifier
classifier = train_rf(pcd_features, labels=label.astype(int), test_size=0.2,
n_estimators=100)
#%% Predict labels for the entire point cloud

def prediction(pcd, model):

    #Open3d to numpy
    points = np.asarray(pcd.points)
    colors = np.asarray(pcd.colors)
    normals = np.asarray(pcd.normals)
    point_cloud = np.hstack((points,colors,normals))

    # Normalize features
    scaler = StandardScaler()
    point_cloud_scaled = scaler.fit_transform(point_cloud)

    # Inference
    predicted_labels = model.predict(point_cloud_scaled)

    return predicted_labels

label_results = prediction(pcd, classifier)
```

The results are shown in Figure 17-13.

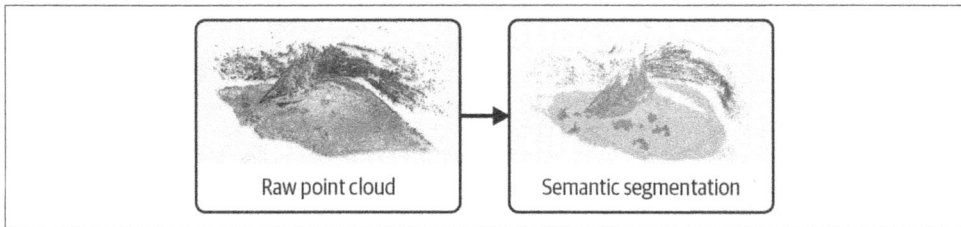

Figure 17-13. The task and results of semantic segmentation on a 3D point cloud

This is an excellent example of the power of 3D machine learning. Of course, the downside is that you need to find labeled 3D datasets, which are hard to come by. We investigated some augmentation strategies and solutions that can reduce friction at this stage.

Then comes the paradigm shift. Handcrafted features give way to deep learning approaches. This is mainly to move away from prior knowledge injection and to find out how best to delineate various semantic classes based on data-driven insights through the provided training dataset.

This took us to the 3D deep learning area and various architectural investigations. While much relevant work in traditional architectures like CNNs, RNNs, and Transformers can be used for 2D datasets, unfortunately, this is not the case when we touch on 3D datasets. This is where we focused on architectures such as PointNet to handle 3D datasets, which is useful for some cases, such as having a literal understanding of the relevant critical factors.

In our experiments with urban environments, we observed that hybrid networks combining both point-based and voxel-based representations outperform single-representation approaches. The key lies in maintaining geometric precision while capturing hierarchical features. This is where the particular interest lies in attention mechanisms that can handle varying point densities—a common challenge in real-world scans often overlooked in academic papers.

3D Object Classification

Very closely related to semantic segmentation, 3D object/scene classification is another excellent addition to your skill set (see Figure 17-14). We consider the complete data as a single study unit, so we need to label it semantically. We developed a PointNet-based solution to determine the class of each single piece of furniture, represented as a 3D mesh (Chapter 16). Is the point cloud a chair? Is it a bed? Is it a cupboard?

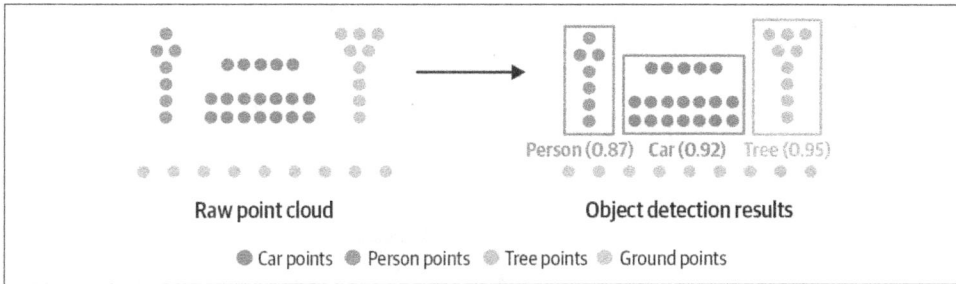

Figure 17-14. 3D object classification principle

To guide you through your code endeavor, you can find here pseudocode that summarizes what we are pushing through our PointNet architecture (if you struggle to understand the values or the organization, I recommend falling back to Chapters 15 and then 16):

```
class pointnet_architecture():
    # Input transformation net
    def tnet(points):
        # Learns 3x3 transformation matrix
        net = shared_mlp(points, [64, 128, 1024])
        net = max_pool(net)  # Global feature
        transform = mlp(net, [512, 256, 9])
        return transform

    # Main PointNet
    def pointnet(points):
        # Input transform
        transform = tnet(points)
        points_transformed = apply_transform(points, transform)

        # Feature extraction
        features = shared_mlp(points_transformed, [64, 64])

        # Feature transform
        feature_transform = tnet(features)
        features_transformed = apply_transform(features, feature_transform)

        # Global feature extraction
        net = shared_mlp(features_transformed, [64, 128, 1024])
        global_feature = max_pool(net)

        # Classification
        net = mlp(global_feature, [512, 256, n_classes])
        return softmax(net)

    return pointnet
```

This returns a label for each 3D model, as illustrated in Figure 17-15.

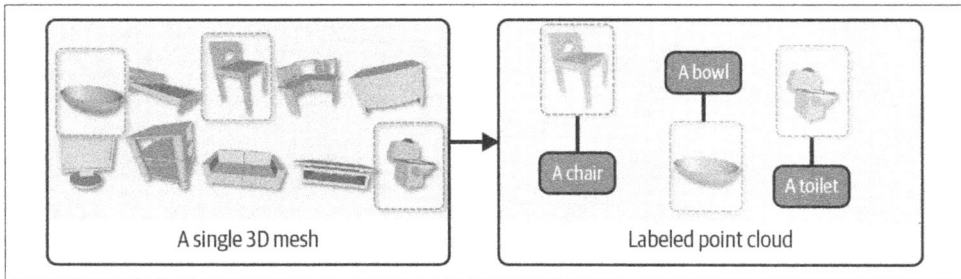

Figure 17-15. 3D object detection results (bounding box and label)

While global descriptors are key in many workflows that need a meta-level understanding (especially in controlled indoor environments), there is great potential to leverage multiscale approaches. This is an excellent track for successfully implementing hierarchical feature learning networks that can identify objects across different scales—from individual architectural elements to entire building structures.

I have found that the real challenge is developing these individual techniques and creating robust pipelines that can handle the variability of real-world data. In the various implementations from Chapters 4 to 16, we learned that the most successful approaches combine multiple semantic extraction methods, using confidence metrics to determine what is most appropriate for different scene parts.

It is good to know that tremendous potential lies in integrating temporal aspects into semantic extraction. You can already do this for projects that monitor urban infrastructure. For these projects, you can start making methods to recognize objects and track how their shape and meaning change over time. This temporal factor makes 3D scene understanding even more complicated, but it is necessary for tasks like keeping an eye on the health of buildings and finding changes in cities.

Since the release of PointNet, many new architectures (e.g., PointNet++, KPConv, and Point Transformer) have improved the performance/reliability of point-based semantic segmentation. For example, self-driving cars can get around in cities with lots of traffic, robots can do complicated jobs very well, and augmented reality can make the digital and real worlds blend together perfectly.

To achieve this, we must push the boundaries of domain adaptation, multimodal fusion, real-time inference, and semantic extraction. By addressing these challenges, we can ensure the safe and reliable deployment of spatial AI technology based on sound digital environments. But this naturally misses a component. What about 3D data visualization? How did it help us create 3D analytics?

3D Data Visualization and Analysis

In this book, we mainly developed tools and techniques to help us better address specific and crucial challenges, like examining the deviation between a point cloud and a mesh (Chapter 9) or understanding the distribution in our dataset (Chapter 8).

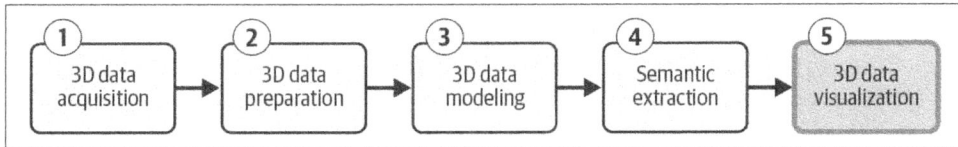

Figure 17-16. Phase 5: 3D data visualization

Specifically, we saw how to develop a multimodal viewer (Chapter 5) to handle any 3D data representation, including meshes, voxels, CAD, point clouds, and depth images. Equipped with this skill, you can address any engineering problem with a sound set of 3D data science tools. Of course, this also means that you eventually need to improve on some modules as new approaches emerge. But they will build on the first principles we have unfolded in the book.

This topic has become increasingly crucial as our ability to capture high-fidelity 3D data outpaces our capacity to analyze it effectively. Therefore, let me share our final insights on 3D data analysis and visualization, drawing from this extensive experience with complex spatial datasets.

3D Shape Recognition

In Chapter 9, we found that 3D shape recognition forms the cornerstone of meaningful spatial analysis. RANSAC has proven particularly powerful, even in its basic form, which most practitioners can use. We have developed adaptive RANSAC variants that dynamically adjust their parameters based on local point density and noise levels.

For instance, we developed a multiscale RANSAC approach that can adapt to local densities, which the pseudocode here and the book's teaching can help you replicate:

```
# Simplified pseudo-code of my adaptive approach
def adaptive_ransac(point_cloud, local_density):
    threshold = compute_adaptive_threshold(local_density)
    min_points = adjust_minimum_points(local_density)
    return detect_shapes(point_cloud, threshold, min_points)
```

While computationally more intensive, region-growing techniques offer superior results for organic shapes. We have found them particularly effective when analyzing geological formations. The key lies in selecting appropriate seed points and growth criteria. You can extend the chapter's project by combining curvature analysis with normal consistency checks to achieve robust segmentation, even in areas with varying point densities.

Finally, by combining both, we were able to develop a fully automatic multiscale shape recognition and reconstruction solution for indoor datasets. While RANSAC excels at identifying basic shapes like planes, cylinders, and spheres, we have found that combining them with region-growing algorithms provides a more robust approach.

This proved very effective in complex projects, where irregular surfaces and weathered elements require more sophisticated detection methods, as well as in-depth data-driven analysis. However, we saw that 3D geometric analysis represents a workflow's mathematically intensive aspect. Through more experience, you will find that combining multiple analysis techniques yields the most robust results.

For instance, when analyzing surface deformation in structural elements, a hybrid approach that combines curvature analysis, normal vector statistics, and local feature histograms can complete the features extracted from the shapes. This multimodal analysis has proven particularly effective in detecting subtle geometric anomalies that single-metric approaches might miss. To carry on these analyses, which tools can we leverage?

3D Data Analytical Tools

Moving to 3D data analytical tools, you have developed a comprehensive framework beyond basic statistical measures. While histograms and threshold-based analyses remain helpful (e.g., Chapter 7), we have found that multidimensional feature spaces provide much richer insights (Chapter 8). This is what we can call "geometric fingerprinting," and it plays on these three simple buckets:

Density-based analysis
　　Implement adaptive kernel density estimation for surface orientation rather than simple point density measures.

Local feature histograms
　　Combine multiple geometric properties (curvature, roughness, normal variation) into multidimensional histograms that better characterize complex surfaces.

Statistical profiling
　　Extend the chapter's implementation methods to generate nonlinear profiles that follow complex geometries. This will help analyze complex geometric structures.

Regarding these 3D data analytical tools, we have observed a significant shift toward integrated analysis pipelines. The combination of NumPy, SciPy, and specialized libraries built on Open3D has revolutionized how we process spatial information, and it is now all yours.

In my work with urban digital twins, I developed analytical workflows that merge traditional statistical analysis with spatial queries, enabling us to correlate building geometry with energy consumption patterns. The key lies in maintaining spatial indexing structures that facilitate rapid neighborhood queries while preserving the ability to perform complex geometric calculations.

Finally, let's talk about the power of visualization for digital environments.

3D Multimodal Python Viewer

The development of 3D multimodal Python viewers has been crucial for making sense of complex spatial data. In our projects, we have moved beyond simple visualization to interactive analytical platforms. By leveraging Open3D (*https://open3d.org*) and PyVista (*https://docs.pyvista.org*), we have created visualization systems that dynamically link geometric properties with semantic information. For example, in our project in Chapter 7, our viewer could simultaneously display point cloud geometry with real-time filtering capabilities based on spatial and semantic queries.

What is often overlooked is the importance of interaction design in 3D visualization, the need to leverage the experiences with a developed set of interaction paradigms that make complex analysis tools accessible to nonexpert users while maintaining the depth needed for detailed technical analysis. This can include context-sensitive measurement tools that adapt to the type of analyzed geometry. Also, interactive filtering mechanisms that operate in both geometric and semantic domains are a great way to provide experts with a combination of useful tools that provide 3D visualization support.

However, the next stage of 3D data analysis and visualization is tightly coupled with integrating machine learning approaches with traditional geometric processing. For example, you could define a system that learns from user interactions to improve visualization parameters and analysis workflows automatically. This represents a shift from static visualization tools to adaptive systems that evolve with use.

Summary

We established a complete library of 3D data science modules by moving from first principles to complete solutions. The critical challenge moving forward lies not in the individual components but in creating cohesive workflows that maintain data fidelity and computational efficiency across all stages. I have found that careful attention to data structures and pipeline processing is crucial for managing the complexity of modern 3D analytical workflows, such as Figure 17-17, that we covered in this book.

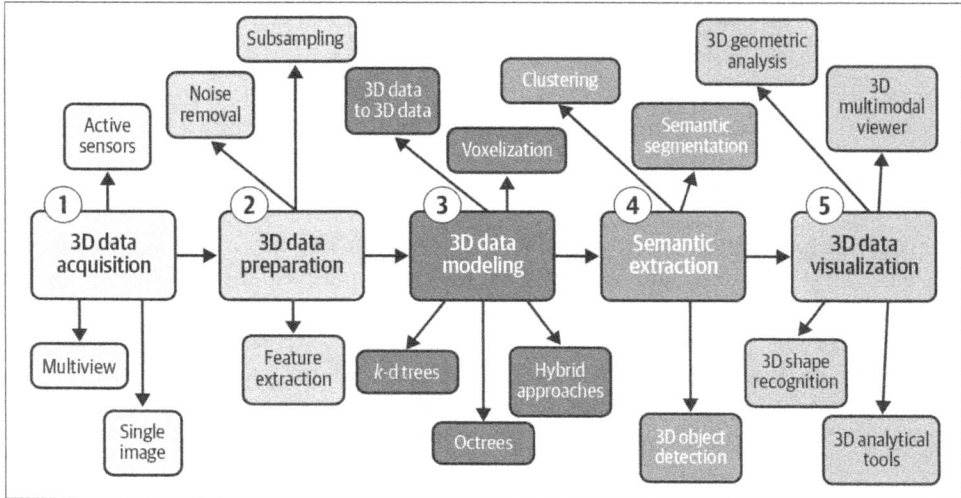

Figure 17-17. A reference 3D data science workflow, with internal cogs that you can now tune to your applications

Therefore, my natural trigger is not to stop here. As lifelong learners, the journey continues. I am thrilled to share with you the subsequent, direct natural paths to extend the learning of this book with the latest innovations to prepare an accurate 3D digital environment.

In the next chapter, I have selected several Python projects that you can investigate directly to improve your chosen 3D data science module. This is another stepping stone toward the integration of spatial AI systems.

From 3D Generative AI to Spatial AI

As experts in cutting-edge applications, you understand the power of 3D data science to solve real-world problems. We can create digital 3D environments using point cloud workflows. However, we need to ask a bigger question: are we close to replicating human perception with our current developments? This chapter looks at how to expand 3D data science to advance to the next step in our journey: 3D spatial AI.

Spatial AI systems are more than just tools; they enhance our ability to perceive and interact with the physical world, reflecting how humans think and understand. This involves helping computers see the world in three dimensions, understand their environment, and make decisions based on what they perceive around them.

Consider how people naturally move through complicated spaces, grasp the connections between different areas, and engage with various objects. Spatial AI seeks to give computers abilities like those of humans. In this chapter, I will explain the main ideas and challenges of building smart spatial systems that can perceive, understand, and reason about 3D environments in two sections, as shown in Figure 18-1.

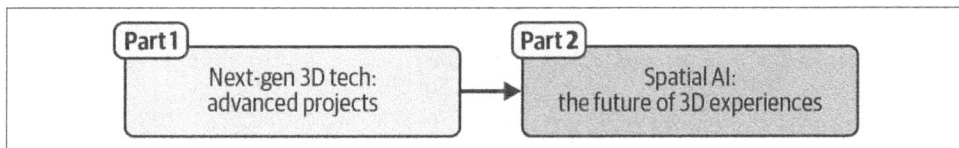

Figure 18-1. The first part of this chapter will focus on five cutting-edge 3D projects, whereas the second part will target perspectives for spatial AI

We first investigate 3D module add-ons that directly extend or improve this book's solutions. The idea is to narrow the gap with cutting-edge research and provide you with five advanced projects that you can build with Python. This includes generative AI for 3D reconstruction, 3D deep point cloud registration, 3D semantic modeling, efficient semantic extraction, and advanced immersive experiences with 3D Gaussian splatting.

Finally, we are going to look into the future of 3D. What are the missing ingredients in our digital model to achieve a complete 3D scene understanding? From this perspective, we lay down the most direct path to unlocking 3D spatial AI systems with advanced reasoning capabilities.

This is our last deep dive; let's ensure these words resonate with you and inspire your subsequent innovative work.

Advanced 3D Projects

I've selected five main projects that will extend the lessons of this book while closing the gap with the most promising research at the time of this writing. These projects build extensively on the concepts that we developed in the rest of the book. Therefore, my last recommendation is to strengthen your knowledge and know-how so that you have a strong foundation. The five projects, shown in Figure 18-2, target the full scope of 3D data science, from 3D acquisition to 3D visualization.

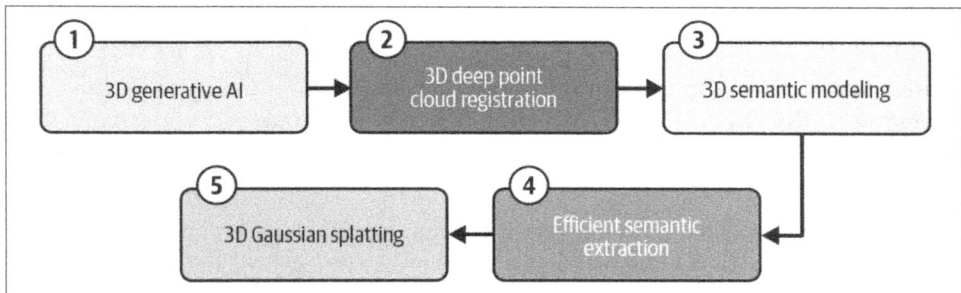

Figure 18-2. The five projects that we will execute in this chapter

Let's start with the first project: generative AI for 3D reconstruction. We will examine how progress in 3D data science and new AI methods can help us create more intelligent machines that can see and interact with the world in a more human-like way.

Generative AI for 3D Reconstruction

In this project, we aim to leverage generative AI to create synthetic 3D datasets from text or image prompts (Figure 18-3). As we saw from Chapters 1 to 5, 3D reconstruction is a compelling solution for converting an idea into an actual 3D data representation. With passive (e.g., cameras) and active (e.g., LiDAR) sensors, we saw how to generate 3D point clouds with relative ease.

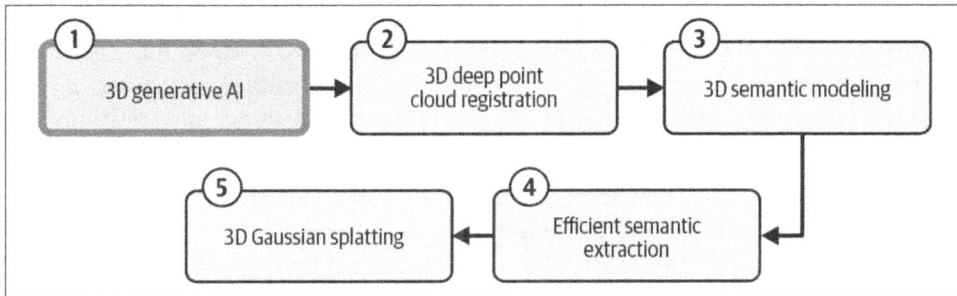

Figure 18-3. 3D generative AI

However, while we see how to turn sensor outputs into 3D data representation, we can also explore techniques that help generate entirely new 3D forms: creative 3D reconstruction (as reviewed in Chapter 3). For a long time, this was exclusively done through pure creativity and manual 3D modeling (with 3D modeling software (*https://oreil.ly/Nbtv0*)) to ensure that ideas could be translated into the most appropriate 3D representation. The product is generically termed a "3D model" (*https://oreil.ly/IIfLp*), originating from a 3D artist or a 3D modeler.

But with 3D generative AI, we can leverage new approaches that allow us to reconstruct 3D data by mainly leveraging NLP (*https://oreil.ly/wrcPp*), which helps us understand and communicate with our systems more naturally. This, specifically for 3D, showcases two main inputs: text or images, as illustrated in Figure 18-4. Let's focus on the first one: text to 3D.

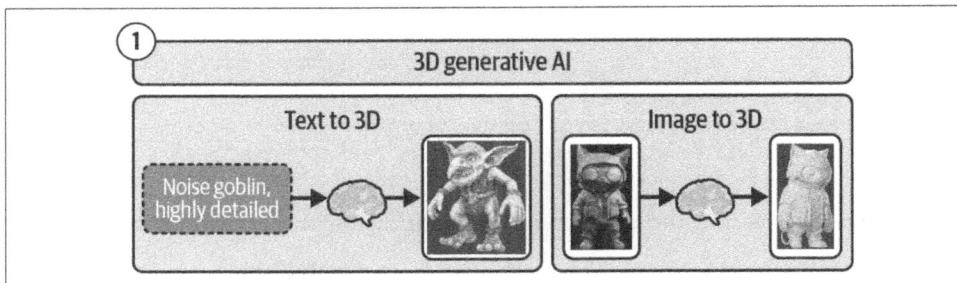

Figure 18-4. 3D generative AI: Text to 3D and image to 3D

Text to 3D

Text-to-3D models generate 3D objects or scenes based on textual descriptions. This can be from your speech or by typing some prompts like "generate a little cat with big eyes and flurry fur." Once we have a prompt, the next stage is to feed it to our 3D GenAI system, which can generate a 3D reconstruction. This 3D reconstruction can then be shaped in the form of point clouds, meshes, voxels, or any 3D data representation seen in Chapter 4.

Let's examine the specifics. The process described is possible due to recent advances in NLP (*https://oreil.ly/wrcPp*) and deep learning techniques. First, NLP breaks down the text input into individual words or tokens and represents them as numerical vectors. These vectors essentially capture the semantic meaning and relationships between words.

Next, the model learns a latent, high-dimensional mathematical space where complex relationships between text and 3D objects can be represented. The text embeddings are mapped to points in this latent space. Finally, a generative model, such as a variational autoencoder (VAE) (*https://oreil.ly/mwXwu*) or a generative adversarial network (GAN) (*https://oreil.ly/jL7UX*), decodes the latent representation into 3D coordinates. The model may iteratively refine the generated 3D structure based on feedback from a discriminator network or reconstruction loss.

For example, in 2023, OpenAI officially released Shap-E (*https://oreil.ly/7xhJp*). Similar to its predecessor Point-E (*https://oreil.ly/sn16q*), Shap-E can generate 3D models based on a text prompt. Shap-E is composed of an encoder and a latent diffusion. The encoder converts the 3D assets into parameters that represent the 3D shape and texture as an implicit function. This means that the resulting implicit function can be interpreted from any angle. Then, the latent diffusion process creates new implicit functions based on the text descriptions. It produces latents that need to be linearly projected to obtain the final implicit function parameters.

You can use the model by following the instructions on GitHub (*https://oreil.ly/z2GiO*). Once installed, you can, for example, create a new Python file and define the import statements:

```
import torch

from shap_e.diffusion.sample import sample_latents
from shap_e.diffusion.gaussian_diffusion import diffusion_from_config
from shap_e.models.download import load_model, load_config
from shap_e.util.notebooks import create_pan_cameras, decode_latent_images
```

Then, we can instantiate the encoder and latent diffusion model as follows:

```
device = torch.device('cuda' if torch.cuda.is_available() else 'cpu')

xm = load_model('transmitter', device=device)
model = load_model('text300M', device=device)
diffusion = diffusion_from_config(load_config('diffusion'))
```

During the initial run, it will download several configuration files and checkpoints (around 4 GB). For the diffusion process to work, we need to define the hyperparameters (see Chapter 15 for a detailed understanding):

```
batch_size = 4
guidance_scale = 15.0
prompt = "a big boat"
```

```
latents = sample_latents(
    batch_size=batch_size,
    model=model,
    diffusion=diffusion,
    guidance_scale=guidance_scale,
    model_kwargs=dict(texts=[prompt] * batch_size),
    progress=True,
    clip_denoised=True,
    use_fp16=True,
    use_karras=True,
    karras_steps=64,
    sigma_min=1e-3,
    sigma_max=160,
    s_churn=0)
```

Finally, you can save the generated latent as single *.ply* file with the following code snippet:

```
from shap_e.util.notebooks import decode_latent_mesh

for i, latent in enumerate(latents):
    with open(f'example_mesh_{i}.ply', 'wb') as f:
        decode_latent_mesh(xm, latent).tri_mesh().write_ply(f)
```

This results in Figure 18-5.

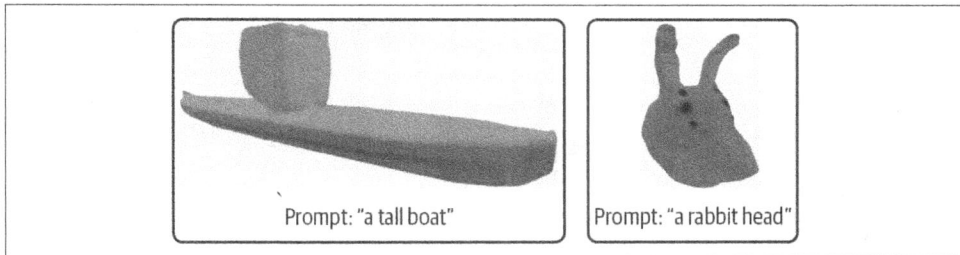

Figure 18-5. Results of text-to-3D modeling with our prompts

Despite significant advancements, challenges remain, such as achieving fine-grained control over the generated 3D objects, rendering them realistically, and training models efficiently. However, independent of that, what powers the system is essentially a combination of the 3D data science concepts you mastered such as 3D reconstruction and 3D deep learning combined with large language models (LLMs) (*https://oreil.ly/MaSfR*) for decoding the prompts.

For example, if you want to generate a 3D reconstruction with generative AI, you could use a video as a prompt. The system would then split that video into a series of frames and then use a monocular depth estimation model to extract the depth information. This is then used as an entry point for your generative AI, which generates a series of 3D reconstructions, combined and exported as a 3D point cloud or a 3D mesh with a specific strategy.

> You can find variants of text-to-3D approaches, such as text-to-3D motion. It involves generating static 3D objects and animating them based on textual descriptions. This allows users to describe dynamic scenes, actions, or behaviors and have the AI generate corresponding 3D animations. While there aren't many off-the-shelf solutions, some research directions include combining text-to-image models with 3D rigging techniques, using large language models to generate animation parameters, or adapting text-to-video models for 3D character animation. Here is an example implementation: generating diverse and natural 3D human motions from text (project page (*https://oreil.ly/gRBGX*) | paper (*https://oreil.ly/d4L9P*) | GitHub (*https://oreil.ly/Agwbk*)).

If you want to explore other approaches, I selected two other notable research works for review in Table 18-1. Now, we can focus on the second central aspect of 3D generative AI: image to 3D.

Table 18-1. Additional text-to-3D architectures

Year	Name	Method	Resources
2022	DreamFusion: Text-to-3D Using 2D Diffusion	Text to 3D	arXiv (*https://oreil.ly/mfiVS*), project (*https://oreil.ly/E2S20*)
2023	Point-E: A System for Generating 3D Point Clouds from Complex Prompts	Text to 3D	arXiv (*https://oreil.ly/8rAMK*), code (*https://oreil.ly/MgpC7*)

Image to 3D

Image to 3D is also super interesting because you can extend the scope we reviewed with monocular depth estimation in the sense that you try to reconstruct what is not seen and not only try to infer the depth from a single image. Remember that you can also go very far with single-image depth estimation, as we saw with foundation models in Chapter 14 such as Depth Anything (*https://oreil.ly/FaCUZ*), MiDaS (*https://oreil.ly/LM38Y*), and Depth Anything V2 (*https://oreil.ly/IbZRm*).

If you want to explore the capabilities of such a system, you can leverage any of the solutions in Table 18-1. For example, as a natural extension from text to 3D, TRELLIS (*https://oreil.ly/4IS0M*) is a native 3D generative model built on a unified structured latent representation and rectified flow transformers for 3D asset creation. Figure 18-6 showcases the results that you can obtain from a single image using the architecture. As you can see, the network recovers details that would be bypassed when using monocular depth estimation.

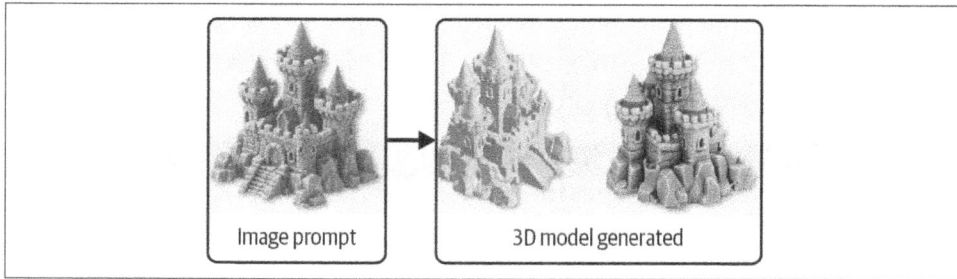

Figure 18-6. 3D model based on an image prompt of a virtual castle using the TRELLIS architecture

To use TRELLIS, you can follow the instructions on the GitHub repository for installation. To use a pretrained model, you can then spin off a Python environment and use the following script:

```
import os
import imageio
from PIL import Image
from trellis.pipelines import TrellisImageTo3DPipeline
from trellis.utils import render_utils, postprocessing_utils

# Load a pipeline from a model folder or a Hugging Face model hub.
pipeline = TrellisImageTo3DPipeline.from_pretrained("JeffreyXiang/
TRELLIS-image-large")
pipeline.cuda()

# Load an image
image = Image.open("your_image.png")

# Run the pipeline
outputs = pipeline.run(image,seed=1)

# outputs is a dictionary containing generated 3D assets in different formats:
# - outputs['gaussian']: a list of 3D Gaussians
# - outputs['radiance_field']: a list of radiance fields
# - outputs['mesh']: a list of meshes

# GLB files can be extracted from the outputs
glb = postprocessing_utils.to_glb(
    outputs['gaussian'][0],
    outputs['mesh'][0],
    simplify=0.95,
    texture_size=1024)
glb.export("sample.glb")
```

If you want to extend your understanding and know-how, I have selected three interesting and complementary approaches with Python in Table 18-2. Now that we have a new solution for creative 3D reconstruction, let's move on to the next stage: data preparation.

Table 18-2. Image-to-3D advanced architectures

Year	Name	Method	Resources
2024	TripoSR: Fast 3D Object Reconstruction from a Single Image	Image to 3D	arXiv (*https://oreil.ly/_8WOG*), model (*https://oreil.ly/_f2Wt*), code (*https://oreil.ly/HEYqK*)
2024	InstantMesh: Efficient 3D Mesh Generation from a Single Image with Sparse-View Large Reconstruction Models	Image to 3D	arXiv (*https://oreil.ly/504xf*), code (*https://oreil.ly/8c7Bm*)
2024	CRM: Single Image to 3D Textured Mesh with Convolutional Reconstruction Model	Image to 3D	arXiv (*https://oreil.ly/wkalE*), demo (*https://oreil.ly/qfhra*), code (*https://oreil.ly/W4A7a*)

3D Deep Point Cloud Registration

In this second project, we tap into 3D data preparation. The idea is to create an automated point cloud registration approach that outperforms the global-to-local registration from Chapter 4 by leveraging deep features (Figure 18-7).

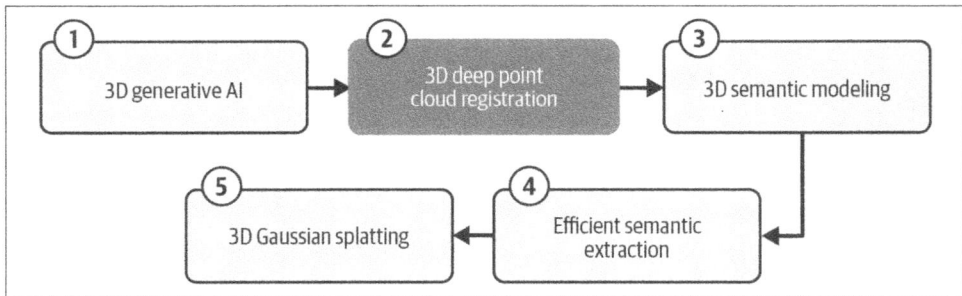

Figure 18-7. 3D deep point cloud registration

If you remember, preprocessing, registration, and feature extraction are three examples of modules that you have a firm grasp on. But what about leveraging deep learning to extend the range of possibilities? If you want to improve the registration module, we could utilize three steps, as illustrated in Figure 18-8.

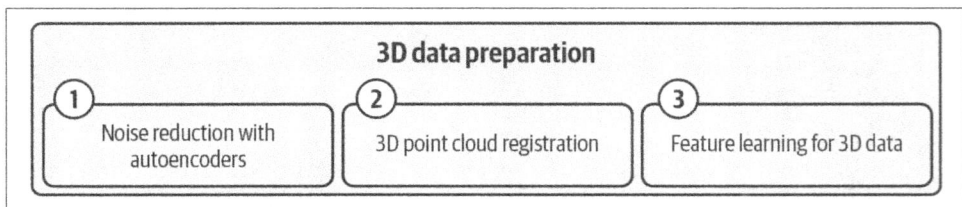

Figure 18-8. The three steps of 3D deep point cloud registration

Let's first start with a noise reduction mechanism. How about sorting out noise from valuable points? Outlier rejection is genuinely a critical step that I believe is often underestimated. It filters erroneous correspondences that can severely distort the final alignment.

To address this challenge, you can leverage supervised and unsupervised methods, including adaptations of RANSAC (*https://oreil.ly/vfyru*) and differentiable consensus maximization. Noise reduction autoencoders (*https://oreil.ly/uzgng*) offer another elegant solution by learning to reconstruct clean 3D data from noisy input.

> Active research involves incorporating learned priors into autoencoder architectures. By incorporating prior knowledge about the underlying 3D structure, autoencoders can achieve superior noise reduction performance, especially in challenging scenarios with high noise levels.

When you are satisfied with your cleaned dataset, we can leverage deep learning's transformative impact on point cloud registration. At this stage, you can categorize deep learning–based registration methods into correspondence-based (*https://oreil.ly/dVn0c*) and correspondence-free (*https://oreil.ly/RZBV4*) methods.

Drawing inspiration from traditional techniques, correspondence-based methods focus on learning robust feature descriptors to improve correspondence matching between the source and target point clouds. Correspondence-free methods, on the other hand, bypass explicit correspondences and directly predict the transformation parameters from global features. This offers potential speed advantages but can be less robust to variations in the input data.

A deep dive into correspondence-based methods reveals the intricacies of their core modules. Feature extraction, the first crucial step, focuses on learning discriminative representations of each point's local geometry. You can leverage PointNet-based architecture and cleverly adapt it for this purpose. This should be a great starting point for improving the registration process. The matching module, responsible for establishing correspondences in feature space, can then employ techniques ranging from nearest-neighbor search to "virtual point" generation and weighted averaging.

If you want to delve deep into some exciting work, I recommend that you check out this paper (*https://oreil.ly/eOMKc*) and its implementation (*https://oreil.ly/NcXi2*), which will guide this 3D deep point cloud registration method to fruition.

> This project is a first-hand extension. But, of course, you also have the capacity to improve by integrating semantic priors into your process to guide the registration. Creativity should guide you toward a perfect blend of processes that can provide robust and more efficient workflows.

At this stage, you have projects that extend learning through 3D reconstruction with generative AI and 3D data preparation with deep registration. As we saw in Chapter 10, we can then use our prepared 3D dataset to create 3D models with various 3D modeling strategies. However, we also saw that we can be limited by the lack of fine-tuning of the geometry locally or in a context. Therefore, let's merge 3D scene understanding concepts (Chapters 12 to 17) with 3D modeling.

3D Semantic Modeling

Let's design a third project: 3D semantic modeling (Figure 18-9). Semantic modeling essentially aims to guide the 3D modeling process with semantic priors to best adapt the geometry to the specific context of the local features. This means that we can combine different 3D modeling strategies based on the semantic decomposition of our scene, which is done through semantic priors.

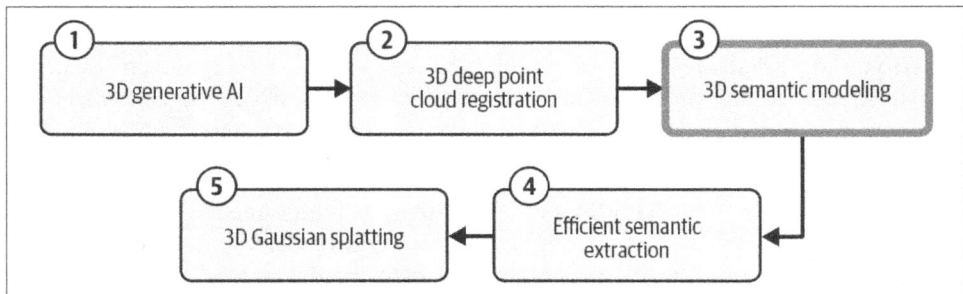

Figure 18-9. 3D semantic modeling

Semantic priors hold additional knowledge about objects in a scene (like their class, shape, and spatial relationships), which can significantly improve the modeling process. By incorporating semantic information into the point cloud data, we can make more informed decisions during modeling. What I propose is that you leverage semantic priors to:

Constrain object shapes
    Knowing an object is a chair, for instance, limits its potential shapes.

Identify and segment objects
    Semantic understanding aids in distinguishing objects from the background.

Guide geometric primitive fitting
    Semantic information can direct the process of fitting shapes to the point cloud data.

For example, if we know an object is a chair, we can infer its likely shape and orientation, even if parts of it are obscured. This information can help you accurately model the occluded parts.

We could even deepen the approach by tying the semantic priors to their information models. This can be done through ontologies (*https://oreil.ly/CYRM_*), a formal way to describe the knowledge of an object for the computer to use in autonomous processes. It provides a structured vocabulary for describing objects and their relationships so that the computer can use it to reason. This is optional, as you can create rules or leverage an LLM system (*https://oreil.ly/JL-dT*) for this, but the ontology guides the modeling process in three main ways.

First, it provides a shared language to describe objects. This vocabulary (RDF (*https://oreil.ly/SQkz2*), a standard model for data interchange on the web) defines basic classes (e.g., furniture, chair) and properties (height, color, and location) and handles the instances.

Then, it defines rules for inferring information about objects. These can be rules for determining whether furniture pieces are adjacent or creating new relationships based on existing data. Finally, it handles the definition of constraints to create validation rules.

Now, depending on how you define your ontology and how your semantic priors are extracted, you can leverage localized 3D shape-matching techniques to achieve part-based 3D modeling. To extend from previous chapters, you could fit entities in the scene to a database of known shapes (e.g., ModelNet40 for the chairs). The best match from the database is then used to create a detailed model.

Moreover, you could generate 3D models at multiple levels of detail, similar to the approach presented in this paper (*https://oreil.ly/DocjV*). This flexibility is crucial for various applications. For instance, a low-detail model might suffice for navigation, while a high-detail model is essential for visualization or simulation.

When you add semantic priors to the 3D modeling process, you can make 3D models from point cloud data much more accurate and faster. As a bonus, based on your task, you can also aim to guide the algorithm (Poisson (*https://oreil.ly/Z5Clh*), ball pivoting (*https://oreil.ly/xPhIj*), marching cube (*https://oreil.ly/Vm4N9*), or Delaunay triangulation (*https://oreil.ly/AXxjA*), as seen in Chapter 10) to a robust and efficient method for meshing point clouds. This also means controlling the meshing process through automatic parameters like voxel size and ISO level, tailoring the output to your needs.

We talked about semantic priors, which means we need semantics! While you have many strategies, let's consider some perspectives on improving your current semantic extraction module transformers.

3D Semantic Extraction with Transformers

This project leverages the Point Transformer (*https://oreil.ly/oLHT_*) architecture to improve the performance of point cloud (or mesh) semantic segmentation tasks (Figure 18-10). 3D semantic segmentation aims to assign a semantic label (e.g., chair, table, wall) to every unit (point in a 3D point cloud, voxel in a voxel dataset, face in a 3D mesh). Think of it like coloring a 3D scan of a room, where each object gets its distinct color. While we took a deep dive into the topic in Chapters 14, 15, and 16, you may want to explore new ways to extend your semantic segmentation module.

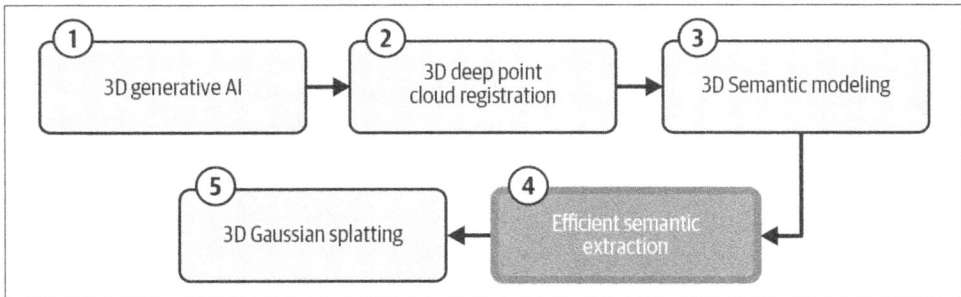

Figure 18-10. Efficient semantic extraction

As seen in Chapter 16, raw point-based methods that directly consume point clouds without requiring any preprocessing are exciting because they increase the robustness of our system solution. Indeed, instead of chaining together many steps, which adds potential failure points to the solution, we leverage a single approach that directly works on the raw dataset.

These methods have gained significant attention recently due to their ability to handle irregular and unordered data. While PointNet is a game changer, it is limited in its ability to capture local features. PointNet++ was created to solve this problem. It uses hierarchical feature learning to gather both local and world data. Researchers have changed the designs of CNNs to work with point clouds after seeing how well they work with 2D images. Dynamic Graph CNN (*https://oreil.ly/n6w1t*), PointCNN (*https://oreil.ly/Crntm*), RGCNN (*https://oreil.ly/f6iRC*), Pointwise CNN (*https://oreil.ly/TQbmL*), PointConv (*https://oreil.ly/DS8u4*), and SpiderCNN (*https://oreil.ly/5Vena*) are examples of CNN-based methods developed for point cloud feature learning. These methods introduce regular representations into the network before applying CNN operations.

However, one of the most promising directions is using autoencoders (AEs) (*https://oreil.ly/Hc9dn*), which can learn latent data representations. Autoencoders actively use transformers. Initially designed for NLP, transformers have revolutionized various fields, including 3D data science. Their ability to capture long-range dependencies and global context makes them well suited for 3D feature extraction. There are many resources online to dive deep into the role of transformers (*https://oreil.ly/ rXADI*) and attention mechanisms (*https://oreil.ly/7OcF1*), with a starting point being the paper "Attention Is All You Need" (*https://oreil.ly/BMlew*).

FoldingNet (*https://oreil.ly/xmaTX*), PPFNet (*https://oreil.ly/WEO3c*), NeuralSampler (*https://oreil.ly/CezX4*), 3dAAE (*https://oreil.ly/6lmeG*), and 3D point-capsule (*https:// oreil.ly/8pMv6*) networks are examples of AE-based point cloud feature learning methods. These methods can encode point clouds' irregularity and sparsity.

The methods described in Figure 18-11 offer significant gains in feature extraction for point cloud data. They permit the acquisition of more detailed and complete features by recording both the local and global features of point clouds. This results in more detailed feature descriptions that better show the 3D structure underneath. Additionally, many of these approaches achieve invariance to transformations such as rotation and translation, ensuring consistent feature representations regardless of the point cloud's orientation. Furthermore, the methods effectively handle point cloud data's irregular and sparse nature. Therefore, we benefit from the same characteristics as PointNet, but largely extend the system's performance when used for 3D semantic segmentation purposes.

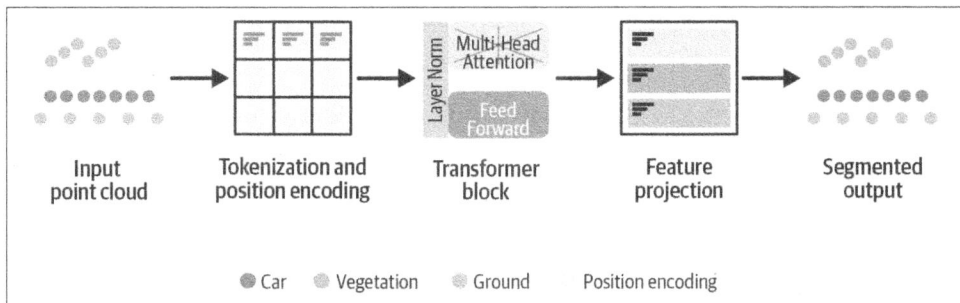

Figure 18-11. Autoencoder example for point clouds

In addition, we observe that the learned features can also help reduce overfitting, leading to more generalizable models. Some methods also offer computational efficiency, enabling faster training and inference, which offers potential for real-time applications.

To guide you in your next steps, I selected an approach you can leverage efficiently: Point Transformer (*https://oreil.ly/3LVFr*). This technique also offers variants and extensions such as Point Transformer V2 (*https://oreil.ly/Br6Ip*) and Point Transformer V3 (*https://oreil.ly/MfSWN*).

Point Transformer is a neural network architecture using a transformer-like approach to understand local and global geometric relationships among points. It differs from earlier point cloud networks that depend on local geometric methods, such as local aggregation or graph convolutions. Instead, Point Transformer employs a self-attention mechanism that enables each point to engage with its surroundings, adjusting the significance of nearby points according to their geometric and feature similarities (Figure 18-12).

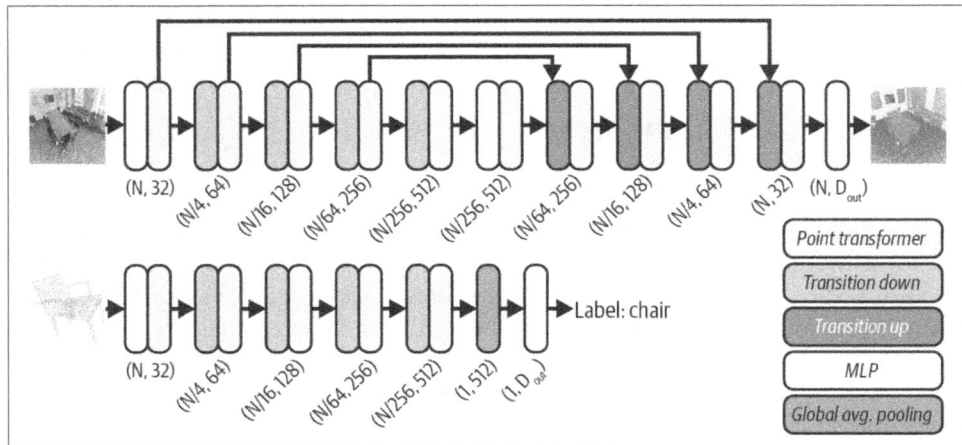

Figure 18-12. The Point Transformer architecture as shown in the original research paper (https://oreil.ly/sBnoB)

The main innovation is in its feature aggregation module, inspired by transformers. It has three main parts: a position-aware feature transformer, a relative position encoding, and a self-attention mechanism. These are achieved within the light gray area from the architecture, which you can design with PyTorch as follows:

```python
class PointTransformerBlock(nn.Module):
    def __init__(self, in_features, out_features, position_dim):
        super(PointTransformerBlock, self).__init__()
        self.fc1 = nn.Linear(in_features, out_features)
        self.point_transformer_layer = PointTransformerLayer(out_features,
        position_dim)
        self.fc2 = nn.Linear(out_features, out_features)

    def forward(self, x, p):
        residual = x # Save

        x = self.fc1(x) # Apply the first linear layer
        x = self.point_transformer_layer(residual, p) # Apply the Point
        # Transformer layer
        y = self.fc2(x) # Apply the second linear layer

        y += residual # Add residual connection

        return y, p
```

Now, let's look at the three components that lie within the `PointTransformerLayer` class.

The position-aware system permits updating each point's feature values by looking at its nearby points. The attention weights are calculated using the distance and feature differences between the points. Therefore, the purpose of the `PointTransformer Layer` is to perform self-attention in the local region of the point cloud. To this end, you can use the following class definition:

```
class PointTransformerLayer(nn.Module):
    def __init__(self, feature_dim, position_dim):
        super(PointTransformerLayer, self).__init__()
        self.gamma_mlp = GammaMLP(feature_dim, feature_dim, feature_dim)
        self.phi = nn.Linear(in_features, out_features)
        self.psi = nn.Linear(in_features, out_features)
        self.alpha = nn.Linear(in_features, out_features)
        self.position_encoder = PositionEncoder(position_dim, feature_dim,
        feature_dim)
        self.rho = nn.Softmax(dim=-1)

    def forward(self, x, p):
        # calculate position embedding
        N = x.size(0)
        delta_p = p.unsqueeze(1) - p.unsqueeze(2)
        delta_p = delta_p.view(-1, delta_p.size(-1))
        position_encoding = self.position_encoder(delta_p)
        position_encoding = position_encoding.view(N, N, -1)

        # calculate attention generation branch
        phi_x = self.phi(x).unsqueeze(1)
        psi_x = self.psi(x).unsqueeze(0)
        gamma_output = self.gamma_mlp(phi_x - psi_x + position_encoding)
        attention_generation_branch = self.rho(gamma_output)

        # calculate feature transformation branch
        feature_transformation_branch = self.alpha(x).unsqueeze(0) +
        position_encoding

        y = torch.sum(attention_generation_branch *
        feature_transformation_branch, dim=1)

        return y
```

The layer takes input features x and their corresponding positions p and computes a transformed representation through multiple key steps. It calculates positional embeddings by computing relative distances between points, which are then encoded using a `PositionEncoder`. This permits relative position encoding to help the network grasp spatial relationships. To design such an encoder, you can use this PyTorch class:

```
class PositionEncoder(nn.Module):
    def __init__(self, input_dim, hidden_dim, output_dim):
        super(PositionEncoderMLP, self).__init__()
        self.fc1 = nn.Linear(input_dim, hidden_dim)
        self.fc2 = nn.Linear(hidden_dim, output_dim)

    def forward(self, p_i, p_j):
        x = p_i - p_j
        x = F.relu(self.fc1(x))
        x = self.fc2(x)

        return x
```

Finally, the self-attention mechanism lets points selectively gather information from their surroundings. The attention mechanism is generated by creating pairwise feature differences and incorporating positional encodings through the `gamma_mlp`, which generates attention weights using a softmax function (*https://oreil.ly/WBAKN*). The feature transformation branch combines the input features with positional information, and the final output is computed by applying the attention weights to the transformed features, effectively allowing each point to selectively aggregate information from its neighborhood based on geometric and feature similarities.

This design helps the network to understand point representations better, giving significant gains in the scores (e.g., for the S3DIS indoor dataset (*https://oreil.ly/_HRYD*), PointNet achieves 49.0 mAcc, whereas Point Transformer achieves 76.5 as mentioned in the original article (*https://oreil.ly/VCSFx*)). This concludes our exploration of these advanced 3D perception tasks.

A Word on Multitask Learning

Point Transformer or PointNet are utilized for single-modality scenarios. Many current approaches use multimodal or multitask learning techniques (*https://oreil.ly/csyp5*). The latter differs from traditional single-task learning, where a separate model is trained for each task. As discussed in previous chapters, object detection and segmentation are crucial components of 3D data science. One application where they shine is self-driving systems.

However, individually developing and deploying these models can be cumbersome, requiring separate data processing, training, and optimization for each task. I've often encountered this challenge in my work, leading to the investigation of more streamlined and efficient approaches. This pursuit led me to HybridNets (*https://oreil.ly/ZhdtW*), a great end-to-end perception network that simultaneously tackles object detection and segmentation.

As its name suggests, HybridNets combines different architectural elements to achieve its multitask objective. The network is designed for object detection, drivable area segmentation, and lane detection and is trained on the BDD100K dataset (*https://oreil.ly/duaGm*). What intrigued me most was its elegant approach to combining these tasks within a single model, achieving interesting object and lane detection results. The architecture of HybridNets is built around a shared encoder and two distinct decoders. This shared encoder, based on EfficientNet-B3 (*https://oreil.ly/GNS3J*) and pretrained on ImageNet (*https://www.image-net.org*), serves as a robust backbone for feature extraction. I appreciate the choice of EfficientNet-B3 for its balance of accuracy and computational efficiency, a crucial consideration for real-time applications like autonomous driving.

Furthermore, incorporating a BiFPN layer (*https://oreil.ly/C_y9x*) in the neck of the network further enhances the model's ability to extract multiscale features, which is crucial for capturing objects and features of varying sizes. This clever combination of a strong backbone and a multiscale feature fusion network allows the model to extract rich and informative representations from the input image.

The two decoders then branch out from this shared encoder, specializing in their respective tasks. Inspired by the YOLO framework (*https://oreil.ly/pIVka*), the detection decoder employs anchor boxes determined through k-means clustering. This allows the model to directly predict bounding boxes and class probabilities, providing a streamlined approach to object detection.

The segmentation decoder, on the other hand, focuses on classifying each pixel into three categories: background, drivable area, and lane lines. The decoder utilizes a multilevel feature fusion approach, combining features from different scales to improve the precision of the segmentation output.

If you want to leverage HybridNets, I would advise looking at the code implementation (GitHub) (*https://oreil.ly/5368e*). Rather than training from scratch, which can be computationally intensive, I encourage you to use pretrained weights and infer on custom data. After setting up a new environment and installing the necessary dependencies, download the pretrained weights and run the inference scripts for images and videos. Even on relatively modest hardware, the results are interesting, showcasing the model's ability to detect objects accurately and segment the drivable area and lane lines.

Of course, the next stage is to make that work in a 3D context; at this stage, it should be easy for you!

After targeting semantic extraction and injection techniques, let's explore a promising new research direction: 3D Gaussian splatting.

3D Gaussian Splatting for 3D Visualization

Our last project is leveraging 3D Gaussian splatting to improve visualization experiences (Figure 18-13). 3D Gaussian Splatting, since its inception in "3D Gaussian Splatting for Real-Time Radiance Field Rendering" (*https://oreil.ly/AhtAH*), has gained a lot of traction. It marks a pivotal moment in the evolution of 3D scene representation to complement 3D reconstruction scenarios. It can also be used for novel view synthesis (*https://oreil.ly/yvW5C*), to generate new viewpoints with high fidelity.

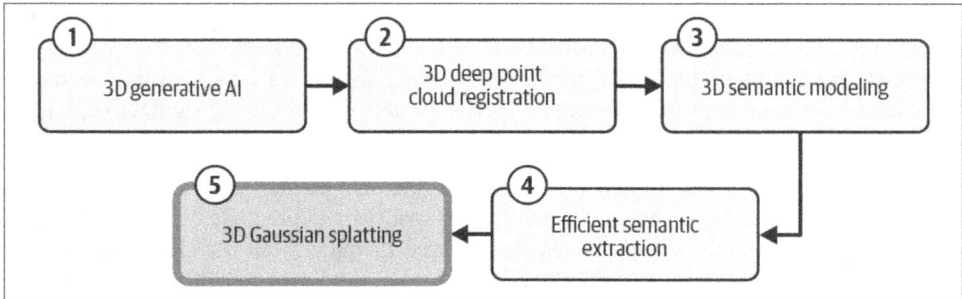

Figure 18-13. 3D Gaussian splatting

Gaussian splatting converts 3D point clouds into visually compelling representations and creates the illusion of solid, tangible 3D objects (see Figure 18-14). However, unlike the 3D modeling techniques studied in Chapter 10, it does not need polygons or meshes; instead, it uses Gaussian distributions to capture and render details.

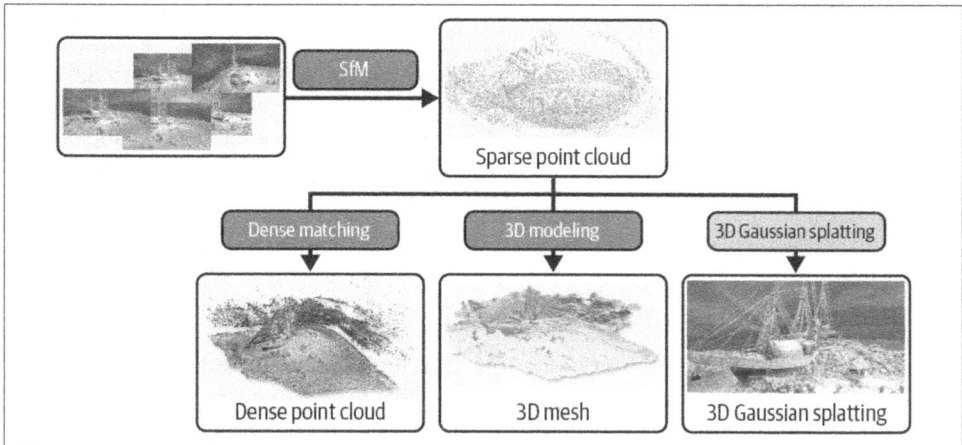

Figure 18-14. Comparing 3D Gaussian splatting to dense point cloud and 3D mesh approaches

Let me give you a quick contextual view of how 3D Gaussian splatting emerged from 3D reconstruction scenarios. As we saw in Chapters 1 to 4, we can leverage photogrammetry to generate dense 3D point clouds or 3D models from a set of photographs, as illustrated in Figure 18-14.

The SfM algorithm (among others) processes large numbers of images, finds common points, analyzes their relative displacement across the images, and generates a point cloud, which can then be transformed into a textured triangulated mesh (Chapters 10 and 11). This method, while expediting 3D modeling work with a fully automatic workflow, has limitations that prevent scanning a variety of objects: transparency (like glass), reflections (such as polished metals), and fine details (like hair). You'll get a lot of unwanted artifacts. However, in other cases, such as high-texture elements and natural objects like rocks, trees, and soil, photogrammetry works well. It's also suitable for industrial objects like concrete, bricks, painted metal, and stone and is widely used for modeling people, especially if hair is hidden.

In 2020, the paper "NeRF: Representing Scenes as Neural Radiance Fields for View Synthesis" (*https://oreil.ly/vCmKq*) brought a new solution that overcomes some of the limitations of photogrammetry: neural radiance fields (NeRFs) (*https://oreil.ly/IZPWe*). Objects created with NeRFs look incredibly realistic, and from certain angles, they appear almost indistinguishable from photographs. The method handles transparency, reflections, and even refractions well. However, there are significant drawbacks. NeRFs are built on a neural network architecture (Chapter 15), where the training phase is computationally intensive.

> The NeRF, with its ability to represent scenes as continuous volumetric functions learned from images, revolutionized novel view synthesis, enabling photorealistic rendering from any viewpoint. However, this fidelity comes at a computational cost. The core of NeRF rendering relies on ray marching—tracing millions of rays through the learned volume for each frame. While it produces stunning results, this process is inherently slow, hindering real-time applications.

On top of that, it is not easy to generalize, as you train a model for each capture, which means retraining the entire model every time. The first-hand identified goal of real-time 3D scene representation was limited to a handful of configurations that need a powerful GPU. Finally, editing is complex, as the models are pretty large, and little can be done to compress them effectively.

This is where, in 2023, Gaussian splatting emerged. 3D Gaussian splatting marks a pivotal moment in the evolution of 3D scene representation. It introduces a powerful alternative to NeRFs, addressing a critical limitation while maintaining impressive visual quality.

Gaussian splatting offers a solution to this performance bottleneck. Instead of an implicit volumetric representation, it utilizes a collection of 3D Gaussians, or "splats," to represent the scene. Each splat, defined by its position, covariance, color, and opacity, acts as a localized, oriented, and colored light point.

The crucial advantage lies in the rendering process. Projecting these Gaussians onto the image plane requires only simple analytical computations, dramatically reducing the computational burden compared to NeRF's ray marching. This allows for real-time or near-real-time rendering, opening doors to interactive applications that demand immediate visual feedback.

The benefits of Gaussian splatting extend beyond its speed. The splat-based representation is inherently more compact than NeRF's implicit volume, requiring less memory and storage (Figure 18-15).

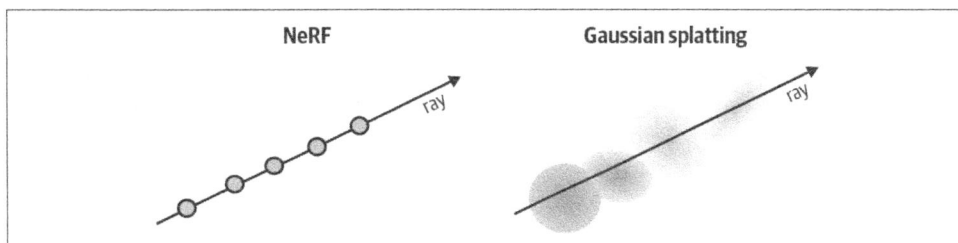

Figure 18-15. How NeRFs trace rays through grids versus how Gaussian splatting calculates how each Gaussian contributes to the light ray using elliptical splats

Furthermore, this explicit representation offers a level of editability that NeRF lacks. Individual splats can be manipulated, allowing straightforward modification of the scene's geometry and appearance.

You can quickly change the color of a car in a scene or adjust the shape of a building, much like you edit point clouds. In essence, a 3D Gaussian splatting scene is a 3D point cloud with many more attributes than only x, y, and z. Spherical harmonics (*https://oreil.ly/2MPir*) are then used by the rendering engine to provide an improved experience.

In the expansion of 3D data science, 3D Gaussian splats provide great new tracks for AR, VR VFX, and scientific visualization applications (among other things). Indeed, if you strip down the Gaussian features and keep only the Euclidean positions, you essentially retrieve the 3D point clouds as a canonical frame of reference. This makes point clouds once again the best candidate for being the digital environment spatial AI systems can use. If you want to see advanced use of point clouds, I encourage you to check the Smart Point Cloud thesis (available online to download) (*https://oreil.ly/4tfRx*).

Like NeRF, Gaussian splatting allows viewpoint-independent rendering, enabling free exploration of the reconstructed 3D scene. Now, to give you an extra edge, let me detail the elegant yet computationally efficient mathematics behind this solution:

- Each Gaussian is defined by a 3D position (μ).
- A covariance matrix, Σ, controls scale and orientation.
- A spherical harmonic coefficient vector handles view-dependent appearance.

The potential applications of Gaussian splatting extend beyond rendering. Its compact and explicit representation makes it a versatile tool for various tasks. Furthermore, individual splats or groups of splats can be assigned semantic labels, enabling scene understanding and object-level interaction (you can use Chapters 12 to 16 for some 3D Gaussian splatting experiments). This opens possibilities for creating virtual environments where understanding the semantic context of the scene is crucial.

To explore a concrete use of Gaussian splatting, I recommend you deep-dive into "3D Gaussian Splatting for Real-Time Radiation Field Rendering," with the precursor work (project (*https://oreil.ly/NUDe3*)) and implementation (GitHub (*https://github.com/graphdeco-inria/gaussian-splatting*)).

Implementing 3D Gaussian splatting involves several key steps. First, you need to take your raw data and make a Gaussian splat representation. Fitting Gaussians to a point cloud and turning a mesh into a set of splats are two ways to do this.

Once you have the Gaussian splat representation, rendering becomes relatively straightforward. For each Gaussian, you project it onto the image plane and compute its contribution to the final pixel color based on its position, covariance, color, and opacity. This involves evaluating a Gaussian function for each pixel, which can be done efficiently using vectorized operations.

Optimizations like hierarchical data structures can further accelerate rendering, especially for large numbers of Gaussians. Using 3D Gaussian splatting on huge scenes means dealing with the problems that come with saving and processing a lot of data. One approach is to use out-of-core techniques, where only the necessary portions of the scene are loaded into memory at any given time. Think of it like streaming a movie—you don't need to download the entire file to start watching, only the current scene.

These technologies are not just improving visualization—they're fundamentally changing how we capture, analyze, and understand 3D data. The ability to maintain high visual fidelity while enabling complex analysis operations marks a significant step forward for accurate digital environments.

What excites me is how these approaches democratize advanced 3D data science. Nowadays, standard workstations can run tasks that used to need special hardware. This means that more students and practitioners can use advanced 3D analysis.

Despite its advantages, Gaussian splatting faces challenges. Accurately capturing complex lighting effects, such as global illumination and subtle shadows, remains an active area of research. Representing dynamic scenes with moving objects and changing lighting conditions also presents a significant hurdle. Finally, developing efficient algorithms for optimizing splat placement and covariance is crucial for achieving high-quality reconstructions with minimal computational overhead.

These challenges, however, also represent exciting avenues for future research. The ongoing development of techniques to address these limitations further solidify Gaussian splatting's position as a powerful and versatile tool for 3D computer graphics and vision. Its unique combination of high quality, real-time performance and compact representation promises to unlock new possibilities in areas ranging from interactive virtual experiences to autonomous navigation and robotics.

These groundbreaking projects in 3D data science can revolutionize how we approach spatial AI. The amount of resources on the internet is sure to keep you busy with new innovations. If I had to sort things out, I would recommend skimming through O'Reilly's resources (*https://oreilly.com*), the 3D Data Academy (*https://learn geodata.eu*), and the research community (*https://oreil.ly/I_-wj*). Then, start scrutinizing scientific conferences and workshops like CVPR (*https://oreil.ly/-YYtc*), SIGGRAPH (*https://oreil.ly/_CYKg*), ICRA (*https://oreil.ly/mB_tl*), ICCV (*https://oreil.ly/wZa5Y*), and ISPRS (*https://isprs.org*) with papers accompanied by code to replicate experiments. However, checking out the approach first and then the implementation is always essential. Identify the gaps and use your newfound expertise to bring and join the two ends toward your goals.

We have answered how 3D data science can be the solution, one application at a time. But what if we ask a bigger question? We have robust point cloud workflows to build 3D digital environments: 3D data science for 3D virtual worlds. Are we close to mimicking human cognition with what we have?

Spatial AI: The Future of 3D Experiences

Let's explore the next frontier in our 3D data science journey: 3D spatial AI. It combines AI techniques and algorithms with spatial data and processes to help computer systems understand and make decisions based on spatial relationships and patterns.

Spatial AI becomes more meaningful when we think about human-centered mechanisms. It will be our personal tools and interfaces for interacting with the physical world, enhanced by a digital layer. We already see virtual reality (VR) (*https://oreil.ly/jbcA8*), augmented reality (AR) (*https://oreil.ly/dXy_W*), and eXtended reality (XR) (*https://oreil.ly/nP94_*) moving in a more human-focused direction.

Here's a quick recap. Virtual reality uses a headset to change our entire view of the world. Augmented reality leverages a headset to add information to what we see. Extended reality combines digital elements with the real world, focusing on hands-on experiences with simple objects to make the overlays feel more real. VR keeps us in one safe spot, but AR and XR help us explore our surroundings and connect with others in both physical and digital ways. VR, AR, and XR are just tools for accessing spatial AI, not spatial AI itself. Spatial AI includes much more than just interactive headsets.

Through our 3D data science journey, we developed the layers that can perceive the world and create digital environments. How can an "intelligent" machine make sense of it? How can we use "spatial AI" reasoning capabilities to generalize? First, let's break down the three main stages to make this work: 3D perception, 3D scene understanding, and 3D reasoning (Figure 18-16).

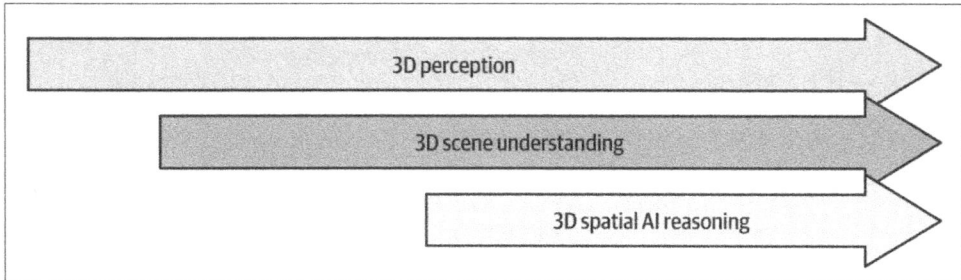

Figure 18-16. 3D perception, 3D scene understanding, and 3D spatial AI reasoning

In 3D perception, we have three main elements: 3D reconstruction, which is the art of creating some kind of 3D data from other modalities. We then review 3D data preparations and see preprocessing, registration, filtering, noise, and 3D data engineering. These are important for extracting features and manipulating and representing any data for higher-level scenarios where we want to gain some understanding. You can already create systems that can leverage this through 3D data science.

This is where we move on to stage 2, which is 3D scene understanding. With 3D data science, we essentially deal with semantic extraction. Some specializations of semantics are also significant to grasp: affordances, functions, and physical properties. These can then be used to form sufficient knowledge about the scene to get a single (or multiple) semantic representation to define a 3D scene with more detail.

Finally, we want to enable AI systems to understand and reason about spatial relationships, including 3D awareness, based on the insights gained from the previous 3D perception and 3D scene understanding stages. By combining these perspectives, we can achieve a richer and more comprehensive understanding, enabling machines to perceive and interact with the world more intelligently and human-like. Examples of such systems include:

- Robots designed to navigate and manipulate objects in various environments
- Autonomous vehicles that require an understanding of their surroundings for safe navigation
- AR and VR systems that need to make it easy for virtual items to fit with real-world settings

As these constitute the entry point to reasoning systems that can act on the data and make decisions, let's develop perspectives that extend the teachings of this book that you can leverage to deepen the depth of your systems.

3D Scene Understanding with Open Vocabularies

To effectively represent world scenes for spatial AI systems, we must mix some necessary components that extend semantic segmentation processes. If you remember our early discussions, the idea is to extract and combine geometry, semantics, and topology for a complete system that AI agents can leverage.

Our 3D data science workflow allows us to achieve this as core modules. We can especially leverage methods that allow us to identify objects and enrich the semantic understanding of the scene. But for spatial AI, we want to tie down two elements: the object categories, i.e., the labels or classes that categorize objects based on their semantic meaning (e.g., "car," "building," "tree"), and the attributes, i.e., the properties associated with objects, such as color, texture, material, function, physical properties, or affordances.

Let's clarify some extra terms: function, physical properties, and affordances. An object's function is its intended purpose or role within a scene. Physical properties describe objects' intrinsic characteristics, such as their size, weight, mass, and material composition. Affordances are the perceived possibilities for action that an object offers, which help determine how an object can be interacted with and what actions it enables.

We then need to get a topological understanding. This should be carried out at the object level and also at the scene level, knowing how each object relates to its external and internal constituents. So, taking things lightly, a complete understanding of a 3D scene involves inferring various aspects of a scene from our 3D data feed and other information sources.

This means we want to understand the semantics, affordances, functions, and physical properties of each constituent in our scene. This is where scene graph analysis and knowledge base integration can be employed. The idea is to make sure to tie in the connections between objects, including spatial relationships (e.g., "next to," "in front of") and semantic relationships (e.g., "part of," "is a").

For instance, given a 3D model of a house, a complete scene understanding should identify surfaces belonging to a fan (semantics), those made of metal (material), those within a kitchen (room type), where a person can sit (affordance), where they can work (function), and those that are soft (physical property). This information is crucial for future robots to interact with their surroundings effectively and for humans to understand a scene through interactive queries and visualizations.

This is where we have another challenge: traditional 3D scene understanding systems are limited by their reliance on labeled datasets for training, resulting in models specific to single tasks and closed sets of classes. This approach struggles with understanding rare objects or queries beyond those in the training data.

We are starting to see some seminal work, such as "OpenScene: 3D Scene Understanding with Open Vocabularies" (*https://oreil.ly/XX3Qj*) that goes in this direction. OpenScene offers an alternative approach to 3D scene understanding by utilizing pretrained text-image embedding models like Contrastive Language-Image Pretraining (CLIP) (*https://oreil.ly/SCKGq*). These models learn a shared feature space for visual and language concepts from massive datasets of captioned images, enabling zero-shot understanding (see Chapter 12 for more details on zero-shot methods (*https://oreil.ly/RhpY3*)).

This zero-shot approach enables OpenScene to identify objects, materials, affordances, activities, and room types without using labeled 3D data. Although not as accurate as fully supervised approaches on limited tasks with sufficient training data, OpenScene excels in tasks with larger classes or rare objects, surpassing even fully supervised methods in such scenarios.

What I like about this approach is its open vocabulary capability. It allows users to segment and identify objects, materials, affordances, activities, and room types using arbitrary text descriptions or images as queries. In addition, OpenScene's open vocabulary capabilities enable innovative applications like:

- Open-vocabulary 3D object search based on text queries
- Image-based 3D object detection using internet images as queries
- Open-vocabulary 3D scene understanding and exploration to identify properties, materials, and activities

Such approaches represent the scene as a graph, where nodes are objects and edges represent relationships between them (e.g., "car is parked on the road"). This provides an additional layer that can be leveraged.

> Without knowing it, particularly in Chapter 13, you have been experimenting with spatial knowledge graphs. The idea is to create a graph structure where nodes represent concepts and actual 3D entities. The edges between these nodes encode not only semantic relationships but also spatial ones—distances, orientations, containment, you name it. The magic happens when we train AI models on these spatial knowledge graphs. We are not just teaching the AI about abstract concepts—we are teaching it to reason in 3D space. This opens up a whole new world of possibilities.

These advances bring new perspectives for spatial AI systems.

3D Spatial AI Reasoning

Once we have a 3D scene with all these components, it becomes the base for the queries of a spatial AI (Figure 18-17) system. This is where we move on to reasoning capabilities for 3D datasets.

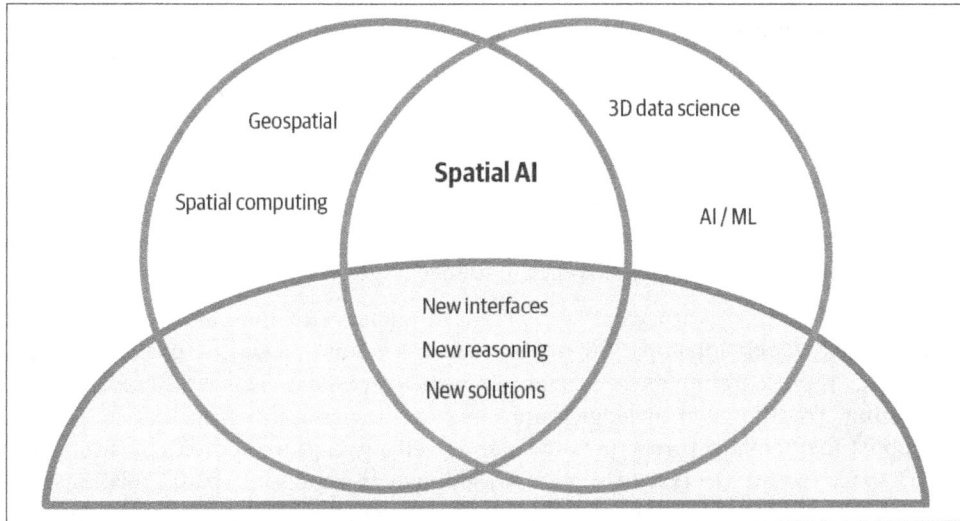

Figure 18-17. Spatial AI and how it relates to other fields

For instance, in a manufacturing context, a spatially aware AI could optimize robotic assembly processes by understanding not just the sequence of operations but also the physical constraints of the workspace. It could predict potential collisions, suggest more efficient movement paths, and even redesign the assembly line layout for better performance.

Or consider urban planning. A spatial AI system could analyze proposed building designs for their impact on the surrounding environment. It could predict how a new skyscraper might affect street-level wind patterns or alter the skyline's visual character from different vantage points.

The key is that we are moving beyond pattern recognition or 3D data science. We are enabling spatial AI systems to reason about space in ways fundamentally similar to how humans do it—but at a scale and level of detail that far surpasses human capabilities.

First, we must recognize that the 3D data science workflow we have established—from acquisition to visualization—gives us an incredibly rich foundation. However, it is not only a foundation as a digital environment. The real power comes when we can layer AI reasoning on top of this spatial data.

Think about it this way: we have intricate 3D models of our world packed with semantic information and topological relationships. But they are still, in a sense, passive. They are waiting for something to breathe life into them, to make them truly interactive and intelligent.

This is where spatial AI reasoning steps in. It is not just about recognizing objects in 3D. We are talking about an AI that can understand the spatial relationships between objects, infer hidden information, and make decisions based on 3D contexts. Let me give you a concrete example.

Imagine we have used our 3D data science workflow to create a highly detailed model of an urban environment. We have buildings, roads, trees, and works. Now, let's say we want to optimize the placement of new 5G antennas. A spatial AI system could reason about line-of-sight obstructions and potential signal reflections off buildings and even predict how seasonal changes in foliage might affect signal propagation.

It does not just work with what is explicitly modeled—it infers and reasons about complex spatial relationships. But to get there, we must tackle the challenge of AI knowledge representation and extraction in a spatial context. This is where things get interesting. Traditional knowledge graphs and ontologies are the first choice. Graphs are robust mathematical structures that can effectively capture complex relationships in 3D space. In the context of spatial AI, graphs can represent spatial relationships between objects, topological connections in 3D structures, and hierarchical organizations of spatial data.

To leverage these ensembles, we can leverage the NetworkX library, which provides robust tools for creating, manipulating, and analyzing graphs, as seen in Chapter 13. Then, implementing automated reasoning with spatial AI is more about pragmatic choices than theoretical frameworks. Here is a concrete example of how you can structure a spatial AI reasoning system:

Data representation
> You can use PostgreSQL (*https://www.postgresql.org*) with the PostGIS (*https://postgis.net*) extension to store and query spatial data (even 3D). It is robust, scalable, and supports complex spatial operations out of the box.

Knowledge formalization
> For ontology development, you can use the Web Ontology Language (OWL) (*https://www.w3.org/OWL*), which is expressive and well supported. I typically use Protégé for ontology design and then export to the Resource Description Framework (RDF) (*https://www.w3.org/RDF*) for use in my systems.

Reasoning engine
> Here is where things get interesting. I have had great success with the RDFLib library (*https://oreil.ly/zgCmq*) in Python for basic SPARQL (*https://oreil.ly/_7qd3*) querying and inference. I have been experimenting with PyCLIPS

(https://oreil.ly/vL4Mz), a Python interface to the CLIPS expert system *(https://www.clipsrules.net)*, for more complex spatial reasoning.

3D data science

This complete book, now yours, is a resource for 3D spatial operations in Python. The techniques we reviewed integrate well with machine learning workflows and can handle complex geometries efficiently.

With this, you can aim to develop unified representations that encompass diversity. This involves creating a common language for representing and integrating information from various sources, enabling seamless and intelligent interaction with the 3D world.

Imagine a universal translator for sensory data—this unified representation would allow different systems and algorithms to communicate and collaborate seamlessly. I can add that it is essential to retain an unsupervised approach to prepare the data for use by many reasoners and, thus, applications. Creating a topological graph between "connected elements" (Chapter 12) is a novelty that provides exciting perspectives concerning the reasoning possibilities that play on relationships between elements.

Conclusion

We are on the cusp of creating truly intelligent spatial systems by combining the rich, detailed 3D environment we can capture and digitize with 3D data science with advanced AI reasoning capabilities. These systems will not just represent the world—they will understand it, reason about it, and help us make better decisions. Human-machine 3D interfacing is a frontier rife with exciting possibilities and formidable challenges. The key is creating AI systems that can reason about 3D space in truly useful ways for human users. It is not just about spatial analysis but about understanding human intent and providing relevant, contextual assistance in 3D environments.

We have the technical capabilities to create vibrant, immersive 3D digital worlds. The challenge is to interface them in a way that is truly intuitive, accessible, and meaningful for users. This multidisciplinary challenge requires us to rethink our assumptions about human-machine interaction and push the boundaries of what's possible in spatial computing.

It is an exciting frontier, and I cannot wait to see how it unfolds. The 3D data science workflow we have developed in this book is the crucial first step. Now, it is time to take that leap into truly intelligent spatial AI.

Index

density-based clustering, 408
Density-Based Spatial Clustering of Applications with Noise (see DBSCAN)
Depth Anything model, 363, 588
depth estimation predictions, 357
depth maps, 112-113
Depth Pro model, 363
descriptive data analysis (DDA), 223
Designing Machine Learning Systems (Huyen), 32
deviation analysis
    meshes, 263-266
    planes, 259-263
DGCNN (Dynamic Graph CNN) architecture, 542, 622
Dice coefficient, 500
difference operation, 350
diffusion models, 355
digital twins, 4, 17, 92, 609
dimensions, 2-3, 41
discover attention, 4
distance metrics, 224, 401
distance threshold, 303, 306
distribution-based clustering, 409
DL (see deep learning)
domain knowledge, 35
downsampling, 153, 174, 194
DreamFusion architecture, 616
DXF (Drawing Exchange Format), 342-345
Dynamic Graph CNN (DGCNN) architecture, 542, 622

E

EDA (exploratory data analysis), 224
edges
    in B-rep systems, 340
    in graphs, 444, 448
EfficientNet-B3 dataset, 627
eigenvalues, 183, 186-190, 209-210, 257
eigenvectors
    analytical app development, 209-210
    principal component analysis, 183, 186-190
elbow method, in k-means clustering, 419
embodied AI, 511
engineering analysis, with ParaView, 70
enrichment of data, 21
ethical considerations, in 3D AI systems, 511
Euclidean clustering, 427, 457-460
Euclidean distance metric, 224, 299, 402

explainability, challenge of, 37
exploratory data analysis (EDA), 224
eXtended reality (XR), 633
extrinsic camera parameters, 359
extrusion, 341, 346

F

F1-score, 500
Facebook, 41
faces, in B-rep systems, 340
false negative (FN) metric, 499
false positive (FP) metric, 499
Farley, David, 32
Fast Point Feature Histograms (FPFH), 194
FEA (finite element analysis) tools, 65
feature extraction, 178-182, 592
    defined, 179
    global, 180
    local, 181
    normal estimation, 178
    principal component analysis, 182-190
    transfer learning, 547
feature perturbation, 549
few-shot learning, 466
filleting, 347
fine registration, 197-199
fine-tuning, 547
finite element analysis (FEA) tools, 65
FN (false negative) metric, 499
focal length, 361
FoldingNet network, 623
Ford, Neal, 31-32
forward propagation, 522
FP (false positive) metric, 499
FPFH (Fast Point Feature Histograms), 194
Fréchet distance metric, 225
FreeCAD, 65
fully connected layers, in neural network architecture, 519, 537
Fundamentals of Software Architecture (Richards and Ford), 31-32

G

GANs (generative adversarial networks), 355, 614
Gaussian mixture models (GMMs), 409
Gaussian splatting, 60, 94, 628-633
GeForce RTX 4060+ GPU, 54
GenAI (generative AI), 612-618

Boolean operations, 349-350
CadQuery, 339-342
file formats, 342-345
modeling various pieces, 350-352
techniques for, 345-349
RANSAC, 286-298
semantic modeling, 620-621
sensor-related 3D modeling, 92
voxels, 331-337, 596-597
creating grid for, 334-335
environment initialization, 332
exporting mesh objects, 336
generating voxel cubes, 335
loading data, 333
structure from motion (SfM), 60, 94, 588, 629
subsampling, 153, 174, 591
SuperPixel framework, 600
SuperPoint framework, 600
supervised learning, 393, 483-490
(see also machine learning)
classification, 488
concepts, 484-488
semantic segmentation, 489-490
training phase, 484-488
SuperVoxel framework, 600
surface models, 124-131
meshes, 125-128
parametric models, 128-131
surface normals, 256
surface reconstruction methods, 595
sweeping, 341

T
Tanh function, 540
tensor processing units (TPUs), 37
TensorFlow, 525
tensors, 526
text-to-3D models, 613-616
Thomas, David, 32
three-dimensional data (see 3D data)
three-dimensional data science (see 3D data science)
three-dimensional data science workflow (see 3D data science workflow)
three-dimensional point clouds (see point clouds)
thresholding
adaptive thresholds, 288
interactive, 216-217

tiling strategy, for k-d trees, 597
TN (true negative) metric, 499
topology
integrating with other concepts, 11
overview of, 9
spatial AI, 41
TP (true positive) metric, 499
TPUs (tensor processing units), 37
transfer learning, 511, 546
transformation of data, 19, 174
Transformers library, 355
transformers, semantic extraction with, 622-627
translation, in data augmentation, 548
Treisman, A. M., 4
TRELLIS, 616
triangle meshing, 125
TripoSR architecture, 618
true negative (TN) metric, 499
true positive (TP) metric, 499
two point clouds, 194, 197, 199

U
undersegmentation, 300, 407, 602
union operation, 349
unit sphere normalization, 236
Unity, 66
Universal Transverse Mercator (UTM) projection, 371
Unreal Engine, 67
unsupervised learning, 393
(see also machine learning)
unsupervised segmentation, 439-479, 600-602
clustering, 396-410
fundamentals of, 397-400
representativity, 401-407
types of algorithms, 408-410
connectivity-based clustering, 440-461
connected components, 455-457
environment setup, 445-447
Euclidean clustering, 457-460
graph analytics, 453
graph theory, 447-453
plotting graphs, 454
principles of, 443-444
DBSCAN, 421-434
experimental setup, 424
k-means versus, 433
multi-RANSAC framework, 429

About the Author

Florent Poux is an esteemed authority in the field of 3D data science. He teaches and conducts research for top European universities. He's also the head professor at the 3D Geodata Academy (*https://learngeodata.eu/3d-data-science-with-python*) and innovation director for FrenchTech120 companies. With a decade of experience, Florent's expertise in Python and its applications to 3D data is unrivaled. His ground-breaking research and real-world contributions have garnered acclaim in academia and industries with several awards. His unique journey merges a deep academic perspective with a pragmatic view of creating innovative 3D products and solutions. Whether you're a coder, data scientist, engineer, or curious researcher, Florent's insights will propel you toward new frontiers and exciting opportunities in this rapidly evolving field.

Colophon

The animal on the cover of *3D Data Science with Python* is a blue viper (*Trimeresurus insularis*). Also known as the white-lipped island pit viper, blue vipers are closely related to the white-lipped pit viper.

As the name suggests, blue vipers are known for their vibrant blue scales. This makes them extremely rare, as most of the snakes in this species are green. The scales can range from a bright, electric blue to a pale, powdery blue and often have an iridescent or metallic appearance.

Blue vipers are native to Komodo Island but can be found in other areas of southeast Asia as well. They prefer forested areas such as monsoon forests, bushlands, and bamboo forests where their blue coloration helps them blend in with their surroundings. Blue vipers are arboreal (tree dwellers) and nocturnal, waiting until nighttime to hunt their prey (rodents, birds, lizards, frogs, and other small mammals).

Although not generally aggressive, blue vipers will fight when provoked. They have hollow fangs which they use to inject venom into their prey. These snakes are poisonous to humans, but there are antidotes.

Many of the animals on O'Reilly covers are endangered; all of them are important to the world.

The cover illustration is by José Marzan Jr., based on a black-and-white engraving from *Lydekker's Royal Natural History*. The series design is by Edie Freedman, Ellie Volckhausen, and Karen Montgomery. The cover fonts are Gilroy Semibold and Guardian Sans. The text font is Adobe Minion Pro; the heading font is Adobe Myriad Condensed; and the code font is Dalton Maag's Ubuntu Mono.